AGRICULTURAL RESEARCH UPDATES. VOLUME 2

AGRICULTURAL RESEARCH UPDATES

Additional books in this series can be found on Nova's website
under the Series tab.

Additional E-books in this series can be found on Nova's website
under the E-book tab.

AGRICULTURAL RESEARCH UPDATES. VOLUME 2

BARBARA P. HENDRIKS
EDITOR

Nova Science Publishers, Inc.
New York

NOTICE TO THE READER

Library of Congress Cataloging-in-Publication Data

ISSN: 2160-1739

ISBN: 978-1-61470-191-0

Published by Nova Science Publishers, Inc. † New York

CONTENTS

PREFACE

This compilation examines agricultural research from across the globe and covers a broad spectrum of related topics. In this book, the authors discuss research including the diet selection of herbivores on species rich pastures; the table olive food supply chain in Mediterranean regions; the invasive Argentine ant and its impact on agriculture and ecosystems; sustainability challenges in seafood supply chains; toxic element contamination in animal feed and agricultural management and land-use practices on pesticide loads.

Chapter 1 - A food supply chain encompasses a series of connections and inter-dependencies, spanning from the farm to the final consumer. In Mediterranean regions, diet entails not only supply of macro- and micronutrients – but is also a valuable piece of a multi-century cultural and gastronomic heritage; more recently, the issue of favourable contribution to health has also been more and more often addressed. Said diet consists on a balanced combination of fresh, local and seasonal foods – and includes several fruits and vegetables, and olives and olive oil in particular as major source of antioxidants and fat. Hence, the Mediterranean diet is characterized by a relative richness in complex carbohydrates and fibre, besides unsaturated fatty acid residues and natural antioxidants.

Table olive manufacture remains one of the most important food industries, especially in southern Europe; it is still carried out following artisanal (and thus empirical) practices, with a moderate degree of technological innovation, despite its regional and overall economic impacts. The uniqueness of fermented olives has indeed made them unavoidable ingredients of gourmet recipes and gastronomic excellence – and nowadays health-aware consumers seek them as well for their natural origin and functional features. However, the transformation sector is rather fragmented (despite the overriding trend for globalization), and exhibits poor profitability margins – thus being particularly vulnerable to year-to-year fluctuations in production supply and market demand.

It is estimated that no less than 25% of the diet in European countries, and up to 60% of the diet in developing countries consists of fermented foods – a realisation derived from the usefulness of fermentation in extending shelf life of (otherwise perishable) foods and in assuring their safety, but also from the improved sensory properties resulting therefrom. Fermented olive farming is indeed critical for several regions in Spain, Portugal, Italy and Greece – which dominate by far the world supply of table olives. Nevertheless, stricter and stricter environmental legislation, further to competition with other unsaturated fats and oils from plant origin worldwide, have created a threatening situation regarding this specific food supply chain. Survival of this sector will require incorporation of dedicated and less conventional technologies, more efficient traceability and control of the associated processes, and diversification of the product portfolio designed for more demanding consumers.

A new generation of natural (and healthy) foods is clearly on the rise – and fermented olives are a nuclear part thereof. Public health concerted policies aimed at efficiently facing budgetary restrictions will require investment on preventive health approaches – especially as human life expectancy is extended; and the aforementioned types of foods are seminal vectors within this strategy. However, manufacture and marketing of fermented olives hinges critically upon the primary production sector; traditional olive orchards already account for a major share of agricultural land in Mediterranean countries, particularly in marginal, less developed areas. In addition to the socioeconomic impact in terms of maintaining local employment and thus avoiding rural exodus, sustainable exploitation is a must – so that the needs of current generations are fulfilled, without compromising the capacity of future generations to meet their own needs.

On the other hand, the adequacy of the food supply in a particular area or country is also dependent on weather, economic and even political factors; resorting to autochthonous flora that have adapted since ancient times to the local ecosystems is a goal to pursue, hence avoiding introduction of genetically modified or exotic species – while rationally intensifying productivity and resistance to pests. Traditional low intensity, non-irrigated groves often contribute to landscape preservation by decreasing desertification and contributing to a greater biodiversity – e.g. via providing shelter and food for the wild fauna. On steep slopes, terraced olive groves can prevent soil erosion; however, when they are left behind, they turn into shrub that enhances the risk of summer fires (one of the chief environmental hazards in Mediterranean regions).

This chapter covers the various points summarized above – and presents general data useful for characterization of the food supply entertained by table olives, while discussing current problems and anticipating future issues in this field.

Chapter 2 - Hydrological field studies have shown that runoff generation processes do not occur uniformly over an entire catchment during storm events. The variable-source area, or VSA, concept encompasses this spatial and temporal variability in rainfall-runoff response at catchment scale. Early on, hydrologists envisioned that by targeting VSAs, the transport of contaminants to streams could be managed, thereby protecting water quality. However, predicting VSAs at catchment scale has remained a daunting task even in intensively studied catchments, let alone transferring the understanding to other catchments or linking runoff generation processes to water quality research.

The application of the VSA concept to pastoral catchments in New Zealand for water quality management is further compounded by open grazing, which is integral to livestock farming in this country. Persistent animal traffic and behaviour such as camping, wallowing, and fence-line pacing significantly deteriorate soil physical quality, resulting, for example, in soil compaction and infiltration-excess runoff. When such compacted areas are connected to streams via animal tracks or farm roads, their potential to contaminate surface waters increases. In addition, the occurrence of animal-induced features has been shown to be randomly distributed. Thus, in pastoral catchments, mapping runoff generation areas to manage contaminant transport needs to take into account animal behaviour and the VSA concept.

In this chapter, we trace the evolution, adaptation, and application of the VSA concept for surface water quality management in grazed, pastoral catchments in New Zealand. Early desk-top studies combined traditional data (streamflow measured at catchment outlets) and few generalised catchment characteristics (topography and soil drainage properties) with rainfall characteristics to predict the dominant runoff generation mechanisms at catchment scale. However, these studies neither extended these predictions to map specific runoff generation areas within the catchments nor linked water quality management to runoff generation processes. Field studies that followed focused on mapping the temporal and spatial dynamics in runoff generation processes and the dominant flow and transport paths that connect land to streams. However, findings from these field studies largely remained within the realms of science. Very little of the new knowledge was applied to manage pastoral catchments, largely because appropriate tools to transfer the findings between catchments were not available. The subsequent development of various indices and concepts, for example, the Phosphorus Index and the Topographic Index, has not provided the desired breakthrough, because they either oversimplified hydrological processes or resulted in solutions that were too complex for practical implementation.

The objective of this chapter is to identify challenges facing the integration of the VSA concept to water quality protection in grazed, pastoral catchments. These challenges range from refining our understanding of current runoff generation concepts to transfer and adaptation of field results across multiple spatial scales. This chapter highlights that understanding runoff generation processes such as saturation-excess and infiltration-excess surface runoff and identifying major overland and subsurface transport pathways that connect land to streams, have to be integrated with knowledge on the key contaminant source and transport areas (e.g., near-stream locations, animal tracks and farm roads) to develop and implement sustainable management practices at catchment scale.

Chapter 3 - In the last 100-150 years, 90 % of Iowa's tall-grass prairies containing many wetlands and some riparian forests has been converted to annual row crops and cool-season grass pastures. Bare soil between crop rows erodes easily and can be transported to streams by overland flow, while grazing in riparian areas can reduce stream bank vegetation, making banks susceptible to erosion. The USDA-NRCS estimated that approximately 12,000 kg ha^{-1} yr^{-1} of soil are eroded from 1/3 of Iowa's land with much of these materials entering streams. As a result, sediment is the most common nonpoint source pollutant and a major contributing factor for stream degradation in agricultural watersheds.

In the agricultural states of the United States including Iowa, conservation practices such as riparian forest buffers and grass filters are being established to improve water quality by reducing sediment delivery to stream channels. These conservation practices provide some financial incentives to farmers because they are part of the United Sstates Department of Agriculture Conservation Reserve Program that began as part of the 1996 Farm Bill. This program subsidizes part of the lost income for establishing perennial plant communities on previous managed row crop or grazed pasture land. When these conservation practices are established in the riparian areas of a watershed, stream water quality can improve while maintaining the largest area of the watershed in agricultural production.

In this chapter results are presented on the effectiveness of riparian forest buffers and grass filters in reducing sediment from surface runoff and streambank erosion that are the major sources of sediment to the stream load. Results are from studies on Bear Creek of central Iowa, a National Restoration Demonstration Watershed as designated by the Interagency Team implementing the Clean Water Action Plan (1999) and a Riparian Buffer

National Research and Demonstration Area as designated by the United Sstates Department of Agriculture (1998) and other Iowa streams.

Chapter 4 - Understanding grazing by domesticated ruminants for animal production is of high economic importance throughout the world and therefore, is paramount in designing management strategies for livestock production. The plant-animal interface is the central feature of these systems. Food quantity and quality are major determinants of animal production. Both food quantity and quality herbivores maintain through selective foraging which alters sward structure, modifies plant species composition and thus produces new patterns of plant biomass production. Therefore, we focus our chapter on the mechanisms of foraging selection which may enable us to have insight into grazing decision-making and processes. The central question is: what are principle drivers in grazing decision processes leading to high selectivity on species rich grasslands and this at different spatial and temporal levels? However, before we answer this question we have to be precise in what we mean under selective grazing and to distinguish it from other terms used in the domain of foraging strategies. We will also summarize known methods and quantifications of grazing selectivity. Only then may we bring a complex view on various factors affecting diet selection strategies of herbivores in species-rich pastures. Finally, we propose management rules in order to use herbivore foraging selectivity to utilize food resources in semi-natural grasslands most efficiently and simultaneously keeping forage production and quality of grasslands from a long-term perspective.

Chapter 5 - Introduction of alien organisms is a major risk that follows international trade and globalization. Ants are among the most harmful groups of invasive organisms, with five species including the Argentine ant *Linepithema humile* listed among the world's 100 worst invasive species by the IUCN. We review the impacts, ecology, and dispersal of invasive ants, with the Argentine ant as a representative. Invasive ants attain high population densities in the introduced range, and cause damage to ecosystems, agriculture, and human well-being by the sheer number. The high densities stem partly from their characteristic social structure 'supercolonies', i.e., aggregations of numerous, mutually cooperative nests. In the Argentine ant, high consistency of their supercolony identities makes supercolony an important unit in inferring dispersal history of the species: colonies originating from a common source colony remain mutually compatible even if they are isolated for a long time. We highlight two topics in the dispersal history of the Argentine ant: 1) formation of an unprecedented intercontinental supercolony by the 150 year international trade; and 2) recent successive introductions to Pan-Pacific region seemingly in accordance with globalization.

Chapter 6 - Common bacterial blight (CBB) is a significant foliar disease of dry bean caused by the pathogen *Xanthomonasaxonopodis* pv. *phaseoli*; a gram-negative bacillus with a genome of approximately 3.9Mb. This disease is endemic to most regions where *P. vulgaris* is cultivated and is annually responsible for millions of dollars of crop loss worldwide. The bacteria are soil residents and initial infections occur predominantly through the stomata of leaves, or through plant wounds. Infected plants generally display symptoms on the leaves first, with the formation of small water soaked lesions appearing 1-2 weeks after infection. The lesions gradually enlarge, and become encircled by a region of yellow tissue. As the disease progresses these lesions become necrotic and extensive defoliation is common in infected plants. The bacteria can migrate throughout the plant, including into the seeds through the vascular system of the pedicle. Treatment options for infected plant tissues are limited and the most effective preventative measure is to grow breeder seeds in regions that are free of the pathogen. Topical application of antibiotics and anti-microbial compounds are used as secondary control measures.

Research efforts to mitigate the damage caused by the pathogen have focused on developing bean germplasm that is resistant or tolerant to the pathogen and integrating these sources of resistance into new varieties. Two main sources of resistance have been discovered, *P. vulgaris*-derived resistance and *P. acutifolius*-derived resistance. CBB resistance in *P. vulgaris* is relatively weak but, interspecific crosses between *P. vulgaris* and *P. acutifolius* have been shown to possess high levels of CBB resistance. Molecular studies have shown that resistance in the interspecific lines is conditioned by several quantitative trait loci (QTL), which interact in various ways. Molecular markers that are associated with the resistance QTL have been used for marker assisted selection and are the starting points for studies to identify resistance genes.

Chapter 7 - Beef burgers were traditionally associated with nutrients and nutritional profiles that are often considered negative including high levels of saturated fatty acids, cholesterol, sodium, and high fat and caloric content. However, meat is a major source for many bioactive compounds including iron, zinc, conjugated linoleic acid, and B vitamins. By selection of lean meat cuts, removal of adipose fat, including oils of vegetal or marine origin (with high content of polyunsaturated fatty acids), and addition of phytosterols, this type of products could become into healthier functional foods. To obtain a product with similar characteristics to the "classical" beef burgers emulsifiers or binding agents are needed, as well as antioxidants to control lipid oxidation. Low-fat beef burgers with high oleic sunflower, deodorized fish oil, and phytosterols were formulated including whey proteins or egg white and natural antioxidants (tocopherols and/or oregano-rosemary extract). Products were characterized and the effect of frozen storage at -20ªC on the quality of the cooked hamburgers was studied. Cooking yield, moisture and lipid retention, press juiciness, texture profile analysis, microstructure, oxidative stability, color, fatty acid profile, phytosterols contents, microbiological counts, and sensory acceptation were determined.

Cooking yields ranged between 79.4 and 84.5 % for all the formulations, and became slightly lower with frozen storage. Press juiciness also diminished with storage time while hardness increased, for both emulsifiers used. Lipid retention was higher than 95 % while water retention was higher than 70 %, for whey proteins or egg white and these parameters did not change during frozen storage.

Whey proteins protected better from oxidation than egg white, and tocopherols demonstrated an adequate antioxidant effect in formulations with egg white. For all the formulations unsaturated/saturated fatty acids ratio was higher than 5.8, showing a good lipid balance in the products.

Global acceptability for all the formulations presented sensory scores higher than 7 (in a 1 to 9 scale). More than 82.3 % of the panelists liked the taste of the products, 78.7 % liked the texture, and over 86.8 % of the panelists liked the products, considering the overall acceptability of the burgers. These results showed that the presence of the high oleic sunflower and deodorized fish oil did not adversely affect the low-fat beef burgers.

The consumption of 100 g of the cooked product would provide 6% of the recommended daily intake of phytosterols to decrease cholesterol and heart disease risk.

Chapter 8 - Principles of sustainable systems are based on a vision where animals should be part of an agricultural system that is environmentally sound, animal friendly and considering the whole system rather than only optimizing its parts. Sustainable systems such as integrated, low-input and organic farming use ecologically sound management strategies with the potential to benefit and respect the physiological and behavioral needs of livestock. This chapter focuses on the most critical obstacles to meet mineral requirements in ruminants in sustainable systems when a high degree of home-grown feed and constrained lower concentrate and mineral supplementation in the ration is promoted; and identifies the effects of different feeding regimes on mineral nutrition such as winter feeding and grazing intensity, fertilizer and pesticide-free pasture management, forage diversity and the evidences that mineral metabolism might be negatively affected by parasite infections that severely affect sustainable systems. The document addresses all these factors that are likely to exert a potentially adverse effect to meet nutritional requirements for animals and current research on strategies to improve animals' mineral nutrition and means of minimizing mineral disorders by specific husbandry practices.

Chapter 9 - Industrial and agricultural activities are responsible for polluting the environment with toxic elements. In terms of potential adverse effects on animal and consequently human health, cadmium, lead, but also mercury and the metalloid arsenic are amongst the elements that have caused most concern; this is because they are readily transferred through food-chains and their exposure can result in adverse effects on a great variety of physiological and biochemical processes. Although episodes of lethal toxicity associated to accidental exposure to very high doses of toxic elements have been largely described in literature, the main concern for livestock is dietary exposure, in agricultural regions diet being the main source of toxic elements for animals. This chapter reviews current data on toxic element concentrations in animal feedstuffs and husbandry practices related to toxic metal exposure, and analyse the effect of toxic element exposure on animal health and residues in animal products.

Chapter 10 - The objective of this study was to evaluate the importance of 25 intrinsic and extrinsic attributes in the purchase of beef, to detect relations between the attributes and to distinguish the existence of consumer segments in the south of Chile. The study was based on a direct survey of 1,200 people in the Maule, Biobío and Araucanía Regions, Chile. Five dimensions were obtained which characterise the relations between the attributes (62.7% variance). Consumers attributed great importance to the price and intrinsic attributes related to the organoleptic quality, harmlessness and health care. Three segments were distinguished, with differences of gender, region and zone of residence, age, family size, education, socio-economic level and ethnic origin. The majority group (42.0%) valued intrinsic and extrinsic attributes, the second group (29.8%) attributed little importance to the production system and the third group (28.5%) gave less importance to the organoleptic quality. The results are discussed in relation to studies conducted in developed countries.

Chapter 11 - The current chapter examined the combined influence of changing row crop production, implementation of agricultural Best Management Practices (BMPs), and enrollment of 112 ha into Conservation Reserve Program (CRP) on pesticide contamination and potential risk to lake aquatic biota in a 914-ha Beasley Lake watershed from 2000-2009. A suite of six current-use herbicides, five current-use insecticides, and two legacy insecticides were measured in lake surface water sampled approximately monthly from 2000-2009. Relative risk of these pesticides to lake aquatic biota was assessed using individual toxicity quotients (TQs), mixture pesticide toxicity index (PTI) scores based upon acute (48-96h) LC/EC50 values, and acute restricted-use pesticide levels of concern (LOCs) (LC/EC50 x 1, 0.5, and 0.1) for freshwater crustaceans (*Daphnia* sp.), insects (*Culex* sp.), fish (*Lepomis* sp.), and algae (*Psuedokirchneriella* sp.). During the ten-year study period, row-crop production shifted from primarily cotton in 2000-2001 to predominantly soybean in 2002-2004, 2006, 2008, and 2009 with milo and corn dominant in 2007. Reduced tillage BMPs were implemented in 2001 and CRP enrollment began in 2003. From 2000-2009, most individual pesticide concentrations were frequently <0.1 µg/L, with the exception of atrazine. Greatest herbicide concentrations occurred for triazine herbicides atrazine and cyanazine. Greatest insecticide concentrations occurred for methyl parathion and bifenthrin. Greatest legacy compound concentrations occurred for the organochlorine insecticide, *p,p'*-DDT. Temporally, peak lake water concentrations of current-use herbicides, current-use insecticides, and legacy compounds occurred during 2000, 2002, and 2002, respectively. Lowest lake water concentrations of current-use herbicides, current-use insecticides, and legacy compounds occurred during 2005. Results of the pesticide risk assessment showed greatest risk would be to crustaceans, primarily from the pyrethroid bifenthrin, during 2000-2002, with decreasing risk to fish and aquatic insects and minimal risk to algae. Although most individual pesticides were below LOCs, PTIs indicated increased risk from pesticide mixtures to aquatic fauna. Temporally, relative risk to lake aquatic biota decreased from greatest potential risk in 2000-2002 to minimal risk in 2005-2006 with infrequent risk to aquatic fauna in 2007-2009. Overall, lake water pesticide contamination decreased annually until 2005-2006 and increased again in 2007-2009 due, in part, to a shift in row crop from reduced tillage soybeans to conventional-till milo and corn in 2007. Concomitantly, changes in land-use with implementation of BMPs, CRP and crop type reduced the frequency and duration of risk of pesticides to lake aquatic biota.

Chapter 12 - In the eastern Amazon Capim River pole, farming is a major economic activity, in which the slash-and-burn system prevails in creating small farms. In the last two decades, some smallholders have changed the landscape of their Family Production Units (FPU) by increasing their small farms and backyards to Agroforestry Systems (AFSs), in order to ensure food safety. Currently, the AFSs, in addition to the numerous environmental services they conduct in the region, already allow some smallholders to enter their products into the consumer market. This paper presents the structural characteristics related to the spatial arrangements of the circuit space and the main product marketing difficulties, which include the difficulty to obtain credit in order to produce, agricultural extension services – there is not enough human material to meet regional demand, the perishable nature of the products, the middleman, and the low prices and lack of specific marketing policies at this pole. From the results of the statistical analyses, as well as by the perceptual maps from expert knowledge this study presents a fuzzy system for the study of the marketing/development of the AFS in the studied area.

Chapter 13 - Discharge of effluents produced in the fish-canning industry contributes significantly to the contamination of the environment in the littoral zones where they are discharged. These effluents have salinity similar to sea water, high organic matter content, and high protein concentration. Firstly, solids and oil are separated by physicochemical methods and, then, anaerobic digestion is generally applied to remove organic matter from these wastewaters. However the generated effluent contains high levels of ammonium concentration due to protein degradation, producing effluents characterized by low C/N ratios.

The post-treatment of these effluents by conventional nitrification-denitrification processes is not economically feasible since the addition of an external carbon source is needed. Therefore, the application of processes such as anammox or autotrophic denitrification, where ammonia and reduced sulphur compounds are used as electron donor, respectively, instead of heterotrophic denitrification can be a feasible alternative to remove nitrogen from these effluents.

Chapter 14 - Orange juice is the predominant juice manufactured by the beverage processing industry worldwide. However, this product is not free from microbiological spoilage problems, especially unpasteurized single-strength juice. This challenge study was undertaken to assess the effects of storage temperature on survival of *Escherichia coli* ATCC 25922 and *Escherichia coli* O157:H7 ATCC 35150 in fresh Sanguinello, Tarocco and Moro orange juice [*Citrus sinensis* (L.) Osbeck] varieties. Standard (ascorbic acid) and sensory (anthocyanins)-influencing quality parameters of these orange juice cultivars were monitored in order to detect the limiting quality factor. Microbial and nutrient analyses were conducted every week. The initial concentration (ca. 5×10^8 cfu ml^{-1}) of *E. coli* gradually decreased by about 3 $\log_{10}$ in 4 weeks in all varieties compared to the control samples. *Escherichia coli* O157:H7 cells were reduced in Sanguinello juice with the same trend as *E. coli* whereas the former decreased by about 5 $\log_{10}$ and 6 $\log_{10}$ in Moro and Tarocco juices respectively. Both strains seem to use ascorbic acid for the their survival, they significantly ($P > 0.001$) reduced the ascorbic acid concentrations in all orange juice varieties with respect to variety controls. The strains significantly also decreased the anthocyanins content in Sanguinello ($P > 0.05$) and Moro ($P > 0.001$) juices but not in the Tarocco juice. The reduction of the bacterial number of each strain was highly correlated ($r > 0.99$) to ascorbic acid and anthocyanin degradation levels. Despite its physical and chemical properties and critical conditions it is subjected to during storage, the fresh orange juice can be considered a food suitable for the survival of *E. coli* and *E. coli* O157:H7. Moreover, the ascorbic acid could be held as a good control marker to monitor microbiological quality and safety of orange juice.

Chapter 15 - While there are many studies reporting on the UPR in plant vegetative tissues, the number of publications investigating the effect of UPR on plant seeds is limited. However, as the few publications already show, the effect of UPR in seeds differs dramatically in the number as well as in the composition of genes that are affected regardless of whether UPR has been induced chemically or by the overexpression of recombinant proteins. Results shown so far indicate that, unlike in vegetative tissue, the UPR in embryogenic tissue affects components of the abscisic acid (ABA)-dependent transcriptome. Persistent UPR in developing seeds could thus have implications on seed maturation and seed stability. Here, we attempt to analyze the possible causes of this difference with respect to the specific physiological conditions that are present in the developing seed. Maize starchy endosperm mutants impaired in the accumulation of zeins, members of the prolamin family of seed storage proteins, exhibit several phenotypic features associated with the UPR, particularly those encoding defective zeins. Analysis of a common bean mutant lacking the 7S globulin phaseolin and major lectins revealed increased levels of cell division cycle protein 48 and ubiquitin, suggestive of enhanced endoplasmic retiruculum (ER) associated degradation. However, levels of the luminal binding protein (BiP) were actually reduced, along with those of rab1 GTPase, consistent with a decreased activity of the secretory pathway. These results indicated that BiP levels in seed may vary according to the rate of secretory traffic, and not necessarily as a component of UPR. A perspective integrating information from the different model systems of UPR in seed is presented.

In: Agricultural Research Updates. Volume 2
Editor: Barbara P. Hendriks

ISBN: 978-1-61470-191-0
© 2012 Nova Science Publishers, Inc.

Chapter 1

TABLE OLIVES IN MEDITERRANEAN REGIONS: ECONOMICS, SUSTAINABILITY AND ENVIRONMENTAL CONSIDERATIONS FOCUSED ON A TRADITIONAL FOOD SUPPLY CHAIN

Cátia M. Peres[1,2], Cidália Peres[2,3] and F. Xavier Malcata[1,4]*

[1] Instituto de Tecnologia Química e Biológica, Universidade Nova de Lisboa,
Avenida da República, Oeiras, Portugal
[2] Instituto de Biologia Experimental e Tecnológica, Avenida da República,
Oeiras, Portugal
[3] INRB/INIA, Avenida da República, EAN, Oeiras, Portugal.
[4] ISMAI – Instituto Superior da Maia, Avenida Carlos Oliveira Campos,
Avioso S. Pedro, Portugal

ABSTRACT

A food supply chain encompasses a series of connections and inter-dependencies, spanning from the farm to the final consumer. In Mediterranean regions, diet entails not only supply of macro- and micronutrients – but is also a valuable piece of a multi-century cultural and gastronomic heritage; more recently, the issue of favourable contribution to health has also been more and more often addressed. Said diet consists on a balanced combination of fresh, local and seasonal foods – and includes several fruits and vegetables, and olives and olive oil in particular as major source of antioxidants and fat. Hence, the Mediterranean diet is characterized by a relative richness in complex carbohydrates and fibre, besides unsaturated fatty acid residues and natural antioxidants.

Table olive manufacture remains one of the most important food industries, especially in southern Europe; it is still carried out following artisanal (and thus empirical) practices, with a moderate degree of technological innovation, despite its regional and overall economic impacts. The uniqueness of fermented olives has indeed made them unavoidable ingredients of gourmet recipes and gastronomic excellence – and nowadays health-aware consumers seek them as well for their natural origin and

---

* Corresponding author

functional features. However, the transformation sector is rather fragmented (despite the overriding trend for globalization), and exhibits poor profitability margins – thus being particularly vulnerable to year-to-year fluctuations in production supply and market demand.

It is estimated that no less than 25% of the diet in European countries, and up to 60% of the diet in developing countries consists of fermented foods – a realisation derived from the usefulness of fermentation in extending shelf life of (otherwise perishable) foods and in assuring their safety, but also from the improved sensory properties resulting therefrom. Fermented olive farming is indeed critical for several regions in Spain, Portugal, Italy and Greece – which dominate by far the world supply of table olives. Nevertheless, stricter and stricter environmental legislation, further to competition with other unsaturated fats and oils from plant origin worldwide, have created a threatening situation regarding this specific food supply chain. Survival of this sector will require incorporation of dedicated and less conventional technologies, more efficient traceability and control of the associated processes, and diversification of the product portfolio designed for more demanding consumers.

A new generation of natural (and healthy) foods is clearly on the rise – and fermented olives are a nuclear part thereof. Public health concerted policies aimed at efficiently facing budgetary restrictions will require investment on preventive health approaches – especially as human life expectancy is extended; and the aforementioned types of foods are seminal vectors within this strategy. However, manufacture and marketing of fermented olives hinges critically upon the primary production sector; traditional olive orchards already account for a major share of agricultural land in Mediterranean countries, particularly in marginal, less developed areas. In addition to the socioeconomic impact in terms of maintaining local employment and thus avoiding rural exodus, sustainable exploitation is a must – so that the needs of current generations are fulfilled, without compromising the capacity of future generations to meet their own needs.

On the other hand, the adequacy of the food supply in a particular area or country is also dependent on weather, economic and even political factors; resorting to autochthonous flora that have adapted since ancient times to the local ecosystems is a goal to pursue, hence avoiding introduction of genetically modified or exotic species – while rationally intensifying productivity and resistance to pests. Traditional low intensity, non-irrigated groves often contribute to landscape preservation by decreasing desertification and contributing to a greater biodiversity – e.g. via providing shelter and food for the wild fauna. On steep slopes, terraced olive groves can prevent soil erosion; however, when they are left behind, they turn into shrub that enhances the risk of summer fires (one of the chief environmental hazards in Mediterranean regions).

This chapter covers the various points summarized above – and presents general data useful for characterization of the food supply entertained by table olives, while discussing current problems and anticipating future issues in this field.

INTRODUCTION

General Considerations

A food supply chain encompasses an intricate matrix of connections and interdependencies, spanning from the farm to the final consumer. In Mediterranean regions, diet often entails not only supply of macro- and micronutrients – but is also a valuable piece of a multi-century cultural and gastronomic heritage; more recently, the issue of favourable contribution to health has also been more and more frequently raised. Said diet consists of a balanced combination of fresh, local and seasonal foods – and includes several fruits and vegetables, and olives and olive oil in particular as major source of antioxidants and fat. Hence, the Mediterranean diet is characterized by a relative richness in complex carbohydrates and fibre, besides unsaturated fatty acid residues and natural antioxidants.

Agricultural practices determine the level of food production and, to a greater and greater extent, the critical interface between human development and environmental sustainability – since they produce goods for satisfaction of basic human needs; a sustainable agriculture is a *sine qua non* condition for actual development [1]. The concept of supply chain is thus nuclear to rationalize the food sector; in a sense, it entails a route 'from farm to fork' – i.e. from agricultural production, via agricultural product transformation, to eventual distribution to the final consumers. Studies on the supply chain focus on how economic activities are to be developed, and duly integrated for the best and most balanced outcome.

Sustainable development is a global objective that aims at responding to the needs of current generations without compromising the abilities of future generations to meet their own needs. Hence, the food supply chain overview has significant implications towards sustainability – encompassing e.g. fulfilment of human needs, provision of employment, maintenance of economic growth, and assurance of health and well-being, as well as favourable impacts upon the environment [2]. Any form of sustainable development of products entails social, ethical, environmental and economic considerations pertaining to any goods and services throughout the supply chain. Said global overview also contributes to an increasing awareness of the outcomes of food production and consumption, upon the natural environment ecosystems in an ever developing world [3, 4].

The concept of sustainability has become increasingly important in supply chain management, as companies tend to respond to external pressure from policy makers and consumers – and to internal pressure from their leaders' values and sense of responsibility [5]. Production and distribution of food must be viewed and mastered as a true supply chain [6] – in that it encompasses a system of stages representing a sequence of economic activities, through which resources, materials and information flow downstream and upstream, for production of goods and services intended for ultimate consumption by a consumer [7].

Typical products – and typical foods in particular, exhibiting special features due to local ingredients and unique traditional manufacture techniques, have driven many studies in rural sociology and geography. Consensus assumptions and recommendations have accordingly emerged regarding the historic and artisanal properties of these food products, and their beneficial contributions to rural development. On the other hand, although food industry has a major impact on our lives and the collective economy, but most food is currently supplied by SMEs – including in special those dealing with traditional products.

In recent years, issues relating to public health and other relationships deriving from the supply chain of agricultural products have received a great deal of attention – and emergence of new dynamics in agricultural commodity markets has been witnessed. The key dimensions of such novel rural development patterns encompass creation, operation and evolution of new or alternative food supply chains; the supply chain dimension has indeed become a key element in attempts to better understand the current patterns of rural development [8].

In particular, the environment prevailing in the Mediterranean basin is seminal for the development of this region; continuous environmental degradation has to be discontinued, owing to its unusually high economic and social costs. Although environmental issues associated with contemporary food systems are well documented, there is yet no consensus on an adequate tool (or set of indices) to measure the sustainability of food supply. As a result, there are several disparities as to the appropriate actions that consumers, businessmen and policy-makers should take in order to reduce the negative impacts associated with food supply and consumption – and eventually contribute to an improved quality of life and ecological restoration; means analysis may become a useful tool in the process of identifying the most environmentally benign options for food provision [9].

The current agricultural dynamics is to be rationalized within the context of a more general transition in rural economies – where the creation, operation and evolution of alternative food supply chains is one of the key dimensions [10]. This is why the European Rural Environmental Policy has focussed on the preservation of natural resources and valuable landscapes, by reducing negative impacts of farming on the natural environment [11].

Traditional Foods

The traditional fermented food market has been experiencing an increased consumer interest, consubstantiated on the demand for authentic foods – with a consequently wider variation in flavour, unlike the increasing standardisation and globalisation in this market. However, a stricter and stricter European legislation steering towards more controlled procedures, workplaces and facilities has hampered the regular operation of local producers of artisan fermented foods, mostly SMEs – which are now in risk of completely vanishing from the market [12].

Traditional practices have been passed from generation to generation – and play an important role in terms of cultural and regional identity, based on thousands of years of culinary and gastronomic heritages; traditional foods have in fact experienced continuous modifications, which reflect the history of a region, or even a country. The foreign occupations over the centuries, particularly if they lasted for long, left traces on the food and food manipulation recipes; however, foreign peoples that came to Europe without actually occupying a country have also influenced the existing traditional foods in many regions. The discovery of the New World, and the development of international trade – with the resulting availability of foods unlike happened previously, have also influenced traditional foods all across Europe [13]. Recall that a traditional food may be defined as a food possessing a specific feature or features, which distinguish it clearly from other similar products of the same category. They receive the influence of such factors as availability of raw materials, agricultural habits and location – complemented by the dietary patterns of its inhabitants [14].

Food safety does not necessarily mean food uniformity. The system for ensuring food safety is essentially common to all EU countries, but it permits diversity – so there is a place for traditional foods and local specialities. EC authorities actively protect distinctive or traditional foods associated with certain regions or certain production methods from being unfairly copied by others. In some member States, quality policies have attempted to justify the protection of names and/or collective brands by arguing that what differentiates a product is its specific mode and place of production [15]. In the case of *Origin Labelled Products*, reference to the mode of production is further reinforced by the realisation that quality is "due to a particular geographical environment with its inherent natural and human factors" (for

PDOs), or that "a specific quality, reputation or other characteristics are attributable to that geographical origin" (for PGIs) [16]. A growing segment of consumers prefer food bearing PDO and PGI labels, besides often stemming from an organic mode of production. Marketing of specific, exclusive varieties of olives also constitutes a way of differentiation – even though further efforts are still necessary to explain the differences between the various products and varietal categories.

Stricter and stricter environmental legislation – further to competition with other food commodities from plant origin worldwide, have meanwhile created a threatening situation regarding the specific food supply referred to above. Survival of those sectors will require incorporation of dedicated and less conventional technologies, more efficient traceability and control of the associated processes, and diversification of the product portfolio designed for more demanding and dynamic consumers. Traditional foods are restricted by the current constraints in the market, but they entertain good perspectives to grow in the future, should a number of challenges be effectively addressed. A significant issue for traditional food production is improvement of its competitiveness – by identifying innovations that guarantee the safety and nutritional quality of the product, and at the same time meeting general consumer demands, and specific consumer expectations and attitudes towards traditional food. The challenge will have to consider communication, legal protection of collective brands and quality assurance – which, as happens with all branded products, are a priority underlying innovation [17]. To increase the value of traditional food products, studies of the social, environmental, human and economic impacts of innovation, and of the positive effects on cultural heritage and strengthening of regional links are in order. Most of these challenges will only be accomplished via efficient and jointly shared efforts by the SMEs that produce each kind of product. Simultaneously, improvements are urged in the marketing and distribution chains for such products.

Mediterranean Diet

The primary role of one's diet is to provide enough nutrients to fulfil the basic human requirements, while giving a feeling of satisfaction and well-being. However, recent discoveries in several aspects of bioscience support the hypothesis that, beyond nutrition, diet may modulate various functions in the body. In particular, the Mediterranean diet has been practiced for thousands of years; despite the relatively recent realization of its unique contribution to health in addition to its appeal in sensory terms [18], the Mediterranean diet is regularly described as the golden standard for healthy eating [19].

However, there is not a single, but a variety of Mediterranean diets in the countries bordering the Mediterranean Sea – and even variations are found from region to region within a given country; the differences in culture, ethnic background, religion, economy and agricultural production account for such a variability, even though there is a common Mediterranean dietary pattern [20]. A strong nutritional recommendation from the 1950s and early 1960s encompassed olives and olive-derived products – which are a nuclear part of the Mediterranean diet, at a moment when the fast food culture did not influence our nutritional habits so strongly. Said diet is based on products derived from wheat, olive and grape – i.e. the Mediterranean triad accounted for bread, oil and wine [21]; however, this overview is somewhat incomplete, as legumes also have a great weight (so a tetrad, rather than a triad ought to be hypothesized).

In Mediterranean regions, diet not only entails supply of macro and micronutrients, but is also a valuable piece of a multi-century cultural and gastronomic heritage. Several reports have indeed elaborated on the relatively good health of Mediterranean people being based on the diet, coupled with their culture, history and lifestyle [22, 23]. More recently, the issue of favourable contribution to health has been been more and more often addressed [24] – especially recalling that such a diet consists on a balanced combination of fresh, local and seasonal foods that includes several fruits and vegetables with a complex set of carbohydrates and fibre [25], besides olives and olive oil as major source of unsaturated fatty acid and natural antioxidants [20].

Therefore, it seems that the healthy Mediterranean diet is essentially related to olive products [26]; their health benefits are known ever since Hippocrates in ancient Rome, who claimed them to be nutritious and provide energy, as well as promote regeneration with subsequent therapeutic benefits. For centuries in a row, the nutritional, cosmetic and medicinal benefits of olive oil have indeed been recognized by the various peoples living in the Mediterranean basin [27]. Note that when olive oil is a part of one's diet, it is greatly assimilated by the human body, and apparently contributes to a high longevity [23]; the assimilation of this 'natural fruit juice' is mainly attributed to its high percent of triolein – but the pigments chlorophyll and pheophytin, as well as the accompanying aroma components, facilitate its absorption.

The fat in this type of diet assures 25-35 % of the total caloric intake, with saturated fat accounting for less than 8 % [21]. Furthermore, the large proportion of fish included in this diet makes it particularly rich in ω-3 fats – known to reduce inflammation that contributes to both heart and cancer risks. On the other hand, an abundant supply of fruits, vegetables and beans provides many different phytochemicals that protect blood vessels and guard cells against carcinogens; and recent studies have shown that olive oil may actually bolster each one of these benefits [27]. In a word: the Mediterranean area has the potential to support a pleasing and healthy diet, which merits further consideration with regard to the underlying food supply chain.

Microbial Biodiversity

On a world basis, the sustainability of agriculture depends on its intrinsic genetic diversity – despite humankind being dependent on a more and more reduced number of agricultural biological sources and (often genetically engineered) cultivars. Beyond the number of species and cultivars used to produce food, the genetic diversity within species and populations – including the wild microbiota, is crucial as well. When food producers invest on clones, cultivars and breeds die out – along with their specialized traits, which have a value towards resistance to diseases that is hard to estimate.

The Mediterranean ecosystems nowadays compete with tropical ecosystems in terms of plant biodiversity; however, conservation of this rich biodiversity is a difficult task in the region owing to bio-geographical and political issues – but is surely of the utmost priority. An efficient assessment of said biodiversity requires a variety of biological, ecological and cultural data, the nature and modes of collection of which unfortunately differ from country to country: e.g. there is a bias in the existing literature on EU countries, with scarce information pertaining to northern Africa, eastern Mediterranean and Balkans [28].

Microorganisms are essential for proper function of Earth; they play many roles both on land and in water – including being the first to colonize and ameliorate both naturally occurring and man-disturbed environments. In particular, artisanal table olives are associated with a rich microbial diversity, with a potential that essentially remains to be explored. Regional know-how and microbial biodiversity are better kept in small artisanal units – at the expense of higher fluctuations in product quality. Conversely, simpler and more standard manufacture methods are used in larger industrial plants, so their final products exhibit a much more constant quality level and homogenous specifications. However, the operations that minimize the risk for product spoilage also constrain strain variation – particularly if starter cultures are used.

Microbial biodiversity can decrease relatively fast owing to the rate of microbial replication, spread and exchange of genetic determinants; however, this tendency may also be reversed. Artisanal fermented processes are carried out in microbial-rich environments, where the prevailing specific technological operations cause a gradual selection of the active populations; this approach is inappropriate for large scale production because of the risk of product spoilage to large numbers and batch sizes. It is then essential to develop a strong understanding of artisanal processing on scientific gounds, so as to efficiently scale-up them up. This should entail institutional support to seek and select model artisanal units, to implement efficient sampling schemes, to isolate and characterize the essential microorganisms involved in each case, and to determine the role of external factors upon fermentations – and the effects of these upon the metabolism of microorganisms. It is also necessary to investigate the result of specific pre-treatments of raw materials upon the fermentation process, and to identify the options more suitable for further processing – and of how these affect the taste and texture of the product; finally, the resulting know-how is to be disseminated. This type of research efforts is thus both capital- and labor-intensive.

On the other hand, predictive microbiology techniques can be useful to control the sequence of microbial populations in scaled-up fermentation processes, by fine-tuning a few controllable parameters. Development of starter cultures, preferentially by exploiting the ecological principle of inoculum enrichment by natural selection, will also aid in this process. A multi-strain dehydrated starter may indeed be developed for each special case (i.e. olive cultivar/region/manufacture process) that can be stored at room temperature; this implies the properties of the starter culture to be well-known in advance, and ease of manipulation of the culture conditions.

Fermented Foods

Various fermented foods are consumed around the world – and spontaneous fermentations precede human history; however, humans have been controlling the fermentation process since ancient times, and each nation has its own types of fermented food (that represent the staple diet and the raw ingredients available in that particular place). As a differentiated technology, food fermentation dates back to at least 6000 years; it probably originated from natural microbial interactions that led to positive outcomes in terms of food preservation (and even flavor). Fermentation is also a major requirement for human survival wherever food preserved for long periods is required [29]. General fermented food is usually based on lactic acid fermentation – which add health benefits to food preservation; furthermore, fermentation can transform the flavor of a product from plain and mundane to a mouth-puckering sourness, as enlivened by colonies of beneficial bacteria and enhanced by micronutrients. It is thus essential to ensure that only the desired bacteria (yeasts or molds, for that matter) multiply and grow on the substrate to sufficient viable numbers. Besides releasing desirable metabolites, this also suppresses other microorganisms which may be either pathogenic or cause food poisoning.

Fermented foods can nowadays be in general described as palatable and wholesome foods, prepared from raw or heated raw materials. They are generally appreciated for such attributes as pleasant flavour, aroma and texture, as well as improved cooking and processing properties. Microorganisms, by virtue of their metabolic activities, contribute to the development of characteristic features, e.g. taste, aroma, visual appearance, texture, shelf life and safety – and enzymes indigenous to the raw materials may play an extra role in enhancing these characteristics [30].

Between 25% to 60% of the world diet consists of fermented foods; being one of the first methods used by Man to preserve foods, the usefulness of fermentation in extending shelf life of otherwise perishable foods (and thus assure their safety) is more and more often complemented with improvement of sensory properties [31]. Louis Pasteur the first to scientifically rationalize fermentation in 1857, when he was able to relate yeasts to the process of fermentation – by demonstrating that the latter is caused by microbial multiplication of these microorganisms, rather than to spontaneous generation, which that claimed that complex living organisms might be generated from plain decay of organic substances [32].

Although traditional fermentation processes have evolved to produce high quality food, new threats to safety and quality may arise – that include those relating to the quality of raw materials and to modifications in the traditional processing technology. (Note, however, that fermentation technologies play an important role in ensuring food security to millions of people around the world – particularly a number of more vulnerable groups.) This is achieved through increasing the range of raw materials that can be used to produce edible food products, and removing anti-nutritional factors to make food safe to eat. Fermentations provide a way to preserve food products – thus enhancing the associated nutritional value, destroying undesirable factors, making safer products, improving appearance and taste, salvaging materials otherwise not usable for human consumption, and reducing the energy required for cooking [33].

During fermentation, microorganisms secrete hydrolytic enzymes into the substrate medium – and assimilate some of the fatty acids, amino acids and simple sugars thus released. Fermentation in food processing typically consists on the conversion of carbohydrates to alcohols, carbon dioxide or organic acids, as catalyzed by yeasts, bacteria, molds or combinations thereof. Most fermented fruit and vegetable products depend on lactic acid fermentation [34]; those microbial consortia normally break down carbohydrates, proteins and lipids of the raw materials, thus increasing the contents of soluble solids – which constitutes a nutritionally desirable event, as the food is partially digested prior to actual consumption.

Besides a better digestibility, fermentation also enhances the nutritional value of a food product through increased vitamin levels – as a consequence of the metabolism of some yeasts able to synthesize thiamin, nicotinic acid and biotin. Vegetables, fruits, legumes and grains subjected to lactic acid fermentation also experience increases in both their macro- and micronutrient profiles; the bioavailability of amino acids (in particular lysine owing to its antiviral effects, and methionine) increases typically with fermentation [35]. In the case of grains, sprouting prior to souring will increase the availability of protein even further. Finally, vegetables that have undergone lactic acid fermentation often witness an increase in the activity of vitamins C and A.

On the other hand, certain microorganisms contain cellulases that cannot be synthesized by human beings, but which will soften the texture of some foods and liberate sugars that would otherwise be unavailable; e.g. microbial cellulases hydrolyse cellulose into sugars, which are then readily digestible by humans; likewise, pectinases make the texture of foods less compact and release sugars for digestion [36].

Since fermentation increases the quantity of soluble proteins in foods, it may concomitantly improve the underlying amino acid profile; besides, it reduces the levels of certain anti-nutritional factors that interfere with digestion, so fermented foods are more efficiently utilized through the human digestive tract. As the raw materials become hydrolyzed, the environment also changes – and sometimes undergoes a pH drop. In addition, the extra peptides and amino acids thus formed may be further converted into smaller volatile molecules – that are odoriferous, and thus improve the flavour characteristics of the fermented foods. A better understanding of the biochemical role of microorganisms actively engaged in fermentation would help explain several changes in flavour – but also in texture of fermented foods; both these attributes are important, so elucidation of their evolution could result in development of more efficient and better controlled fermentation processes.

Preparation of traditional fermented foods is more complex and time-consuming than their unfermented counterparts [37]; in fact, the quality and safety of traditional fermented foods is strictly dependent of the quality of raw materials in the first place, as well as on the strain(s) and the processing steps undergone thereby. Most artisanal fermented foods are based on adventitious strains, whereas others resort to natural or stabilized starter cultures; however, the principles of microbial action are identical, with both the type of microorganism and the environmental conditions playing a role upon the characteristics of the final products. By criteriously manipulating the external conditions, the rate and balance of the microbial reactions can be controlled to produce desirable results. There are several ways of altering the environment so as to encourage preferential growth of certain microorganisms; studies of the microbial ecology of fermented foods are now facilitated because of specific molecular techniques that allow the dynamics of each population to be monitored throughout the fermentation process. Moreover, post-genomic approaches can provide innovative methods to ascertain the behavior of microorganisms in structured food matrices, rather than in plain homogeneous suspensions (that are often a poor representation of the actual fermented foods).

There is a long array of beliefs on the medicinal properties of fermented food products – yet the beneficial health effects of lactic acid bacteria on the intestinal flora have been thoroughly documented on scientific grounds [38]; many compounds in fermented foods have been found to possess a protective effect against e.g. cancer development [39]. This is a reason for the current boom in healthy food demand by educated consumers – for whom a wider variability in sensory terms is also appealing, as a sign of less processed and industrialized products. On the other hand, fermentation can improve the flavour and appearance of food – which is helpful in "spicing" an otherwise dull diet, especially in the case of the elderly.

Meanwhile, the European legislation has been oriented toward stricter procedures, workplaces and facilities; hence, local producers of artisan fermented foods have faced a stronger challenge by regulatory authorities that will worsen in the near future. Many of these producers bring about spontaneous fermentations that harbour a unique ecosystem – or resort to back-lopping, whereby a new fermentation is initiated with a small portion of an already fermented material. Such naturally fermented foods are a rich source of potentially interesting microbial strains, owing to their useful metabolites; hence, disappearance of traditional food manufacturers will result in a concomitant loss of microbial diversity, besides the obvious loss of gastronomic specialties. This is why it is essential to safeguard such foods – not only for their role as part of the cultural heritage of a nation, but also as a rich source for technological innovation in the food industry at large.

Although fermentation of foods has been in use for quite a long time, the exact microbial and enzymatic processes responsible for the transformations are still unknown. Without a more fundamental scientific knowledge of the art of traditional food fermentation, rational optimization will be hampered and control will be difficult. It is thus essential to know not only which microorganisms are present and active during food fermentation, but also how their physiology and metabolism are affected by physicochemical (abiotic) conditions. The rapid progress in biological sciences, in both basic and applied aspects, made it possible for one to gain a better understanding of fermentation processes – especially in terms of the microorganism(s) involved; often times, starter and nonstarter cultures have been designed and manufactured to large scale therefrom, which will aid toward more economical and faster traditional fermentations.

A particular interest with regard to food fermentation encompasses table olives, which constitute the basis of one of the most important food industries in southern Europe; despite its local and overall economic impacts, most of it is still carried out following artisanal practices with a moderate degree of technological innovation. The uniqueness of fermented olives has indeed made them unavoidable ingredients of gourmet recipes and gastronomic excellence – but nowadays health-aware consumers seek them also for their biological origin and functional features. However, the transformation sector is rather fragmented despite the overriding trend for globalization; and it exhibits poor profitability margins, thus being particularly vulnerable to year-to-year fluctuations in production supply and market demand.

Healthy Foods

One of the strongest food trends is healthy eating; faced with changing lifestyles and urban globalization, it has become a priority for many families today. The concept of bioactivity – as well as of benefits and safety of traditional foods, constitute interesting topics; advances in food and medical science, coupled with changing consumer demand and demographics have encouraged growth of this market. Consumers began indeed to view food from a radically different perspective back in the 1990s; this 'changing face' of food has evolved into an exciting area where food and nutrition sciences meet to give rise to functional foods. Its significant potential has already attracted large investments [40]; and the industry is well-positioned to respond with nutraceuticals and functional foods to emerging healthcare trends – including personalized medicine and incentives to reduce medical costs via prevention of incidence of diseases [41].

The worldwide market of functional foods has been growing exponentially over the latest years; they already account for 5% of the total food production, with consumers more interested than ever in fortified foods – even those with a variable overall nutrient profile. In the USA only, the current sales of those foods amount to 18.9 billion US$, and have increased at an average yearly rate of 7.2% within 1999-2004 – whereas in Europe they are 5.0 billion US$ and 7.4%, respectively. The role of healthiness in the food choice has increased continuously [42]; driven by science and technology, novel products do regularly reach the market stage. Despite their diversity in nature, they share something: prior to approval for marketing, they must pass a thorough safety assessment process. To assess the safety of new food products, the EC has introduced a new regulation on Novel Foods and Novel Food Ingredients [43].

One measure of the timeliness and impact of this trend comes from realization that 25-30% of all patents recently issued in Europe in the food sector pertain to nutraceuticals and functional foods. Furthermore, the specific item of functional snack foods – as is the case of fermented olives and the like, has undergone a market increase from 1.6 in 2001 to (an expected) 4.0 billion US$ in 2010 [44]. Despite the consistently strong interest since 1998, market studies have revealed little familiarity among US consumers with terms commonly used to describe the concept of functional foods [45].

As population grows older, and thus becomes more susceptible to disease, an excessive burden will be put on the State budgets for health support – especially when the active population is decreasing at an even higher rate. Preventive medicine is surely the key – which possesses the much more important advantage of assuring a good quality of life throughout such longer life expectancies. On the other hand, taking active principles directly obtained from nature and including them in the diet is psychologically more appealing – and, in practice, more effective than systematically taking a pill or another form of medicine with a concentrated active principle (so often obtained from chemical synthesis) [46]. But the outstanding barriers upon consumer acceptance of the relation between diet and health will only be fully overcome if foods are rationally designed, and if sound scientific evidence of their health-promoting features is made available *a priori* [47]. Consumer adhesion to the concept of functional foods, and a better understanding of its determinants are widely recognized as key success factors for marketing strategies, consumer product development and successfully negotiation of market opportunities.

Recall that traditional foods are often the result of agricultural practices aimed at preserving and enhancing rural environments. In fact, their production is very much in line with current EC thinking on rural development, preservation of biodiversity and sustainability. Although a small number of large companies are responsible for over 50% of the sector turnover, the food and drink industry is characterised by a predominance of small and medium capacity. These SMEs, usually experiencing low internal R&D investment, may however become genuine innovators since they often have an expansive business profile and are continuously looking for opportunities of development. Traditional foods are restricted by a few unfavorable conditions in the market, but they have good perspectives to grow in the future should a few challenges be duly met. A significant one is to improve competitiveness by ensuring that proper labelling will underlie quality assurance of the product – by communicating what is different in terms of health claims, while at the same time abiding to general consumer demands, and specific consumer expectations and attitudes towards traditional food. A key aspect of the innovation challenge is thus to assess the adequacy, and transfer dedicated and suitable technology into the traditional food industry [17].

Therefore, the future looks encouraging regarding demand for traditional foods – besides guaranteeing low-environmental impact production, favourably contributing to consumers' health, and perpetuating gastronomic and cultural values, consumers appear to be more and more directly attracted by such food products; furthermore, socio-demographic trends are also in favour of functional foods [48]. Dietary factors and habits are indeed a major tool to aid in reduction of incidence of diet-related diseases and health-conditions [46]; a more clear understanding of the biological and physiological relationships between nutrition and health should take advantage of nutrigenomics and systems biology, and might lead to reformulation of food processes as well as development of novel foods able to target specific consumer groups. This will likely boost the food products themselves, while opening up new markets in the expanding field of nutraceuticals and functional foods [49]. However, this approach should develop along with the commitment taken by EC of an holistic approach described by the motto 'from fork to farm' – in that the primacy of the consumer should override the food chain, while providing safe foods for everyone. The main Cooperation Programme currently active under the 7th EC Framework Program of Research and Development includes specifically this idea; key aspects covered are food quality, safety and consumer concerns, considered throughout the whole food supply chain, while also promoting further research on areas of relevance for the traditional food sector.

Specific scientific information on valuable active ingredients found in a wide variety of traditional plant products and ingredients has proven successful in enhancing the continued use of such crops in local communities [50]. However, new findings will permit entrance in complementary (and higher-added value) markets, thus providing opportunities for creation of jobs and economic growth in a number of depressed rural regions [51].

FERMENTED TABLE OLIVES

Supply and Demand

Market failure to deliver a level of safety sufficient to meet public health requirements and consumer demands constitutes economic grounds for public policy intervention [52]; and olive-related issues surely are covered by this statement. The distinctive geographical, economic and social features of olive growth have been thoroughly highlighted by the International Olive Oil Council (IOOC) – and proposed lines of action encompass expanding global trade, and drawing up standards and procedures for international trading of e.g. olive oil and table olives, among other products [53]. In the Mediterranean basin, olive-related businesses are equivalent to the rural industries of northern Europe; and this conceptual equivalence can be extended, if not to the volume of income, at least to the number of people involved. Olives account in fact for a major fraction of the commercial exchange between North and South.

Olive-related activities constitute the main source of employment and economic activity in many Mediterranean regions, and they have been shaping the landscape and the actual farming in these countries over many centuries. From an overall perspective, European olive production and export are quite impressive, and assure a major world market share, yet the production the sector is extremely fragmented: it is characterized by a large number of small holdings, generally with rather low profitability margins, and thus particularly vulnerable to annual variations in production. The costs of implementing the most recent European environmental legislation – which has imposed stricter regulation upon the agricultural sector, also had a significant impact upon olive farmers; additionally, this sector faces a serious threat arising from an ever growing international competition. To be able to overcome these risks in the coming future, the olive sector needs innovation – via application of novel technologies and development of novel products.

There are ca. 850 million olive trees on the planet, which cover more than 10 million hectares of land – but 98% of them are grown in the Mediterranean area, where they play a major role on environment and rural economy [54]. Moreover, this region accounts for 99% of the world production of olives, and 85% of table olives: more than one million hectares are indeed used to produce the latter. The total production of olives exceeds 18 million tons per year: 90% thereof is used to make oil, and 10% is processed into table olives. The major European producers are Spain (499.000 ton), Greece (108.000 ton), Italy (80.000 ton) and Portugal (19.200 ton, in the 2006/07 campaign) [53].

Of a total of 2,088 million ton of olives, 677,900 ton go for table olive manufacture – and Spain is again the world's leading table olive producer, with a share of ca. 475,000 ton. Over the latest 10 crop years, the world consumption of table olives has risen by 6% – and the EU consumption of table olives accounts for 88% of its production. Despite this self-sufficiency, EU imports and exports represent 14.6% and 36.47%, respectively, of its production [53]; Morocco comes after EU with 23%, and Turkey with 8% of exports, whereas Russia stands out within the top importing countries – followed by USA (30%), Brazil (16%) and Canada (6%) [55].

There is an increasing market for fermented foods in Europe at large due to consumers' realisation that fermentation plays a beneficial role upon human nutrition and health, besides contributing to safety [56]. However, consumers are more and more interested in higher-quality products – for which they know the geographical origin and the traditional mode of manufacture. Hence, it is crucial that olives can abide to consumers' expectations [57]. In fact, traditional foods are often considered healthy and wholesome – and this public perception in terms of nutrition and health has boomed the demand for traditional foods. However, transparency is a must, so consumers have to be informed in full, yet concisely.

The uniqueness of fermented olives has indeed made them unavoidable gastronomic ingredients – and nowadays health-aware consumers seek them as well for their natural origin and functional features. However, the sector faces a serious competitive threat of growing international competition, and is thus particularly vulnerable to annual variations in production supply and market demand from an international perspective; survival of this sector at a European scale will require changes in processing and management, as well as incorporation of new technologies and more efficient control of the processes – together with adaptation of its products to consumers, and launching of new products.

Olive Groves

The adequacy of food supply in a particular area or country depends on weather, economic and even political factors. Classical low intensity, non-irrigated groves often contribute to landscape preservation by decreasing desertification and contributing to a greater biodiversity; they actually provide shelter and food for the wild fauna. Furthermore, terraced olive groves can prevent soil erosion on steep slopes; however, when they are left behind, they turn into shrub that enhances the risk of summer fires (one of the major environmental hazards in Mediterranean regions).

The wild olive tree, mainly *Olea europea* subsp. *sativa*, is an important crop in the Mediterranean basin (with recent successful introduction in USA and Australia) – and possesses a strong historical, symbolical and economical relevance, as an icon of the agriculture practiced in unfavored regions. The domesticated olive tree is of an ancient origin, probably by the dawn of agriculture; it likely originated in the Middle East, and spread south- and west-bound to the remainder of the Mediterranean basin through movement and trading activities of Phoenicians and ancient Greeks. European olives are in fact as old as the civilizations that encircle the Mediterranean Sea, so olives have historically been an essential foodstuff for inhabitants on these regions. The domestic olive belongs to *Olea europaea* subsp. *europaea*, and includes a great number of cultivars and semi-wild forms that have evolved via hybridisation and selection from the original species.

As part of the traditional food supply chain throughout the Mediterranean zone, the olive production area and volume have been steadily increasing – so olive trees have played for ages an important role in rural development in southern Europe, northern Africa and Near East; nowadays, olive cultivation is still the basic tree cultivation in the Mediterranean basin, and dominates its rural landscape. Olive-related businesses have constituted one of the major sources of income and employment there [58], and have helped constrain the rural exodus in those relatively poor areas – thus reinforcing conservation of natural resources. The olive sector accounts indeed for a great percentage of the total employment in the agriculture sector therein, so it is sometimes called a "social crop" – even more than in such other dynamic agricultural subsectors as horticulture. Simultaneously, soil erosion is one of the worst environmental problems in those zones, with severe downstream effects; erosion in olive production areas is currently high, especially on sloping land where several erosion risk factors are present – viz. erosivity of rainfall, erodibility of soils, steep slopes and insufficient ground cover because of clean weeding. Preservation of olive trees is thus compatible, and even actively helpful in preventing environmental decay.

Domesticated olive trees also contribute to maintain biodiversity at high levels; recall that biodiversity is the variability among living organisms from all sources, including diversity within species, between species and of ecosystems themselves [59]. Hence, biodiversity applies not only to plants and animals on Earth, but also to microorganisms that naturally contaminate them – and live under relationships of mutual benefit, commensalism or even predation/competition; their interplay as a whole assures important functions in element cycles that are required to sustain the natural patterns. On the other hand, olive cultivation and tourism are strongly linked to each other in the Mediterranean area; such an interactive and mutual dependence is apparent in the population, the culture, the rural landscape, the economy and the natural environment [60]. Since olive tree cultivation is part of the traditional set of food resources, the associated agricultural practices can preserve and enhance rural environments – so they are in line with current EU policy on rural development that entails preservation of biodiversity and sustainability as major focus. This trend counterbalances previous orientations toward intensifying olive tree production in the past – which have caused unwanted environmental damage [61].

Technological issues are critical in the final costing of the olive products, along with land and labour costs; competitiveness actually hinges on a balance between these factors and world demand. As happens with many other typical Mediterranean crops, olive growing has followed basically the same procedures for centuries – but new planting, irrigating, harvesting and processing techniques have been under development and implementation, and adopted to higher and higher levels. Therefore, three major types of olive growing technologies in EU may be pinpointed: traditional groves, managed traditional plantations and intensive plantations.

Traditional olive growing in groves and extensive plantations is an intrinsically low-intensity production system – encompassing usually old trees, grown at low densities, producing low yields, and receiving poor inputs of labour and materials. Though such systems are environmentally sustainable, their economic viability has been questioned; EU orientations have been in favour of more intensive and thus more competitive systems. Olive groves that have not undergone intensification seem to be threatened by the recent reform of the EU olive and olive oil policy, since income support has been decoupled from production [62]. The causes and consequences of abandonment of this traditional mode of production have revealed that such systems are barely sustainable from an economic point of view: their viability is only assured if reduced opportunity costs for family labour exist, or if olive growing is a part-time activity. However, preservation of traditional olive groves is benign for the environment, owing to the systems they represent and protect; in addition, they contribute positively to maintain the levels of soil erosion low [18] – a particularly important issue since groves are frequently on sloping land, i.e. the most susceptible to erosion.

Managed traditional plantations involve a larger use of inputs, including more mechanization and irrigation. Under these circumstances – and despite some variability between regions, harvesting accounts for at least 50% of the production costs, as it heavily depends on the labour costs. It should be noted that many olive producing areas in Europe are mono-varietal, so the majority of the olives ripen at essentially the same time, thus leading to critical seasonal labour shortage. Dense groves allow for a greater use of mechanization, yet mechanization is only justified on holdings of a minimum size and yield per hectare. On the other hand, the intended final use for olives makes a difference: table olives, in particular, require careful picking.

Finally, Super High Density (SHD) olive groves are a fairly new development, and are found mainly in southern Spain – but are also expanding in other countries [63]. They are generally irrigated via a drip system, and are heavily pruned; this means that only very few varieties are adapted to this cultivation system, which is estimated to reduce harvesting costs by as much as 80%. Another advantage is that the olives can be moved faster from the tree to the processing plant, thereby reducing the chance for spoilage. Adequate cultivars generally start producing within 2 years of plantation, and reach their full maturity potential by the fourth or fifth year. However, there are several limitations to SHD production – the first and most significant being the high costs of installation. Consequently, there is a mere 25,000 hectares of SHD plantation, as compared with the total of 17.9 million hectares of olives planted around the world. Secondly, this intensive system accommodates only a limited number of varieties, and requires irrigation. On the other hand, it does not work if the olives are planted on a slope. New players in the production of olives, e.g. California, Chile, Argentina, South Africa and Australia, are better positioned to adopt SHD technologies because they are not as dependent on traditional production systems. In addition, many well-established agricultural industries are looking for new profitable crops – as is typically the case of California, where grape producers are diversifying their investments [64]. Eventually, the major obstacles are moving from production to processing: lack of processors, bottlers and marketers of olive oil appear to have hampered successful exploitation in the New World.

In areas characterized by poor agricultural structures and limited employment opportunities, the economic, social and environmental factors of sustainability seem to favour stability – although a right balance between these systems is mainly dependent on the economic factor [65]. The recent possibilities in olive grove exploitation will surely contribute to put the traditional system in danger, with obvious negative consequences upon the environment and the local economies; this problem is particularly acute when the possibility of changing the use of land into pastures is considered. This is why a concerted effort to preserve traditional olive groves is deserved, owing to their environmental relevance – particularly in southern Europe, where the ecosystems are more critically dependent on olive groves. This obviously requires low intensity olive farming, and consequent reduced use of agrochemicals, as well as combination of old olive trees with semi-natural herbaceous vegetation [66].

Desertification Problems

The production of olives, particularly in southern Europe where olive trees are particularly well-suited to the prevailing harsh and dry conditions, requires a complex analysis of environmental impacts – of which desertification is a good example.

Traditional low intensity, non-irrigated groves often contribute to landscape preservation by decreasing desertification – and contribute to a greater biodiversity, by providing shelter and food for the wild fauna. Furthermore, terraced olive groves can prevent soil erosion, especially on steep slopes. However, when these olive groves are left behind, they turn into a form of shrub – which, if not duly maintained, constitutes a chief environmental hazard in regions characterized by particularly dry summers. In addition, such groves meet the social objectives of keeping people living in remote areas. As a result of their particular plantation characteristics and associated farming practices, low input traditional plantations have potentially the highest natural value – in terms of biodiversity and landscape value, as well as most positive effects – e.g. water management in upland areas.

The intensified traditional, and even more the modern intensive systems are inherently of a lesser value to nature – and may in practice pose greater negative environmental impacts, particularly in the form of soil erosion, run-off to water bodies, degradation of habitats and landscapes, and depletion of scarce water resources. In large areas of olive monoculture, as happens in Andalucia (Spain), olive production has already caused appreciable problems of soil erosion [67]. Biodiversity decay has chiefly resulted from the increased use of fertilizers and herbicides that contaminate water systems; and high levels of cultivation lead to excessive use of water – it is estimated that 300 million m^3 per year of water is already consumed to irrigate olive farms [68]. In regions where water supply is limited, the environmental problems raised by irrigation are harder – and the associated waste run off contains usually high levels of organic substances. Processing table olives also produces highly polluting liquid waste, because of a high organic content and traces of sodium from the brines used in processing.

Agriculture has for some time been recognised as an important vector of soil degradation, especially in slopping areas and others constituted by poor quality soils. Mediterranean regions are accordingly recognised among the most threatened areas worldwide – and Spain, Portugal, southern Italy, Greece and Cyprus are particularly endangered. More specifically, the main causes claimed to promote desertification are overgrazing, combined or not with fire occurrence and current agricultural practices – mainly tillage that favours soil erosion. In addition, irrigation with low quality water may also lead to detrimental effects on soil structure [69].

Desertification is currently recognised as one of the most severe environmental traits; it causes decreases in food production, famines, increased social costs, and decline in water resource availability and degradation of their quality – further to decreased soil productivity [70]. Olive trees have been reported to contribute to desertification mainly on groves cultivated under intensive and semi-intensive systems [71, 72]. Restoration of the original vegetation is an effective way to control desertification, while rehabilitating degraded land. The use of native plants, adapted to local environmental conditions, constitutes a successful solution. Note that olive trees tolerate drought, salinity and poor soils, so they are particularly adequate for degraded soils in Mediterranean climates; hence, *O. europea*, in conjunction with other native species, are potential candidates to upgrade otherwise spent areas [73]. However, implementation of consistent programs and adoption of common policies, both at State and EU levels, is required to reverse the impacts of intensive olive tree cultivation upon the environment – knowing that market mastering based on the high current prices of olive oil may encourage olive tree plantation temporarily, but may fail to support efficient strategies on the long run.

Environmental Impacts

Under pressure of the prevailing economic situation in agriculture and the irreversible trend toward globalization, olive groves tend to carry over from traditional inclined, shallow, arid and barren areas into plain and irrigated areas – where mechanization will be possible. Consequently, cultivation of olive trees is currently being pushed out to the most fragile and poor areas of the Mediterranean basin – where a higher probability of becoming deserts exists, with devastating consequences for the environment and the economy of those areas. The importance of olive groves as an element of identity and conservation of life in that area thus justifies its preservation – and farmer multiactivities may positively contribute toward this deed, while promoting sustainable rural development. In this way, olives may continue to offer the gifts as they have done for so long [18].

One of the most severe menaces comes from trade liberalization: increased agricultural activities accelerate resource depletion in most regions, besides enhanced use of fertilizers – and are unmotivated to apply regulatory measures that cope with future environmental degradation. But cultivation of olive trees thus stems out of environmental considerations, well beyond human alimentary needs [18]; and the centuries during which it dominated the Mediterranean landscape have provided a strong evidence for their supporting the equilibrium of Mediterranean ecosystems, and their intrinsically rich biodiversity. Thanks to the conservation of soils and the limited inputs required, olive cultivation contributes clearly to the sustainability of existing natural resources.

Although the importance of olive groves in the life of the Mediterranean basin is unquestionable, such an ecological richness has been constrained in recent times due to modernisation of agricultural practices – with recent relevant changes in the olive sector arising from generalization and intensification of farming. Hence, olive groves tend to carry over from traditional to plain and irrigated areas, where its mechanization is possible, with concomitant marginalisation of low-input farms [18]. Consequently, cultivation of olive trees has been moving away from the most fragile and poor areas of the Mediterranean basin, and several problems related to environmental sustainability have been on the rise in the areas left behind – not only of a socio-economic nature, but also as unfavourable impacts upon the local ecosystems (e.g. erosive phenomena, water pollution or biodiversity losses – including hampering of the natural regeneration of vegetation, with decrease of land coverage by woody species).

As a result of the aforementioned changes in olive exploitation around the world, the world area cultivated with olive trees has never been so large; note that the EU States, under the framework of the Common Agricultural Policy (CAP), encouraged enlargement and intensification of olive systems during recent decades – with resulting higher density of plantation and introduction of irrigation. Such a rapid expansion has caused sustainability problems [74, 75, 76, 77]; and intensification has also brought about extra water requirements, with consequent overexploitation of water resources – coupled with a reduction in the quality of water, because of regular use of agrochemical products (mainly herbicides and fertilizers).

One of the main features associated with traditional olive groves is the high biodiversity associated with the crop: trees and shrubland provide an assorted habitat similar to meadows, where a vast number of insects, birds, reptiles and mammals find shelter and feed. However, olive farm intensification has changed this situation, since it has led to disappearance of said vegetable cover, pollution of surface waters, use of insecticides and soil erosion; coexistence of olive trees with other crops, e.g. pastures, vineyards or cereals, has also been compromised.

Traditional Fermented Foods

Traditional fermented foods of artisanal manufacture have earned a status of gastronomic specialties – and are in most cases produced by SMEs with a strong link to the "terroir", and possessing strong cultural, historical and lifestyle-related ties. They are produced from raw materials, from animal or plant origin, usually by spontaneous (or minimally controlled) fermentations brought about by the adventitious microbiota – sometimes stabilized in specific starter/nonstarter cultures with commercial dimension; the latter are associated chiefly with dairy products. Europe harbours a large variety of such traditional fermented foods, reflecting artisan and region-dependent handling – duly labelled as 'Protected Designation of Origin' (PDO) and 'Protected Geographical Indication' (PGI) [78]. That microflora harbour a wide biodiversity of lactic acid bacteria and acetic acid bacteria, besides yeasts and moulds – and contributes to Europe's cultural food richness. The market share of traditional fermented food products has been growing throughout Europe as a consequence of consumers' increasing awareness of the gastronomic quality, unique features and healthy properties of such foods [17]; and this trend is expected to be reinforced with the joining of new member States, given their many traditional fermented food products produced locally.

However, two unfavourable issues may interfere with this trend, thus compromising the market share and the contribution to the local biodiversity. First, the modern European consumer has more stringent requirements concerning freshness, wholesomeness, safety, taste, novelty and healthiness (viz. low-contents of sugar, fat and salt) of foods, despite the revival of consumers' interest on traditional foods. Second, the tightened European legislation on food safety – including such items as Hazard Analysis of Critical Control Points (which has a severe impact upon small-scale artisan producers), results in lower production flexibility and higher production costs [79]. In the worst case scenario, the regulatory issues posed by EC concerning hygienic manufacturing and handling procedures may lead to disappearance of regional and artisanal products, e.g. traditional beers, wines, cheese and fermented meat products – and obviously fermented vegetables. However, integrating traditional technologies with modern quality-controlled facilities may provide a way to create new fermentation processes, and to build on existing market opportunities; dedicated R&D efforts will in particular permit a deeper knowledge of olive matrices, and thus lead to a more rational optimization of their manufacture processes.

Table Olive Processing

Table olives are currently one of the major fermented products consumed worldwide; recall that fermented olives are prepared from the sound fruits of suitable varieties of cultivated olive trees. A large diversity of recipes, based on both ancient tradition and modern creativity, will eventually lead to strengthening of this type of product in the market [80].

Table olive processing involves transformation of bitter (and thus essentially inedible) olives into an edible foodstuff. Even at full maturation, olives are hardly edible as such because of their bitter-tasting compounds – primarily oleuropein, which are to be eliminated prior to consumption. Processing also contributes to keeping olives away from spontaneous deterioration, so that table olives can be stored and consumed gradually throughout the year at will. Besides being palatable, the processed olives should be safe upon ingestion, and able to retain most of their nutritional attributes. Complementary preservation techniques, e.g. pasteurisation and heat sterilisation, have also been employed for some packaged olive products [81].

Olive cultivars particularly suited for use in table olive preparation should be larger in size and lower in oil content than olive cultivars intended for oil expression. Because overall appearance affects the preference of consumers at the point of selling, harvesting is always carried out manually to avoid fruit wounding. There is quite a number of olive products available worldwide, with each olive growing country producing several local and regional styles, further to various processing protocols; these depend on olive variety, degree of ripeness, processing technology, cultural factors and empiric experience. Although in principle a great many olive cultivars can be processed by the methods available, from a commercial point of view specific varieties are preferred because of technological and organoleptic convenience, as well as features that eventually dictate consumer preference [81].

Recal that olive products are appreciated not only for their nutritional quality, but also for their health benefits. Among edible oils, olive oil is indeed the only one that can (and must, on legal grounds) be produced solely by physical methods from fresh fruits, and which may be consumed immediately after pressing (as happens with most fruit juices). Likewise, a simple process of treatment using either dry salt or salted brine is required to manufacture table olives. The flesh – also called the mesocarp, of the fresh olive is rather juicy; depending on the variety and maturation level, it ranges from yellow green to deep purple black in colour. The skin – also known as epidermis or epicarp, provides a protective barrier against access to the internal nutrients. It is mostly a continuous structure, except for numerous small openings (or lenticels) that allow the olive fruit to transpire. The olive skin also serves as docking support to a rich microbial population of bacteria, yeasts and fungi – which proliferate during spontaneous fermentations, whereas some may cause food spoilage, and more seldom poisoning. The olive fruit is thus a drupe containing a bitter component – oleuropein, low sugar concentration (2.6-6.0%) and high oil content (12-30%); these values usually change with the degree of maturity and the variety [82]. Such characteristics prevent olives from being consumed directly from the tree, and have as well promoted appearance and improvement of several processes to improve their edibility that are considerably distinct from region to region.

The basic commercial preparations of table olives in the international market are dependent on the maturation stage of the fruit; the most important preparations are: the green Spanish style; the olive ripening by alkaline oxidation – the so-called Californian style; and the naturally black olives, also known as Greek style [82, 83, 84]. Recently, such other products as seasoned green table olives have earned the favour of consumers, within a sensory diversification of the product portfolio [85]. The international regulatory agencies have realised the progressive importance of those products by including them under the heading 'Specialties' in the current 'Trade Standard Applying to Table Olives' [53]; nevertheless, most common commercial preparations are green olives prepared according to the Spanish style, or ripened as black olive brining.

The Spanish-style green olives in brine is indeed one of the most important commercial preparations available in the international market, and has undergone a steady increase; the overall production in EU is ca. 500,000 ton/year, most of which (ca. 60%) is produced in Spain [53, 86]. However, most fermented olives on the market still result from spontaneous fermentations, which will likely lead to variations in the final sensory and physicochemical profiles. In the case of green olives prepared by the Sevillian style, fruits are harvested at their maximum size – and just before colour change; and debittering afterwards is achieved by lye treatment with a diluted solution of NaOH, which increases permeability of the fruit by breaking cell walls, thus releasing their contents. The dissolution of NaOH in water is exothermic, so it warms up the processing medium that facilitates diffusion; the alkaline treatment hydrolyzes oleuropein, via conversion to glucose and oleuropein aglycone – and this one is, in turn, converted to simpler non-bitter compounds, e.g. elenolic acid and β-3,4-dihydroxy-phenyl ethyl alcohol, and other compounds able to inhibit microbial fermentation [87, 88].

The growth of microorganisms and their metabolic activity are affected by extrinsic and intrinsic factors prevailing in the fermentation process – viz. oxygen limitation, low glucose and high salt concentration, and low pH [89, 90, 91, 92, 93, 94, 95, 96]. It should be stressed, however, that the physicochemical changes in the main substrates (e.g sugars) and end products (e.g. organic acids, volatile compounds and biogenic amines) during spontaneous fermentation of green table olives has not been fully elucidated to ate [97]; in particular, the type and concentration of main metabolic products responsible for its typical profile – e.g. D- and L-lactate, acetate, formate, ethanol, free amino acids and biogenic amines, and their implication in the final flavor are not known in detail. If this knowledge existed, it would be helpful to highlight situations where evolution of the fermentation process would be termed abnormal [98]. The manufacture of table olives is thus an extremely difficult task that is often affected by costly losses of raw material due to faulty fermentation; on the other hand, proper fermentation would produce a high quality, unspoiled, consistent and safe product – but requires a deeper knowhow.

In the Spanish-style method, fruits are then placed in a brine solution of ca. 10 % NaCl, where nutrients diffuse from the fruit pulp to the brine – while NaCl diffuses into the olives, until equilibration is reached prior to the onset of fermentation. Brenes-Balbuena et al. [99] demonstrated that the concentrations of tyrosol, p-coumaric acid and vanillic acid remain constant throughout processing, whereas the concentrations of caffeic acid and hydroxytyrosol decline markedly after lye treatment; this observation was attributed to differences in chemical structure, and the fact that the latter two compounds possess both an o-diphenol group.

According to IOOC, the designation of "natural olives" may be applied to fruits at various maturation stages – green, turned colour and black; and which have not undergone any alkaline treatment. Natural olives are harvested when the fruits are green, or semi- or fully ripe; depending on the region, such olives can be reddish black, violet black, deep violet, greenish black or deep chestnut, with both skin and flesh being coloured [80].

After brief sorting and cleaning, olives are immersed in fresh water for a gentle washing for 6 to 8 d, with several water changes; it will give place to brine containing 8-10 % NaCl, where olives undergo fermentation – the final results of which depend on the prevailing physicochemical conditions, cultivar, temperature and chiefly salt content. The fruits are kept in this solution until they lose (partially or totally) their original bitterness. As per market demand, olives are sorted, graded and packed; and in some commercial presentations, they can be broken or cut along their longer longitudinal diameter, and/or seasoned with natural products or their flavours [80]. Microbial communities will then establish and start multiplying, although the contents in polyaromatic compounds and tannins (1.5 to 2.5%) can prove inhibitory – coupled with low nutrient concentrations. Equilibrium of pulp and brine nutrients takes place within 2-3 mo; and containers are kept full and tightly closed during fermentation, to avoid growth of oxidative organisms. Salt is periodically added to the brine to help in selection of the desired microbial population, as pH gradually decreases.

Most black ripe olives are processed by placing them directly in brine after picking, or submitting them to repeat soaking in water – where they undergo spontaneous fermentation; the finished product retains some fruity and bitter flavours. During processing, olives do often lose their intense black purple pigments, thus resulting in pale to dark brown coloured olives – but the original colour can be (partially) restored by exposing them to air after processing.

They are preserved in brine, following sterilisation or pasteurisation, or else by addition of a suitable preservative.

Semi black ripe olives can also be processed by an initial alkaline treatment with light lye, followed by oxidation under particularly oxidizing conditions; they are then preserved likewise in brine. Olives intended for production of ripe olives by alkaline oxidation are previously preserved in an aqueous solution – e.g. brine or acidic water, and darkened throughout the year. Darkening consists of several treatments with dilute NaOH solutions, with water washes between them; and during the oxidation process, air is passed throughout the suspension of olives in the liquid. Once the olives attain the proper colour ring around the outer surface, this is fixed by immersion in a lactate or gluconate solution containing iron cations. These olives are usually packed in light brine, and their commercial presentations span plain or whole, pitted, sliced and even olive paste [82].

Note, finally, that there are many other traditional and industrial ways of processing table olives, which differ in their fermentation conditions – temperature, level of salt and type of acid, or raw material – green, turning colour or black olives; a comprehensive account thereof is provided elsewhere [81, 82, 83].

Lactic Acid Fermentation

Fermented foods are produced worldwide according to various manufacturing techniques, using distinct raw materials and microorganisms [100]. Fermentation is rather useful because it is a form of biopreservation, in that it extends the shelf life of the food by minimizing unwanted bacterial contamination – based on ecological principles (or outrunning in viable numbers), as well as active antimicrobial action (via release of e.g. bacteriocins or organics acids). Despite the primary purpose of preservation, fermentation also enhances the sensory attributes of the processed product [101].

Table olives are probably the most popular fermented vegetable in the western World, as well as a core ingredient of the so-called Mediterranean diet together with olive oil; despite their underlying economic importance, the degree of innovation in this sector is very limited [80, 102]. The need to design appropriate control processes for table olive fermentation is often forgotten, yet shifting from craft to industrial technology would surely avoid undesirable fermentations – thus also avoiding unwanted losses in product quality and economic value, while leading to higher (and more predictable) yields [103, 104, 105].

In general, the manufacture of table olives encompasses a spontaneous fermentation, and generates a sensory profile that depends on the cultivar and the native strain – besides the degree of maturity at harvest and the steps of postharvest processing. During fermentation, native LAB release lactic and acetic acids, as well as small molecules responsible for flavour (e.g. aldehydes) and other antimicrobial compounds [106, 107]. Despite this general, well-established picture, a more fundamental understanding on the biochemical and microbiological aspects of table olive fermentation is relatively recent – even though the art of preparing the aforementioned gourmet snacks lies somewhere in Antiquity [87, 108]. As happens with other fermented vegetables, traditionally fermented olives harbour a unique biodiversity [109] – particularly of LAB and yeasts [110], which produce metabolites against, and ecologically dominate over contaminating microflora that would otherwise eventually pose eventual sensory and health hazards [111].

The action of lactic acid bacteria (e.g. *Lactobacillus* spp.), as well as yeasts upon such fermentable substrates as sugars released from olives during soaking, constitutes a major issue; although those beneficial bacteria influence fermentation in a variety of ways, their most important role is the production of lactic acid, which causes a drop in pH and a concomitant increase in acidity – thus ensuring microbiological stability of the product as storage time elapses. The combination of high salt contents and low pH values greatly reduces the risk of microbial spoilage of table olives on the long run [112]. However, care is to be exercised to reduce the risk of overgrowth of undesirable or harmful microorganisms, which might lead to product decay and even toxinfections upon ingestion. Process control usually focuses on keeping the salt and acid levels at preset values, via criterious adition of sodium chloride and food acid(s), respectively. It is desirable that the fermentation will eventually attain predominance of the *Lactobacillus* genus, and preferentially of the *Lactobacillus plantarum* species.

Experimental studies developed encompassing the ecosystem olive/brine under commonly encountered conditions have indicated that special strains of adventitious lactic acid bacteria can use phenolic compounds as the only carbon source, when in the absence of sugars [113, 114, 115]. One representative example is oleuropein, the dominant phenolic compound of olives: it usually undergoes acid hydrolysis, but is also susceptible to the action of β-glucosidase synthesized by *L. plantarum* strains [116]. More recent studies have shown that a relatively high fraction of strains tolerate, and even degrade oleuropein: use of oleuropein and X-Gluc as substrates by lactic acid strains associated with the natural fermentation process of green olives has been proven elsewhere [117]; however, it is not known whether lactic acid bacteria can breadown oleuropein via any route besides acid hydrolysis.

Yeasts play a double role in olive fermentation: they influence their flavour and texture [118, 119], but they also appear to be active in synthesising vitamins, amino acids and purines, or breakdown complex carbohydrates – all of which are essential deeds to promote growth of *Lactobacillus* sp., which are nutritionally fastidious [120]. Thiamine (vitamin B_1), nicotinic acid, pyridoxine (vitamin B_6) and pantothenic acid are among the vitamins and other enzyme cofactors accumulated and/or synthesized by yeasts *in situ* [121].

Health Promotion

Fermentation is among the oldest methods used for food preservation – and lactic acid fermentation has for centuries been the predominant biological preservation method for vegetables [122]; as emphasized before, fermentation also contributes to unique flavor development, and lactic acid bacteria are the major actors therein. The association of these beneficial bacteria with the human environment, and their numerous beneficial interactions – both in food and in the human intestinal tract, combined with the long tradition of including lactic fermented foods in many cultures, led to granting of these foods with a 'generally recognised as safe' (GRAS) status. The *Lactobacillus* genus is a heterogeneous group; the ability of its various species and strains to colonize a variety of habitats is a direct consequence of their wide metabolic versatility – with lactobacilli implicated beyond doubt in starter cultures for dairy products, fermented vegetables, fish and sausages, as well as silage inoculants [123].

Lactic acid bacteria do not constitute a family in taxonomic terms; they are instead a physiologically diverse group of Gram-positive, non-spore-forming bacteria, which can be further described as anaerobic – and with lactic acid being the major product of carbohydrate metabolism. They are aerotolerant, but generally lack catalase (although a pseudocatalase can sometimes be found); they are chemo-organotrophic, with high nutritional needs. Two major pathways are used thereby for energy production: homolactic species utilize hexoses by the glycolytic pathway, whereas heterofermentative species follow the 6-phosphogluconate pathway [124].

Due to their stringent nutritional needs, lactic acid bacteria exist in nutrient-rich ecological niches, including such fermentable vegetables as table olives, sauerkrout, pickles and wine, meat and fish products, dairy products and cereals. In those habitats, a variety of microorganisms compete with each other for survival – and have adapted to eventually constitute a unique microflora. The competitive ability of LAB is a result of main primary metabolites – chiefly lactic and acetic acids, ethanol and carbon dioxide, as well as other antimicrobial compounds, viz. formic and benzoic acids, hydrogen peroxide, diacetyl, acetoin and bacteriocins [125, 126, 127, 128, 129].

The antimicrobial effect of organic acids in the culture medium is associated not only with pH reduction, but also with its undissociated form [130]. The extent of dissociation decreases as the medium pH fals down; the undissociated form of acids can diffuse passively across the membrane, owing to their lipophilic character. Once inside the cell, the acid will easily dissociate at the higher cytoplasmic pH: it acts by collapsing the electrochemical proton gradient, or by altering the cell membrane permeability – thus leading to disruption of the substrate transport systems [131, 132, 133].

Different organisms vary considerably in their sensitivity to lactic acid – particularly the more toxic undissociated form: pH 5.0 is inhibitory to spore-forming bacteria, but essentially ineffective against yeast and moulds [134]. Lindgren and Dobrogosz [135] showed that the minimum inhibitory concentration (MIC) of the undissociated acid molecules is different against *Clostridium tyrobutyricum*, *Enterobacter* spp. and *Propionibacterium freudenreichii* ssp. *shermanii*, depending on the pH range at stake. Acetic and propionic acids are also produced by certain strains via heterofermentative pathways; their antimicrobial efficiency is higher than that of lactic acid, due to their higher pK_a values. Acetic acid may also act synergistically with lactic acid in lowering pH, thus increasing its own antimicrobial capacity [136].

Under certain environmental conditions, polyaromatic compounds can result from lactic acid bacterium metabolism – e.g. benzoic acid, 5-methyl-2,4-imidazolidinedione, tetrahydro-4-hydroxy-4-methyl-2H-pyran-2-one and 3-(2-methylpropyl)-2,5-piperazinedione; those microorganisms can also produce small amounts of ethanol and carbon dioxide through the lactic heterofermentative pathway. The former results from decarboxylation of pyruvate to acetaldehyde, which is further reduced to ethanol; if this second step is incomplete, acetaldehyde may accumulate in the medium. Hydrogen peroxide is produced in the presence of oxygen, as a result of the action of flavoprotein oxidases or NADH peroxidase. Its antimicrobial effect lies on oxidation of sulphydryl groups in proteins – thus denaturing several types of enzymes, as well as on peroxidation of membrane lipids that lead to increased membrane permeabilities. H_2O_2 may also be a precursor of bactericidal free radicals, such as superoxide (O_2) and hydroxyl (OH·), which can compromise DNA chemical integrity [137].

Diacetyl is produced by citrate fermentation; it affects arginine utilization in bacteria by reacting with the arginine-binding protein. Jay (1986) [138] showed that Gram-negative bacteria are more sensitive to diacetyl than their Gram-positive counterparts; however, diacetyl may act synergistically with other antimicrobial compounds.

Reuterin was found during anaerobic growth of *Lactobacillus reuteri*, and a glycerol dehydratase apparently catalyzes its synthesis from glycerol [139]. It is chemically defined as 3-hydroxypropanal (or β-hydroxypropionaldehyde) – a highly soluble and neutral compound, which usually exists in equilibrium between their hydrated monomeric and cyclic dimeric forms [140]. According to these authors, reuterin is active against *Salmonella*, *Shigella*, *Clostridium*, *Staphylococcus*, *Listeria*, *Candida* and *Trypanossoma* spp.

The selection of strains for food fermentations usually entails characterization of their potential to produce and release bacteriocins [141]. Some of them may in addition be able to survive passage through the gastrointestinal tract, and accordingly act against pathogens installed therein (e.g. in the stomach or in the small bowel) and eventually colonize the large bowel [142, 143, 144]. Besides brined table olives, another matrix bearing a high potential is olive pastes fermented with adventitious *L. plantarum* and *Lactobacillus pentosus* [107]. Evidence for some form of probiotic activity has been made available – viz. release of bacteriocins by those strains cultivated *in vitro* [106].

Despite their fastidiousness [145], *Lactobacillus* spp. have been found to act against pathogens [146] – mostly via synthesis of major products of primary metabolism (e.g. organic acids, ethanol and carbon dioxide), or by-products and other minor compounds (e.g. bacteriocins, cyclic dipeptides, reuterin and reutericyclin, hydrogen peroxide, acetaldehyde, diacetyl and acetoin, and such 3-hydroxylated fatty acids and polyaromatic compounds as phenyl-lactic and benzoic acids) [147, 148] and pyroglutamic acid [149]. Preliminary data encompassing wild strains of *Lactobacillus* spp. in fermented olives have pointed specifically at inhibition of clinical strains of *Helicobacter pylori* [150]. This Gram-negative pathogen causes peptic ulcers and stomach cancer [151], and chronically infects more than 50% of the human population; in the latter case, resistance to multiple antibiotics has been detected [152]. Other possibilities already referred to in the literature encompass antimicrobial potential against such food-borne pathogens as *L. monocytogenes*, *Salmonella enteritidis*, *Campylobacter* spp., *Staphylococcus aureus* and *Escherichia coli*, besides spoilage microorganisms, e.g. *Pseudomonas aeruginosa* [153, 154].

Bioactive Compounds

In spite of the widely proclaimed health benefits of olive oil, much fewer data exist on the nutritional value of table olives. Olives have indeed suffered from an undeserved reputation of a fattening food – and although olives do have a higher fat content than most drupes, they are relatively low in calories. Olives supply indeed ca. 2,500 kcal/kg, i.e. less than half of what is normally found in chips (5,600 kcal) or dark chocolate (5,340 kcal). To eliminate such myths (including a high salt content), table olives are currently promoted as healthy foods in Spain, based on comprehensive scientific studies. Good quality olives are obviously required for high quality olive products – and quality starts at the very olive grove, where the farmer needs to understand the specifications and requirements of the olive tree and its fruits. Poor selection of varieties and growth under suboptimal conditions can unfavourably influence the properties of raw olives; this matter is of high relevance because most producers lack efficient technical support.

The olive fruits contain a wide variety of phenolic compounds that are potent antioxidants – and play an important role in the chemical, organoleptic and nutritional properties of virgin olive oil and table olives [155, 156, 157, 158]. In terms of composition, ca. 82 %(w/w) of olives is accounted for by oleic acid, 13% by linoleic acid (ω-6) and 3% by linolenic acid (ω-3) – of a total of 15-30% of the fruit, mainly in the forms of triglycerides and waxes. Studies on changes in the unsaponifiable matter, sterols and fatty and triterpenic alcohol changes during olive processing indicated that the values of most of these parameters by the moment of consumption lie within the limits established by EU Directives. However, statistically significant effects of cultivars and processing steps have been claimed on unsaponifiable matter, β-sitosterol, 5-avenasterol, total sterols, 1-docosanol, 1-tetracosanol, erythrodiol and erythrodiol +uvaol [159].

The mineral composition of table olives also depends on the cultivars and preparation style. As expected for a brined food product, Na is the most abundant element – but olives are also a good source of Ca, K, Mg and P. Fe concentrations are also high in ripe olives, but relatively low in green and directly brined ones – whereas the microelements Cu, Zn and Mn appear at levels similar to other plants. Processing affects the mineral content of table olives (except Ca and Mg) between green (Spanish style), directly brined and ripe olives; the most discriminating elements appears to be Fe, K, Na, Mn, Cu and P between styles, and these plus Mg and Ca between cultivars [160].

The levels of vitamins in olive pulp are important; the dominant ones are the water-soluble ascorbic acid (vitamin C) and thiamine (vitamin B_1), and the oil-soluble tocopherols (vitamin E) and carotenes (pro-vitamin A) [161]. In the traditional Spanish-style green olive fermentation, the four vitamins essential for growth of *L. plantarum* are required: nicotinic and pantothenic acids, biotin and vitamin B6 are available in the fermentation brines within the first few days of processing, and their levels throughout the fermentation process remain well above those required by *L. plantarum* for growth at the maximum rate possible [162]. Various yeast strains isolated from fermentation media were found to produce those vitamins *in vitro* to levels several-fold those required by *L. plantarum* – so some yeast species may play a symbiotic role with *L. plantarum* in this mode of fermentation.

Fibre determines texture and digestibility of the olive fruit – and relatively high contents in dietary fibre have accordingly been recorded, viz. 2-3 % (w/w) [163].

The content of antioxidant polyphenols is also noteworthy in table olives, as compared with other fruits. Note that polyphenols may contribute to fruit quality in a number of ways – including sensory attributes, such as colour and flavour; this is the case of some specific phenolics, in particular oleuropein [164 – given its intense bitterness. Other bitter phenolics found in the fruit include glucosides, salidroside, nuezhenide and nuezhenide oleoside, as well as two secoiridoid glucosides of uncertain structure containing tyrosol, elenolic acid and glucose moieties. In Portuguese olive varieties, various analyses have identified and quantified seven polyphenolic compounds: hydroxytyrosol, tyrosol, 5-*O*-caffeoilquinic acid, verbascoside, luteolin 7-*O*-glucoside, rutin and luteolin [165]. The highest concentrations of such antioxidants compounds were observed in 'Galega' natural black olives, and the lowest contents pertain to chemically oxidized olives (Californian style) – irrespective of variety. Hydroxytyrosol, and such flavonoids as rutin and luteolin-7-glucoside are the major phenolic compounds in Portuguese olive varieties, whereas verbascoside, luteolin-7-glucoside and oleuropein are abundant in leaves [166].

Olive fermentation plays an important role in phenol compound degradation, with inhibition of lipid peroxidation [167] in addition to several physiological activities [168]. The safety of hydroxytyrosol appears to be excellent, so it has been labelled as GRAS; and other phenolic constituents of olives have been associated with potent biological activities *in vitro*, including (but not limited to) antioxidant action. Such olive phenolics as tyrosol and hydroxytyrosol are absorbed by humans after ingestion in a dose-dependent manner, and they are excreted in the urine as glucuronide conjugates [169]. In particular, hydroxytyrosol and oleuropein scavenge free radicals, e.g. superoxide, and exert other biological activities encompassing inhibition of platelet aggregation and potentiation of nitric oxide-mediated macrophagic immune response. Furthermore, an increase in the dose of phenolics administered apparently increases the proportion of conjugation with glucuronide. Hydroxytyrosol is one of the most actively investigated natural phenols, and is endowed with interesting pharmacological activities – many of which have already been demonstrated *in vivo*; given its lack of toxicity, future use as human supplement is conceivable [169]It Finally, hydroxytyrosol and related olive phenols have been tested, as supplements, in humans – and thus retain their antioxidant activity after ingestion; the human metabolic pathway has been elucidated, and it suggests extensive glucuronidation and subsequent urinary excretion – with a protective role from second hand smoke-induced oxidative damage, inhibition of platelet aggregation, increase of brain cell resistance to oxidation and mitochondrial membrane potential [169]. Further experiments have confirmed the anti-thrombotic potential of hydroxytyrosol, coupled with its ability to ameliorate osteoarthritis [171]; and synergies with other olive phenols cannot at present be ruled out. The contribution of excessive free radical formation to the onset of certain pathologies (e.g. atherosclerotic heart disease and cancer) strongly recommends higher dietary intake of fruits and vegetables – i.e. foods with large contents of antioxidant vitamins, flavonoids and polyphenols [170].

Several nutritional allegations for olive oil have been proposed, and are currently under evaluation in EU [172, 173]; however, nutritional allegations for table olives may be more difficult to be granted, due to their intrinsically high level of salt – so a compromise with their monounsaturated oil composition and high levels of dietary fibre, vitamins and antioxidants has to be duly publicized.

FUTURE ISSUES

General Considerations

Because of an increasing and elder global population, and the emergence of economies of once peripheric and underdeveloped countries, the food supply chain has been facing serious challenges; however, it has also gained access to sophisticated, accurate and trustworthy analytical techniques and processing strategies, which may help mitigate most underlying risks. Note that the nuclear issue is to respond to a rising demand with at least a quality as ever, while protecting the environment and leaving resources available and useful for the generations to come [174]. This issue of security and safety – coupled with sustainability, requires multidisciplinary approaches that are powerful and ingenious enough to guarantee feasibility and efficacy.

The aforementioned trends have also contributed to new dynamics within the agricultural commodity markets: such dynamics cover from a more gradual transition in rural economies characterized by changes of the productivity food regime [175], to establishment of a new 'rural development paradigm' [176] – but, in either case, the evolution of existing, or the rising of alternative food supply chains are key dimensions that need to be addressed.

Food is the second largest manufacturing sector in EU; it already accounts for 13.5 % of the total revenue of the manufacturing industry, in addition to ranking first in exports – thus accounting for ca. 20 % of the global value. Nevertheless, such a world leadership has been eroded by up to 5% over the last 10 years [177] – and one reason for that derives from the highly fragmented nature of production and transformation (more than 99% of the companies are SMEs), and the low technological input therein, unlike the extremely strong concentration of food retail – which smashes prices down, and makes it more and more difficult to compete in bulk products, especially given the much lower labor costs in less developed countries [178].

The past three generations have witnessed a focus of European food supply on security, which goes along with low prices to the consumer – whilst seeking to reduce the environmental impact, and maintain reasonable economic returns to rural communities. The recent expansion of EU has led to an increasing diversity of food production systems, thus affording an unprecedented opportunity to take advantage of this diversity to create and support more sustainable systems. Given the strong ties between food production and sustainability, a holistic view over European food production and supply systems is in order – which must come along with strengthening of competitiveness entertained by all stakeholders; and novel foods bearing features well beyond plain nutrition are surely the way, where traditional products may provide a source of inspiration and experience.

Demand of Traditional Foods

Food is an integral part of all cultures, and has evolved everywhere in parallel to human civilisation; however, traditional production is to abide to modern expectations and constraints associated with food production, viz. convert raw materials into edible, safe, wholesome and nutritious food products – with desirable physicochemical properties, extended shelf-life, and optimal features for palatability and convenience of preparation and eating. A rapid urban growth has turned most traditional foods into species on the verge of extinction, while a few were able to retaining their prominence and even gain in market share following criterious adaptations and improvements – especially when differentiated sensory features and proactive contribution to health are at stake [179].

Although traditional foods are particularly sensitive to the current behavior of the world markets, they entail good perspectives for expansion in the near future – should a number of challenges be adequately addressed; these include wider communication, legal protection, quality assurance and acceptable innovation [180]; and olive-related products fall well within this list. Production and marketing of less usual products derived from traditional ones requires adjustments in the traditional food supply chain – so as to preserve their identity, while responding adequately to consumer demand and expectations; and the emergence of novel food circuits in globalised agricultural economies should be analyzed versus the background associated with a number of crucial changes within the various links within and departing from the agro-food chain.

The unique (and often irreplaceable) meaning of several traditional food products should be discussed *vis a vis* with the delicate situation of a global market where consumers require a great interchangeability and availability of nutritional and nonexpensive bulk foods. These limitations have promoted production and distribution of traditional food stuffs by large industrial and commercial corporations, with the purpose of assuring consumer satisfaction, while making the food choices more and more uniform on an pan-European basis – despite the misleading large diversity of brands and displays in each country and region [179]. The art of manufacturing traditional food products needs to be improved and tuned, thus permitting incorporaton of objective methods of process control and some degree of standardisation of the final products – without losing their desirable attributes, but along with improving quality, taste and nutritional profile [181].

Feasibility of industrial, essentially collective and of artisanal, essentially individual manufacture is central to the whole planning exercise. Feasibility is an integrated concept, determined by world price and enterprise positioning – and encompasses collaboration between competitors, along with product portfolio expansion, complemented with readiness for investing and establishing positions of influence in global supply chains [182]; a recent wave of interest on consumer satisfaction has also stimulated several thoughtful interpretations of the causes and effects thereof. In the specific case of the olive-based sector, the major issues are cost-effective harvesting for small to medium growers, water quantity and availability under drought conditions, and possibility to access new markets.

Besides creating more sustainable niche markets for local food products, fermented traditional foods can bring about significant social, environmental and economic benefits for those involved – and thus lead to greater overall gains, which may undoubtedly lead to sustainability of mainstream agriculture and international supply chains. Consequently, links between food business people, farmers, researchers and public authorities are important to raise the baseline for near-commodity and commodity food supply chains. These links have also to build on the olive grove as an element of identity and conservation of life (at least in the Mediterranean basin) – which thus necessitates active preservation; farmer multi-activities may positively contribute to maintenance of traditional olive groves, and so to promotion of sustainable rural development [183].

Demand of Functional Foods

A new generation of natural (and healthy) foods has been clearly on the rise – and fermented olives are a central part thereof. Recall that concerted policies of public health aimed at efficiently facing budgetary restrictions will require investment on preventive health approaches – especially as human life expectancy is extended; and the aforementioned types of foods are seminal vectors towards this strategy. However, manufacture and marketing of fermented olives hinges critically upon the primary production sector; traditional olive groves already account for a major share of agricultural land in Mediterranean countries – particularly in marginal, less developed areas. In addition to the socioeconomic impact in terms of maintaining local employment and avoiding rural exodus, sustainable exploitation is a must – so that the needs of current generations are fulfilled, without compromising the capacity of future generations to meet their own needs.

The rising income and changing consumption patterns towards healthier and safer products favour the quantitative and qualitative increase of consumption in and away from home; most food companies already consider that product innovations are an opportunity to increase profits and market share – and functional foods are particularly interesting, because their added value lies between that of a bulk food and that of a pharmaceutical active principle. A functional food can be a natural food, a modified traditional food or a novel food – as defined in the germane EU Regulation; however, the perspective on functional foods is clearly different between west and east – in the former, they are viewed as a revolutionary approach (and thus represent a fast growing segment of the food industry), whereas functional foods have been a part of the eastern culture for centuries [184]. Since countries are been faced with a number of major health challenges arising from an ageing population and an increase in lifestyle-derived diseases, functional foods can make a positive contribution to address those challenges [185]. Development of functional foods requires, however, that a functional component is either created/introduced or optimized/made more available – but, in either situation, the level of complexity increases, and so does the need for accurate monitoring [186].

In the past years, it has become apparent that some naturally occurring components of traditional foods can help maintain a state of health through optimisation of human functions, or reduction of the risk of incidence of chronic diseases [187]; this issue is particularly important because most countries are experiencing a considerable rise in the proportion of elderly people, and proper nutrition will not only contribute to a longer life but also to a better life [188]. This expectation has made consumers increasingly motivated toward purchase of functional foods – often traditional foods (or based thereon), provided that they also are environmentally-friendly and do not violate one's ethical principles. Synergies should accordingly be generated between economic growth, environmental protection and social fairness – with multidisciplinary scientific and technical approaches, and integration between public and private research being utterly recommended.

In addition, many technological innovations have arisen from realisation of the important role of food in promoting and sustaining health; the food industry is redirecting its strategy to new product development containing nutraceuticals, i.e. natural food components that play favourable physiological roles upon health. Hence, nutraceutical-containing (or functional) foods add to the classical primary and secondary roles of foods, viz. nutrition and sensory preference – as wel as the capacity to modulate physiological systems with beneficial effects provided that they are ingested regularly and as part of a balanced diet. The interactions between food, nutrition and health constitute indeed quite a new challenge for both food science and technology; however, it is important that claims on health-promoting effects have sufficient scientific support – so health authorities should establish criteria for mandatory qualification of functional foods. Within the wide range of functional foods, those containing live bacteria (mainly certain lactobacilli and bifidobacteria) that provide beneficial health effects deserve a special mention – and are currently traded under the label of probiotics [189]; other functional foods include prebiotics, i.e. compounds that facilitate growth of said beneficial bacteria in the intestinal tract of the consumer (if both are present in a given food, one has a synbiotic food).

Development of a functional food should not be only a marketing feature able to attract more and wealthier consumers, but also possess scientific validation – and this creates complex scientific challenges. In fact, the simple presence of a bioactive compound in a food does not necessarily ensure that it will be biologically active at the moment of ingestion, and afterwards within one's organism. Therefore, the true benefits of functional foods will not be realized unless scientifically sound and non-misleading messages are provided to consumers [190]. Food will then be said to behave as a functional one if it has been satisfactorily demonstrated that it beneficially affects one or more target functions in the body beyond adequate nutritional effects, in a way that is relevant to either the state of well-being, or the health, or else the reduction of risk of incidence of a disease [187]. Hence, growth of the functional food sector brings about not only significant benefits to public health, but also offers new opportunities for processing and manufacturing companies [191]; larger added-value and higher margin products thus provide a key impetus for food manufacturers.

Demand of Probiotic Foods

As emphasized above, some foods can play a third role besides responding to nutritional needs and providing organoleptic pleasure – which is gaining a renewed interest worldwide: conveying extra protection against degenerative diseases, and delaying incidence and severity of chronic health conditions (upon regular inclusion in a balanced diet). One major class of said items is probiotic foods – the constituent microorganisms of which can survive passage through the gastrointestinal tract, and eventually establish and act at the colon, where they exert antimicrobial roles against local pathogens, further to exhibiting antitumor and hypocholesterolemic features. Probiotic foods have to date been restricted almost exclusively to dairy products, and have encompassed mainly *Bifidobacterium* spp. – but vegetable matrices constitute a rich alternative source of probiotic microorganisms, mostly belonging to the *Lactobacillus* genus.

WHO/FAO [192] has defined probiotic organisms as "live microorganisms, which, when administered in adequate amounts, confer a health benefit to the host". Most probiotics belong to the *Bifidobacterium* and *Lactobacillus* genera, which are considered as GRAS ingredients; their documented health-promoting effects include regulation of intestinal flora and body cholesterol levels, immuno-modulation and cancer prevention [146]. Intake of probiotics stimulates growth of beneficial microorganisms, reduces the amount of pathogens and strengthens the body's natural defences, thus helping boost the immune system [193, 194, 195, 196] and consequently lowering the risk of gastrointestinal diseases. In fact, when the intestinal microflora is altered, administration of probiotic bacteria not only re-establishes its normal equilibrium, but also improves the microbial balance and properties of the endogenous flora. A positive role of probiotics in prevention of food allergies and intolerances has been also claimed [197]. After antibiotic-based clinical strategies, probiotics have been prescribed to avoid the related intestinal incidents, besides useful in the treatment of candidiasis. Probiotic bacteria can also help relieve the symptoms of inflammatory bowel diseases, constipation, irritable bowel syndrome, colitis and alcoholic liver disease; probiotic consumption will as well reduce the risk of colon, liver and breast cancers [198].

Therefore, several pieces of evidences exist that support clinical applications of probiotics upon prevention and treatment of diseases of the gastrointestinal, respiratory and urogenital tracts [199]; but probiotics go well beyond, in that they can provide a number of potential health benefits, via maintaining a good balance and an appropriate composition of the intestinal flora – thus helping increase the body ability to resist invasion of pathogens, and maintain the host's well-being [200]. Although several other benefits are yet to be thoroughly proven, the tested ones point at the growth and action of probiotic strains to be the direct cause of alleviation of symptoms and health improvement [201].

Recall that the primary role of diet is to provide enough nutrients to fulfil body-building and energy requirements, besides pleasing the sensory organs and assuring convenience – while providing a feeling of social satisfaction and well-being. Recent scientific evidence has in addition proven that diet may modulate several functions in the human body – as well as delay, or even decrease incidence of several chronic diseases and health conditions [202]. Health-proactive ingredients – or nutraceuticals, and matrices that incorporate them – or functional foods, are accordingly the major food trend of the 21st century, especially because of a global population growing older and of the State health budgets (suffering from the dramatically increasing magnitude of health-related items). Usually sold at higher prices than classical food items, novel nutraceuticals have been sought and novel functional foods have been increasingly formulated by the food industry; however, the associated health claims are often not supported by scientifically solid and statistically validated data. This has led to stricter and stricter enforcement of regulatory issues by the food and health authorities Europe-wide. Consequently, the food industry has been urged to abide to such requirements, while facing market competition – e.g. via design of novel food matrices for use as vectors to deliver probiotic strains.

Processing of microbial systems aimed at manufacturing probiotic foods can be organized into processing of starter cultures, processing of products with desired functional characteristics, and processing of foods containing probiotics. Technological issues related to microorganisms in functional foods are usually complex and diverse; furthermore, limited data are available on the impact of processing, distribution and storage on cell viability of those specific strains. Technological challenges include the need to obtain high initial productivity and viability as starter cultures, as well as high stability, viability and productivity as probiotic strains [203]. Future studies are obviously required to collect data so as to definitely assess (by *in vivo* studies) whether the living bacteria and the food matrix own beneficial components significantly improve the health of the consumer; if so, allegations can be proposed – as starts to be the case of certain types of table olives, which may be eventually be classified as functional foods.

Over recent decades, development and consumption of probiotic foods has increased along with awareness of their beneficial effects in promoting gut health, as well as in disease prevention and therapy [194, 204]. As a result of the natural tendency to diversify supply, food industries are now focusing on non-dairy foods that can contribute also to regular ingestion of probiotics by individuals having lactose intolerance, or be incorporated in diets lacking milk-derived products. Interest in the field of non-animal originated probiotics has consequently been on the rise, in parallel with the renewed interest on microbial ecology of the gut – powered by the availability and generalization of randomised, blind or double-blind human trials. Demand for new probiotic products has led to new scientific achievements, together with a strong utilization of improved and scientifically-based selection criteria [196, 205, 206].

Table Olives as Functional, Probiotic Foods

A new generation of natural and healthy foods is on the rise – and fermented olives are clearly a nuclear part thereof. Public health concerted policies aimed at efficiently facing budgetary restrictions will require investment on preventive health approaches – especially as human life expectancy is extended; and the aforementioned types of foods are seminal vectors in this strategy, as applied in the Mediterranean basin (but also elsewhere). In generation of functional foods, LAB with diverse physiological and metabolic traits are to be combined; their metabolic and technological properties are often different from those of traditional starter cultures, so appropriate production processes are envisaged for eventual development [207].

Since positive association to health issues permits selling of food products at higher prices – and consumers prefer traditional products, innovation in fermented traditional foods should be targeted at identification of health-proactive food matrices to carry unusual probiotic strains, while keeping their traditional character and uniqueness. In this regard, it has been shown that probiotic strains from either intestinal (animal) or crop (plant) origin show essentially identical metabolic and functional properties [143, 208]. Traditional olives thus seem an obvious choice – and a target product with a good potential; furthermore, some of the components of the fruit itself may function also as prebiotics, viz. carbohydrate polymers (also known as dietary fibers), besides such nutraceuticals as essential fatty acids (of the ω-3, -6 and -9 types) and such antioxidants as sterols, polyphenols and glycosides. In particular, recent work has claimed a number of beneficial effects of its dietary fibers toward prevention of human pathologies [209], while other authors have refered to the anti-cancer activity of some of the aforementioned polyunsaturated fatty acids (present chiefly in the olive skin) [210].

Some of the adventitious lactic acid bacteria of natural table olives may be able to survive passage through the gastrointestinal tract – thus resisting digestion and surviving in the gut, and accordingly having a chance to act against pathogens installed therein (e.g. in the stomach or in the small bowel). Moreover, they produce metabolites that are inhibitory to *H. pylori* [211]. Although the probiotic strains of *L. plantarum* available commercially and traded as protiotics are all of intestinal origin, Haller et al. [208] reported that similar useful metabolic and functional properties can also be found in strains contaminating spontaneously fermented vegetables and fruits. This realisation is expected to increase the added value of table olives in a dramatic fashion – with a special economic impact, given that such products originate in less developed regions. On the other hand, the market share of traditional fermented food products in Europe is steadily growing, as consumers become more and more aware of their cultural heritage, unique properties and gastronomic quality – further to their functional status; and said market already encompasses a considerable share accounted for by fermented table olives.

The olive phylloplane – and the fruit surface in particular, is suitable for survival of microbial populations, especially lactic acid bacteria [212, 213] – that are actively involved in performance of their lactic fermentation [82]. On the other hand, table olives are suitable as both a biological carrier of useful probiotics and a tasty, nonrefrigerated matrix that is appealing to consumers at large.

Recent attempts to use table olives as vehicle to incorporate probiotic bacterial species, viz. *Lactobacillus rhamnosus, Lactobacillus paracasei, Bifidobacterium bifidum* and *Bifidobacterium longum* [214], have met with success; fermented olives were immersed in a suspension of probiotic bacteria, and their survival was monitored for up to 3 mo – with *L. rhamnosus* showing a good survival on the pericarp of drupes, followed by bifidobacteria (which actually tended to slightly decline throughout storage). A commercial patent application covering this aspect has meanwhile been issued [215].

Probiotic Strains from Olives for Olive Processing

Microorganisms bearing a health impact will likely remain important functional ingredients in the coming future; new strains will probably be identified, and foods will accordingly be developed to fulfil the needs of an ever increasing number of specific consumer groups. Increased understanding of the interactions prevailing between the gut microbiota, the diet and the physiological conditions prevailing in the host will open up new possibilities of producing new ingredients and nutritionally optimised foods – both of which are expected to promote consumer health via microbial activities in the gut.

Manufacture (and trading) of olives fermented with probiotic bacteria isolated among the native microbiota colonizing the olive surface constitutes indeed a new avenue to enrich the olive-based product portfolio. The selected probiotic bacteria may thus be deliberately introduced into brines, at the onset of fermentation, to act as starters – and will eventually dominate and ensure proper fermentation outcomes, while inhibiting growth of undesirable microorganisms. The goal is thus to obtain a probiotic food, containing probiotic bacteria in sufficient amounts to play a significant role, and consequently improve consumers' health without hampering the sensory characteristics of the final fermented olives. Commercial success will obviously require the consumer to be informed about the nutritional value of the fruit; moreover, consumer acceptance studies will be essential to more adequately design and plan the launching of those types of food items in the global market. Simultaneously, a better control of the fermentation process – with the possibility to prevent and detect faulty fermentation and spoilage at early stages, will also demand higher quality of the raw materials, i.e. well-cared trees capable of producing large quantities of robust and intact fruits.

Recall that efforts pertaining to probiotic foods have to date focussed mainly on dairy matrices, so only few data are available on alternative carriers – including olives in particular; however, inclusion of probiotic strains will reinforce the already strong functional behaviour associated with unsaturated oil and antioxidant compounds. Furthermore, the European olive production and export numbers are already substantial – although competition involving low-added value, traditional agricultural commodities entertained by peripheric countries has been on the rise. From the point of view of production, the sector is extremely fragmented and characterized by a large number of small businesses – typically with very low profitability margins, and hence particularly vulnerable to annual variations in production. The costs of adhering to the new European environmental legislation that imposes stricter regulations upon the agricultural sector will also have a significant impact on olive farmers, since they constitute mainly SMEs. Survival of this sector at a European scale will require significant incorporation of new technology, more efficient control of processes, and adaptation of its products to the consumers and their preferences and expectations.

Therefore, the important market share of olive products in terms of snacks worldwide will hardly be maintained unless high-added value products from olives are meanwhile developed; this is the case of olive matrices as probiotic strain carriers, the features of which are to be convincingly validated before approaching the market. Although table olives are inherently safe – no relevant toxinfection outbreaks have indeed been ever reported, the capacity of this food subsector to respond effectively to the world trend of health-oriented foods is still incipient.

Market Opportunity for Probiotic Olives

Table olive production remains a huge conglomerate of handicraft, small industries in most EU countries possessing a tradition in this type of food product; ancient technology leading to very limited profits is still the rule. In addition, the sector is undergoing a progressive competition arising from countries with lower production costs. Solutions to conveniently respond to this situation may entail a number of approaches – such as basic, pre-competitive and applied research that permits deeper scientific insight into the fermentation process (including both processing and packaging steps), in order to improve product quality and safety, as well as consumer acceptability thereof. Moreover, it will likely allow a more extensive standardization of processing, as prediction of the fermentation outcomes will be more feasible if starter cultures are used – but which are necessarily based on the native microflora. This will complement the market with new added value products, for which a suitable response is expected on the consumers' side: consumers are more frequently seeking for new preparations based on olives that can be found in delicatessen shops (e.g. olive paste, or olives stuffed with almonds or peppers); and they are more worried about their own health, knowing that natural products will more effectively address such a worry. Hence, the olive sector should take advantage of the market opportunities, but should also seek entrance in and reinforcement of presence in adequate niches; adaptation to new consumers will require a major effort in innovation in product and redesign of current processes, adjustment of traditional products to new specifications, and development of new ones on a consumer-oriented concerted effort [216].

Lactobacilli will be particularly important players in this regard, as they ecologically dominate over contaminating microflora that may eventually pose organoleptic and health hazards – while producing metabolites that can act against those microorganisms. Recent data originating from various European teams indicate that adventitious wild strains of *Lactobacillus* spp. in table olive brines possess antimicrobial character, and can resist conditions similar to those prevailing in the gastrointestinal tract. Hence, a strategy to oppose the decrease in production and export of native cultivars of table olives is to conceive and manufacture probiotic olive brines and pastes bearing a market added-value, and which take advantage of the underlying cultural heritage and unique gastronomic character of olives. Since most of the probiotics available are dairy products, consumption by dairy-sensitive individuals is not recommended at all – so it will also be advantageous to make available food products that allow administration of probiotic bacteria without causing allergies or intolerances associated with their matrix – and which can be stored for longer times after opening [217, 218]. Note that probiotic dehydrated foods have been obtained on an experimental scale by vacuum drying fruits soaked in probiotic microorganisms [219], while some oat-based products and fruit juices containing probiotic bacteria are already available on the market [198, 220]. Likewise, it will be of great importance to identify new probiotic bacteria initially isolated as a part of the natural microflora of food products that are not related to common allergies. Application of such types of novel technologies will convey a significant marketing advantage to European fermented table olives worldwide, which is rather appealing to the consumer.

Diversification of Olive Product Portfolio

A significant challenge in traditional food production is to improve its competitiveness, by identifying innovations that guarantee safety – while abiding to general consumer demands, as well as specific consumer expectations and attitudes towards traditional foods. The integration of the rich traditions of European cuisine with the innovation-driven consumer market is a particularly challenging task for food SMEs; ensuring that the European food industry remains innovative through the use of advanced technologies, tuned specifically for traditional food processing, is a main driver toward competitiveness.

The emergence of new technologies for processing of table olives opens new ways for the valorisation of certain less relevant cultivars – particularly those that are hardly adequate for oil production, or even for typical table olive consumption. At the same time, those technologies will make it possible to upgrade several valuable compounds in otherwise waste by-products, so production losses can be reduced. A good example is the production of specialty products, e.g. olive paste and tapenade prepared from processed olives, as alternative products for the global market. This may indeed constitute a solution to process olives bearing physical defects, and thus not adequate for commercialization as table olives – despite their organoleptic features, as well as to use the olive paste obtained after the first oil extraction. However, stuffed olives or olives in marinades containing fresh material (e.g. Fetta cheese, anchovies, vegetables, herbs and spices) exhibit generally a reduced shelf life, and are thus labile to spurious contamination.

Possible additives – viz. garlic, chilli, basil and oregano, should indeed be in a dry state and free of microorganisms; but processed table olives should retain some level of bitterness and fruitiness. The final salt and pH levels of packed olives depend normally on the style, variety and ripeness of the fruit – but all should meet standard safety requirements. In all cases, processed table olives should be checked at random and proven to be microbiologically safe, by an accredited laboratory, before being made available for sale.

Antiseptics from Olive Cultivars

Hand sanitation is a common concern in the western World, and is a critical procedure in healthcare and food processing facilities. Currently, most sanitizers are alcohol-based – and constitute an excellent alternative to hand washing; they are actually more effective than soap and water in killing bacteria and viruses that cause diseases. A good sanitizer should accordingly be able to eliminate Influenza virus, *Candida albicans, Salmonella* sp., *Clostridium difficile, S. aureus, Bacteroides fragilis, Haemophylus influenzae, Proteus mirabilis, P. aeruginosa, E. coli, Bacillus* sp., *Streptococcus* sp., *Shigella* sp. and *Klebsiella* sp. from one's hands. The activity of alcohol-based sanitizers is affected by concentration and type(s) of alcohol included in the formulation – among other ingredients; and their concentration is a key factor for activity against pathogens be displayed, at least *in vitro*. For instance, isopropanol has a greater activity than ethanol against bacteria, fungi and viruses. Despite their performance, such sanitizers are tolerated but not preferred by consumers, because alcohols cause skin dryness. Conversely, consumers recognize natural antimicrobials as beneficial, particularly plant-derived substances.

The leaves of *O. europea* contain natural antimicrobials in their composition, which are a part of the plant defence systems against diseases; the main active compound is oleuropein, a bitter secoridoid; it is hygroscopic, freely soluble in ethanol and NaOH solutions, but only moderately soluble in water. Moreover, oleuropein is edible, despite scarce data so far on its toxicity and eventual metabolism. Olive leaf extracts have been already commercialized as food supplements [221]; that compound is indeed a known antimicrobial, and plays a key role in table olive fermentation by inhibiting spoilage and pathogenic microorganisms. It has been shown to inhibit growth of *E. coli, Klebsiella pneumoniae* and *S. aureus* – besides several yeasts and molds, [222], and to possess antiviral activity against some viruses and retroviruses [223], including anti-HIV activity by blocking the HIV virus entry into host cells [224]. Oleuropein has been used in a number of medical treatments [223]; its medicinal use dates back to the early 1800s, when it was used in liquid form for treatment of malarial infections. Hydroxytyrosol is another phenolic compound of olive leafs; it was recently deemed effective against clinical human pathogenic strains of *H. influenzae, Moraxella catarrhalis, Salmonella typhi, Vibrio parahaemolyticus* and *S. aureus* [222, 225]. Olive leaf extracts are already in use as active ingredients in cosmetics and pharmaceuticals; they are immediately perceived by consumers as fully natural, because of their origin and association with olive oil (and the healthy properties thereof).

Consumer preferences of alcohol-based hand rubs vary considerably with regard to their consistency (gel, foam or rinse), scent, drying characteristics and cost. Although efficient, they have the disadvantage of causing dryness to skin – particularly if they are repeatedly used along the day. This major disadvantage can be overcome by reducing the alcohol concentration, and concomitantly including other disinfectant in their formulation. For hand sanitizers, the Food and Drug Administration (FDA) [226] has actually recommended low levels of ethanol or isopropanol.

Sanitizers based on olive leaf extracts appear feasible for large scale manufacture and use: the active principles are (sorted by decreasing concentration) ethanol, isopropanol, oleuropein and other phenolic compounds [227]. This type of antimicrobial additive may be obtained by a one-step solid-liquid extraction from olive leafs, using an ethanol/isopropanol/water solution; a high content of oleuropein was indeed obtained using ethanol as extraction solvent [228]. The concentration of each active component may easily be adjusted, and the process will not involve expensive unit operations (or high temperatures, for that matter) – and may thus be well-adapted to gel formulation [222].

Antimicrobials from Olive-Borne Wild Strains

New natural antimicrobial substances, particularly bacteriocins synthesized by a few lactic acid bacteria, seem to play a role against specific microorganisms in some foods [203, 229]; bacteriocins are indeed one of the most abundant and diverse family of microbial defence systems – and consist of an ill-defined group of extracellular proteinaceous compounds active specifically against some families of bacteria [230]. A bacteriocin is assembled in ribosomes in the form of an inactive prebacteriocin – which its N-terminal leader sequence, to be eventually cleaved during transportation throughout the cytoplasm. The essential genes include information for its precursor peptide(s), an ATP-binding cassette transport system, a self-immunity protein, a gene encoding an accessory protein useful for export, and regulatory genes (e.g. a two-component sensory system) [231, 232]. Nisin is the most widely studied bacteriocin – and, as happens with various other (linear) antimicrobial peptides and or pheromones, it is detected by bacteria via a quorum-sensing mechanism, which enhances in turn bacteriocin synthesis and expression when a certain threshold is attained [233, 234].

According to Riley & Gordon [235], bacteriocins are produced chiefly under stress conditions that cause the rapid elimination of neighbour sensitive cells, and play an important role in mediating intra-species interactions and modulating population dynamics. Although bacteriocins were first studied in Gram-negative bacteria, lactic acid bacteria (as Gram-positive bacteria) have been the focus of comprehensive research in more recent years, and accordingly exploited as a reservoir of antimicrobial peptides with potential food applications. Despite their heterogeneity in chemical and biological features, most bacteriocins are characterized by sensitivity to various proteases, a bactericidal mode of action, high thermal stability, higher activity and stability under acidic pH, biosynthesis via a precursor peptide consisting of a leader sequence and a characteristic proteolytic cleavage site, and organisation of the structural and regulatory genes in an operon – as well as existence of immunity proteins that protect the producer from the lethal action of its own products, and the cytoplasmic membrane acting as their primary target [236, 237]

At this stage, it is important to differentiate between 'classical' peptide antibiotics (viz. gramicidins and polyketides) and bacteriocins: the former are synthesized by multi-enzyme complexes from building blocks provided by a variety of cellular processes [238] – more specifically, by a multiple-carrier thiotemplate mechanism where peptide synthetases assemble amino acids to form the antibiotic molecule [239]; whereas bacteriocins are ribossomally synthesized [141]. As the active molecule is encoded by one structural gene, active sites and structure-function relationships can be more simply examined than in antibiotics [239]. In addition, peptide antibiotics are not associated to host cell immunity, they have specific intracellular targets, and they may show side effects. The target cell's resistance mechanisms are also quite different. The resistance to antibiotics is genetically transferable and affects different sites, depending on the mode of action – whereas the resistance to bacteriocins relies on adaptation mechanisms that affect membrane composition. Whereas many antibiotics disable or kill pathogens over a series of days by inhibiting essential enzymes, most bacteriocins kill microorganisms rapidly by destroying or permeating the microbial membrane, and thus impairing their ability to carry out any metabolic process. These peptides are therefore unlikely to face the same antimicrobial resistance mechanisms that limit current antibiotic use [239].

The most obvious application of bacteriocins released by lactic acid bacteria is as food preservatives, since most of them are active against spoilage and food-borne pathogens [240]; however, applied studies are still needed because the physicochemical conditions prevailing within the food matrix itself can have a significant influence upon the activity of the bacteriocin. It should be emphasized that consumers at present demand more 'natural' and 'minimally processed' foods; as a result, there has been a major scientific interest on naturally produced antimicrobials (as is the case of bacteriocins) [239].

Despite the huge potential of bacteriocins in food processing, only nisin has been employed to a significant extent by the food industry – and it is identified on product labels by the code E234; it is produced by *Lactococcus lactis* ssp. *lactis*, but possesses a few shortcomings because of its lack of stability at neutral pH [241]. This clearly unfolds a window of opportunity for research directed to identification of less conventional lactic acid cultures, able to produce antimicrobial compounds active at neutral pH conditions and thus susceptible of a wider use [241]. Among the processing advantages of nisin, its sensitivity to K-chymotrypsin, heat stability at low pH and nontoxic nature have promoted its application. Nisin has in fact been in current use for more than 30 years as a dairy food biopreservative, and it was approved by FDA in April 1989 to prevent growth of botulism spores in pasteurized process-cheese spreads – and has obviously been classified as GRAS (21CFR184.1538); use of nisin as a food preservative has also been approved by the European Directive for clotted cream, ripened and processed cheese, as well as Semolina, Tapioca and similar puddings.

Some bacteriocins produced by lactic acid bacteria have gained further interest because they show an impressive activity against bacteria that are resistant to conventional antibiotics [242]. This property, coupled with the peculiar characteristics of resistance to low pH and high temperature, high productivity under such stress conditions as high sodium chloride content [106, 150], and wide bioactivity spectrum – including several spoilers and pathogens, have promoted their application in food preservation, and more and more often in the health area itself. Nevertheless, the applicability of each bacteriocin as food preservative is to be assessed in advance for each particular food. A few lactic acid bacteria found in table olive brines can also secrete bacteriocins, and in addition be labelled as probiotics – i.e. they will beneficially affect the host animal by improving its intestinal microbial balance [243], as a consequence also of their good adhesion properties. Recently, a wider definition of 'biotherapeutic agent' has prompted inclusion of non-viable probiotics, cell components and active molecules [244] – yet little is known about them, despite their potential advantages in terms of improved stability and safety.

As mentioned above, promising evidence is available on inhibition of clinical strains of *H. pylori* by metabolites of *Lactobacillus* sp. isolated from olive brines; recall that *H. pylori* is a Gram-negative, helicoidally-shaped, flagellated, human pathogen that causes peptic ulcers and stomach cancer [152]. *In vitro* studies and animal models have indeed demonstrated that they can suppress growth of *H. pylori*, possibly by a combination of biochemical mechanisms. On the other hand, lactic acid present in the supernatant and an intracellular protein also contributed to the observed bactericidal effect of *L. acidophilus* against *H. pylori* [245]. Specifically, a *Weissella confusa* strain was shown to have a dual inhibitory action: its supernatant had bactericidal activity, and (viable and non-viable) cells prevented *H. pylori* from binding to human gastric cells by competitive adhesion to mucosal glycolipids [246].

Following the practical interest of using bacteriocins as biopreservatives in foods, research efforts have been directed to obtain information about their properties and mode of action – which have already led to purification and characterization of the corresponding bioactive moieties. Classically, bacteriocins are produced in commercially available culture media, where they accumulate in small amounts; the ability to obtain a concentrated crude bacteriocin mix via optimisation of growth parameters greatly simplifies recovery thereof in subsequent purification steps – since it is well-known that dramatic losses of activity do frequently occur during downstream processing. To optimise bacteriocin production, the relationships between growth, bacteriocin synthesis and gene expression ought to be understood.

Bioactive Molecules from Olive Byproducts

Nowadays, two of the main environmental problems are the high rate of waste generation and energy consumption. The current situation requires a change in waste and energy management models – with a greater focus on waste reduction and recycling. Supply of agricultural products is crucial to human existence, as well as to a balanced and sustained quality of life. However, the intensive agricultural practices aimed at greatly increasing the global food supply have inadvertent, yet detrimental impacts on the environment – thus urging more sustainable agricultural methods; this is surely the case of olive groves. All processing issues are to be undertaken so that environmental impacts are minimised; and methods that use less energy and water, and generate lower volumes of wastes – viz. natural brine fermentations, are more favourable than those involving lye treatments or heat. Olives treated with lye require in fact multiple washes, and can use up to 5-fold the amount of drinking water demanded by natural methods; even so, lye-based methods have a number of advantages, including shorter processing times and specific sensory features of the final product.

Therefore, environmental pollution control constitutes a major challenge for the olive industry today. Relatively large amounts of olive waste waters are discharged into (large) ponds, but this is far from being a solution – because off-odours and groundwater contamination result, and natural evaporation is actually slow. Moreover, the physicochemical treatment of such wastewaters reduces by only 0-30% its organic contents, whereas appropriate biological treatments would eliminate up to 75-85% [247]. In any case, the chlorides remaining in solution represent a major concern for industrial operation; fermentation brines can be ultrafiltered and partly reused in canning, yet a large amount of dilute brine generated in the desalting process remains – for which no practical reuse exists. Salt-free fermentation could be a solution, but it is only available for Californian-style black olives – since Spanish-style green olives tend to spoil at low salt concentrations. A preservation method has recently been developed for wash waters resulting from the Spanish-style process, based on lactic acid fermentation [248]; but even if chlorides were eliminated from table olive processing, a high volume of contaminated wastewater would still remain – for which a new method, based on evaporation technology, has been proposed for implementation in olive factories [247].

Byproducts of plant food processing usually represent a major disposal problem for the industry concerned – but they may also constitute promising sources of compounds, with a market based on their favourable technological or nutritional properties. Spent olives, olive brines and olive oil wastewaters are indeed rich in polyphenols, such as tyrosol and hydroxytyrosol – which are valuable antioxidants with a pharmacological value; but many other valuable components can be extracted from olives and related products. Olive extracts and olive leaf extracts are known for their antimicrobial features; and there is a growing interest in finding phytochemicals as alternatives to compounds used for processing – and obtained from chemical synthesis routes, in the food, pharmaceutical and cosmetic industries [249]. Disposal of such materials usually represents a problem that is further aggravated by legal restrictions. The possibility of using these wastes and byproducts for further exploitation, in terms of production of food additives or supplements with high nutritional value, have earned an increasing interest – because they contain high-value added products, and their recovery may even be attractive from an economic standpoint.

The wastes generated during some of the stages of olive processing are rich in specific compounds with both nutritional and functional features; and it is widely accepted that byproducts from olives represent an important source of sugars, minerals, organic acids and bioactive compounds – e.g. phenolics, which entail a diverse group of secondary plant metabolites that includes phenolic acids, lignans, lignins, coumarins, flavonoids, stilbenes, flavonolignans and tannins [250]. Many of these compounds have shown strong antioxidant properties as oxygen scavengers, peroxide decomposers, metal chelating agents and free radical inhibitors [251]; the large range of useful activities attributed to these compounds raises an interest on new forms of use of these byproducts towards production of food additives or supplements.

On the other hand, the market demand for olive-derived cosmetics is another way to upgrade olive byproducts and wastes; one illustrative example is the use of low quality olive oils in saponification reactions. It is also possible to extract natural flavours, texture modifiers and colour stabilizers for eventual uses in food formulation.

Olive Grove Safety

The great economic and social importance of the olive crop, and the possible benefits to be derived from utilization of any of its products and byproducts have already been comprehensively discussed [252, 253]. Following the increase in consumption of olive oil in the last decade – based mainly on an impressive list of health-favourable properties, there has been a rapid expansion of olive groves, both in traditional producer countries and in emergent countries.

However, the aforementioned expansion has also promoted dissemination of plant diseases; one of the most important is olive anthracnose (commonly designed by 'gafa'). The incidence and severity of this disease has been on the rise, particularly in southern Portugal; its causal agent is a fungus (*Colletotrichum acutatum*, and *Colletotrichum gloeosporioides* to a lesser extent) that occurs primarily on leaves and twigs, and which induces premature fruit drop thus reducing olive oil quality. Although olive anthracnose symptoms in most cases only arise at fruit maturity, it is not exactly known where or under which form the fungus survives during the rest of the year. It has been hypothesized that the fungus may be present in fruits prior to maturity in the form of latent infections, or develops on the surface of host tissues. It the exact moment when the fungus establishes contact with the host were known, the factors that trigger infections and the initial steps of infection/colonization would allow an educated and effective intervention in this string of events. Fungicides are a way to fight this disease, but they should be used only to supplement a cultural control program. Copper-based solutions are one of the conventional practices aimed at reducing the incidence of olive diseases; although useful, their systemic characteristic raises an inconvenient for human health.

Another problem in olive groves that compromises product quality is the olive knot disease, caused by *Pseudomonas syringae* subsp. *savastanoi*; it is characterized by hyperplasia formation on the stems and branches of olive plants, and occasionally on the leaves and fruits. The incidence can cause severe damage in olive groves, mainly when weather conditions favour survival of epiphytic populations of that pathogen and entry thereof into the bark. This is a serious disease that has also been spreading fast, promoted by a combination of circumstances – including highly susceptible cultivars and predisposal to its causal agent. Olive knot control measures are usually of a preventive nature. Attention should be paid to cultural practices, so that pruning, destruction of infected plant material and alternative harvesting methods will not harm the tree. Although control of the olive knot disease is rather difficult, copper compounds have also been traditionally employed to reduce symptoms; treatment therewith may reduce the disease, but will not eliminate the underlying bacterial population – which will soon multiply back to previous levels. The diffuse resistance to copper bactericides among pathovars of *P. syringae* calls for development of alternative control methods of bacterial pathogens, such as biological control products – to lower toxic pesticide residues on fruits and vegetables, and to avoid environmental accumulation of chemicals (and consequent development of resistance among pathogens) [254].

On the other hand, many lactic acid bacteria adventitious on olive fruits produce potentially useful inhibitory compounds, such as bacteriocins (as already discussed) and smaller molecules such as reuterin [139], besides different types of aromatic compounds [255]. But their defence systems may simultaneously rely on lactic acid, antifungal cyclic peptides and aromatic compounds [256, 257]. The findings to date on this issue have prompted a new interest in bacteriocins produced by lactic acid bacteria, which can be considered as an alternative biocontrol system to reduce the hazard associated with synthetic pesticides. Valid formulations to ensure effectiveness of the bactericide under natural environmental conditions should thus be pursued; among the mould-related agents, there are species belonging to the *Aspergillus, Penicillium, Fusarium* and *Claviceps* genera that justify public safety concerns because they can produce mycotoxins – which are able to elicit strong intoxication in livestock and humans. However, not all species of said genera are mycotoxin-producers, and there is a wide variability of toxin synthesis ability within a toxigenic species. Antagonistic tests with olive fermentation fungi and bacteriocins produced by lactic acid bacteria from olive brines have unfolded the efficiency of the later against fungi [258, 259].

The use of bacterial metabolites instead of chemical antimicrobial agents appears interesting – but the scarcity of information on strains and their ecology, and the difficulty in obtaining the regulatory approval required for their formulation and application constrain a wide application in biological control on the short run. A new strategy is necessary, based on biologically safe exploitations – and thus prone to increase yields, together with reduction in use of toxic fungicides and increase of olive quality; economic competitiveness will eventually be promoted if useful results are generated via search for novel antibiotics with specific antifungal activity against various kinds of phytopathogenic microorganisms. This approach will indeed reduce the need to use environment-unfriendly, undesirable chemical compounds – which are at present the only effective method to protect olive groves from pathogen attack. Furthermore, it is necessary to assess the efficacy of olive natural compounds against olive disease agents, and compare their performance to those that are most widely used in commercial fungicides.

Recommended Strategies for Future Work

Olea europaea L. has been widely studied as part of gastronomic recipes and Mediterranean-type diets: the fruits and the oil are important nutritional components of a considerable part of the world population, whereas leaves are important because of their secondary metabolites. Their extracts consist in fact of an extremely complex mixture of organic components – with most of them still not identified; however, several antibacterial features of oleuropein have already been demonstrated *in vitro* [223, 260, 261, 262, 263]. *In vitro* studies have also made apparent that olive leaf extracts strongly inhibit the classical pathway of complement system; since the aglycone is composed of elenolic acid bound to β-3,4-dihydroxyphenylethyl alcohol and the latter compound is not inhibitory, elenolic acid may be the inhibitory part of the aglycone molecule [264].

Several properties of oleuropein have been thoroughly studied; it appears that they inactivate bacteria by dissolving their outer lining, thus interfering with amino acid production, and thus neutralizing production of essential enzymes – while stimulating phagocytosis [113, 114]. In addition, an acid hydrolyzed extract of oleuropein inhibited growth of eight species of bacteria – a realization that can be explored in attempts to inhibit germination and spore formation of microorganisms on olive trees. Other promising substances from olive tissues include esters, multiple iridoids, rutin, apigenin, luteolin and phospholipid complexes – which work synergistically, and provide an effective year-round defence system for the olive tree. A long way exists to investigate on various types of olive leaf substrates (extracted by both aqueous and organic solvents), coupled with optimization of the extraction processes (using e.g. temperature and time of extraction as manipulated factors). Furthermore, preservation methods for the extracts, viz. drying, demy-evaporation and lyophilization, should be studied regarding their stability and efficiency. Likewise, the best conditions for the antagonistic activity, water activity, pH, natural adjuvants and type of treatment should be sought, so that the antagonistic effect regarding each microorganism ecology can be maximized: in a first stage, assays should be conducted under laboratory conditions, but the second stage will require field experimentation in actual olive groves.

CONCLUSION

The concept of supply chain is nuclear in attempts to rationalize the food sector, as it entails the pathway 'from farm to fork' – i.e. from agricultural production, via food product transformation, and eventually to distribution; and it encompasses an intricate matrix of inter-dependencies along the way. Educated overviews on the food supply chain are useful and have significant implications upon sustainability – encompassing e.g. fulfilment of human needs, provision of employment, maintenance of economic growth, assurance of health care and well-being, and favourable impact upon natural ecosystems. The emergence of unusual dynamics in agricultural commodity markets in recent years has promoted creation of novel or alternative food supply chains, as well as different forms of operation thereof.

Typical products – and traditional foods in particular, exhibit special features owing to their ingredients available only locally, coupled with their unique manufacture protocols. There is a general consensus on their beneficial contributions to rural development, along with preservation of historic and gastronomic heritages. Hence, the traditional fermented food market has earned an increased consumer interest, based on the demand for authentic foods – with consequently wider variation of flavour, in contrast with the overwhelming standardisation brought about by globalisation in the food sector. However, survival of that growing niche market will require incorporation of dedicated and less conventional technologies, more efficient traceability and control of the associated processes, and diversification of the final product portfolio – designed for more demanding and dynamic consumers.

Traditional Mediterranean diets have been practiced for thousands of years, and are characterized by a rich mix of complex carbohydrates and fibre, unsaturated fatty acid residues and natural antioxidants; recently, the issue of proactive contribution to human health has also been more and more frequently addressed. Scientific evidence has it that, beyond nutrition, diet may modulate various physiological functions. The healthy components of the aforementioned Mediterranean diets are directly related to olive-based products – including fermented table olives. Being one of the first methods used by Man to preserve foods, the usefulness of fermentation has also encompassed improvement of sensory properties – and, more recently, the probiotic character of its adventitious microflora. The fermentation of table olives remains one of the most important food industries in southern Europe – and their uniqueness has indeed made them unavoidable ingredients of gourmet recipes and gastronomic excellence, which is now complemented with health claims that motivate health-aware consumers particularly.

A new generation of natural healthy foods is thus on the rise – and fermented table olives are clearly a part of it. In said generation of functional foods, lactic acid bacteria – possessing diverse physiological and metabolic traits, besides technological features often different from classical commercial starter/nonstarter cultures, play a major role. New added-value food products are accordingly to be designed, for which a favourable response is expected on the consumers' side; however, new consumers will not be gained unless a major effort is developed and consistently implemented, toward product and process innovation, coupled with tuning of traditional olive matrices to meet unusual specifications.

Less conventional approaches to process table olives open up new avenues of R&D – yet functional ingredients present spontaneously in the original matrix (e.g. sanitizers in olive leaf extracts, as well as antimicrobial agents) are to be considered, together with the possibility to upgrade byproducts of processing (e.g. olive residues and brines, and olive oil wastewaters) by taking advantage of a number of valuable antioxidants therein. Despite the current food situation prevailing in the world, the future of the unique food supply chain encompassing olives holds by all means a promising potential.

REFERENCES

[1] Convay, G. R. & Barbie, R. E. B. 1990. After the green revolution: Sustainable agriculture for development, Earthscan Publications, London.

[2] Roth, A. V., Tsay, A. A., Pullman, M. E. & Gray, J. V. 2008. Unravelling the food supply chain: strategic insights from China and the 2007 recalls, *Journal of Supply Chain Management*, 44, 1: 22-39.

[3] Maloni, M. J. & Brown, M. E. 2006. Corporate social responsibility in the supply chain: an application in the food industry. *J Business Ethics*, 68, 1: 35-52.

[4] Matos, S. & Hall, J. 2007. Integrating sustainable development in the supply chain: the case of life cycle assessment in oil and gas and agricultural biotechnology. *J Operations Manag*, 25: 1083-102.

[5] Seuring, S. & Müller, M. 2008. From a literature review to a conceptual framework for sustainable supply chain management, *J Cleaner Prod*. 16: 1699-1710.

[6] Marsden, T., Harrison, M. & Flynn, A. 1999. Creating competitive space: exploring the social and political maintenance of retail power. *Environ Planning A*. 30: 481-498.

[7] Stevens, J. 1989. Integrating the supply chain. *Int J Physical Distribution Mat Manag*. 19, 8: 3-8.

[8] Marsden, T. K., Banks, J., & Bristow, G. 2000a. Food supply chain approaches: exploring their role in rural development. *Sociologia Ruralis*. 40: 424-438.

[9] Andy, J. 2002. An environmental assessment of food supply chains: a case study on dessert apples. *Environ Manag*. 30, 4: 560–576.

[10] Schucksmith, M. 1993. Farm household behaviour and the transition to post-productivism. *J Agric Econ*. 44, 466-478.

[11] CAP 2003/2004, European Commission. 2003a). Working Paper of the Directorate - General for Agriculture: The olive oil and table olive sector. Brussels; European Commission (2003b) and eventually accomplishing a sustainable agricultural model for Europe through the reformed Common Agricultural Policy, CAP – including the tobacco, olive oil, cotton and sugar sectors, among others (Communication from the Commission to the Council and the European Parliament, CON (2003) 554 final, Brussels. European Commission (2004); European Union citizens and agriculture from 1995 to 2003. Special Eurobarometer survey, September 2004, Brussels).

[12] ftp://ftp.cordis.europa.eu/pub/fp7/kbbe/docs/traditional-foods.pdf.

[13] Parasecoli, F. 2005. Introduction. *In*: Culinary cultures of Europe. Identity, diversity and dialogue (C. Goldstein and K. Merkle, eds). Council of Europe, Verlagsgruppe Lübbe, Germany, pp 11-37.

[14] Krystallis, A., Chryssochoidis, G. & Scholderer, J. 2007. Consumer-perceived quality in 'traditional' food chains: the case of the Greek meat supply chain. *Appetite*, 48, 1:54-68.

[15] Allaire G., & Sylvander. 1996. Qualité spécifique et innovation territoriale, conférence introductive au séminaire Qualification des produits et des territoires, Toulouse, 2-3 October 1995, *Actes et Communications*, Cahiers d'Economie et Sociologie Rurales, 44, 30-59.

[16] EU Regulation 2081/92.

[17] Jordana, J. 2000. Traditional foods: challenges facing the European food industry. *Food Res Int.* 33, 3-4: 147-152.

[18] Loumou, A. & Giourga, C. 2003. Olive groves: the life and identity of the Mediterranean. *Agric Human Values.* 20: 1, 87- 95.

[19] http://www.oldwayspt.org/traditional-mediterranean-diet.

[20] Trichopoulou, A, & Lagiou, P. 1997. Healthy traditional Mediterranean diet: an expression of culture, history, and lifestyle. *Nutr Res.* 11: 383–389.

[21] Willett, W. C., Sacks, F., Trichopoulou, A., Drescher, G., Ferro-Luzzi, A., Helsing, E., & Trichopoulos, D. 1995. Mediterranean diet pyramid: a cultural model for healthy eating. *Am J Clin Nutr.* 61:1402S–1406S.

[22] Nestle, M. 1995. Mediterranean diets: historical and research overview. *Am J Clin Nutr.* 61:1313S–1320S.

[23] Trichopoulou, A, & Vasilopoulou, E. 2000. Mediterranean diet longevity. *Braz J Nutr* (Supl 2): S205-S209.

[24] Wahrburg, U., Kratz, M. & Cullen, P. 2002. Mediterranean diet, olive oil and health. *Eur J Lipid Sci Technol.* 104: 698- 705.

[25] Shai, I., Dan Schwarzfuchs, M. D., Yaakov Henkin, M. D., Danit, R., Shahar, R.,D., Shula Witkow, R. D., Ilana Greenberg, R. D., Rachel Golan, R. D., Fraser D., Bolotin A., Vardi, H., Tangi-Rozental, O., Zuk-Ramot, R., Sarusi, B., Brickner D., Schwartz, Z., Sheiner, E., Marko, R., Katorza, E., Thiery, J., Martin Fiedler, G., Blüher, M., Stumvoll M., & Stampfer, M. J. 2008. Weight loss with a low-carbohydrate, Mediterranean, or low-fat diet. *Nat Engl J Med.* 3, 359: 229-241.

[26] http://www.suite101.com/content/composition-and-health-benefits-of-olive-oil.   Read 21.12. 2010.

[27] Leonhäuser, I. U., Dorandt, S., Willmund, E. & Honsel, J. 2004. The benefit of the Mediterranean diet. Considerations to modify German food patterns. *Eur J Nutr* (Suppl 1) 43: I/31-I/38.

[28] Willett, W. C., Sacks, F., Trichopoulou, A., Drescher, G., Ferro-Luzzi, A., Helsing, E., & Trichopoulos, D. 1995. Mediterranean diet pyramid: a cultural model for healthy eating. *Amer J Clin Nutr.* 61, 6: 1402S–1406S.

[29] Vogiatzakis, I.N., Mannion, A.M. & Griffiths, G.H. 2006. Mediterranean ecosystems: problems and tools for conservation. *Prog Phys Geography* 30, 2: 175-200.

[30] Prajapati, J. B., and B. M. Nair. 2003. The history of fermented foods, p. 2-23. *In* E. R. Farnworth (ed.), Handbook of Fermented Functional Foods. CRC Press, New York.

[31] Hammes, W.P., 1990. Bacterial starter cultures in food production. *Food Biotechnol.* 4: 383– 397.

[32] Stiles, M.E. (1996) Biopreservation by lactic acid bacteria. *Antonie van Leeuwenhoek* 70: 331-345

[33] Bamforth, C. W. 2005. Food, Fermentation, and Microorganisms. Oxford: Blackwell Science.

[34] Paredes-Lopez, O., and G. I. Harry. 1988. Food biotechnology review: traditional solid-state fermentations of plant raw materials. Application, nutritional significance, and future prospects. *Crit Rev Food Sci Nutr.* 27:159-187.

[35] Beuchat, L. R. 1978. Traditional fermented food products. Pp. 224- 253 *in*: Food and Beverage Microbiology, L. R. Beuchat (Ed.), Westport, CT: AVI Publishing.

[36] Odunfa, S. A., Adeniran, S. A., Teniola, O. D. & Nordstrom, J. 2001. Evaluation of lysine and methionine production in some lactobacilli and yeasts. *Int J Food Microbiol.* 22: 159-163.

[37] Parades-Lopez, O. 1992. Nutrition and Safety Considerations in Applications of Biotechnology to Traditional Fermented Foods, report of an Ad Hoc Panel of the Board on Science and Technology for International Development, National Academy Press, Washington D.C., USA.

[38] Scott, R. & Sullivan W. C. 2008. Ecology of Fermented Foods. *Human Ecol Rev.* 15, 1: 25-31.

[39] Motarjemi, Y., Nout, M. J. R., Adams, M., Bosman, L., Fuchs, R., Harcourt, D., Hastings, J. W., Von Holy, J. W., Holzapfel, A., Kouthon, G., Lee, C. H. & Liebenerg. 1996. Food Fermentation: a Safety and Nutritional Assessment, Bulletin of the World Health Organisation, Switzerland.

[40] Frohlich, R. H., Kunze, M. & Kiefer, I. 1997. Cancer Preventive Impact of Naturally Occurring, Non-nutritive Constituents in Food, Acta Medica Austriaca, Austria.

[41] Hasler, C. M. 2000. The changing face of functional foods. *J Amer Coll Nutr,* 19, 90005: 499S-506S.

[42] www.newgenerationfoods.com.

[43] Biacs, P. A. 2007. Regulations and claims of functional foods. In Proceedings of the fourth international FFNet meeting on functional foods.

[44] Regulation EC 258/97.

[45] American Dietetic Association. 2009. Position of the American Dietetic Association: Functional foods. *J Amer Diet Assoc.* 109, 4:735-74.

[46] IFIC. 2002. The consumer view on functional foods: yesterday and today. Food Insight (May/June).

[47] Muanda, F., Koné, D., Dicko, A., Soulimani, R. & Younos, C. 2009. Phytochemical Composition and Antioxidant Capacity of Three Malian Medicinal Plant Parts. *eCAM* 2009; pp 1-8.

[48] Verbeke, W. 2005. Consumer acceptance of functional foods: socio-demographic, cognitive and attitudinal determinants. *Food Qual Preference.* 16, 1: 45-57.

[49] Jordana, J. 2000. Traditional foods: challenges facing the European food industry. *Food Res Int.* 33, 3-4: 147-152.

[50] Istva´n Siro, Emese Ka´polna, Bea´ ta Ka´polna, Andrea Lugasi. 2008. Functional food. Product development, marketing and consumer acceptance – a review. *Appetite,* 51:456–467.

[51] Bruulsema, T.W. 2002. Nutrients and Product Quality. *Better Crops,* 86, 2: 18-19.

[52] Zhao, J. 2007. Nutraceuticals, nutritional therapy, phytonutrients, and phytotherapy for improvement of human health: a perspective on plant biotechnology application. *Recent Patents Biotechnol. 1:* 75-97.

[53] Prio, R. L., & Cao, G. 2000. Antioxidant phytochemicals in fruits and vegetables: diet and health implications. *Horticultural Sci.* 4: 588–92.

[54] Garcia-Martinez, M. & Poole, N. 2004. The development of private fresh produce safety standards: implications for developing Mediterranean exporting countries. *Food Policy.* 29: 229–255.

[55] International Olive Oil Council (IOOC). 2008. Table Olive Production. www.internationaloliveoil.org.

[56] http://europa.eu.int/comm/agriculture/publi/fact/olive/index_en.htm.

[57] International Olive Council, Olive Products Market Report Summary. No 35 – January 2010. Working paper of the Directorate General for Agriculture; The Olive Oil and Table Oil Sectors.
http://europa.eu.int/comm/agriculture/markets/olive/reports/rep_en.pdf, 2004. p. 19.

[58] Steinkrauss, K. H. 1994. Nutritional significance of fermented foods. *Food Res Int.* 27: 259-267.

[59] Campbell-Platt, G. 1994. Fermented foods – a world perspective. *Food Res Int.* 27: 253-257.

[60] Graaff de, J. & Eppink, L. A. A. J. 1999. Olive oil production and soil conservation in Southern Spain in relation to EU subsidy policies. *Land Use Policy* 16: 259–267.

[61] Convention on biological diversity, Rio de Janeiro, 1992.

[62] Guzman Alvarez, J. R. 1999. Olive cultivation and ecology: the situation in Spain. *Olivae.* 78: 41–49.

[63] Beaufoy, G., 2001. EU policies for olive farming unsustainable on all counts. Worldwide Fund for Nature (WWF) and Bird Life International, Brussels: 7-8.

[64] CAP. 2003. Analysis of the 2003 CAP Reform,
http://www.oecd.org/dataoecd/62/42/32039793.pdf.

[65] Loumou, A, & Giourga, C. 2003. Olive groves: the life and identity of the Mediterranean. *Agric Human Values* 20: 87-95.

[66] Johnson, H. D. 2010. Density: Life For Olive Trees Getting Harder, www.oliveoiltimes.com/features/high-density-olive/2571.

[67] Vossen, P. & Devarenne, A. 2004. California Olive Oil Industry Survey Statistics 2004. California Olive Oil Survey Statistics. http://cesonoma.ucdavis.edu/files/71873.pdf.

[68] Giourga, C., Loumou, A. Tsevreni, I. & Vergou, A. 2008. Assessing the sustainability factors of traditional olive groves on Lesvos Island, Greece (Sustainability and traditional cultivation). *GeoJournal* 73:149–159.

[69] Beaufoy, G. & Cooper, T. 2008. Guidance Document to the Member States on the Application of the HNV Impact Indicator, European Evaluation Network for Rural Development, Bruxelles.

[70] Pastor, M. & Castro, J. 1995. Soil management systems and erosion. *Olivae*, 59, International Olive Oil Council, Madrid.

[71] http://www.igreens.org.uk/eu_olive_subsidies.htm.

[72] Tedeschi, A. & Dell'Aquila, R. 2005. Effects of irrigation with saline waters, at different concentrations on soil physical and chemical characteristics. *Agric Water Manag* 77: 308-322.

[73] UNCCD. 2004. Preserving our common ground. *UNCCD 10* years on United Nations Convention to Combat Desertification, Bonn, Germany.

[74] Metzidakis, I., Martinez-Vilela, A. Nieto, G. C., Basso, B. 2008 Intensive olive orchards on sloping land: good water and pest management are essential, *J Environ Manag.* 89: 120–128.

[75] Xiloyannis, C. Martinez, A. R. Kosmas, C. & Favia, M. 2008. Semi-intensive olive orchards on sloping land: requiring good land husbandry for future development, *J. Environ. Manag.* 89: 110–119.

[76] Stroosnijdera, L., Mansinho, M.I. & Palese, A.M. 2008. OLIVERO: The project analysing the future of olive production systems on sloping land in the Mediterranean basin, *J Environ Manag.*, 89: 75-85.

[77] Barea, F. & Ruiz-Avilés, P. 2009. Estrategias de futuro para el sector oleícola andaluz. In Gómez Calero, J.A. (ed.) *Sostenibilidad de la producción de olivar en Andalucía*, Consejería de Agricultura y Pesca-Junta de Andalucía, Sevilla.

[78] García-Brenes, M. D. 2007. Los impactos ecológicos del cambio estructural. El olivar andaluz. En: Sanz, J. (ed.) *El futuro del mundo rural*, Síntesis, Madrid, pp. 223-241.

[79] Gómez-Calero, J. A. (ed.) (2009) Sostenibilidad de la producción de olivar en Andalucía, Consejería de Agricultura y Pesca-Junta de Andalucía, Sevilla.

[80] Consejería de Agricultura y Pesca. 2008. El sector del aceite de oliva y la aceituna de mesa en Andalucía, Consejería de Agricultura y Pesca - Junta de Andalucía, Sevilla, Spain.

[81] The European Commission – Agriculture - Quality Policy. Protected Designation of Origin / Protected Geographical Indication – Table olives. http://europa.eu.int/comm/agriculture/qual/pt/pgi_02pt.htm.

[82] WHO. 2005. World Health Organisation: Hazard Analysis Critical Control Point System (HACCP). See http://www.who.int/foodsafety/fs_management/haccp/en/.

[83] IOOC. 1990. Table Olive processing. Collection: Technical handbooks. International Olive Oil Council. Madrid.

[84] Díez, M., Ramos, R., Fernandez, A., Cancho, F., Pellisó, F., Vega, M., Moreno, A., Mosquera, I., Navarro, L., Quintana, M., Roldán, F., Garcia, P., & Gómez-Millán, A., 1985. Biotecnologia de la aceituna de mesa. Instituto de la Grasa y sus Derivados. CSIC. Sevilla. Spain.

[85] Garrido Fernández, A., Fernández Díez, M. J., & Adams, M. R., 1997. Physical and chemical characteristics of the olive fruit, p. 67-109. *In* A. Garrido-Fernández et al. (ed.), Table Olives. Chapman & Hall, London, UK.

[86] Sanchez-Gómez, A. H. Garcia-Garcia, & P, Rejano-Navarro, L. 2006. Elaboration of table olives. *Grasas y Aceites.* 57: 86-94.

[87] Panagou, E. Z, Schillinger, U., Franz, C. M. A. P., & Nychas, G. J. E. 2008. Microbiological and biochemical profile of cv. Conservolea naturally balck olives during controlled fermentation with selected strains of lactic acid bacteria. *Food Microbiol.* 25: 348-358.

[88] Arroyo-López, F.N., Durán Quintana, M.C., & Garrido-Fernández, A. 2006. Use of the generalized z-value concept to study the effects of temperature, NaCl concentration and pH on *Pichia anomala*, yeast related to table olive fermentation. *Int J Food Microb.* 106: 45–51.

[89] Beumer, R. R. 2001. Microbiological hazards and their control: bacteria, p. 141-158. *In:* M. R. Adams and M. J. Robert Nout (ed.), Fermentation and Food Safety, Aspen Publishers, Maryland, USA.

[90] Tassou, C.C. 1993. Microbiology of olives with emphasis on the antimicrobial activity of phenolic compounds. Ph.D Thesis, Univ. of Bath, Bath, UK.

[91] Medina, E., Brenes, M., Romero, C., Garcia, A., de Castro, A., 2007. Main antimicrobial compounds in table olives. *J Agric Food Chem.* 55: 9817-9823.

[92] Bobillo, M., & Marshall, V. M., 1991. Effect of salt and culture aeration on lactate and acetate production by *Lactobacillus plantarum*. *Food Microbiol.* 8: 153-160.

[93] Bobillo, M., & Marshall, V. M., 1992. Effect of acidic pH and salt on acid end-products by *Lactobacillus plantarum* in aerated, glucose limited continuous culture. *J Appl Bacteriol.* 73: 226-233.

[94] García-García, P., Durán-Quintana, M. C., Brenes-Balbuera, M., & Garrido-Fernández, A., 1992. Lactic fermentation during the storage of 'Alorena' cultivar untreated green table olives. *J Appl Bacteriol.* 73: 324-330.

[95] Fernández-González, M. J., García-García, P., Garrido-Fernández, A., & Durán-Quintana, M.C. 1993. Microflora of the aerobic preservation of directly brined green olives from Hojiblanca cultivar. *J Appl Bacteriol.* 75: 226-233.

[96] Montaño, A., Sánchez, A. H., & de Castro, A., 1993. Controlled fermentation of Spanish-type green olives. *J Food Sci.* 4: 842-844.

[97] Castro, A., Montaño, A., Casado, F. J., Sánchez, A. H., & Rejano, L. 2002. Utilization of *Enterococcus casseliflavus* and *Lactobacillus pentosus* as starter cultures for Spanish-style green olive fermentation. *Food Microbiol.* 19: 637-644

[98] Tassou, C. C., Panagou, E. Z., & Katsaboxakis, K. Z. 2002. Microbiological and physicochemical changes of naturally black olives fermented at different temperatures and NaCl levels in the brines. *Food Microbiol.* 19: 605-615.

[99] Álvarez, D. M. E., Sánchez, A., & Lamarque, A. L. 2003. Naturally black olives: comparison of three processes for fermenting cv. 'Farga' olives. *Olivae.* 97: 47-51.

[100] Montaño, A., Sánchez, A.H., Casado, F.J., de Castro, A., & Rejano, L., 2003. Chemical profile of industrially fermented green olives of different varieties. *Food Chem.* 82: 297-302.

[101] Hornero-Mendez, D., & Garrido-Fernandéz, A. 1994. Rapid high-performance liquid chromatography analysis of biogenic amines in fermented vegetable brines. *J Food Prot.* 60, 4: 414-419.

[102] Brenes-Balbuena, M., García-García, P., & Garrido-Fernández, A. 1992. Phenolic compounds related to the black color formed during the processing of ripe olives. *J. Agric Food Chem.* 40: 1196-1196.

[103] Soni S. K., & Sandhu, D. K. 1990. Indian fermented foods: microbiological and biochemical aspects. *Indian J Microbiol.* 30:135-157.

[104] Ross, R. P., Morgan, S., & Hill, C., 2002. Preservation and fermentation: past, present and future. *Int J Food Microbiol.* 79: 3-16.

[105] Isolauri, E, Sütas, Y, Kankaanpää, P, Arvilommi, H, & Salminen, S. 2001. Probiotics: effects on immunity. *Am J Clin Nutr* 73 (suppl.): 444s-450s.

[106] Durán Quintana, M. C., García-García, P., Garrido-Fernández, A., 1999. Establishment of conditions for green olive fermentation at low temperature. *Int J Food Microbiol.* 51: 133-143.

[107] Castro, A., Montaño, A., Casado, F. J., Sánchez, & A. H., Rejano, L. 2002. Utilization of *Enterococcus casseliflavus* and *Lactobacillus pentosus* as starter cultures for Spanish-style green olive fermentation. *Food Microbiol.* 19: 637-644.

[108] Chorianopoulos, N. G., Boziaris, I. S., Stamatiou, A., Nychas, G.-J. N., 2005. Microbial association and acidity development of unheated and pasteurized green table olives fermented using glucose or sucrose supplements at various levels. *Food Microbiol.* 22: 117-124.

[109] Delgado, A., Brito, D., Peres, C., Noé-Arroyo, F. & Garrido-Fernández, A. 2005. Bacteriocin production by *Lactobacillus pentosus* B96 can be expressed as a function of temperature and NaCl concentration. *Food Microb* 22: 521- 528.

[110] Peres, C., Delgado, A. & Catulo, L. (in press). What perspectives for the Portuguese table-olive industry? *OLEA*-FAO Olive Network.

[111] Balatsouras, G. D., 2004. Table Olives: Varieties, Chemical Composition, Commercial Preparations, Quality Characteristics, Packaging, Marketing. Agricultural University of Athens Edition (in Greek).

[112] Macedo, A. C., Venâncio, A. & Malcata, F. X. 2003. Biotecnologia dos alimentos. *In* Lima, N. & Mota, M. (Eds.), Biotecnologia: Fundamentos e Aplicações. LIDEL, Lisboa, Portugal: pp. 429-472.

[113] Marquina, D., Peres, C., Caldas F. V., Marques, J. F. Peinado, J. M. & Spencer Martins, I. 1992. Characterization of the yeast population in olives brines, *Lett Appl Microbiol* 14: 279–283.

[114] Schnürer, J. & Magnusson, J. 2005. Antifungal lactic acid bacteria as biopreservatives. *Trends Food Sci Technol* 16, 1-3: 70-78.

[115] Hurtado, A., Reguant, C. Esteve-Zarzoso, B., Bordons, A. & Rozès, N. 2008. Microbial population dynamics during the processing of *Aberquina* table olives. *Food Res Int.* 41: 738–744.

[116] Rozès, N., & Peres, C. 1996. Effect of oleuropein and sodium chloride on viability and metabolism of *Lactobacillus plantarum*. *Appl Microbiol Biotechnol.* 45: 839-843.

[117] Rozès, N. & Peres, C. 1998. Effects of phenolic compounds on the fatty acid composition of *Lactobacillus plantarum*. *Appl Microbiol Biotechnol.* 49: 108-111.

[118] Oliveira M., Brito D., Catulo L., Leitão F., Gomes L., Silva S., Vilas Boas L., Peito A.; Fernandes I., Gordo F., & Peres C. 2004. Biotechnology of olive fermentation of Galega Portuguese variety. *Grasas y Aceites.* 55, 3: 219-226.

[119] Ciafardini, G., Marsilio, V., Lanza, B., Pozzi, N., 1994. Hydrolysis of oleuropein by *Lactobacillus plantarum* strains associated with olive fermentation. *Appl Environ Microbiol.* 60: 4142–4147.

[120] Ghabbour, N., Lamzira, Z., Thonart, P., Cidalia, P.,Markaoui, M., & Asehraou, A. 2011. Selection of oleuropein-degrading lactic acid bacteria strains isolated from fermenting Moroccan green olives. *Grasas y Aceites.* 62, 1: 84-89.

[121] Hernández, A., Martín, A. Córdoba, M. G. Benito, M. J. Aranda E. & Pérez-Nevado, F. 2008. Determination of killer activity in yeasts isolated from the elaboration of seasoned green table olives. *Int J Food Microbiol.* 121: 178–188.

[122] Arroyo-López, F. N. Querol, A., Bautista-Gallego, J. & Garrido-Fernández, A. 2008. Role of yeasts in table olive production. *Int J Food Microbiol.* 10:189-196.

[123] Viljoen, B. C. 2006. Yeast ecological interactions. Yeast–yeast, yeast bacteria, yeast–fungi interactions and yeasts as biocontrol agents. *In*: A. Querol and H. Fleet, Editors, *Yeasts in Food and Beverages*, Springer–Verlag, Berlin (2006), pp. 83–110.

[124] Abbas, C. A. 2006. Production of antioxidants, aromas, colours, flavours, and vitamins by yeasts. *In*: A. Querol and H. Fleet, Editors, *Yeasts in Food and Beverages*, Springer–Verlag, Berlin (2006), pp. 285–334.

[125] Prajapati, J. B., & Nair, B. M. 2003. The history of fermented foods, p. 2-23. *In* E. R. Farnworth (ed.), Handbook of Fermented Functional Foods. CRC Press, New York.

[126] Giraffa, G., Chanishvili, N.,· & Yantyati, W. 2010. Importance of lactobacilli in food and feed biotechnology. *Res Microbiol.* 161, 6: 480-487.

[127] Hammes, W. P., Weiss, N. & Holzapfel, W. 1992. The genera *Lactobacillus* and *Carnobacteria. In* The Prokaryotes, Balows, A., Trüper, H.G., Dworkin, M., Harder, W. and Schleifer, K. (Eds.), pp.1535-1544, Springer-Verlag: New York, NY, USA.

[128] Buckenhüskes, H. J., 1993. Selection criteria for lactic acid bacteria to be used as starter cultures for various food commodities. *FEMS Microbiol Rev.* 12: 253-272.

[129] Fleming, H. P., Mc Feeters, R. F., & Daeschel, M. A. 1985. The lactobacilli, pediococci and leuconostoc from vegetable products. *In*: Bacterial Starter Cultures for Foods. Ed. Gilliland, S. E. CRC Press, Boca Raton FL, USA.

[130] Piard, J. C., & Desmazeaud, M., 1991. Inhibiting factors produced by lactic acid bacteria. 1. Oxygen metabolites and catabolism end products. *Lait.* 71: 525-541.

[131] Piard, J. C., Desmazeaud, M. 1992. Inhibiting factors produced by lactic acid bacteria. 2. Bacteriocins and other antibacterial substances. *Lait.* 72: 113-142.

[132] Ray, B., 1992. Bacteriocins of starter culture bacteria as biopreservatives, an overview. *In*: Food Biopreservatives of Microbial Origin. Ed. Ray, B. & Daeschel, M. A. CRC Press, Boca Raton FL, USA.

[133] Podolak, P. K., Zayas, J. F., Kastner, C. L., Fung, D. Y. C., 1996. Inhibition of *Listeria monocytogenes* and *Escherichia coli* O157:H7 of beef by application of organic acids. *J Food Prot.* 59: 370-373.

[134] Earnshaw, R. G. 1992. The antimicrobial action of lactic acid bacteria: natural food preservation systems. *In*: The Lactic Acid Bacteria in Health and Disease. Ed. Wood, B. J. B. Elsevier Applied Science. London, New York.

[135] Kashket, E. R. 1987. Bioenergetics of lactic acid bacteria: cytoplasmic pH and osmotolerance. *FEMS Microbiol Rev.* 46: 233-244.

[136] Smulders, F. J. M., Barendsen, P., Van Logtestijn, J. G., Mossel, D. A. A., & van der Marel, G. M. 1986. Lactic acid: considerations in favour of its acceptance as meat decontaminant. *J Food Technol.* 21: 419-436.

[137] Woolford, M. K. 1975. Microbiological screening of food preservatives, cold sterilants and specific antimicrobial agents as potential silage additives. *J Sci Food Agric.* 26: 229-237.

[138] Lindgren, S. E. & Dobrogosz, W. J. 1990. Antagonistic activities of lactic acid bacteria in food and feed fermentations. *FEMS Microbiol Rev.* 87: 149-164.

[139] Adams, M. R., & Hall, C. J. 1988. Growth inhibition of foodborne pathogens by lactic and acetic acids and their mixtures. *Int J Food Sci Technol.* 23: 287-292.

[140] Kong, S., & Davidson, A. J. 1980. The role of interactions between O_2, H_2, OH·, e·, O_2^- in free radical damage to biological systems. *Arch Biochem Biophys.* 204: 13-29.

[141] Jay, J. M. 1986. Modern Food Microbiology. 3rd ed. van Nostrand Reinhold, New York, USA.

[142] Talarico, T. L., Casas, I. A., Chung, T. C., & Dobrogosz, W. J., 1988. Production and isolation of reuterin, a growth inhibitor produced by *Lactobacillus reuteri. Antimicrob Agents Chemother.* 32: 1854-1858.

[143] Axelsson, L., Chung, T. C., Dobrogosz, W. J., Lindgren, S. E., 1989. Production of a broad spectrum antimicrobial substance by *Lactobacillus reuteri. Microb Ecol Health Dis.* 2: 131-136.

[144] Klaenhammer, T. 1993. Genetics of bacteriocins produced by lactic acid bacteria. *FEMS Microbiol Rev.* 12: 39-86.

[145] Madureira, A. R., Soares, J. C., Pintado, M. E., Gomes, A. M. P., Freitas, A. C. & Malcata, F. X. 2008. Sweet whey cheese matrices inoculated with the probiotic strain *Lactobacillus paracasei* LAFTI® L26. *Dairy Sci Technol.* 88: 649-665.

[146] Madureira, A. R., Pereira, C. I., Truszkowska, K., Gomes, A. M., Pintado, M. E. & Malcata, F. X. 2005. Survival of probiotic bacteria in a whey cheese vector submitted to environmental conditions prevailing in the gastrointestinal tract. *Int Dairy J.* 15, 6-9: 921-927.

[147] Kongo, J. M., Gomes, A. M. & Malcata, F. X. 2006. Manufacturing of fermented goat milk with a mixed starter culture of *Bifidobacterium* animalis and *Lactobacillus acidophilus* in a controlled bioreactor. *Lett Appl Microbiol.* 42, 6: 595-599.

[148] Gomes, A. M. P., & Malcata, F. X. 1998. Use of small ruminants' mik supplemented with available nitrogen as growth media for *Bifidobacterium lactis* and *Lactobacillus acidophilus*. *J Appl Microbiol.* 85, 5: 839-848.

[149] Gomes, A. M. P. & Malcata, F. X. 1999. *Bifidobacterium* spp. and *Lactobacillus acidophilus*: biological, biochemical, technological and therapeutical properties relevant for use as probiotics. *Trends Food Sci Technol.* 10, 4-5: 139-157.

[150] Schnürer, J. & Magnusson, J. 2005. Antifungal lactic acid bacteria as biopreservatives. *Trends Food Sci Technol.* 16, 1-3: 70-78.

[151] Haller, D., Colbus, H., Ganzle, M. G., Scherenbacher, P., Bode, C. & Hammes, WP. 2001. Metabolic and functional properties of lactic acid bacteria in the gastrointestinal ecosystem: a comparative *in vitro* study between bacteria of intestinal and fermented food origin. *Syst Appl Microbiol.* 24, 2: 218-226.

[152] de Keersmaecker, S. C. J., Verhoeven, T. L. A., Desair, J., Marchal, K., Vanderleyden, J. & Nagy, I. 2006. Strong antimicrobial activity of *Lactobacillus rhamnosus* GG against *Salmonella typhimurium* is due to accumulation of lactic acid. *FEMS Microbiol Lett.* 259: 89-96.

[153] Delgado, A., Noé-Arroyo, F., Brito, D., Peres, C., Fevereiro, P. & Garrido-Fernández., A. 2007. Optimum bacteriocin production by *Lactobacillus plantarum* 17.2b requires absence of NaCl and apparently follows a mixed metabolite kinetics. *J Biotechnol.* 130, 2: 193-201.

[154] Montecucco, C., & Rappuolli, R. 2001. Living dangerously: how *Helicobacter pylori* survives in the human stomach. Nature Reviews - *Mol Cell Biol.* 2, 6: 457-466.

[155] Lopes, A. I., Oleastro, M., Palha, A., Fernandes, A. & Monteiro, L. 2005. Antibiotic-resistant *Helicobacter pylori* strains in Portuguese children. *Pediatric Inf Dis J.* 24, 5: 404-409.

[156] Mättö, J., Fondén, R., Tolvanen, T., von Wright, A., Vilpponen-Salmela, T., Satokari, R. & Saarela, M. 2006. Intestinal survival and persistence of probiotic *Lactobacillus* and *Bifidobacterium* strains administered in triple-strain yoghurt. *Int Dairy J.* 16, 10:1174-1180.

[157] Bertazzoni, E. M. & Benini, A. 2008. Relationship between number of bacteria and their probiotic effects. *Microb Ecol Health Disease.* 20, 4: 180-183.

[158] Owen, R.W., Giacosa, A., Hull, W. E., Haubner, R., Spiegelhalder, B., & Bartsch, H., 2000. The antioxidant/anticancer potential of phenolic compounds isolated from olive oil. *Eur J Cancer.* 36: 1235–1247.

[159] Visioli, F., Romani, A., Mulinacci, N., Zarini, S., Conte, D., Vincieri, F. F., & Galli, C. 1999. Antioxidant and other biological activities of olive oil mill waste water. *J Agric Food Chem*. 47: 3397–3401.

[160] de la Puerta, R., Ruíz-Gutiérrez, V., & Hoult, J. R., 1999. Inhibition of leukocyte 5-lipoxygenase by phenolics from virgin olive oil. *Biochem Pharmacol*. 57: 445–449.

[161] Auroma, O. I., Deiane, M., Jenner, A., Halliwell, B., Kaur, M., Banni, S., Corongiu, F.P., Dess, M. A., & Aesbach, R., 1998. Effects of hydroxytyrosol found in extra virgin olive oil on oxidative DNA damage and on low-density lipoprotein oxidation. *J Agric Food Chem*. 46: 5181– 5187.

[162] López-López, A., Rodríguez Gómez, F. Cortés-Delgado, A. & Garrido-Fernández, A. 2009. Changes in sterols, fatty alcohol and triterpenic alcohol during ripe olive processing. *Czech J Food Sci*. 27: S225-S226.

[163] López-López, A., García, P. & Garrido-Fernandéz, A. 2008. Multivariate characterization of table olives according to their mineral nutrient composition. *Food Chem*. 106, 1: 369-378.

[164] Garrido-Fernandéz, A. & Lopez-Lopez, A., 2008. Revalorización nutricional de la aceituna de mesa. Oral communication. II Jornadas Internacionales de la Aceituna de Mesa. Sevilla. 26-27 Março 2008.

[165] Ruiz-Barba, J. L. & Jiménez-Díaz, R. 1995. Availability of essential B-group vitamins to *Lactobacillus plantarum* in green olive fermentation brines. *Appl Environ Microbiol*. 61, 4: 1294–1297.

[166] Guillén, R., Heredia, A., Felizoón, B., Jiménez, A., Montaño, A., & Fernández-Bolanños, J. 1992. Fibre fraction carbohydrates in *Olea europeae* (Gordal and Manzanilla var.). *Food Chem*. 44: 173–174.

[167] Amiot, M. J., Fleuriet, A., & Macheix, J. J. 1986. Caracte´risation des produits de de´gradation de l'oleuropein (Technological debittering process of olives: characterization of fruits before and during alkaline treatment). *Groupe Phenolic*. 364-369.

[168] Pereira, J. A., Pereira, A. P. G., Ferreira, I. C. F. R., Valentão, P., Andrade, P. B., Seabra, R., Estevinho, L. & Bento, A. 2006. Table olives from Portugal: phenolic compounds, antioxidant potential, and antimicrobial activity. J Agric Food Chem. *54*, 22: 8425–8431.

[169] Silva, S., Gomes, L., Leitão, F., Coelho, A. V., & Vilas-Boas, L. Phenolic compounds and antioxidant activity of *Olea europaea* L. fruits and leaves. *Food Sci Technol Int*. 12, 5: 385-396.

[170] Teissedre, P. L., Frankel, E. N., Waterhouse, A. L., Peleg, H., & German, J. B. 1996. Inhibition of *in vitro* human LDL oxidation by phenolic antioxidants from grapes and wines. *J Sci Food Agric*. 70, 1: 55-61.

[171] Cody, V., Middleton, E., Harborne, J. B., & Beretz, A., (ed.) 1988. Plant Flavonoids in Biology and Medicine. II - Biochemical, Cellular and Medicinal Properties, *Prog. Clin. Biol. Res.*, 280, Alan R. Liss, New York, USA

[172] Visioli, F., Galli, C., Bornet, F., Mattei, A., Patelli, R., Galli, G. & Caruso D. 2000. Olive oil phenolics are dose-dependently absorbed in humans. *FEBS Lett*. 468, 2-3: 159-160.

[173] Visioli, F, & Galli, C. 1998. The effect of minor constituents of olive oil on cardiovascular disease: new findings. *Nutr Rev*. 56, 5:142-147.

[174] Visioli, F. & Galli, C. 2002. Biological properties of olive oil phytochemicals. *Crit Rev Food Sci Nutrition*. 42, 3: 209-221.

[175] Ryan, D. & Robards, K. 1998. Phenolic compounds in olives. *Analyst*. 123: 31R–44R.

[176] MADRP, 2007a. Ministério da Agricultura do Desenvolvimento Rural e das Pescas. Gabinete de Planeamento e políticas. Anuário Vegetal 2006. Crop Production yearbook. Ed. MADRP 2007. Olivicultura.

[177] MADRP, 2007b- Ministério da Agricultura do Desenvolvimento Rural e das Pescas. Guia dos Produtos de Qualidade. DOP/IGP/ETG/DOC/IPR/AB/PI. Ed. MADRP 2007.

[178] Brookes, G., Craddock, N. & Kneil, B. 2005. The Global GM Market: implications for the European food chain. An analysis of labelling requirements, market dynamics and cost implications, p. 106. http://www.pgeconomics.co.uk/.

[179] Ilbery, B., & Kneafsey, M. 1999. Niche markets and regional speciality food products in Europe: towards a research agenda *Environ Planning A* 31: 2207-2222.

[180] van der Ploeg, J. D., Renting, H., Brunori, G., Knickel, K., Mannion, J., Marsden, T. K., de Roest, K., Sevilla-Guzma¨n, E., & Ventura, F. 2000. Rural development: from practices and policies towards theory. *Sociologia Ruralis*. 40, 4: 391-408.

[181] Confederation des Industries Agro-Alimentaires de l'UE. 2009. The Competitiveness of the EU Food and Drink Industry, *Fact and Figures*, September 2009, p. 2.

[182] European Commission, EU. *Food Market Overview*, http://ec.europa.eu/enterprise/sectors/food/eu-market/index_en.htm.

[183] Jordana, J. 2000. Traditional foods: challenges facing the European food industry. *Food Res Int*. 33, 3-4: 147-152.

[184] EU Regulation 2081/92. The European Commission – Agriculture - Quality Policy. Protected Designation of Origin / Protected Geographical Indication – Table olives. http://europa.eu.int/comm/agriculture/qual/pt/pgi_02pt.htm.

[185] Broerse, J. E. W., & Visser, B. 1996. Assessing the Potential, *In*: Biotechnology: Building on Farmers' Knowledge, Ed. Bunders, J., Haverkort, B. and Hiemstra, W., (1996), Macmillan Education, UK.

[186] Renting, H. 2003. Understanding alternative food networks: exploring the role of short food supply chains in rural development. *Environ Planning A*. 35, 393-411.

[187] Loumou, A., & Giourga, C. 2003. Olive groves: the life and identity of the Mediterranean. *Agric Human Values*. 20: 87–95.

[188] Shi, J., Ho, C.T., & Shahidi, F. 2005. In *Asian Functional Foods*, CRC Press, Boca Ratton FL, USA.

[189] Subirade, M. 2007. Report on functional foods. Food Quality and Standards Service (AGNS), Food and Agriculture Organization of the United Nations (FAO), November, 2007.

[190] Scientific Concepts of Functional Foods in Europe: Consensus Document. 1999. *British J Nut*. 81: S1–S27.

[191] Roberfroid, M. B. 1999. What makes food functional? *in*: Lásztity, R., Pfannhauser, W., Simon-Sarkadi, L. & Tömösközi, S. (Eds.) Proceedings of EURO FOOD CHEMX: Functional Foods - A new challenge for the food chemists, 22-24 September 1999, Budapest., Budapest, Hungary, Vol.1, pp. 3-10.

[192] Pszczola, D. E. 1999. It's never too late: ingredients for the aging. *Food Technol*. 53, 5: 60-68.

[193] Sanders, 1999. Sanders, M. E. (1999) Probiotics. *Food Technol*. 53, 11: 67-77.

[194] Milner, J. A. 1999. Functional foods - The US perspective. *in*: Lásztity, R., Pfannhauser, W., Simon-Sarkadi, L. & TömösközI, S. (Eds.) Proceedings of EURO FOOD CHEMX: Functional Foods - A new challenge for the food chemists, 22-24 September 1999, Budapest. Budapest, Hungary, Vol. 1, pp. 46-55.

[195] Jones, P.J., & Jew, S. 2007. Functional food development: concept to reality. *Trends Food Sci Technol.* 18: 387-390.

[196] Joint FAO/WHO. 2002. Working Group Report on Drafting Guidelines for the Evaluation of Probiotics in Food. Guidelines for the Evaluation of Probiotics in Food.

[197] Salminen, S., Ouwehand, A. G., & Isolauri, E. 1998. Clinical applications of probiotic bacteria. *Int Dairy J.* 8:563-572.

[198] Saarela, M., Lähteenmäki, L., Crittenden, R., Salminen, S., & Mattila-Sandholm, T. 2002. Gut bacteria and health foods – the European perspective. *Int J Food Microbiol.* 78: 99-117.

[199] Drago, L., Gismondo, M. R., Lombardi, A., de Haen, C. & Gozzini, L. 1997. Inhibition of *in vitro* growth of enteropathogens by new *Lactobacillus isolates* of human intestinal origin. *FEMS Microbiol Lett.* 153: 455-463.

[200] Cross, M. L. 2002. Microbes versus microbes: immune signals generated by probiotic lactobacilli and their role in protection against microbial pathogens. *FEMS Immunol Med Microbiol.* 34: 245-253.

[201] Jahreis, G., H., Vogelsang, G., Kiessling, R., Schubert, C., Bunte, & Hammes, W. P. 2002. Influence of probiotic sausage (*Lactobacillus paracasei*) on blood lipids and immunological parameters of healthy volunteers. *Food Res Int.* 35:133-138.

[202] Prado, F. C., Parada, J. L., Pandey, A., & Soccol, C. R., 2008. Trends in non-dairy probiotic beverages. *Food Res Int.* 41, 111-123.

[203] Gardiner, G. E., Bouchier, P., O'Sullivan, E., Kelly, J., Kevin Collins, J., & Fitzgerald, G. 2002. A spray-dried culture for probiotic Cheddar cheese manufacture. *Int Dairy J.* 12, 9: 749-756.

[204] d'Aimmo, M. R., Modesto, M., & Biavati, B. 2007. Antibiotic resistance of lactic acid bacteria and *Bifidobacterium* spp. isolated from dairy and pharmaceutical products. *Int J Food Microbiol.*115, 1: 35-42.

[205] Rasic, J. L. 2003. Microflora of the intestine probiotics. *In* B. Caballero, L. Trugo, & P. Finglas (Eds.), Encyclopedia of Food Sciences and Nutrition, pp. 3911–3916. Oxford: Academic Press.

[206] Newell-McGloughlin, M. 2008. Nutritionally improved agricultural crops. *Planta Physiol.*147: 939-953.

[207] Knorr, D. 1998. Technology aspects related to microorganisms in functional foods. *Trends Food Sci Technol.* 9, 8-9: 295-306.

[208] German, B., Schiffrin, E. J., Reniero, R., Mollet, B., Pfeifer, A., & Neeser, J. R. 1999. The development of functional foods: lessons from the gut. *Trends Biotechnol.* 17: 492-499.

[209] Tannock, G.W. 1997. Probiotic properties of lactic acid bacteria: plenty of scope for fundamental R & D. *Trends Biotechnol.* 15: 270-274.

[210] Marteau, P. R., de Vrese, M., Cellier, C. J., & Schrezenmeir, J. 2001. Protection from gastrointestinal diseases with the use of probiotics. *Am J. Clin Nutr.* 73S: 430-436.

[211] Cross, M. L. 2002. Microbes versus microbes: immune signals generated by probiotic lactobacilli and their role in protection against microbial pathogens. *FEMS Immunol Med Microbiol.* 34, 245-253.

[212] Oberman, H., & Libudzisz, Z. 1998. Fermented milks. In B. J. B. Wood (Ed.), Microbiology of fermented foods, Vol. 1 (pp. 308–350). London: Blackie Academic & Professional.

[213] Haller, D., Colbus, H., Ganzle, M. G., Scherenbacher, P., Bode, C., & Hammes, W. P. 2001. Metabolic and functional properties of lactic acid bacteria in the gastro-intestinal ecosystem: a comparative *in vitro* study between bacteria of intestinal and fermented food origin. *Syst Appl Microbiol.* 24: 218-226.

[214] Rodríguez, R., Jiménez, A., Fernández-Bolaños, J., Guillén, R., & Heredia, A., 2006. Dietary fibre from vegetable products as source of functional ingredients. *Trends Food Sci Technol.* 17: 3–15.

[215] Juan, M. E., Wenzel, U., Ruiz-Gutierrez, V., Daniel, H., & Planas, J. M. 2006. Olive fruit extracts inhibit proliferation and induce apoptosis in HT-29 *Human Colon Cancer Cells J Nutr.* 136: 2553–2557.

[216] Brito, D., Serrano, C., Pereira, A., Delgado, A., Oleastro, M., Monteiro, L. & Peres, C. 2007. Evaluation of the inhibitory activity of *Lactobacillus* sp. from table-olives against *Helicobacter pylori* Abstrat no: PO91. *Helicobacter.* 12: 438.

[217] Lavermicocca, P., Gobbetti, M. Corsetti, A. & Caputo, L. 1998. Characterization of lactic acid bacteria isolated from olive phylloplane and table olive brines. *Ital J Food Sci.* 10: 27-39.

[218] Nychas, G.-J.E., Panagou, E. Z., Parker, M. L., Waldron, K. W., & Tassou, C. C. 2002. Microbial colonization of naturally black olives during fermentation and associated biochemical activities in the cover brine. *Lett Appl Microbiol.* 34:173-177.

[219] Lavermicocca, P., Valerio, F., Lonigro, S. L., de Angelis, M., Morelli, L., Callegari, M. L., Rizzello, C. G., & Visconti, A., 2005. Study of adhesion and survival of lactobacilli and bifidobacteria on table olives with the aim of formulating a new probiotic food. *Appl Environ Microbiol.* 71: 4233-4240.

[220] Lavermicocca, P., Lonigro, S. L., Visconti, A., De Angelis, M., Valerio, F., & Morelli, L. 2007. Table olives containing probiotic microorganisms. International Patent WO2005053430 (A1); US2007086990 (A1); EP1843664 (A0); CA2546776 (A1).

[221] Panagou, E. Z., Tassou, C. C. & Katsaboxakis, K. Z.. 2002. Microbiological, physicochemical and organoleptic changes in dry-salted olives of Thassos variety stored under different modified atmospheres at 4 and 20 °C. *Int J Food Sci Technol.* 37: 635-641.

[222] Saxelin, M., Grenov, B., Svensson, U., Fondén, R., Reniero, R., & Mattila-Sandholm, T. 1999. The technology of probiotics. *Trends Food Sci Technol.* 10: 387-392.

[223] Molin, G., 2001. Probiotics in foods not containing milk or milk constituents with special reference to *Lactobacillus plantarum* 299v. *Am J Clin Nutr.* 73S: 380-385.

[224] Betoret, N., Puente, L., Diaz, M. J., Pagan, M. J., Garcia, M. J., Gras, M. L., Martinez-Monzo, J., & Fito, P., 2003. Development of probiotic-enriched dried fruits by vacuum impregnation. *J Food Eng.* 56: 273-277.

[225] Johansson, M. L., Nobaek, S., Berggren, A., Nyman, M., Bjorck, I., Ahrne, S., Jeppsson, B., & Molin, B. 1998. Survival of *Lactobacillus plantarum* DSM 9843

(299v), and effect on the short-chain fatty acid content of faeces after ingestion of a rose-hip drink with fermented oats. *Int J Food Microbiol.* 42: 29-38.

[226] at: http://en.wikipedia.org/wiki/Olive_leaf.

[227] Markin, D., Duek, L., & Berdicevsky, I. 2003. *In vitro* antimicrobial activity of olive leaves. *Mycoses.* 46:132–136.

[228] Benavente-García, O., Castillo, J., Lorente, J., Ortuño, A., & Del Rio, J.A. 2000. Antioxidant activity of phenolics extracted from *Olea europae* L. leaves. *Food Chem.* 68: 457-462.

[229] Bao, J., Zhang, D. W., Zhang, J. Z. H., Huang, P. L., & Lee-Huang, S. 2007. Computational study of bindings of olive leaf extract (OLE) to HIV-1 fusion protein gp41. *FEBS Lett.* 581: 2737- 2742.

[230] Bisignano, G., Tomaino, A., LoCascio, R., Crisafi, G., Uccella, N. & Saija, A. 1999. On the *in vitro* antimicrobial activity of oleuropein and hydroxytyrosol. *J Pharm Pharmacol.* 51: 971–974.

[231] Food and Drug Administration. Topical antimicrobial products for over-the-counter use; tentative final monograph for healthcare antiseptic drug products. Federal Register. (1994). 59: 31221–2.

[232] *Hibbard, J. S. 2005.* Analyses comparing the antimicrobial activity and safety of current antiseptic agents: a review. J. Infusion Nursing. *28: 194-207.*

[233] Altıok, E., Bayc, D., Bayraktar, O., & Semra, U., 2008. Antibacterial properties of silk fibroin/chitosan. *Sep Purif Technol.* 62: 2: 342-348.

[234] Abee, T., KrocheL, L. & Hill, C. 1995. Bacteriocins: modes of action and potentials in food preservation and control of food poisoning. *Int J Food Microbiol.* 53: 43-52.

[235] Riley, M. A., & Wertz, J. E. 2002. Bacteriocins: evolution, ecology, and application. *Annu Rev Microbiol.* 56: 117-137.

[236] Chatterjee, C., Paul, M., Xie, L., & van der Donk, W. A. 2005. Biosynthesis and mode of action of lantibiotics. *Chem Rev.* 105: 633-683.

[237] Stephens, S. K., Floriano, B., Cathcart, D. P., Bayley, S. A., Witt, V. F., Jiménez-Díaz, R., Warner, P. J., & Ruiz-Barba J. L. 1998. Molecular analysis of the locus responsible for production of plantaricin S, a two-peptide bacteriocin produced by *Lactobacillus plantarum* LPCO10. *Appl Environm Microbiol.* 64: 1871-1877.

[238] Kleerebezem, M., Quadri, L. E. N., Kuipers, O. P., & de Vos, W. M. 1997, Quorum sensing by peptide pheromones and two-component signal-transduction systems in Gram-positive bacteria. *Mol Microbiol.* 24: 895-904.

[239] Kuipers, O. P., Beerthuyzen, M. M., de Ruyter, P. G. G. A., Luesink, E. J., & de Vos, W. M. 1995. Autoregulation of nisin biosynthesis in *Lactococcus lactis* by signal transduction. *J Biol Chem.* 270: 27299-27304.

[240] Riley, M. A., & Gordon, D. M. 1999. The Ecological role of bacteriocins in bacterial competition. *Trends Microbiol.* 7: 129-133.

[241] Muriana, P. M. & Luchansky, J. B. 1993. Biochemical methods for purification of bacteriocins. *In Bacteriocins of Lactic Acid Bacteria* ed. Hoover, D.G. and Steenson, L.R. pp. 41–61. Academic Press, San Diego CA, USA.

[242] de Vuyst, L. Callewaert R. & Crabbé, K. 1996. Primary metabolite kinetics of bacteriocin biosynthesis by *Lactobacillus amylovorus* and evidence for stimulation of bacteriocin production under unfavourable growth conditions. *Microbiology.* 142: 817– 827.

[243] Keating, T. A., & Walsh, C. T. 1999. Initiation, elongation, and termination strategies in polyketide antibiotic biossinthesis. *Cur Opin Chem Biol.* 3: 598-606.

[244] Cleveland, J., Montville, T. J., Nes, I. F., & Chikindas, M. L. 2001. Bacteriocins: safe, natural antimicrobials for food preservation. *Int J Food Microbiol.* 71: 1-20.

[245] Leroy, F., & de Vuyst, L. 1999. The presence of salt and a curing agent reduces bacteriocin production by *Lactobacillus sakei* CTC 494, a potential starter culture for sausage fermentation. *Appl Environ Microbiol.* 65: 5350-5356.

[246] Enan, G., el-Essawy, A. A., Uyttendaele, M. & Debevere, J. 1995. Antibacterial activity of *Lactobacillus plantarum* UG1 isolated from dry sausage: characterization, production and bactericidal action of plantaricin UG1. *Int J Food Microbiol.* 30: 189-215.

[247] Moll, G. N., Konings, W. N., & Driessen, A., 1999. Bacteriocins: mechanism of membrane insertion and pore formation. *Antoine van Leeuwenhoek.* 76: 186-198.

[248] Fuller, R. 1989. Probiotics in man and animals. *J Appl Bacteriol.* 66: 365-378.

[249] Salminen, J.-P., Ossipov, V., Loponen, J., Haukioja, E., & Pihlaja, K. 1999. Characterisation of hydrolysable tannins from leaves of *Betula pubescens* by high-performance liquid chromatography–mass spectrometry. *J Chromatogr A.* 864: 283–291.

[250] Felley, C., & Michetti, P. 2003. Probiotics and *Helicobacter pylori*. *Best Pract Res Clin Gastro.* 17: 785-791.

[251] Nam, H., Ha, M., Bae, O., & Lee, Y., 2002. Effect of *Weissella confusa* strain PL9001 on the adherence and growth of *Helicobacter pylori*. *Appl Envir Microbiol.* 68: 4642-4645.

[252] Brenes, M., García, P., Romero, C., & Garrido, A. 2000. Treatment of green table olive waste waters by an activated-sludge process. *J Chem Technol Biotechnol* 75:459-63.

[253] Brenes, M., Romero, C., & de Castro, A. 2004. Combined fermentation and evaporation processes for treatment of washwaters of the Spanish-style green olive processing. *J Chem Technol Biotechnol.* 79, 3: 253-259.

[254] Djilas, S., Čanadanović-Brunet, J., & Ćetković, G. 2009. By-products of fruits processing as a source of phytochemicals. *Chem Industry Chem Eng Quart.* 15, 4: 191-202.

[255] Dewick, P.M. 2002. Medicinal Natural Products, Wiley, Chichester, UK, p. 121.

[256] Nijveldt, R. J., van Nood, E., van Hoorn, D. E. C., Boelens, P. G., van Norren, K. & van Leeuwen, P. A. M. 2001. Flavonoids. A review of probably mechanisms of action and potential application. *Am J Clin Nutr.* 74: 418–425.

[257] Guinda, A., Albi, T., Camino, M. C. P., & Lanzón, A. 2004. Supplementation of oils with oleanolic acid from the olive leaf (*Olea europaea*). *Eur J Lipid Sci Technol.* 106: 22–26.

[258] Tabera, J., Guinda, A., Ruiz-Rodriguez, A., Senorans, J. F., Ibanez, E., Albi, T., & Reglero, G. 2004. Counter current supercritical fluid extraction and fractionation of high-added-value compounds from a hexane extract of olive leaves. *J Agric Food Chem.* 52: 4774–4779.

[259] Wilson, M., & Backman, P. A. 1999. Biological control of plant pathogens, USA

[260] Niku-Paavola, M. L., Laitila, A., Mattila-Sandholm, T., & Haikara, A., 1999. New types of antimicrobial compounds produced by *Lactobacillus plantarum*. *J Appl Microbiol.* 86: 29-35.

[261] Lavermicocca, P., Valério, F. Evidente, A., Lazzaroni, S., Corsetti, A., & Gobbetti, M., 2000. Purification and characterization of novel anti-fungal compounds from the sourdough *Lactobacillus plantarum* strain 21B. *Appl Environ Microbiol.* 66: 4084-4090.

[262] Ström, K., Sjögren, J., Broberg, A., & Schnürer, J. 2002. *Lactobacillus plantarum* MiLAB 393 produces the anti-fungal cyclic dipeptides cyclo (L-Phe-L-Pro) and cyclo (L-Phe-trans-4-OH-L-Pro) and 3-phenyllactic acid. *Appl Environ Microbiol.* 68: 4322-4327.

[263] Asehraou, A, Lamzira, Z, Brito, D, Faid, M. & Peres, C. 2006. Interacções entre estirpes de *Lactobacillus plantarum* e *Pichia anomala* e *Debaryomyces hansenii* isolados de uma salmoura de azeitona. *Melhoramento.* 41: 224-231.

[264] Brito, M. D., Serrano, C., Delgado, A. & Peres, C. 2006. Inibição de fungos por bacteriocinas produzidas por bactérias lácticas de azeitona de mesa. *Melhoramento.* 41: 268-273.

[265] Gordon, M. H.; Paiva-Martins, F.; & Almeida, M. 2001. Antioxidant activity of hydroxytyrosol acetate compared with that of other olive oil polyphenols. *J Agric Food Chem.* 49: 2480–2485.

[266] Paiva-Martins, F., Gordon, M. H., & Gameiro, P. 2003. Activity and location of olive oil phenolic antioxidants in liposomes. *Chem Phys Lipids.* *124*: 23–36.

[267] Meirinhos, J., Silva, B., Valentão, P., Seabra, R. M., Pereira, J. A., Dias, A., Andrade, P. B., & Ferreres, F. 2005. Analysis and quantification of flavonoidic compounds from Portuguese olive leaf cultivars (*Oleae europeae* L.) can be found in the literature, but the investigation of those volatiles is mostly correlated to their antioxidant properties. *Nat Pro Res. 19*: 189-195.

[268] Bisignano, G., Tomaino, A., Lo Cascio, R., Crisafi, G., Uccella, N., & Saija, A. 1999. On the *in vitro* antimicrobial activity of oleuropein and hydroxytyrosol. *J Pharm Pharmacol. 51*: 971–974.

[269] Markin, D., Duek, L., & Berdicevsky, I. 2003. *In vitro* antimicrobial activity of olive leaves. *Mycoses.* 46: 132–136.

In: Agricultural Research Updates. Volume 2
Editor: Barbara P. Hendriks
ISBN: 978-1-61470-191-0
© 2012 Nova Science Publishers, Inc.

Chapter 2

ADAPTATION AND APPLICATION OF HYDROLOGICAL APPROACHES FOR WATER QUALITY MANAGEMENT: PASTORAL AGRICULTURE IN NEW ZEALAND

M.S. Srinivasan[1] and K. Müller[2]

[1] National Institute of Water & Atmospheric Research Limited,
Christchurch, New Zealand
[2] The New Zealand Plant & Food Research Institute, Ruakura Research Centre
Private Bag, Waikato Mail Centre, Hamilton, New Zealand

ABSTRACT

"..[P]ronounced variations in hydrological processes can occur from point to point over what, in surface appearance, is an extremely homogenous area."

Pilgrim et al. 1978, *p. 327*

Hydrological field studies have shown that runoff generation processes do not occur uniformly over an entire catchment during storm events. The variable-source area, or VSA, concept encompasses this spatial and temporal variability in rainfall-runoff response at catchment scale. Early on, hydrologists envisioned that by targeting VSAs, the transport of contaminants to streams could be managed, thereby protecting water quality. However, predicting VSAs at catchment scale has remained a daunting task even in intensively studied catchments, let alone transferring the understanding to other catchments or linking runoff generation processes to water quality research.

The application of the VSA concept to pastoral catchments in New Zealand for water quality management is further compounded by open grazing, which is integral to livestock farming in this country. Persistent animal traffic and behaviour such as camping, wallowing, and fence-line pacing significantly deteriorate soil physical quality, resulting, for example, in soil compaction and infiltration-excess runoff. When such compacted areas are connected to streams via animal tracks or farm roads, their potential to contaminate surface waters increases. In addition, the occurrence of animal-induced features has been shown to be randomly distributed. Thus, in pastoral catchments, mapping runoff generation areas to manage contaminant transport needs to take into account animal behaviour and the VSA concept.

In this chapter, we trace the evolution, adaptation, and application of the VSA concept for surface water quality management in grazed, pastoral catchments in New Zealand. Early desk-top studies combined traditional data (streamflow measured at catchment outlets) and few generalised catchment characteristics (topography and soil drainage properties) with rainfall characteristics to predict the dominant runoff generation mechanisms at catchment scale. However, these studies neither extended these predictions to map specific runoff generation areas within the catchments nor linked water quality management to runoff generation processes. Field studies that followed focused on mapping the temporal and spatial dynamics in runoff generation processes and the dominant flow and transport paths that connect land to streams. However, findings from these field studies largely remained within the realms of science. Very little of the new knowledge was applied to manage pastoral catchments, largely because appropriate tools to transfer the findings between catchments were not available. The subsequent development of various indices and concepts, for example, the Phosphorus Index and the Topographic Index, has not provided the desired breakthrough, because they either oversimplified hydrological processes or resulted in solutions that were too complex for practical implementation.

The objective of this chapter is to identify challenges facing the integration of the VSA concept to water quality protection in grazed, pastoral catchments. These challenges range from refining our understanding of current runoff generation concepts to transfer and adaptation of field results across multiple spatial scales. This chapter highlights that understanding runoff generation processes such as saturation-excess and infiltration-excess surface runoff and identifying major overland and subsurface transport pathways that connect land to streams, have to be integrated with knowledge on the key contaminant source and transport areas (e.g., near-stream locations, animal tracks and farm roads) to develop and implement sustainable management practices at catchment scale.

NOTE

Throughout this chapter, the term variable-source area, or VSA, has been used to describe the dynamic participation of catchments in runoff generation processes. The description of the term, VSA, follows from Freeze (1974) and Pearce et al. (1986). According to them, the variable-source-area concept encompasses partial-area Hortonian overland flow concept (infiltration-excess overland flow) as forwarded by Betson (1964) and others, saturation-excess overland flow from saturated areas, a concept described by Dunne and Black (1970) and others, and subsurface flow to channel in soils with high infiltration capacities, a concept advanced by Hewlett (1961) and others.

ABBREVIATIONS

IE	Infiltration Excess
SE	Saturation Excess
VSA	Variable Source Area
SSF	Subsurface Flow

CSA	Critical Source Area
BMP	Best Management Practice
FDE	Farm Dairy Effluent
P	Phosphorus
TP	Total Phosphorus
N	Nitrogen
TN	Total Nitrogen
DRP	Dissolved Reactive Phosphorus
TDP	Total Dissolved Phosphorus
TPP	Total Particulate Phosphorus
NO_3-N	Nitrate-N
TDN	Total Dissolved Nitrogen
TPN	Total Particulate Nitrogen
DIP	Dissolved Inorganic Phosphorus
SS	Suspended Sediments
CN	Curve Number
PI	Phosphorus Index
DD	Drainage Density
TI	Topographic Index
SWR	Soil Water Repellency

1. Introduction

1.1. Pastoral Agriculture and the State of Soil and Water Resources in New Zealand

In New Zealand, open grazing is integral to pastoral agriculture. In 2008, low and high producing grasslands covered 29 and 22% of the country's land area, respectively (Figure 1a). In other words, half of New Zealand's land area is grassland, and most of it is used as grazing land for sheep, cattle (beef and dairy), or deer (MfE, 2010). A combination of mild weather conditions with ample rainfall and the availability of large expanse of pastures have made year-round grazing an attractive and cost-effective livestock farming practice. Dairy and beef cattle, deer, and sheep are the most commonly grazed animals in New Zealand. Recent statistics (2010) indicate that6 million dairy cows, 3.9 million beef cattle, 1.1 million deer, and 32.5 million sheep are grazed on 5.8 million ha of pasture lands across New Zealand (http://www.stats.govt.nz/browse_for_stats/industry_sectors/agriculture-horticulture-forestry/AgriculturalProduction_HOTPJun10prov.aspx). Milk powder, butter, and cheese are the country's most important export commodities at New Zealand $8.8 million followed by meat and edible offal valued at New Zealand $5 million (http://www.stats.govt.nz /browse_for_stats/Corporate/Corporate/nz-in-profile-2011.aspx).

Open grazing systems and the associated animal traffic across landscape can result in significant changes to the state of soil and water resources. Almost all grazed-pastoral catchments of New Zealand bear the scars of grazing - deer wallows, cattle camp sites, zig-zagging animal tracks that are often connected to waterways, and compacted fence lines from fence pacing by deer (see Figure 2 for some of these features). In addition, farm tracks that allow stock and vehicle movement are also visible features in pastoral catchments. Many of these changes have significantly altered the hydrology of pastoral catchments and impacted surface water quality.

Deterioration of surface water quality is an issue of growing concern in New Zealand as elsewhere in the world.For example, Figure 1b shows recent total phosphorus trends at Regional Council Sites and National River Water Quality Network Sites in New Zealand. In New Zealand, there is growing evidence that the declining water quality of streams and lakes is linked to pastoral agriculture(McColl et al., 1977; Quinn and Stroud, 2002), and its increasing intensification (Cooke and Petch, 2007; Larned et al., 2004; Vant and Smith, 2004). For example, 90% of nitrogen (N) and 78% of phosphorus (P) entering streams in the Waikato region, a region dominated by grazed-dairy farms, has been related to pastoral land use (Cooke and Petch, 2007). Other surface water quality concerns related to livestock farming include siltation, pathogen, and herbicide contamination (McDowell et al., 2008). The loss of nutrients and contaminants from pastoral catchments can occur in particulate and dissolved forms. For example, in grazed-pastoral systems, most P loss occurs as dissolved (<0.45 μm) reactive P (as determined by molybdate-based colorimetry; DRP) in surface runoff (Nash et al., 2000). Nitrogen is mainly transported to surface waters as nitrate-N in subsurface flow (McDowell et al., 2008).Grazing livestock has also been identified as the dominant source of faecal contamination of New Zealand's freshwaters (Wilcock et al., 1999& 2006a; Donnison et al., 2004; Parliamentary Commissioner for the Environment, 2004), and, often, unimpeded access to waterways to grazing animals can result in direct faecal contamination of surface waters (e.g., Collins et al., 2007; McDowell, 2009).

Hydrologically, grazing and animal traffic can alter, and often aggravate, the occurrence of runoff generation,soil erosion and sediment transport processes, and alter the connectivity between land and water. The impact of grazing on soil and water quality has been recorded in many field studies across various physiographic and climatic regions of New Zealand – from sub-tropical Northland to temperate Southland (e.g., Gradwell, 1968;Edmond, 1974; Warren et al., 1986a & 1986b; Dormaar et al., 1989; Proffitt et al., 1993; McDowell, 2009).

Generally, the extent of damage to soil and water resources is directly associated with grazing intensity, implying the higherthe number of animals grazing per unit area, the greater the traffic and, hence, the more severe the damage. However, grazing intensity, expressed as number of animal units per ha, does not fully capture the intensity of the actual grazing practice. For instance, nationally, the average grazing intensity for New Zealand farms is 2.8 cows ha^{-1}, and across the country it ranges from 2.3 (Northland; low intensity dairying) to 3.3 (Canterbury; high intensity dairying) (http://www.dairynz.co.nz/file/fileid/34190). However, as animals are grazed on a rotational basis, and,are moved from paddock to paddock every few days to ensure maximum pasture utilisation, the actual grazing intensities can be as high as 300 to 600 cows ha^{-1},especially during winter periods (Singleton et al., 2000). Such intensive grazing during wet, winter months, even for a short period (2-3 days) may cause significant soil damage (Singleton et al., 2000), and contamination of waterways as a result of surface runoff of suspended sediments and nutrients (Smith et al., 1993; Quinn et al., 1994). Grazing paddocks immediately following wet periods has been shown to result in soil structural degradation in the form of compaction and remoulding of soil by cattle hooves to a depth of 20 cm (Climoand Richardson 1984; Burgess et al., 2000).

Many field and plot scale studies have found evidence for the deterioration of soil physical propertiesfollowing grazing events, and this change in state can influence hydrological and hydraulic processes (e.g., McDowell et al., 2004). A summary of results from a selection of these studies are presented in Table 1. Often grazing impacts have been found to be transient, although the features resulting from repeated grazing events such as camp sites and wallows can remain part of the landscape for long periods. Singleton et al. (2000) reported the decline in soil physical conditions such as hydraulic conductivity and aggregate size with increased treading, and found these effects remained measurable for as long as 18 months after the event. Drewry and Paton (2000) based on a field study in Southland, New Zealand, showed that following an exclusion of dairy cow grazing,macroporosity increased by 70% within four months, air permeability increased by over two orders of magnitude in 18 months, and saturated hydraulic conductivity increased by 200% to the 10-cm soil depth. Nguyen et al. (1998), in a plot scale study, observed that grazing effects disappeared six months following the events. However, Muirhead (2009) indicated that in grazed pastures, faecal material reservoirs can be a big source of pollutant, even six months after the dairy cows had been removed.

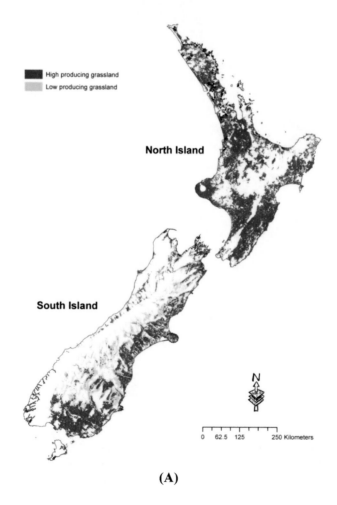

(A)

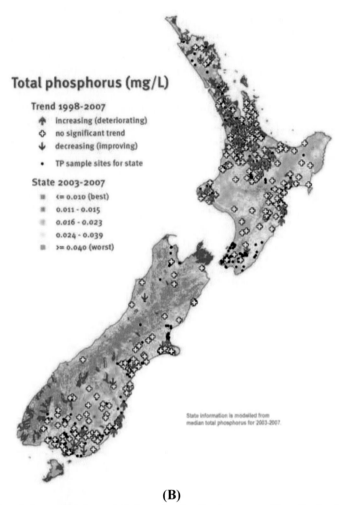

Total phosphorus (mg/L)

Trend 1998-2007
- ↑ increasing (deteriorating)
- ◇ no significant trend
- ↓ decreasing (improving)

- • TP sample sites for state

State 2003-2007
- ■ <= 0.010 (best)
- ■ 0.011 - 0.015
- ■ 0.016 - 0.023
- ■ 0.024 - 0.039
- ■ >= 0.040 (worst)

State information is modelled from
median total phosphorus for 2003-2007.

(B)

Data source: A) Land Cover Database, Ministry for the Environment, New Zealand, B) Adopted with permission from the Ministry for the Environment, New Zealand

Figure 1. A) Extent and distribution of grazing land in New Zealand, B) Recent trends in total phosphorus concentrations at Regional Council sites and National River Water Quality Network sites, 1998–2007, for New Zealand.

Few studies identified grazing behaviour of animal types and its impact on soil and water resources. Finlayson et al.(2002), who developed a mathematical model to predict the treading damage of pasture due to grazing, concluded that there is a need to study animal behaviour to predict the spatial pattern of the damage. Following an observational study over two summers and a spring season, Bagshaw (unpublished data) found that dairy cattle spent 99.1% of time in the paddocks, 0.7% in the riparian zone, and 0.1% in the stream. McDowell (2009) indicated that red deer *(Cervuselaphus)* have a natural instinct to seek out water and wallow it, which could result in deteriorating soil and water qualities. Willat and Pullar (1983) observed that cattle exert greater loads on soil than sheep, so soils on dairy farms would be expected to be more compacted than on sheep farms. Betteridge et al. (2008 & 2010) used GPS (global positioning system) collars on sheep and cattle to track their movement within paddocks, andreported that sheep camped at high zones in hill country paddocks, while such distinction was not easily observable among cattle. Betteridge et al. (2008) observed that in hill country, animals typically seek shade and shelter to camp on flat rather than on steep areas, and to generally exhibit different behaviours than animals grazing on flat land. Many studies (Collins et al., 2007, Betteridge et al., 2010) have shown that streams are attractive areas for animals, resulting in a constant traffic to and from streams and well-established animal tracks.

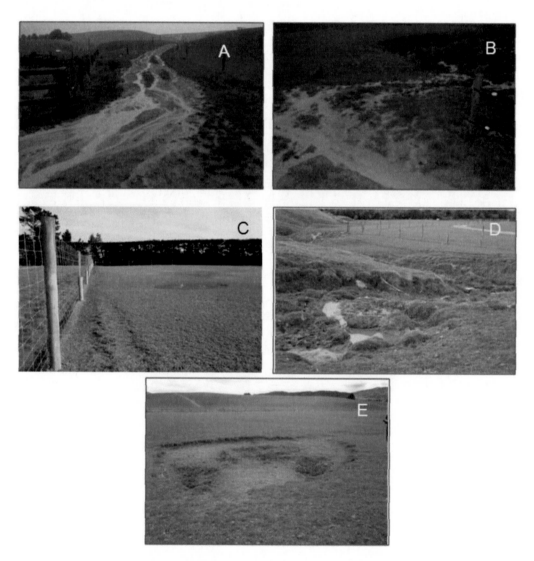

Figure 2. Grazing-induced landscape features in pastoral catchments of New Zealand; (A): Infiltration-excess runoff from farm track, (B): Infiltration-excess runoff from animal track, (C): Compacted, bare fence line and an emerging deer wallow, (D): Well-developed deer wallow, (E): Cattle camp site.

Table 1. Summary of results from selected field studies from New Zealand catchments on the impact of grazing on soil and water quality

Literature reference	Geographical location	Animal type grazed	Observations
Nguyen et al., 1998	Whatawhata Research Centre, Hamilton, Waikato, North Island	Dairy cattle	On steep slopes (28-39°), grazing damage resulted in 46% less infiltration and 87% more sediment, 89% more N, and 94% more P in runoff than from undamaged areas.
Environment Southland, 2000	Southland, South Island	Deer	Concentrations of SS, $N-NH_4$, and faecal bacteria were 20-30 times higher downstream of a deer wallowing site than upstream of that site.
Rodda et al., 2001	Modelling study	Deer	Deer farming can result in 4.5 times greater sediment loss than from land under forestry. Sediment loss can be halved, if deer farming were limited to slopes <20%.
Davies-Colley and Nagels, 2002	Waikato, North Island	Deer	Measured concentrations of faecal bacteria were 2- to 10-fold greater downstream of two large deer farms than upstream. Wallowing by deer enhanced erosion of the stream bed and bank and near-channel areas, resulting in reduced water clarity as well as promoting faecal contamination.
Pollard and Drewry, (2002); McDowell et al.(2003)	Southland, South Island	Deer	Fence line pacing resulted in soil compaction, decreased soil pore space and pasture growth, and increased the likelihood of sediment, nutrient, and microbial transport in surface runoff.

Literature reference	Geographical location	Animal type grazed	Observations
Collins et al., 2007	Hamilton, Waikato, North Island	Dairy cattle	Yield of *E. coli* was strongly correlated (negatively) with the time elapsed since the hillside was last grazed. Farm tracks in hill country readily generated surface runoff due to soil compaction and de-vegetation was broadly similar to that occurring with livestock treading damage (can be expected to deliver microbes and other pollutants to waterways).
McDowell, 2008b	Inchbonnie, West Coast, South Island	Dairy cattle	As much as 0.6 kg P ha^{-1} was lost via lanes and stream crossings in dairy farms.
McDowell, 2008a & 2009	Otago/Southland, South Island	Deer	Hydraulic connection between wallows and water bodies can result in significant contamination of water bodies. Recommendation to use "safe" wallows to reduce contamination.
Houlbrooke et al., 2009	Southland, South Island	Dairy cattle	Macroporosity of grazed plots was significantly ($P < 0.05$) greater than on ungrazed plots.

1.2. Variable Source Area (VSA) Hydrology and Its Adaptation to New Zealand Catchments

Hydrological field studies from 1960s and 70s introduced the concept that runoff generation at catchment scale could be limited to small parts, while rainfall on the rest of the catchment may not directly contribute to stormflow (e.g., Hewlett, 1961; Betson, 1964; Hewlett and Hibbert, 1967; Betson and Marius, 1969; Dunne and Black, 1970; Freeze, 1974). Since then, several variations of this concept have been forwarded, although the central idea, the partial participation of catchments, has been widely observed and recognisedin many forested and agricultural catchments across the world (e.g., Maimai Catchment, New Zealand; Mahantango Catchment, United States of America). Studies differed in describing the processes that result in runoff generation. Freeze (1974) and Pearce et al. (1986) grouped all these variations on partial participation under one umbrella, the variable-source area, or VSA, concept. Bonell (1993, *p. 218*) indicated that "by hydrological standards, the introduction of the variable-source area concept was close to being a 'Copernican revolution' in terms of how the runoff component of the water balance is viewed."

Engman (1974) suggested that understanding this catchment behaviour (partial participation in runoff generation) could be very important for water quality, as management practices could be limited to these "participating" or "contributing" areas. Pionke et al. (1996) coupled this concept of runoff generation and sediment and nutrient transport processes at catchment scale and introduced the concept of Critical Source Areas, or CSAs. They described CSAs as overlap of high source and potential transport areas at catchment scale. The high source areas are areas with large nutrient and erodible-sediment concentrations, and potential transport areas are areas with high potential to generate runoff that can reach waterbodies overland. Many field (e.g., Zollweg et al., 1996; Needelman et al., 2004) andmodelling studies (Gburek et al., 2000; Sharpley et al., 2004; Srinivasan et al., 2005) have been based on the CSA-concept forwarded by Pionke et al. (1996).

In New Zealand, water quality studies at catchment scale that explicitly account for different hydrological processes in the transfer of nutrients and sediments from land to water are few in numbers (McDowell and Wilcock, 2004; McDowell and Wilcock, 2007; Stenger et al., 2005; Stenger et al., 2008), and catchment studies with particular emphasis on VSA hydrology are even rarer (McDowell and Srinivasan, 2009; Mülleret al., 2010a). It has been recognisedearly on, however, that effective mitigation of surface water contamination requires the determination of the location of the main source areas in a catchment which "ma[k]e[s] a direct contribution to streamflow"(McColl and Gibson, 1979).

The application of the VSA concept in New Zealand has been largely influenced by land management and the nature of pastoral systems. Climatic conditions allow year-round grazing almost all through the country. Pastures are grazed on average 14-16 times a year (McDowell et al., 2008). These grazing practices leave traces behind on landscape and soil quality. Persistent animal traffic and behaviour such as camping, wallowing, and fence-line pacing, significantly deteriorate soil physical quality, resulting, for example, in soil compaction and infiltration-excess runoff during rainfall events. When such compacted areas are connected to streams via animal tracks and farm roads, their potential to transport contaminants increases (McDowell and Srinivasan, 2009). In addition, the occurrence of animal-induced features are randomly distributed (McDowell et al., 2009; Tian et al., 2002). Thus, in grazed-pastoral catchments, mapping runoff generation areas and managing contaminant transport also needto account for animal behaviour. Traditional approaches to identify contaminant source areas such as soil P tests on a field-to-field basis thus may not be sufficient for open grazing systems.

2. APPLICATION OF VSAHYDROLOGYTO GRAZED-PASTORAL CATCHMENTSIN NEW ZEALAND

Historical application of VSA hydrology and its adaptation to water quality management in pastoral catchments of New Zealand are discussedin this chapter under two broad sections, desktop studies *(section 2.1.1)* and detailed catchment studies *(section 2.1.2)*. Desktop studies primarily focus on the introduction of VSA hydrology to New Zealand catchments, pastoral and non-pastoral, description of runoff generation processes and areas, and delineation of flow paths that connect land to water. The detailed catchment studies subsection focuses at the application of VSA hydrology to water quality management in grazed-pastoral catchments.

2.1.1. Desktop Studies

The desktop studies described here analysed conventional catchment-scale records such as time-series flow and precipitation data along with catchment physical characteristics such as soils, land cover, topography, and landscape position to infer the dominant runoff generation processes, and areas where these dominant processes could be occurring within catchments; all within the context of empirical generalisations published by earlier studies (e.g., Freeze, 1972 & 1974; Dunne, 1978; Dunne et al., 1975). Based on intensive field studies in the humid northeast United States, Dunne et al. (1975) and Dunne (1978) concluded the following:

1) in humid regions, infiltration excess (IE) runoff rarely occurs;
2) steep hillslopes with convex toe aspect, incised channels, narrow valley bottoms and permeable soils generally result in subsurface flows (SSF) during storm events;
3) hillslopes with a concave toe aspect and wide valley bottoms, thin soils, and gentle terrains generally result in saturation excess (SE) runoff;
4) during small rainfall events, the majority of flow is generated via SE runoff from near-stream areas (< 5 m from stream); and
5) during large rainfall events, the SE runoff areas extent beyond near-stream areas (> 5m from stream).

While the above might appear to be a simplistic generalisation of catchment behaviour during rainfall events, nevertheless they sowed the seeds for many hydrologic field studies around the world, including New Zealand. These generalisations provided a framework for many interferential as well as detailed field and modelling studies that followed.

In New Zealand, one of the earliest inferential hydrological field studies with a focus on process understanding of transport processes was reported by Hayward (1976). His 2-year field study in the Torlese catchment in the South Islandwas aimedat understanding the hydrological processes eroding the mountain catchment. He concluded that this catchment seldom produced IE runoff, as the soils had high infiltration capacities. Based on estimates of soil properties such as hydraulic conductivity and infiltration capacity, he concluded that SE runoff from riparian lands and direct precipitation on stream channels contributed the bulk of the stormflow. He estimated that the infiltrating rainfall would have moved 2-3 m by the time stormflow travels from the head of the perennial channel to the catchment outlet, a distance of 1500m, and thus, rainfall contributions via SSF paths could be negligible. Overland flow is generally *known and accepted*, according to Hayward (1976), to move at velocities in the order of 4000 to 8000 m d^{-1}. Thus, he logically concluded that SSFis not an important component of stormflow. However, he concluded that under conditions where hydraulic conductivities are large, as in case of gravel beds of stream channels, SSF can become a part of the stormflow. He applied Hewlett and Nutter's (1970) model of expanding channel network, and indicated that dry channel beds with very high conductivity possibly expanded upslope of the head of the perennial channel to contribute significantly to the stormflow. By comparing the hydraulic conductivities of soil matrix with those of gravel channel beds, he computed that travel times in the dry channel beds, headward of perennial channels, can be 80 times that of soil matrix, and hence can potential become contributors.

Pearce and McKerchar (1979) conducted an inferential study in seventeen headwater pastoral and non-pastoral catchments across the country. A brief summary of these catchments and the key findings from Pearce and McKerchar (1979) are available in Table 2. Using previously recorded rainfall and runoff (streamflow) data from a 'typical' year, Pearce and McKerchar (1979) applied the empirical generalisations from Dunne (1978) and Dunne et al. (1975) to describe the runoff generation processes in these catchments. However, unlike Dunne (1978), Pearce and McKerchar (1979) did not exclude the occurrence of IE runoff in New Zealand catchments.

Apart from catchment physical characteristics data and data on rainfall and runoff, they consulted published infiltration data for the soils in the catchments studied as well as the estimated extent of permanently or seasonally saturated areas. In general, they concluded that for events that generated small flows, the majority of flows came from SE runoff. They used the extent of saturation areas and rainfall intensity data to compute these flows. Pearce and McKerchar (1979), in the absence of any specific observations on the occurrence of SE runoff or SSF, correlated the extent of the saturated areas and stormflows in these catchments. For most of the small storms, they were able to simply relate the extent of saturation areas and rainfall on these saturated areas resulting in SE runoffto account for the recorded stormflows. When the recorded stormflows exceeded the SE runoff estimates, they accounted the excess stormflowlargely to SSF, and to a lesser extent to IE runoff.

Pearce and McKerchar (1979) drew the following conclusions based on their inferential studies:

1) IE might not be a major runoff generation mechanism inNew Zealandcatchments,
2) in most catchments, storm runoff from frequent events (return period, days to weeks) can be best explained as rainfall on channel and saturated areas,
3) for large events (return period, months to years), flows depend on soil hydraulic properties, landscape position, and slope forms. Concave slopes with low to moderate hydraulic conductivity produced IE runoff. On steep slope with shallow soils,SSF dominated.

Table 2. Inferring catchment-scale runoff generation processes based on measured rainfall, flow, and catchment physical characteristics data. Adapted from Pearce and McKerchar (1979) with permission from the Journal of Hydrology (New Zealand). Catchments are spread across the country

Catchment	Area (ha)	Rain-fall (mm)	Runoff (mm)	Quick flow[1] (mm)	Soils	Topography	Land use	Dominant runoff generation processes/areas
Pukewaenga	38.9	1380	570	359	Northern yellow-brown earth (silt and clay loams)	Rolling to steep hill country	Grazed pasture	Small events - SE Large events - IE[2]
Manukau	30.1	1150	450	84	Manurewa silt loam + hill soils variant (Northern yellow-brown earth)	Gentle to rolling terrain	Grazed pasture	Small events - SE from near-stream poorly drained soils Large events - IE
Purutaka	22.5	1460	360	<10	Oruanuisilty sands + hill soils (yellow-brown pumice soils)	Rolling to steep hill country	Grazed pasture	Small events - SE from impermeable Waikokomuka soils[3] Large events - SSF
Ngahere	52.0	2690	1760	425	Steepland soils related to central yellow-brown earth; some pumice soil	Steep hill and mountain lands	Forest (beech)	Small events - SE from near-stream areas Large events - SSF
Makara (Makara 11)	7.4	1190	330	131	Makarasteepland soils from central yellow-brown earth	Steep hill country	Poor pasture and scrub - lightly grazed	Small events - SE from near-stream areas Large events - SSF
Moutere	7.0	1120	270	112	Rosedale hill soils (sand silt and silt loam) – central yellow-brown earth	Moderate to steep dissected hill country	Grazed pasture	Small events - SE from foot-slope wet areas Large events - IE

Table 2. (Continued)

Catchment	Area (ha)	Rain-fall (mm)	Runoff (mm)	Quick flow[1] (mm)	Soils	Topography	Land use	Dominant runoff generation processes/areas
Big Bush (4 catchments)	4.8 - 20.2	1800	800	220	Hope hill soils (stony, slightly podsolised yellow-brown earth)	Steep, dissected hill country	Forest (beech, podocrop, hardwood)	Small events - SE from near-stream wet areas Large events - SSF
Maimai (6 catchments)	1.6 - 4.6	2600	1550	1014	Blackball hill soils (stony, podsolised yellow-brown earth)	Steep, short slopes	Forest (beech, podocrop, hardwood)	Small events - SE from near-stream wet areas Large events - SSF
Hut creek	22.1	1500	750	275	Bealey hill soils (high country yellow-brown earth)	Moderate foothill slopes of Craigieburn Range	Newly established pasture, scrub, mountain beech, conifer mixture	Small events - SE from near-stream areas and foot slopes Large events - SE from concave slope and SSF from convex slopes

[1] Portion of net rainfall (rainfall that passed through the canopy storage) that became stormflow.

[2] Small events are rainfall events with return periods from days to weeks; large events have return periods from months to a few years. SE - Saturation-excess runoff; IE - Infiltration-excess runoff; SSF - Subsurface flow. Also, Pearce and McKerchar (1979) defined those storm events that produced less than 5 mm of quick flow as small and greater than 5 mm as large events.

[3] Impermeable Waikokomuka soils formed by hydrothermal alteration of Taupo Pumice (Rijkse and Bell, 1974).

Taylor and Pearce (1982) studied the runoff generation mechanisms in six steep, headwater catchments in the CraigburnRange of the South Island. Applying Dunne's empirical generalisationslisted above, they differentiated the six catchments as SE runoff and SSF dominated, and concluded that the SE-runoff dominated catchments were positively influenced by storm sizes and antecedent catchment conditions, and the SSF dominated catchments were influenced only by storm sizes. Freeze (1974) believed that quick SSF to storms are possible only from deeply incised streams, with convex lower slopes. Taylor and Pearce (1982) used this as basis to separate stormflow sources and catchment types. However, Taylor and Pearce (1982) concluded that during storm events, both groups of catchments received flows from SE and SSF mechanisms, and that a practical separation of flow sources was impossible. In such catchments, they indicated that the application of topography or soil-property based indicators might not be a useful guide to describe the runoff generation processes and sources.

Pearce et al. (1984) studied the runoff generation sources in three small catchments of the Glendhu State Forest in the South Island and found that less than 30% of annual flow came as stormflow (according to Pearce et al. (1984), the corresponding numbers were 50% at the Hut Creek and 65% at the Maimai catchments – both in the South Island). They concluded that IE runoff did not occur in this catchment, and that wet areas accounted for 10 to 20% of the total catchment area and SE runoff from these areas dominated stormflow. They hypothesisedthat SSF can also be significant in these catchments. During storm recession periods, they observed that a sharp decrease in flows indicating that more than one mechanism could be contributing flow during storm events, SE runoff dominating early, quick flows and SSF occurring later during storm event. The general perception, as earlier indicated by Hayward(1976) and Freeze (1972), that SSF cannot be as quick as surface flow, could have led to this conclusion.

Despite the importance Pearce and McKerchar (1979) imparted to SSF contributions during storm events, Freeze (1972) believed that subsurface translatoryflows (flow through soil matrix or micropores) can never be a significant component of stormflow. He indicated that SSF can occur from convex, steep hillslopes with shallow soils, but even under those conditions significant contributions are only possible if the saturated hydraulic conductivity is sufficiently large. Flows via macropores or preferential flow paths have to be significantly large and well connected to dominate stormflow. Hayward's (1976) conclusions at the Torlese catchment and the observations by Pearce et al. (1986) overall tend to support Freeze's (1972) observations that SSFsare rarely a part of the quick flows during storms. Hayward (1976) concluded only under conditions where the soil hydraulic properties allow large flows to pass through soil matrix, as in case of dry stream beds, SSF can be large and dominating during storm events. Thus, the early desktop studies emphasised the occurrence of SE runoff from near-stream wet areas and the occurrence of SSF, where large and conductive flows paths are available.

Elsewhere, in North America, it was well recorded and recognised (e.g., Whipkey, 1965) in the 1960s that subsurface contributions to stormflows can be significant where, (a) the land is sloping; (b) the surface soil is permeable; (c) an impeding layer is present closer to the surface, resulting in lateral SSF; and (d) the soil is saturated. Whipkey (1965) showed the importance of SSFs during storm events using a trench study ("Whipkey" trench) in the Allegheny Plateau region of the eastern United States. A similar trench study conducted in the Maimai catchment of the South Island of New Zealand by Mosley (1979) reported significant

SSF during storm events. Woods and Rowe (1996) used *"Whipkey"* trenches to quantify the SSF contributions within a small sub-catchment of the Maimai catchment and found that SSF can be spatially and temporally variable even at the hillslope scale with no specific pattern observable.

The two key conclusions from the desktop studies set the direction for many detailed catchment studies that followed:

(i)   occurrence of SE runoff from near-stream areas that described the stormflow contributions for small events or events not big enough to simulate SSFs; and

(ii) occurrence of SSF when soil hydraulic conductivities (horizontal) are highand for events that produced large flows that could not be primarily accounted as SE runoff from near-stream or riparian saturated areas.

2.1.2. Detailed Catchment Studies

The application of VSA hydrology to water quality research at catchment scale started with experimental studies that focused on mapping the temporal and spatial dynamics of runoff generation processes and identifying the dominant flow and transport paths that connect land to streams. Transport of agrichemicals from land to water was inferred by (1) sampling various flow paths (surface runoff, drain flow, etc.), measuring concentrations of agrichemicals in these samples and estimating loads using measured flow volumes, (2) extrapolating measurements made at plot scale to entire catchments taking into account the catchment topography and soil properties, and (3) measuring agrichemical concentrations in streamflowand linking those to stormflows using hydrograph separation techniques. It was observed that dominant flow and transport pathways within a catchment were not necessarily identical. In this context, the importance of the contact time for interactions between soil and agrichemicals for the mobility of agrichemicals was highlighted. IE runoff water has shorter contact times with soils than SE runoff water. It was also discussed that streams can act as sink and/or source for agrichemicals during different seasons. Based on early field studies, best management practices(BMPs) to optimise surface water quality were recommended, including establishing buffer strips along waterways and excluding stock from seasonally saturated areas. Table 3 summarises a selection of catchment studies with their key findings and conclusions, which are discussed below.

2.1.2.1. Catchment-Scale Water Quality Studies based on VSA Hydrology

One of the earliest water quality studies at catchment scale in New Zealand that quantified the relative importance of different flow pathways for P losses to streams and took into account the variable natureof runoff generation across a catchment was conducted by Sharpley and Syers(1979). They separated P losses considering different phosphorus forms in surface, accelerated subsurface (tile drainage), and subsurface runoff (via soil matrix). Flow and P contributions via different pathways to overall streamflowwere measured and P loads were estimated usinghydrograph separation techniques. Loads were estimated by measuring concentrations in tile drainage and by extrapolating measurements from surface runoff plots to the 20-ha subcatchment, assuming that two-third of the catchment would contribute to surface runoff based on topography (Table 3). Although surface runoff only contributed about 10% of the streamflow, its contribution to P-losses dominated the total annual DIP and PP loading compared with the loads through subsurface runoff.

Table 3. Summary of results from key catchment-scale water quality studies from grazed-pastoral catchments of New Zealand

Literature reference	Study catchment information	Study duration (months)	Study focus	Approach	Key observations and conclusions
Bargh, 1978	Tuapaka (180 ha), North Island. sheep & cattle-grazed pasture	12	Determine quantity and quality (TP, TDP, TPP, NO₃-N, TN, TDN, TPN, SS) of surface water discharged from an agricultural catchment	Measured precipitation and streamflow; weekly stream sampling & intensively sampling for few storms	1. Deep percolation was minimal 2. Erosion from farm tracks and stream banks were important sources of SS 3. 76 and 15% of TP and TN, respectively, were transported in particulate form; significant correlations between flow and TPP, and SS and TPP
Sharpley and Syers, 1979	Massey University (20 ha); Manawatu, North Island. dairy cattle-grazed pasture	36	Estimate the relative contribution of surface runoff, accelerated SSF (tile drain) and SSF to the P-loading of stream draining a pastoral catchment	Measured precipitation and streamflow; accelerated SSF and SSF estimated by hydrograph separation; 2 surface runoff plots with/without manual P-application to verify estimates of runoff amounts and contaminant (DIP, TDP, PP, SS) losses of entire catchment (up-scaling)	1. Surface runoff equalled enhanced SSF volumes 2. Soil and plant material delivered significant amounts of TP in surface runoff; fertilisation considerably increased these losses (especially DIP by 32%) 3. Losses in enhanced SSF smaller due to sorption processes; losses in SSF even lower than in runoff (TDP remained constant) 4. Stream bank erosion and re-suspension of stream sediment contributed >70% of TP, sediment and PP
McColl et al., 1985	Judgeford catchment (4.2 ha), North Island. sheep-grazed pasture	18	Investigate the role of catchment scale saturation areas on water quality	Measured streamflow and precipitation; estimated runoff volumes from storm hydrograph; mapping of saturated areas in the field (weekly); soil mapping; infiltration measurements; dye tracer experiments	1. Small and medium rainfall events (1-25 mm) - flow generated by SE runoff from permanently and seasonally saturated areas 2. Large rainfall events (>25 mm) - SE from temporarily-saturated areas and spatially distributed IE contributed to stormflow
Cooke and Dons, 1988	Scotsman Valley (16 ha), Waikato, North Island. sheep & cattle-grazed pasture	18	Record dominant runoff generation processes on a seasonal basis	Measured precipitation and streamflow; 16 surface runoff collectors to measure flow rates and times; 2 Whipkey throughflow trenches	1. Throughflow and surface runoff dominated autumn (throughflow slightly greater than surface runoff) 2. Surface runoff dominated winter and spring 3. Surface runoff and throughflow levels were similar in summer 4. Channel precipitation was significant in summer and in early autumn

Table 3. (Continued)

Literature reference	Study catchment information	Study duration (months)	Study focus	Approach	Key observations and conclusions
Cooke, 1988	Scotsman Valley (16 ha), Waikato, North Island. sheep & cattle-grazed pasture	18	As above & influence of runoff on phosphorus (TP, TDP, DRP, PP) and sediment transport from land to water	As above & a stormflow activated Manning sampler to sample stormflows; piezometers and suction lysimeters (shallow groundwater & soil water at 30 cm depth, respectively)	1. Stormflow was dominated by SE runoff but throughflow was also significant 2. Concentrations in rain and shallow groundwater were generally similar and very low; TP-concentrations in permanently saturated zones 10–100 times higher & concentrations in surface runoff 2 order of magnitude higher than concentrations in rain (dominated by DRP and by PP during winter, PP-concentrations related to % of saturated area 3. 20% of annual P-export due to direct aerial input during fertilisation
Cooke and Cooper, 1988	Scotsman Valley (16 ha), Waikato, North Island. sheep & cattle-grazed pasture	18	As above & influence of runoff on nitrogen (NO_3-N; reduced N (TKN); and Cl as tracer) transport from land to water	As above	1. Organic N inputs through SE runoff peaked in winter (observation of treading damage under saturated topsoil conditions) & autumn. 2. NO_3-N inputs were dominantly through SSF (concentrations peaked in autumn & spring)
Smith, 1987	Tauwhare (16 ha), North Island. sheep-grazed pasture	20	Investigate seasonal and spatial variations in surface runoff water yields and chemistry	Measured precipitation; 8 surface runoff collectors (SRC) (volume, rates) installed within 5 m of the stream bank: 4 and 3 SRCs located in known and expected surface runoff channels, respectively; 1 SRC on steep hill; 1.6% of total runoff was collected for analysis of TP, TDP, TKN, DKN, NO_3-N, SS	1. High variability in runoff among SRCs for any given storm events (4 to 333 m³ in 20 months) and extreme channelization 2. SE runoff (predominantly from Tauwhare soils) during winter and spring events dominant 3. SRCs with highest runoff volumes had also highest nutrient and sediment losses with the exception of nitrate-N 4. Losses peaked in winter and early spring in accordance with runoff volumes 5. For nitrate-N exposition was important
Petch 1988	Upper Mangawhara Valley, Waikato, North Island.	24	Compare landscape forming processes in pasture and forested hillslope and the role of saturation areas	Spatial measurement of water table status at hillslope scale	1. Under similar topographic, soil, and precipitation conditions, larger areas of pasture saturated than forest

Literature reference	Study catchment information	Study duration (months)	Study focus	Approach	Key observations and conclusions
					2. Probability of occurrence of saturation areas decreased as moved away from stream for both land cover types
Bonell et al., 1990	Glendhu Catchments (2 catchments - 218 and 310 ha.), South Island.	Selected storm events between February and April 1988	Partition stormflow as "old" (soil water) and "new" (incoming rainfall) water	Combination of isotopic, hydrometric measurements and hydrograph separation techniques	1. Old water responds to hydrograph earlier than new water 2. Despite large wet areas in the catchments, response from the unconfined groundwater was quicker than the SE flow from wet areas
Bowden et al., 2001	Subcatchment of larger Glendhu Catchment (subcatchment area, 3.64 ha.), South Island.	< 24	Separate SE and SSFs during stormflows using hydrograph separation techniques	Soil moisture tension data from 12 spatial locations collected using recording tensiometers; Water table dynamics data from 7 wells; two transects of 3 throughflow pits each; time series runoff data at catchment outlet	1. Near-surface organic horizons contributed the majority of SSFs during storm event
McDowell and Wilcock, 2004	Bog Burn (8,760 ha) South Island; plantation forestry in headwaters; rest pasture, mainly dairy	12	Relate the bioavailability and concentration of P & SS along a stream draining an agricultural catchment to the overall P loss dynamics	Measured streamflow; biweekly analysis of ^{137}Cs-concentrations in trapped sediments to identify its source (topsoil, subsoil, stream bed, and bank sediment); biweekly stream water sampling at 4 sites for SS, DRP, and TP analysis; analyses of SS, BAP, TP in sediment samples collected from stream flow	1. Speculations which parts of the catchment were dominated by subsurface or surface runoff based on topography, land use and management 2. SSF mainly from topsoil via tile drain or to a lesser extent via surface runoff 3. Seasonal patterns observed in DRP and TP-concentrations in streamflow: highest in summer and lowest in winter (opposite true for loads)
Stenger et al., 2005; Stenger et al., 2008	Toenepi (1,500 ha), Waikato, North Island. dairy-grazed pasture	36	Investigate pathways for N to enter stream draining a pastoral catchment	Measured streamflow, monthly measurements of groundwater level along 7 transects (4-6 wells per transect, mostly 2.5-3 m depth) & tile drain flows:	1. Groundwater often within 0.5 m from surface; 75-85% of streamflow from groundwater 2. Inorganic N concentration in groundwater far less than that in stream; denitrification below root zone evidenced; reducing conditions in shallower groundwater; nitrate-N in subsurface drain

Table 3. (Continued)

Literature reference	Study catchment information	Study duration (months)	Study focus	Approach	Key observations and conclusions
				monthly sampling of wells and streamflow; suction cups in 4-5 depths installed in 2 soil types (6 reps) for NO_3-N, NH_4-N-analysis; NO_2-N, DKN, DIC, DTC, dissolved Fe &Mn (lab), & DO, pH, EC in 18 groundwater samples taken in May 2007. Modelled concentrations of N in leachate using OVERSEER.	
Gillingham and Gray, 2006	Waipawa: 2 pastoral catchments with different Olsen-P levels (6 and 25 mg kg^{-1}); (12.6 & 12.8 ha); Hawke's Bay, North Island.	~ 12	Measure P movement and losses in runoff water; assess usefulness of small scale measurements	Measured streamflow; 12 surface runoff collectors (total volume; sampled for DRP, TDP, and TP analysis); initial 300 mL of runoff from 7 0.55 m^2 runoff plots (with various P-levels; simulated rain), time–proportional stream samples from storm events (DRP, TDP, TP); flow simulations with TOPMODEL to delineate areas in the catchments contributing to surface runoff (SE & IE); modelling used to assess usefulness of small scale measurements	1. Most surface runoff occurred during summer & autumn and was negatively correlated to topsoil moisture 2. TOPMODEL predicted flows correctly but contributing runoff areas were restricted to near-stream areas (not observed at other scales; soil water repellency) 3. DRP concentrations in catchment flow were ~10% of runoff concentrations (runoff collectors); potential explanations: dilution of runoff with SSF or surface runoff did not reach stream
McDowell and Wilcock, 2007	Waiokura (2,100 ha), North Island. dairy-grazed pasture	12	Determine origin and bioavailability of SS in a stream	Measured streamflow; monthly stream water (SS, DRP, TP) analysis and time-integrated SS sampling using in situ samplers (SS, OC, BAP, TP)	1. SS mainly from stream banks during winter & spring (relatively steeply sloped banks with access for cattle) 2. P-enriched surface runoff from laneways or topsoils during summer & autumn

Literature reference	Study catchment information	Study duration (months)	Study focus	Approach	Key observations and conclusions
Davies-Colley et al., 2008	Toenepi (1,500 ha), Waikato, North Island. dairy-grazed pasture	12	Investigate transport of *E. coli* from land to water	Used continuous turbidity measurements as proxy for background *E. coli* concentrations; measured streamflow; storm hydrographs sampled bi-hourly for *E. coli*	1. Storm flow made up 24% of annual flow 2. 95% of total estimated *E. coli* export during stormflows
McDowell and Srinivasan, 2009; Srinivasan and McDowell, 2009	Invermay (9 ha), Otago, South Island. half deer-grazed and half sheep & beef-grazed Two sub-catchments (~20 ha) of the Glenomaru catchment, Southland, South Island. One catchment was deer-grazed pasture; and the other, sheep & beef-grazed pasture	16	Record the occurrence and dynamics of saturation areas and SE runoff in grazed pastures	Measured precipitation &streamflow, transects of wells (<1 m; 5–40 m distance from stream); 20 localised surface runoff samplers (LR); spring flow measured and sampled; regular biweekly and flow-proportional stream sampling during storm events; sub-samples from LRs positioned in strategic IE or SE-runoff areas; manual samples from surface runoff during one storm event; all samples analysed for SS, TP, FRP, TFP, PP, FURP; topsoil moisture measurements to delineate catchment moisture conditions and permanently saturated areas (transport area mapping); estimating surface runoff using empirical and process-based models (Srinivasan and McDowell, 2007)	1. Occurrence of SE limited to <10m from stream during the majority of storm events 2. Occurrence of IE from deer wallows, fence lines, and other impervious areas can be significant in grazed pastures 3. Storm flow accounted for more than 75% of P and SS loss (the majority of these losses occurred during 3 winter, early spring storms); Glenoamru sheep & beef-grazed catchment: ~20% of stream load from IE runoff

Table 3. (Continued)

Literature reference	Study catchment information	Study duration (months)	Study focus	Approach	Key observations and conclusions
Müller et al., 2010a	Upper Toenepi (192 ha), Waikato, North Island. dairy-grazed pasture	28	Investigate the occurrence and dynamics of saturation areas and SE runoff; assess point sources in dairy-grazed catchment	Measured precipitation &streamflow, 5 wells (<1.5 m; 2–30 m distance from stream at various hillslope positions); tile drain flow; localised surface runoff samplers (LR); monthly samples from stream, tile drain, and wells analysed for SS, TP, DRP and PP; LRs used to ascertain IE or SE runoff; modelling surface runoff with different empirical and process-based approaches (Srinivasan and McDowell, 2007)	1. SE runoff during storm events (based on LRs & groundwater levels); 2. Significant SSF contributions during base and storm flows 3. P loss was dominated by effluent ponds; SE runoff and subsurface contributions possible equally important for P transport
Müller et al., 2010a	Kiwitahi (8.7 ha), Waikato, North Island. dairy-grazed pasture	12	Record the occurrence and dynamics of saturation areas and SE runoff in grazed pastures	Measured precipitation &streamflow, 5 wells (<1.5 m; 2–25 m distance from stream); 6 localised surface runoff samplers (LR); 4 runoff plots; monthly and flow-proportional stream sampling during storm events; monthly well samples; all samples analysed for TP and DRP, LRs used to ascertain IE or SE runoff; mapping of permanently saturated and impervious areas (<0.1% of total catchment area); modelling surface runoff with different empirical and process-based approaches (Srinivasan and McDowell, 2007)	1. Limited surfacing of shallow water table indicated that flow during events was not restricted to surface runoff 2. SSF dominated both storm and base flow periods 3. Baseflow accounted for 42% of total stream flow, and contributed 37 and 52% to the DRP and TP loads annually

According to Sharpley and Syers (1979), sorption of P from soil water to soil particles and the longer contact time between P and soils compared with surface processes explained the lower concentrations in subsurface than surface runoff. Stream-bank erosion and resuspension of streambed material were found to dominate PP, TP, and sediment loads in the stream. Similarly, Bargh(1978) observed in a 1-year catchment study that farm tracks and stream bank erosion were important sources for suspended sediment to waterbodies.

Table 4. Linking vegetation to soil saturation

Species throughout the pasture area	*Loliumperenne* L.
	Trifoliumrepens L.
	TrifoliumdubiumSibth.
Species apparently restricted to drier area	*Cerastium* sp.
	Trifoliumsubterraneum L.
Species of drier areas apparently tolerating	*Taraxacumofficinale* Weber
temporary waterlogged conditions	*Plantagomajor* L.
	CynosuruscristatusL.
	HypochaerisradicataL.
	MenthapulegiumL.
	Rumexsp.
Species restricted to seasonally saturated areas	*JuncusarticulatusL.*
	JuncusbufoniusL.
	JuncusgregiflorusL. Johnson
	Juncus spp. (tall rushes)
	Blechnum minus (R.Br.) Allan
Species of permanently saturated areas	*EpilobiuminsulareHaussk.*
	Myosotis sp.
	Veronica sp.
	Nasturtium sp.
	Juncus spp.
	Eleocharis sp.
	SchoenusmaschalinusRoem. Et Schult

Table reproduced from McColl et al. (1985), with permission from the Journal of Hydrology (New Zealand).

McColl et al. (1985) mapped saturation areas at catchment scale to estimate the contribution of SE runoff to stormflow. This study was conducted in a sheep-grazed pastoral catchment, the Judgeford catchment in the North Island. They concluded that the extent of saturation areas was greatly influenced by season and antecedent rainfall. McColl et al. (1985) observed that at a catchment scale, the saturation areas varied from 5% (summer) to as much a 15% (winter) due to differences in evaporative losses. The permanently saturated areas were limited to the riparian zones.Several seepage areas and seasonally-saturated areas were also found on the valley side. Based on the extent of saturation areas, they concluded, that the distribution of vegetation was related to long-term soil moisture conditions (see Table 4, *reproduced with permission* from McColl et al., 1985). McColl et al. (1985) indicated that SSF might not be a major source of stormflow. They did not measure SSF directly, and when the measured stormflow could not be explained by SE runoff from permanently saturated areas and IE runoff , they assumed the unaccounted flow could have come from temporarily saturated areas. These temporarily saturated areas, they concluded, did not remain saturated long after storm events. Petch (1988) reported a hillslope comparison study in the Waikato region of the North Island. He compared a steep pasture and a steep forested hillslope to investigate the role of hillslope form and development of saturation zones as influenced by vegetation. The probability of saturation was the greatest in riparian zonesand decreased when moving upslope for both vegetation types. His observation agrees well with the modelling concept, TOPMODEL (Beven and Kirkby, 1979), which relates the upslope drainage area to saturation propensity.

Cooke and Dons (1988) applied the simple straightforward soil saturation mapping method developed by McColl et al. (1985)(see Table 4) for their hydrological water quality study in the 16-ha hill pasture catchment, Scotsman Valley, Waikato, North Island. In order to assess the relative importance of SE and IE runoff, they mapped soil saturation on a fortnightly basis, and measured infiltration and hydraulic conductivity ofthe topsoil in summer and winter. They also recorded various hydrologic variables: throughflow via subsurface soil horizons at three landscape positions head, shoulder, and toe slopes using "*Whipkey*" trenches, overland flow at several locations along the stream in the riparian zones, streamflow and precipitation (Table 3). In addition, they conducted dye tracer experiments at the throughflow sites to assess the influence of macropores. Based on the infiltration data, they concluded that IErunoff rarely occurred in the catchment. Despite an even distribution of rain throughout the year, they reported that winter (May – July) accounted for 65% of annual flows (0.5% of annual flow recorded in February and 12.5% in June). Cooke and Dons (1988) concluded that the major contribution to stormflow came from surface runoff but throughflow was also significant, especially during autumn and early winter. Throughflow can result from an expansion of saturation areas in the upslope areas, which can exert pressure on the downslope saturation areas, resulting in rapid seepage from the base. Anderson and Burt (1982) termed this as "saturated wedge". The saturated wedge and groundwater ridging result in the same effect – a rapid outflow in the downslope areas of seepage zones. Based on the definitions on groundwater ridging (Gillham, 1984) and saturated wedge (Anderson and Burt, 1982), it appears that the groundwater ridging is described in one dimension (vertically), while the saturated wedge is defined in two dimensions, vertical as well as horizontal. Skalsh et al. (1986) did not present data on the upslope expansion of saturation areas when groundwater ridging was observed. However, Srinivasan (2000) has recorded groundwater ridging in a hillslope adjacent to the stream without any occurrence of saturation areas upslope. Cooke and Dons (1988) quantified the surface and subsurface components ofstormflowbased on hydrometric and hydro-chemical observations, and identified stormflow generation processes for the catchment under observation, which are listed below in their order of significance for the catchment. They concluded that the occurrence and significance of the different flow processes weredependent on the distribution of soil type and physiographic position,

1) SErunoff from saturation areas occurred on 'gleyed' soils (TypicDystrochepts and AericHaplaquepts)
2) SSF resulting from saturation wedges also occurred on 'gleyed' soils
3) return flow occurred on soils from weathered greywacke developed on steeper convex slopes (TypicDystrandepts)

4) IE runoff was thought to be possible for the AericHaplaquepts during high-intensity, mid-winter storms.

Cooke (1988) extended the hydrological field study in the Scotsman Valley catchment (Cooke and Dons, 1988) to a water quality study by determining different P-forms in the runoff components, soils and stream sediments to understand the relative importance of transport pathways for nutrient losses (Table 3). More precisely, surface and throughflow samples were analysed for various P forms (TP, PP, TDP, DRP) and suspended sediment. He found that, similar to flows, surface runoff was the major source of P to the stream during storm events. He found that stream sediments had a higher sorption capacity than the soils in the catchment, and concluded that the stream was a sink for P during baseflow and low-intermediate intensity storm events. Moreover, he observed that about 20% of the annual total P-export was caused by direct aerial fallout during fertiliser application, and proposed that this could be remediated by low-level helicopter application of the fertiliser and by establishing an unfertilised buffer zone around streams. In addition, storm samples from all surface and throughflow of the above catchment study were also analysed for NO_3-N, NH_4-N and TKN (Cooke and Cooper, 1988). An estimated 7 kg N ha^{-1} was exported from the catchment during 1981 of which 86% was in reduced forms (Kjeldahl-N, TKN) and the remainder as nitrate-N (NO_3-N) (Cooke and Cooper, 1988). Cooke and Cooper (1988) found that NO_3-N concentrations were 5-10 times greater in surface flows than in throughflows (Table 3). This provided a key for differentiating flows occurring via surface and subsurface pathways. Virtually all of the reduced N inputs came from SErunoff, whereas NO_3-N inputs were dominantly subsurface derived. A TKN balance for eight stormflowevents showed that except for large floods, the stream system was a net sink for TKN. When discussing water quality management, Cooke and Cooper (1988) indicated owing to surface flow paths being more dominant, riparian controls were critical. They suggested a riparian buffer of 10-20 m. They also suggested that removal of stock from seasonally saturated areas during periods of saturation could significantly reduce soil loss and hence TKN input to the stream.

Smith (1987) similarly concluded that IErunoff rarely occurred in a 16-ha pastoral sheep-grazed catchment located in the North Island, based on measurements of hydraulic conductivities of topsoils and rainfall intensities. She measured surface runoff rates and volumes using surface runoff collectors installed within 5 m of the stream bank during a 20-month study (Table 3). In accordance with the VSA-concept, she observed a huge variability in total runoff volumes amongst the runoff collectors. Surface runoff was extremely channelled along the hillslope, and was mainly generated as SE runoff, limited mainly to winter and early spring seasons. Moreover, return surface runoff was found to be a significant component of total runoff (runoff continuing after rainfall had ceased; exfiltration flow). This spatial variability of surface runoff was correlated to nutrient and sediment loads with the exception of NO_3-N. In a companionpaper (Smith, 1989), she showed that riparian pasture retirement in the same catchment significantly reduced sediment, phosphorus, and nitrate concentrations in surface runoff during a subsequent investigation period of 22 months.

2.1.2.2. Shift to Catchment Studies that Integrate Specifics of Pastoral Agriculture

More recent water quality catchment studies based on VSA hydrology(McDowell and Srinivasan, 2009; Müller et al., 2010a; Srinivasan and McDowell, 2009) were designed to integrate specific features and management procedures of pastoral agriculture that impact on runoff generation and water quality. For example, animal traffic resulting in soil compaction and IE runoff, tile drainage resulting in acceleratedSSF were shown to be significant for stream water quality. Some of these catchment studies also took into account typical management operations linked to year-around grazing of dairy cows such as treating and handlingfarm dairy effluents (FDE, excreta mixed with water produced from washing down yards andmilking parlours). FDEs are often stored in open ponds for subsequent irrigation onto pastoral land (often irrigated on a daily basis during the lactation period). It was highlighted that P release of effluent ponds in pastoral dairy catchments needs to be controlled, especially as the ponds are mainly active during the warmer months (cows are dried of during winter), when P is most detrimental (Müller et al.,2010a). The impact of the methods of treating and disposing of FDE (discharge of effluents into streams vs. effluent irrigation onto land; various effluent pond systems, effluent application techniques, consideration of soil water content in FDE irrigation systems, adding of nitrification inhibitors to FDE) on stream water quality has been discussed widely in former research conducted in New Zealand (Houlbrooke et al., 2004a & 2004b; McDowell et al., 2008; Wilcock et al., 1999 & 2006a).

Srinivasan and McDowell (McDowell and Srinivasan, 2009; Srinivasan and McDowell, 2009) conducted a hydrology-based water quality study in three headwater catchments located in the South Island (Table 3). This study was based on the hypothesis that during storm events, the near-stream areas saturate either due to a rising water table, expanding stream channel or a combination of both, resulting in saturation and subsequent SE runoff. They recorded the water table dynamics in near-stream areas (5-40 m from the stream), and found that topsoil saturation for the majority of rainfall events (days to weeks return period) was limited to within 5 m of the stream corridor, and the bulk of the stormflow came from these areas and from other permanently saturated areas adjacent to the stream. During large storms (return period of months to years), the size of the near-stream saturation extended as far as 40 m and beyond from the stream. During large storm events, they also observed surface runoff from areas as far as 150 m from the stream. They also reported the occurrence of several ephemeral streams, headward of the perennial stream during large rainfall events, effectively connecting the far off surface runoff areas to the stream. Such expansion of stream channels headward was previously reported by Hayward (1976), who indicated that high conductivity of dry channel bed could have resulted in such conditions. Srinivasan and McDowell (2009) also observed the runoff generation dynamics of many small hollows disconnected from the perennial water source and reported that these hollows did not contribute to small and medium sized rainfall events. However, they observed exfiltration or return flows during large storms. Generally, large storms occur infrequently, and in these three catchments, they only occurred in winter and early spring during the investigation periods. Even though they accounted for most of the P lost during the investigation period.The authors concluded that managing these VSAs might be (i) too complex due to the observed temporal and spatial dynamics of the VSAs; and (ii) not economical due to the fact that these P-losses only occurred when their potential impact on periphyton growth in surface water was minimal due to low temperatures.

All the three catchments monitored were grazed by beef cattle, sheep, and deer. Srinivasan and McDowell (2009) monitored the fence lines that were compacted by deer pacing, a deer wallow (Figure 2), and several animal tracks, and reported that they all contributed IE runoff during the majority of rainfall events. Earlier, Cooke and Dons (1988) and McDowell et al. (2003) had indicated, based on field measurements, that pugging and fence-line pacing can significantly reduce water infiltration at the soil surface. Cooke and Dons (1988) suggested that reducing pugging, by keeping animals away from saturated areas, is a key management practice for water quality protection. However, in none of the studies earlier to that by Srinivasan and McDowell (2009), the frequency of runoff contributions from such semi-pervious had been recorded. McDowell and Srinivasan(2009) used localised runoff samplers to measure surface runoff, TP and SS losses from these areas. They estimated that TP and SS from IE areas constituted about 20% of the stream load. The importance of such small IE areas caused by heavy animal traffic to the overall P loss from catchments had been shown by other researchers outside of New Zealand(Edwards and Withers, 2008; Hively et al., 2005). As the areas that produce IE runoff are less dynamic, contribute disproportionately large amounts of P and SS to stream loads and can relatively easy be identified in a catchment, McDowell and Srinivasan(2009) recommended to target mitigation strategies to these areas.

In one of the subcatchments of the Glenomaru catchment, Srinivasan and McDowell (2009) observed that the presence of a subsurface tile drain at 60-90 cm from the surface prevented the occurrence of surface saturated areas (absence of water table at the surface). However, they also reported that the responsive SSF system quickly transferred the incoming rainfall via tile drains, showing that the presence of extensive subsurface drainage network can alter the catchment response. Based on a multi-scale experiment on P-transport in the United Kingdom, Heathwaiteand Dils (2000) highlighted the importance of evaluating thresholds for activating different hydrological pathways in a catchment. The importance of the connectivity of hydrological and transport processes for water quality of receiving waterbodies was also stressed by Gburek et al. (2001).

Müller et al. (2010a) conducted a study similar to that of Srinivasan and McDowell (2009) within a dairy-grazed headwater catchment, the Toenepi catchment and the nested headwater catchment Kiwitahi in the North Island. Apart from measuring the shallow (<1.5 m) water table dynamics, they also analysed(shallow) well water samples for DRP (Table 3). They extended the CSA-concept and hypothesised that, in grazed-pastoral catchments, the transport of P is a function of both SE and IE runoff under stormflow conditions and of SSF under baseflow conditions. Similar to the results of McDowell and Srinivasan(2009), in the Kiwitahisubcatchment, during storm events, IE contributions were significant, highlighting the importance of mapping impervious areas in pastoral catchments. In the Upper Toenepi catchment contributions from SE runoff appeared significant. First-flush DRP-concentrations in surface runoff samples collected with localised surface runoff samplers in the Kiwitahi catchment were very high compared with average DRP-concentrations of stream water samples collected during storm events (Figure 3). However, the absence of the water table at the surface and the presence of a water table gradient towards the stream indicated that P transport during storm events was not limited to surface flow pathways. The P transport during baseflow appeared equally important as P losses during stormflow. The close resemblance in P levels between groundwater and stream samples during baseflow demonstrated the importance of shallow groundwater for streamflow (Figure 3). Furthermore, they demonstrated that the dynamics of the groundwater table and the occurrence of SE runoff areas were influenced by proximity to the stream and hillslope positions. Based on a catchment study conducted in Susannah Brook catchment in Western Australia, Ocampoet al.(2006)hypothesised that the hydrological connectivity between different zones across a hillslope is an important driver for SSF to streams, and concluded that the impact of hydrological connectivity on flow, transport, and reaction has to be captured for predicting runoff generation and chemical transport. The results from Müller et al. (2010a) highlight the need to expand the definition of source areas to include both baseflow and stormflow conditions in catchments. The study by Müller et al. (2010a) also highlighted the impact of point sources (effluent pond systems) for water quality in dairy-grazed catchments in New Zealand.

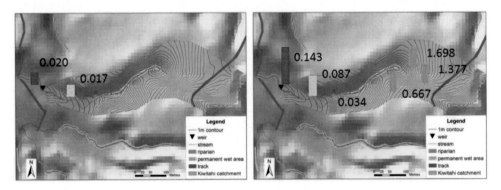

Figure 3. Measured concentrations of dissolved reactive phosphorus (DRP; mgL^{-1}) in stream water (red bars) and well water samples from five shallow (1.5 m deep) wells (blue bars) collected (a) during baseflow conditions betweenDecember 2007-2008; (b) during a storm event (147 mm; 19 April 2008) in the dairy-grazed pastoral catchment Kiwitahi, New Zealand. Green and purple bars represent first flush SE and IE, respectively, runoff samples for that April 2008 storm event. The study is outlined in Müller et al. (2010a).

Gillingham and Gray (2006) conducted a multi-scale (from micro-plots of 0.5 m^2 area to 12-ha catchments), hydrology-water quality study in a seasonally dry hill-country pasture, Waipawa, Hawke's Bay, North Island, to quantify the proportion of surface-runoff transported P to total P losses (Table 3). In contrast to former catchment studies reported in New Zealand (Cooke and Dons, 1988; Smith, 1987; McDowell and Srinivasan, 2009; Srinivasan and McDowell, 2009), they reported that most surface runoff occurred during early summer to late autumn and was negatively correlated to antecedent soil moisture conditions. Thus, SE runoff was excluded. One of the interesting conclusions of this study was that IE runoff may be induced by soil water repellency (SWR). The authors, however, did not attempt quantifying degree and persistence of SWR, nor did they investigate the temporal and spatial variability of SWR.

Soil water repellency is a transient physical soil property that inhibits or completely prevents the infiltration of water into soil (Doerr et al., 2000). In a hydrophilic soil, water infiltrates across the entire cross section of the soil surface, and the rate of infiltration is influenced by soil texture (e.g., higher in coarse textured soils) or structure (e.g., reduced infiltration in compacted soils). In soils where SWR occurs, water typically infiltrates only across a fraction of the soil surface. The reduction of the water infiltration rate into water repellent soilscouldrangeby a factor of six to as high as 25 when compared to hydrophilic control soils (DeBano, 1971; Müller et al., 2010b; Wallis et al., 1990). A recent survey of pastures across North Island of New Zealand demonstrated the potential for a wide distribution for soil water repellent conditions (Carter et al., 2010; Müller et al., 2010b). This survey of the occurrence of SWR in the top 4 cm of soils across 50 sites (10 soil orders x 5 drought proneness classes) under pastoral land use in the North Island showed that 49 out of 50 sites (98%) will become hydrophobic when they dry out, and that 35 out of 50 sites (70%) were hydrophobic at the time of sampling in summer 2009/10. The degree of SWR was found to be positively correlated to the soil organic carbon and nitrogen content of the top 4 cm and negatively correlated to bulk density (Müller et al., 2010b). Another survey in the United Kingdom (Doerr et al., 2006) investigated the occurrence of SWR in topsoils (0-5 cm) and subsurfacesamples (10-15 and 20-25 cm) under different land-use. Of a total of 41 sites, nine were on shrublands, five on permanent pasture, seven on conifer woodland, six on broad-leafed woodland, and 14 on arable land. Discarding the sites under arable land-use, 96% of the top-soils were potentially hydrophobic. The authors concluded "repellency is common for many land-use types with permanentvegetation cover in humid temperate climates irrespective of soil texture." (Doerret al., 2006; *p. 741*)The impact of SWR on IE runoff and transport of nutrients from the soil surface under grazed pasture is currently investigated for different soil orders across the North Island (Carter et al., 2010, ongoing research).

2.1.3. Summary of Findings from Water Quality Field Study

As discussed above, very few water quality field studies have been reported from the pastoral catchments of New Zealand. Forested catchments received the most attention from hydrologists (e.g., Maimai and Glendhu), indicating the lead taken by forest hydrologists in understanding the VSA hydrology in New Zealand, as has been elsewhere. Some of the major conclusions that can be derived from thepastoral catchment studies are,

1) IE runoff rarely occurs except in soils that have been compacted by animal movement or exhibit seasonal SWR. In grazed-pastoral catchments nutrient losses in particulate forms through IE runoff have to be considered even though the flow pathway may be relatively insignificant. Also IE runoff from farm tracks, milking parlours, and feed pads can be important for nutrient transport from land to water in pastoral catchments. Mapping of areas in the catchment where infiltration is inhibited is recommended. The transient character of SWR(Buczko et al., 2005; Wallis and Horne, 1992) and its spatial variability (Deurer et al., 2007; Gerke et al., 2001; Lemmnitz et al., 2008) complicate prediction and mappingof soil water repellent areas at any one time. This is related to the lack of understandingof the fundamental causes and genesis of SWR(Doerr et al., 2006; Hallett, 2008; Horne and McIntosh, 2000), especially at the soil surface where SWR actually appears most often and directly impacts on water infiltration and generation of IE runoff.

2) SE runoff from near-stream saturated areas appear to be the major contributor for the majority of the small and medium events (return period, days to weeks); however, in steep catchments, with shallow soils, subsurface contributions can be significant even for small and medium storm events. Nutrient transport via this pathway generally is smaller than those via IE runoff due to the longer contact time between soil and water enabling sorption processes as indicated by Sharpley and Syers (1979).

3) Expansion of saturation areas during storm events could be happening first headward of perennial channels before occurring lateral to the stream, as has been defined by Hewlett and Nutter (1970).

4) Subsurface flow is a significant component of stormflow during large events. SSF contributions can be highly variable; though can generally be greater at the converging hillslopes than diverging hillslopes. Subsoil discontinuities such as gravel patches within the soil horizon, macropores, and drain pipes can significantly increase subsurface contributions. Nutrient transport via this pathway depends on the nutrient concentration in the shallow groundwater which in turn is determined by nutrient contents in the soil, soil type, shallow groundwater dynamics,and others.

5) Groundwater ridging and saturation wedges at the near-stream locations can result in rapid outflow of subsurface flows in the seepage zones.

6) The importance of stream bank erosion and re-suspension of stream sediment for TP, SS and PP was shown.

7) Management practices in dairy-grazed pastoral catchments including FDE treatment, storage, and irrigation techniques are important factors influencing water quality in pastoral catchments.

2.1.4. Models Developed to Integrate Findings from Field Studies into Management Tools

VSA hydrology provides a sound theoretical basis for identifying hydrologically active areas in the landscape (Qiu et al., 2007) that could be used as foundation for developing spatially distributed, event-based catchment water quality empirical and mechanistic models (Rodda et al., 1998). A brief overview on various modelling approaches, that mostly have been developed outside of New Zealand, and have been modified and tested in some New Zealand catchments, is provided below.

Caruso (2001) developed a risk-based approach to identify subareas and farms within a catchment that are most likely to contribute to nonpoint source pollution to surface waters. His approach uses ranking methods based on expertknowledge and a general understanding of the variable contribution of different catchment areas to the overall pollution of a surface water body. His approach has two stages: firstly, identifying subareas in a catchment and ranking according to their potential to contribute to water pollution based on different combinations of the following criteria (i) catchment characteristics (including vegetation, predominant geology, soil type, slope class, type and severity of potential erosion, proximity to the lake, and bank erosion, and (ii) measured TP concentrations and loadings in receiving waterbodies. Secondly, farm data including intensity of sheep and cattle grazing, fertiliser usage, proximity to surface water bodies, slope/runoff and observed bank erosion, are used to rank the farms located within a catchment according to their risk to contribute to the TP input to the surface water body. Caruso(2000 & 2001)applied this approach to describe the TP pollution of LakeHayes, a lake located in a high country pastoral catchment in the South Island,and concluded that best management practices should focus on the identified high-risk subareas and farms.

The Basin New Zealand (BNZ) model (Cooper et al., 1992) based on CREAMS (Knisel, 1980) was designed to predict runoff, soil erosion, and nutrient loss from pastoral catchments in New Zealand, using the Soil Conservation Service Curve Number method (Soil Conservation Service, 1972). A decision support system, known as the Catchment Decision Support System (CDSS) was developed, which linked the BNZ model to national databases of climate, hydrological and land use information, within an ARC/INFO^{TM} GIS interface(Rodda et al., 1999). The CDSS was applied to theToenepicatchment in the North

Island. Soil, land use, and topographic data on a grid cell basis were used to model average annual sediment and nutrient loadings using climate data for the period 1980-1987. A comparison between modelled and measured data has not been published. While the CDSS model does not consider VSA hydrology, it is mentioned here, because in the meantime the runoff Curve Number approach has been modified to take into account VSA hydrology, and has been successfully applied to two catchments in the Catskill Mountains region of New York State, USA, and to one catchment in south-eastern Australia(Lyon et al., 2004).

Rodda et al. (1998) were the first to attempt process-based modelling of nutrient transport from CSAs to surface water in New Zealand on an event basis. They combined a modified version of the physically-based TOPMODEL (Beven and Kirkby, 1979) with a simple P-transport model based on the P-cycling routines, derived from GLEAMS (Leonard et al., 1990), and applied the model to the sheep-grazed,3-ha Pukemangahill-country catchment, Whatawhata, North Island. They employed a GIS which enabled illustrating the spatial extent of the surface runoff and associated P loads. However, they reported that the performance of this uncalibrated model at catchment scale was rather poor, possibly due to the simplified equation using Olsen-extractable soil P as indicator for DRP in runoff.

Others focused on modelling VSA hydrology and applied a modified verson of TOPMODEL to different catchments in New Zealand to delineate the location of IE and SE runoff areas, and to simulate runoff volumes at catchment scale (Elliott and Ibbitt, 2000a & 2000b), however, also only with limited success. In particular, the identification of runoff generating processes and areas was poor. The poor modelling results were explained, *inter alia*, by seasonal variability in soil properties not considered in the model, lack of routines incorporating re-infiltration of runoff, inappropriate representation of baseflow generation and deep groundwater processes (Elliott and Ibbitt, 2000a) and the fact that SWR is not considered in TOPMODEL (Elliott and Ibbitt, 2000b). Another physically-based catchment model that incorporates VSA hydrology is the Soil Moisture Routing Model (SMoRMod, Zollweg et al. 1996), which has not been tested in New Zealand. Srinivasan et al. (2003) employed a modified version of this model, Soil Moisture Distribution and Routing (SMDR; Soil and Water Laboratory, 2002), in a small agricultural catchment in central Pennsylvania, USA. Based on catchment-scale soil and topographic data and observed rainfall data, they developed seasonal and event-based soil saturation maps that could be used as a proxy for SE runoff areas (see Figure 4). The saturation maps depicted in Figure 4 provide an averaged soil moisture status across the catchment for the months shown. Srinivasan et al. (2003) concluded that such maps, when overlaid with field boundary maps can be an effective visual guide for planning nutrient application practices. However, they reported that SMDR does not handle SSFs, which could be an important transport mechanism during large rainfall events.

Parameterising process-based models is very complex, as the models are extremely data-hungry. Moreover, data on the temporal and spatial dynamics of CSAs during rainfall events to validate and calibrate these models are largely not available. Therefore, validation and calibration as well as uptake of these models has remained very limited. On the other hand, it has to be noted that some of the existing physically-based water quality simulation models like SWAT (Soil Water Assessment Tool; Arnold et al., 1998) and AGNPS (Agricultural Non-Point Source model; http://www.wsi.nrcs.usda.gov/products/w2q/h&h/ tools_models/agnps/index.html) are well established and have scientifically sound nutrient cycling and transport routines even if the hydrological components do not incorporate VSA hydrology.

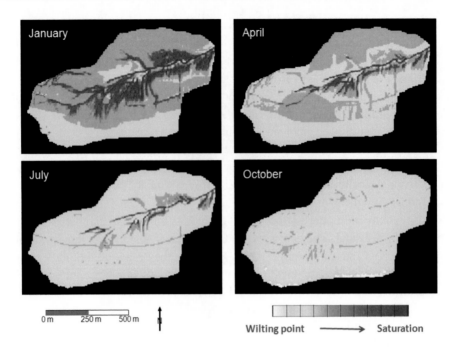

Figure 4. Mapping seasonally saturated areas using process-based models. An example adapted fromSrinivasan et al. (2003). Maps depict averaged monthly soil moisture status for a headwater catchment in central Pennsylvania, USA. Maps were developed using the SMDR (Soil Moisture Distribution & Routing) model (Soil and Water Laboratory, 2002).

A different avenue was chosen in the United States with the empirical Phosphorus Indexto identify CSAs (Gburek et al., 2002; Gburek et al., 2000; Sharpley et al.,2004). The development of isiwas closely based on the CSA concept forwarded by Pionke et al. (1996) stipulating that near-stream areas are active P and sediment transport areas. Field observations on spatial and temporal changes in runoff areas at near-stream locations within headwater catchments in Pennsylvania, United States, by Zollweg et al. (1996), Needelman et al. (2004), and Srinivasan et al. (2002) supported the concept that near-stream areas exhibit pronounced hydrological activity during rainfall events. Recent New Zealand field studies corroborate these observations (Müller et al., 2010a; Srinivasan and McDowell, 2009). A defined distance on either side of the stream has been assumed to be the major contributor of runoff and nutrients to streams, and this assumption has been adopted in many P management strategies in the United States (Sharpley et al., 1999).

In New Zealand, Srinivasan and McDowell (2007) compared the two modelling strategies: empirical and process-based models, to map the hydrologically active areas at catchment scales and tested the model performance by applying them to grazed-pastoral catchments. The five hydrological approaches proposed by Srinivasan and McDowell (2007) assumed that P transfer from land to stream occurs via surface runoff processes. The five approaches were: 1) the curve number (CN) method, 2) the Phosphorus Index (PI), 3) the drainage density (DD) model, 4) the topographic index (TI), and 5) the SE-IE combination model. A detailed description of these approaches can be accessed at Srinivasan and McDowell (2007). The empirically-based CN approach uses hydrologic soil group and land use information to estimate runoff depths at catchment scale, and thus, the transport areas delineated are reflective of land use and soil group combination. The PI approach was based on Gburek et al. (2000). This approach assumes that stormflow generation is limited to fixed near-stream areas. Gburek et al. (2000), based on field research in a small agricultural catchment in central Pennsylvania, USA, assumed that 50 m on either side of the stream is hydrologically most active during storm events, and Srinivasan and McDowell (2007) assumed the same distance. The DD approach combines the CN and PI approaches. This approach derives runoff depths based on CNs. The derived runoff is assumed to be generated from near-stream areas but unlike the PI approach, which assumes a fixed distance for all storm events, the DD model varies this distance based on runoffdepth estimated and antecedent catchment moisture conditions. Both PI and DD approaches assume that stormflow is generated by SE mechanism. The TI approach is based on the TopographicIndex developed by Beven and Kirkby(1979). It is a physically-based approach, and incorporates local slope, upslope drainage area, and soil transmissivity characteristics to rank catchment areas with high propensity to generate SE runoff. The fifth approach (SE-IE combination) is a modification on the TI model wherein Srinivasan and McDowell (2007) included the management-induced IE areas such as animal tracks, farm tracks, and wallows to represent both SE and IE runoff areas.

Figure 5, reproduced from Srinivasan and McDowell (2007), presents a comparison of results from the five approaches for a storm event, 25.4 mm that occurred on a moderately wet catchment, AMC II, in a grazed-pastoral headwater catchment in the South Island, New Zealand. For this particular storm event, a comparison of observed and predicted flow volumes (presented as "deviation of runoff volume" in Figure 5), the empirical approaches fared better than the process-based approaches. However, Srinivasan and McDowell (2007) applied these five approaches to other rainfall events and in other catchments in the region and concluded that the process-based approaches generally resulted in better prediction of measured flows than the empirical approaches. Deviation of runoff volumes indicates the percent difference between measured and predicted stormflow volumes. A positive value indicates under-prediction of measured flow volumes, and a negative value signifies over-prediction. For example, a value of +5% signifies that the observed flow volumes were underpredicted by 5%. The key difference between the empirical and process-based approaches was the spatial extent and spread of the predicted CSAs. From a management perspective, the CSAs predicted by DD and PI approaches could be easily implemented as opposed to the CSAs predicted by the two process-based approaches. This evaluation was also confirmed in later applications to dairy-, deer- and sheep and beef-grazed catchments (McDowell and Srinivasan, 2009; Müller et al., 2010a; Srinivasan and McDowell, 2009). The implementation of results from a process-based model into management practices might be very complex and not practical because of the predicted scatter of CSAs resulting from IE and SE runoff. Srinivasan and McDowell (2009) tested the validity of these approaches using field data, and reported that storm flow generation was often limited to less than 5 m wide near-stream areas. They also reported the measured stormflows could only be accounted by a combination of SE from near-stream areas, IE from near- and far-stream compacted areas connected to the streams, and SSF. Even though SSF contributions were not measured directly, they deduced the potential SSF contributions based on SE and IE contributions and measured flow during storm events.

McKergow and Tanner (2011) developed a procedure to identify key runoff generation processes at landscape and farm scales for the selection and implementation of appropriate nutrient and sediment management practices. This procedure combined soil, landscape, and geologic information to describe the dominant runoff generation mechanisms and nutrient and sediment transport pathways at farm scale. Soil information includes surface and subsoil characteristics such as compaction and presence of low or impermeable zones that can result in quick flows. Landscape scale details include information on the presence of wet areas, seeps and springs, and the geologic data describe the presence of permeable parent material. Their procedure combined this information with farmers' local knowledge of the landscape (e.g., information on the seasonality of water table status and saturation areas) to describe the dominant runoff generation process(es). For managing IE runoff from impervious areas, practices such as grassed filter strips and farm ponds that collect, retain, and eventually infiltrate into the soil, were recommended. Similar recommendations were made for SE areas (exclusion of livestock from wet areas) and SSF areas (riparian buffers downslope of areas where SSF emerges to the surface).

In summary, an appropriate model to predict diffuse pollution of streams taking into account VSA hydrology that could transferthe findings from the field studiesconducted to other catchments is not yet available. The few models that are available have not been fully tested using field data. This means that to date very little of the new knowledge on CSAs and nutrient transport has been applied to manage grazed-pastoral catchments.

3. ADAPTATION OF VSA HYDROLOGY FOR WATER QUALITY IN NEW ZEALAND

Pearce (1990) highlighted the need for more focussed, objective-oriented field studies for understanding the VSA hydrology in New Zealand and Australian catchments. In New Zealand, a vast majority of process-defining field studies relating to VSA hydrology have been reported from forested hillslopes and headwater catchments. The importance of VSA hydrology in pastoral catchments has been increasingly recognised though studies have been far and few. Even within pastoral catchments, the reported field studies focussed at plot, hillslope, and headwater catchment scales, because hydrologic controls can be easily measured and monitored at these scales (e.g., measurement of throughflow using *"Whipkey"* trenches by Cooke and Dons (1988); manual mapping of saturation areas using hand-held soil moisture probes by Srinivasan and McDowell (2009)). Below, the key points of the VSA hydrology concept and its necessary extensions to protect the water quality in New Zealand pastoral catchments, as introduced in this book chapter, are summarised.

3.1. VSA Hydrology and SE Surface Runoff

The early VSA studies in New Zealand and elsewhere recognised that near-stream corridors arethe primary source areas of stormflow during storm events (e.g., Dunne et al., 1975; Gburek, 1990; Gburek et al., 2000), and hence, described these areas as criticalto nutrient and sediment transport (e.g., Pionke et al., 1996; Gburek et al., 2001). Cooke and Dons (1988), McColl et al. (1985), and many others in New Zealand indicated that near-stream corridors need to be fenced off from animals as a first step of water quality protection.

Loss of land from production and high fencing costs have deterred the adaptation of stream fencing in the farming community. However, a few regional governments have started mandating this practice (e.g., Otago Regional Council; http://www.orc.govt.nz/Information-and-Services/Farming-and-land-management/Waterway-protection-programme/Waterway-protection-FAQ/). Based on the assumption that the near-stream corridors are critical source areas,Srinivasan and McDowell (2007),using a simple hydrologicalmodelling approach, showed that by reducing nutrient input to the near-stream areas, farmers can more effectivelycontrol nutrient movement to surface waters than by fencing off streams. However, they did not account for direct faecal deposition in streams, which stream fences can effectively preclude.

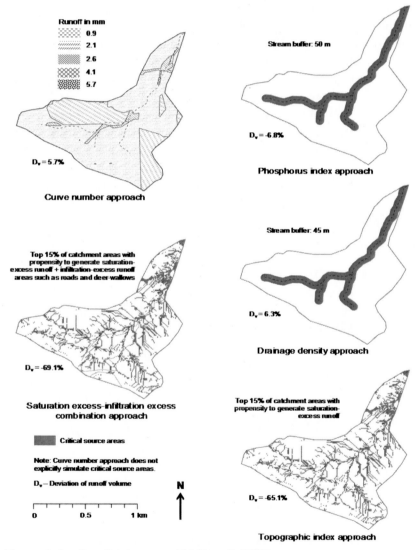

Adopted with permission from Srinivasan and McDowell (2007).

Figure 5. Comparison of empirical and process-based hydrologic approaches in predicting critical source areas at catchment scale. Shown here is an example application in a grazed-pastoral headwater catchment in the South Island of New Zealand.

3.2. IE Surface Runoff Related to Year-Round Open Grazing Systems

Recently, McDowell and Srinivasan (2009), Müller et al. (2010a), and Srinivasan and McDowell (2009) based on detailed field studies in grazed-pastoral headwater catchments indicated that near-stream corridors might not always be the primary sources of stormflowduring storm events. Srinivasan and McDowell (2009) indicated that animal camps, animal tracks, wintering pads (winter confinement systems for livestock where the animals are grazed off over winter and spend winter on pads bedded with woodchips or rubber mats and receive supplementary feed), and deer wallows connected to streams could result in significant IE runoff and sediment and nutrient transfer during storm events. They indicated that these IE areas might be critical to water quality even if their flow contributions are not significant. Müller et al. (2010a), in the absence of significant SE runoff from near-stream areas despite the presence of the water table at the surface, concluded that SSFand IE runoff could be significant sources of water, sediment, and nutrients during storm events.

Thus, inNew Zealand, grazing adds another dimension to the identification, mapping, and management of VSAs. Several field studies (see Table 1) highlight the impact of grazing on soil resources, but very few tools have been developed to generalise and transfer that knowledge. The model developed by Finlayson et al. (2002) attempted to simulate the treading effects of cattle on soil resources at field and paddock scales.Validation and transferability of this model could only be realised by including more data on animal behaviour during grazing, which Finlayson et al. (2002) concluded are not readily available . Betteridge et al. (2008 & 2010) suggested that such data are difficult to collect as animal behaviour, specifically cattle behaviour, is difficult to describe and categorise in steep pastoral catchments.

3.3. IE Surface Runoff Related to Soil Water Repellency

Many studies have shown that surface runoff generally increases with an increase in SWR(Burch et al., 1989; Buttle and Turcotte, 1999; Frasier et al., 1998; Gomi et al., 2008; Leighton-Boyce et al., 2007; Miyata et al., 2009; Pierson et al., 2009; Scott and Van Wyk, 1990; Valeron and Meixner, 2010; Walsh et al., 1994). Estimates ofthe increase in surface runoff due toSWR ranged from three (Burch et al., 1989) to 16 times (Leighton-Boyce et al., 2007). In most studies indirect methods (e.g., correlation analysis) were used to attribute the increase of surface runoff to an increase in SWR. Most of these studies focused on the impact of forest fires. Data on the impact of SWR on surface runoff generation under pastoral land use are very scarce.

3.4. Contributions of SSF to Stormflow

Significant levels of SSF have been reported in forested catchments inNew Zealand, despite many early desktop studies largely discounted the role of SSF during storm events (e.g., Hayward, 1976; Pearce et al., 1986). Mosley (1979), based on a field study in a forested catchment, indicated the potential of substantial contributions via SSF paths during storm events. Even though later studiescontradicted with Mosley's observations on the extent of the

SSF source areas (entire catchment versus near-stream areas),they generally agreed on the potential of SSFas a source of stormflow (e.g., Pearce and McKerchar, 1979; Woods and Rowe, 1996). Srinivasan et al. (2002), based on a hillslope-scale field study within an agricultural catchment in the northeast US, concluded that, even under complete saturation conditions, the majority of transport of water and nutrients could be occurring via SSF paths rather than as SE runoff. Müller et al. (2010a) made similar observation in the Toenepi catchment, New Zealand. These observations highlight the need to investigate the role of SSF during storm events.

Monitoring and quantifying SSF via tile drains have been reported in many field studies in pastoral catchments (e.g., Hedley et al., 2005; Monaghan and Smith, 2004; Srinivasan and McDowell, 2009). For example, Srinivasan and McDowell (2009) monitored a tile drain in a grazed-pastoral headwater catchment in Otago, South Island, and concluded that SSF via tile-drains temporally can be as responsive as surface flows. Similar results have been reported with water quality constituents. Houlbrooke et al. (2004c) and Monaghan and Smith (2004) reported that under wet conditions, *E. coli* concentrations in the drain flowsapproached those in the irrigation effluent applied. McDowell and Srinivasan (2009) reported similar quick transfer of phosphorus via tile drains under rainfall conditions. However, measurement and quantification of SSF contributions in catchments with no artificial drains is a challenging task. Cooke and Dons (1988) and Woods and Rowe (1996) used trenches to collect and quantify SSF contributions from various soil zones. Woods and Rowe (1996) highlighted the spatial and temporal variability of SSF during and between rainfall events. This variability in response can be quantified by increasing the number of observation points (trenches), covering multiple landscape positions and rainfall events, and by applying geostatistical methods. However, such trenching is not practical (loss of productive land) and safe (for grazing animals) in grazed-pastoral catchments. Use of tracers has been demonstrated as an effective method for mapping SSF paths (e.g., Cooke and Dons, 1988; Mosley 1979), however, the controls on SSF during rainfall events could be very different from those during baseflow periods. Ridging and headward and sideward expansion of stream channels during rainfall events could initiate the connectivity of SSF pathways from streams toward land. Hayward (1976) described the possibility of such rapid expansion of headwater stream channels and subsequent SSF contributions via porous stream beds. Non-destructive field techniques to measure SSF and nutrient contributions to flow and loads during stormflow conditions in grazed-pastoral catchments as well as methods to transfer these field scale observations to higher scales of interest are needed.

3.5. Tools for Linking VSA Hydrology to Water Quality Management

Tool development, be it maps (e.g., seasonally saturated areas), best management practices (e.g., exclusion of animals from permanently saturated areas), or simple indices, similar to the Phosphorus Index (Gburek et al., 2000), is discussed at two different spatial scales – thecatchment scale, where best management practices are implemented to protect water quality, and the regional scale, where water quality standards are set, monitored, and reported. At the catchment scale, tools are designed to specifically consider management practices and their control on runoff generation and contaminant transport, while at the regional scale these tools and indices provide an overview of controls on runoff hydrology.

3.5.1. Catchment-Scale Tools

Below, we discussapproaches to integrate VSA hydrology with water quality management practices in grazed-pastoral catchments. Key research and management challenges that need to be tackled for expanding the capacity to identify CSAs are highlighted. Even though the approaches discussed are not exhaustive,they are chosen to stimulate research and technology transfer in this area.

3.5.1.1. VSA Hydrology and SE Surface Runoff

*Mapping saturation areas using process-based models:*Hydrologic simulation models allow mapping of soil moisture and saturated areas at catchment scale.Srinivasan et al. (2005) used a process-based model, SMDR to map and delineate surface saturation areas based on rainfall data and published infiltration data from a soil survey database. These maps can be combined with shallow groundwater data to predict the probability of the occurrence of surface saturated areas at any locations within a farm (Srinivasan et al. 2005).

*Use of vegetation to infer on soil saturation:*McColl et al. (1985) suggested that vegetation might be a good indicator of saturation areas even in grazed catchments. A list of vegetation that is linked to the occurrence of saturation areas is given in Table 4 (McColl et al., 1985). So called 'vegetation signatures' have been used to delineate wetland areas in the past in different contexts (Qiu et al., 2007; Guntenspergen et al., 2002). However, the difficulty with such indices is similar to those of animal induced features, i.e.the mapping of these features.

Use of remote sensing technology to identify saturated areas: As already indicated in the early literature on VSA hydrology (Van De Griend and Engman, 1985), remote sensing might have potential for identifying saturated areas at catchmentscale. Remote sensing methods for deriving soil moisture data have recently been reviewed by Petropoulus et al. (2009). In New Zealand, Wilson et al. (2003) applied microwave remote sensing methods in the Mahurangi River Catchment, and pointed out that spatial distributions of soil moisture in the upper few centimetres of soil could be readily interpreted from the signals. Hutchinson (2003) showed that ERS-2 data may be capable of monitoring near-surface soil moisture conditions over even extremely dense natural grassland vegetation. de Alwis et al. (2007) developed a methodology to determine the spatial variability of saturated areas using a temporal sequence of remotely sensed images. The Normalized Difference Water Index (NDWI; Gao, 1996) was derived from medium resolution Landsat 7 ETM+ imagery that was collected over a seven-month period in the Town Brook watershed in the Catskill Mountains (New York State, USA). The data were used to characterise saturated areas by the soil surface water content when the vegetation was dormant and by the vegetation's leaf water content during the growing season. The results indicate that remote sensing could be successfully used for capturing the spatial and temporal variations in surface water content under a single land use. Furthermore, differences in vegetation and temperature might also be identified with remote sensing (Van De Griend and Engman, 1985).

3.5.1.2. IE Surface Runoff Related to Year-Round Open Grazing Systems

*Inclusion of farm-specific data related to year-round open grazing of livestock:*Srinivasan and McDowell (2009) indicated the importance of deer wallows, animal tracks, and animal camping sites for generating IEsurface runoff during storm events. Connected to waterbodies via animal tracks, these areas can be significant contributors of stormflow. However, such data are not readily available, nor could be mapped from aerial photos. Thus, there is a need to develop a farm-scale database with periodical updates.

3.5.1.3. Contributions of SSF to Streamflow

*Inclusion of subsurface drain data to better represent the subsurface flow processes in drained catchments:*Collins et al. (2007) indicated that many dairy pastures on poorly drained soils have been artificially drained, and suggested that these drains may reduce soil saturation and the propensity to generate SE runoff. However, drains act as bypass flow pathways and can quickly transfer microbes (Collins et al., 2005) and nutrients (McDowell and Srinivasan, 2009) to surface waterways during and immediately after rainfall events. Thus, in artificially drained catchments, SSF could be a dominant transport mechanism. However, like in many other regions of the world, in New Zealand, the installation of artificial drains has not been mapped during the installation. Thus, in regions, where artificial drains are common, we might need simple, automated tools to map their existence. For example, Naz and Bowling (2009) described an approach to map drain lines and their density using aerial photos and remotely sensed images. Such an approach might be needed in New Zealandcatchments, specifically for those in Southland, where soils are poorly drained and the dairy industry has been expanding rapidly.

3.5.1.4. Modelling Approaches

Linking physically-based distributed hydrological models and agrichemical transport models: Hydrologic simulation models that are based on VSA hydrology (e.g., TOPMODEL, SMDR) need to be incorporated into existing widely used water quality simulation models (e.g., SWAT, AGNPS). While these agrichemical transport models include sophisticated and extensively tested agrichemical transport and fate algorithms, the mathematical description of variable participation of catchment areas in runoff generation processes is incompelte.

Empirical indices based on digitalised data: Available digital data resources could be used much more efficiently to develop and apply simple indices for (a) assessing the location of runoff generating areas and the timing of runoff risks in a catchment (e.g., using the topographic index; distance to the stream etc. as, for example, suggested by Srinivasan and McDowell,2007); and (b) intersecting this empirical VSA knowledge with land use information (e.g., grazing animals, effluent application areas etc.).

3.5.1.5. IE Surface Runoff Related to Soil Water Repellency

Based on the current understanding of SWR and our still very sparse observations of the impacts of SWR on water dynamics, SWR models have been developed. Functional approaches (Dekker and Ritsema, 1994), mechanistically complex simulation models (Bachmann et al., 2007; Deurer and Bachmann, 2007; Karunarathna et al., 2010b) and empirical tools predicting the effect of changing soil water contents on SWR(Karunarathna et al., 2010a; Regalado and Ritter, 2005) have been proposed. However, all these models are either too simplistic or too complex to be useful for assessing the impact of SWR on soil moisture dynamics at field scale. Moreover, as long as we are unable to predict when SWR will occur or disappear (Doerr et al., 2007), the usefulness of models to predict the impact of

SWR on water infiltration is questionable. Research is needed to improve our understanding of the causes of SWR under different soil, climatic, and land use conditions.

3.5.2. Regional Scale Tools

The catchment-scale tools and studies discussed in this chapteroffer a detailed understanding of hydrologic and transport processes. However, theuptake of these tools are also constrained by the need for detailed field data at appropriate scales, and often, observations over long periods. Also, the transferability of catchment-specific observations to other scales and regions has been a challenge. Below, we offer few ways to effect this transfer. We see the regional scale tools discussed below have the potential to extrapolate catchment-specific results within the context of large scale data such as soils, topography, climate, and geology. Such transfer is important not only for science but also for management and policy development. Here, we draw parallel to the successful development of the Phosphorus Index (Gburek et al., 2000) in the United States. We conclude this section with few thoughts on one of the emerging issues, surface water-groundwater interaction, in New Zealand. We see this is as an emerging area of research in New Zealand, with potential to influence both water quality and quantity management in many agricultural regions.

3.5.2.1. Regionalisation

Regionalisation, an effective tool for knowledge transfer at large spatial scales, allows groupingof areas with similar characteristics or values. In hydrology, regionalisation can be implemented in many different ways depending on the dominant controls on catchment response during storm events. Also, regionalisation can be varied to represent a specific hydrologic variable (e.g., flood forecast, McKerchar and Pearson, 1989; low flow forecast Laaha and Blochel, 2006). In New Zealand, regionalisation technique has been successfully applied to infer ecological characteristics and values of rivers and streams. Snelder and Biggs (2002) developed a river environment classification system, or REC, for New Zealand's river and streams. The REC classes are geographically independent and available from headwater streams to large rivers, and provide a quick link between stream orders and ecological systems.

Mosley (1981) explored the potential to use the regionalisation technique for New Zealand catchments to describe their hydrologic response to storm events and arrived at twogeneral conclusions:

1) In the South Island, climatic regimes, as modified by topography, are the major control on flood hydrology.
2) In the North Island, no clear factor emerged as a control on the hydrologic systems. A combination of climate, lithology, soil types, and topography controlled the hydrology of North Isalndcatchments.

Mosley (1981) indicated that the regionalisation did not necessarily mean grouping the catchments in certain locations together, but the identified catchments could be spread across a large region and still be grouped based on a common characteristic. He highlighted his approach with an example of the Waikato River in the North Island, which flows across a wide range of climatic regimes, lithologies, vegetation, soil types, and land use. Thus, the catchments of this river would be grouped in more than one region.

Other similar hydrologic variables could be considered to group catchments. By the process of regionalisation, catchments across New Zealand could be grouped based on the controls on runoff generation processes. The examples shown below can be readily developed based on result from studies conducted in New Zealand.

1) *Grouping based on catchment scale stormflow response:*Pearce et al. (1984) studied runoff generation in three small catchments of the Glendhu State Forest in South Island, and found that less than 30% of the annual flow wasstormflow (the corresponding numbers were 50% at the Hut Creek and 65% at the Maimai catchments – both in the South Island). They concluded that IE runoff did not occur in this catchment, and that SE runoff from wet areas, accounting for 10 to 20% of the total catchment area,was the dominant flow generation mechanism.

2) *Grouping based on linking hydrologic response and seasons:*McColl et al. (1985), based on a runoff generation study in the Scotsman Valley catchment, North Island, indicated the existence of large seasonal differences in hydrologic response. They found that this catchment generated 65% of flow in winter, even though rainfall was uniformly distributed throughout the year. Similar variations in seasonal stormflow responses were reported in other studies (e.g., Müller et al., 2010a; Cooke and Dons, 1988).

3) *Grouping based on dominant flow generation mechanism:* Taylor and Pearce (1982) grouped some headwater catchments of New Zealand as SSF and SE runoff dominated. While their grouping was limited to headwater catchments, such grouping scaled up to higher order catchments, could be an useful attribute for regionalisation.

3.5.2.2. Water Quality Risk Assessment Indices

Process complexity has been the hallmark of VSA hydrology. However, to realisethe transfer of knowledge/ process understanding to the management of water quality at farm scale, certain generalisations, similar to those derived by Dunne (1978), are necessary. For instance, in the United States, when the Phosphorus Index was developed as a tool for predicting the potential of P pollution, the entire hydrology associated with P transport from land to water was distilled into one factor, the transport factor (Gburek et al., 2000). Even this factor was empirical, and only varied with the proximity to the stream. A large transport factor was assumed for near-stream areas and a small factor for far-stream areas. Also, the entire transport was assumed to be dominated by SE runoff. The transport factor was indicative of the P transport potential from varying landscape locations to waterbodies. Beegle et al. (2000) indicated that even such simplifications might not be time and cost effective enough for farm managers and nutrient management plan writers.

McLeod et al. (2005) developed a relative risk index to describe microbial transport from land to waterways. They developed a series of maps describing the risk of surface runoff (overland transport), bypass flow through soil (preferential flow), and flow via the vadose zone. They used basic soil properties such as drainage class, depth to layers of low permeability, hydraulic conductivity, particle-size class, parent material, toprock, soil structure, and slope to arrive at a risk classification. However, these maps often do not include explicit information on catchment scale runoff generation processes. A GIS-based riskassessment approach for delineating the spatial occurrence of runoff generation processes and areas using digitalised data of agricultural land use, surface water bodies, soil data, and DEM could be developed. It has to be noted that there is another level of complexity with the seasonal variability of saturated areas. Therefore, the risk assessment approach might have to incorporate long-term climatic data and operate like a probability index.

3.5.2.3. Surface Water-Groundwater Interaction

Apart from catchment-scale controls on runoff generation and nutrient transport, the hydrology of catchments can also be influenced by large scale geology. For example, the lowland catchments of Canterbury and Hawke's Bay (eastern side of New Zealand) receive a large influx of groundwater from upland catchments(Martin and Williams, 2006). Many upland rivers and streamslose significantly to groundwater due to porous bedrock, and this groundwater resurfaces as springs, sustaining flows in many lowland catchments (Martin and Williams, 2006). For example, based on observed rainfall,estimatedetd evaporative loss, and measured streamflow data within a lowland catchment, the Harts Creek Catchment in Canterbury, Meriano et al. *(in preparation)* reported that rainfall accounted for only one-seventeenth of the flows measured in the creek; the remaining streamflow was accounted for by groundwater discharge. Under these conditions, the sources of flows (specifically baseflows) can be groundwater recharged in upland catchments. Depending on the travel times and distances between the recharge (upland) and discharge (lowland) streams, contaminants within the groundwater transport zone can potentially act as nutrient source to lowland streams. While such large scale processes are difficult to manage (with the exceptionof managing nutrient applications in the upland catchments), knowing of such source areas can be useful for water quality management. However, currently the nature of the connectivity between upland and lowland streams is not fully understood. Thus, including upland streams and subsurface nutrient source areas as potential source areas for lowland streams in predictive models cannot be fully comprehended.

CONCLUSION

Hydrological field studies across New Zealand concurred with results and observations from elsewhere – runoff generation and transport processes are limited to small portions of the catchment. Adaptation of this concept, the variable-source area hydrology, for water quality management in New Zealand has not been complete like elsewhere in the world. The definition of critical source areas in New Zealand has been limited to saturated areas in a catchment, resulting from a rising or a perched water table, and the generation of saturation-excess runoff from these saturated areas. The majority of early desktop hydrological studies concluded that SE runoff is the major runoff generation mechanism, and later this was adopted by many water quality studies. However, many field and plot scale studies from grazed-pastoral catchments highlighted changes to the quality of soil and water resources following grazing events, which could influence the hydrologic and hydraulic (e.g., connectivity between land and water) behaviour of grazed-pastoral catchments.

In the New Zealand context, the definition of critical source areas needs to be expanded to include two more important flow generation processes and pathways during storm events: 1. inclusion of subsurface flow contributions; and 2. inclusion of infiltration-excess contributions. The former has been recognised worldwide from various field studies and hence, we are not elaborating on it here. However, the latter is very important to grazed-pastoral catchments in New Zealand. Infiltration excess runoff from compacted soil surfaces and animal induced features such as wallows, camp sites, and cattle tracks, and from hydrophobic soils need to be included into the CSA definition in New Zealand. While some results are available on IE runoff from animal induced features (e.g., McDowell and Srinivasan, 2009), more research is needed in describing and defining IE runoff from water repellent soils. Questions remain unanswered on the extent (spatial and temporal) of these surface soil properties, on the role of seasons, and importantly, on their contributions to catchment scale water quantity and quality. It is unclear how these areas could be 'managed' for water quality, since our knowledge on the reasons for their occurrence is incomplete.

While hydrology has remained the main driver for water quality management, we would like to distinguish the focus of those two disciplines – hydrology and water quality sciences. It is important to recognise that the dominant pathways and mechanisms of flow transfer from land to water are not the same as those of water quality constituents. McDowell and Srinivasan (2009) and Srinivasan and McDowell (2009) indicated that some of the pathways (e.g., IE excess from wintering pads) may carry disproportionately, and consistently, more nutrients than flow contributions. This means that in some cases the nutrient concentrations in surface runoff or SSF rather than the flow volumes might dominate the nutrient loads reaching surface water resources. In this context, it is important to know the potential for nutrient losses of soils. This will be determined, *inter alia*, by soil type, land use, and management practices. For example, McDowell and Condron(2004) proposed simple equations to estimate the potential of DRP concentrations in surface runoff and SSF from recently grazed pastoral soils employing easily measurable parameters, Olsen-P, P-retention and P-sorption index. Validation of such regression equations for different soil types is needed.

From a management perspective, the adaptation of VSA hydrology for water quality management in New Zealand catchments has to occur at farm scale. With the ongoing intensification of dairying in New Zealand, it is necessary to develop simple-to-use tools and easy-to-implement practices that can be transferred to the agricultural community (e.g., McKergow and Tanner, 2011). Such tools are already available in the nutrient management sector (e.g., OVERSEER, http://www.overseer.org.nz/). OVERSEER is widely used among the farming community to develop nutrient management plans, and is well supported by the science community and funded by the industry. The development of VSA tools needs to be inclusive of all sectors of farming, from farmers, to fertiliser industry to water resource managers and regulators to be successful and effective. It has to be communicated to the farming community that water quality management practices based on VSA hydrology are not going to replace existing best management practices such as stream bank fencing, but that these would be complementing measures. Existing BMPs might be extended by targeted management practices based on VSA hydrology. Field research is needed to demonstrate changes in water quality parameters following the implementation of such targeted management strategies. Furthermore, the cost-effectiveness of the new measures has to be evaluated in a range of catchments.

REFERENCES

Anderson, M.G., Burt, T.P. 1982. The contribution of throughflow to stormflow: An evaluation of a chemical mixing model. *Earth Surface Processes and Landforms*, 7:565-574.

Arnold, J.G., Srinivasan, R., Muttiah, R.S., Williams, J.R. 1998: Large area hydrologic modelling and assessment. Part I. Model development. *Journal of American Water Resources Association*, 34:73-89.

Bachmann, J., Deurer, M., Arye, G. 2007. Modeling water movement in heterogeneous water-repellent soil: 1. Development of a contact angle-dependent water-retention model. *Vadose Zone Journal*, 6:436-445.

Bargh, B.J. 1978. Output of water, suspended sediment, and phosphorus and nitrogen forms from a small agricultural catchment. *New Zealand Journal of Agricultural Research*, 21:29-38.

Beegle, D.B., Sharpley, A.N.,Gburek, W.J.,Weld, J. 2000. Integrating the phosphorus index and existing databases for use in nutrient management planning. Department of Agronomy, The Pennsylvania State University, University Park, Pennsylvania, USA.

Betson, R.P. 1964. What is watershed runoff? *Journal of Geophysical Research*, 69:1541-1552.

Betson, R.P.,Marius, J.B. 1969. Source areas of storm runoff. *Water ResourcesResearch*, 5:574-582.

Betteridge, K., Costall, D., Balladur, S., Upsdell, M., Umemura, K. 2008. Urine distribution and grazing behaviour of female sheep and cattle grazing a steep New Zealand hill pasture. *Animal Production Science*, 50:624-629.

Betteridge, K., Hoogendoorna, C., Costall, D.,Carter, M., Griffiths, W. 2010. Sensors for detecting and logging spatial distribution of urine patches of grazing female sheep and cattle. *Computers and Electronics in Agriculture*, 73:66-73.

Beven, K., Kirkby, M.J. 1979. A physically based, variable contributing area model of basin hydrology. *Hydrological Science Bulletin*, 24:43-69.

Bonnell, M. 1993. Progress in the understanding of runoff generation dynamics in forests. *Journal of Hydrology*, 150:217-275.

Bonell, M., Pearce, A.J., Stewart, M.K. 1990. The identification of runoff-production mechanisms using environmental isotopes in a tussock grassland catchment, Eastern Otago, New Zealand. *Hydrological Processes*, 4:15-34.

Bowden, W.B., Fahey, B.D., Ekanayake, J., Murray, D.L. 2001. Hillslope and wetland hydrodynamics in a tussock grassland, South Island, New Zealand. *Hydrological Processes*, 15:1707-1730.

Buczko, U., Bens, O., Huttl, R.F. 2005. Variability of soil water repellency in sandy forest soils with different stand structure under Scots pine (Pinussylvestris) and beech (Fagussylvatica). *Geoderma*, 126:317-336.

Burch, G.J., Moore, I.D., Burns, J. 1989. Soil hydrophobic effects on infiltration and catchment runoff. *Hydrological Processes*, 3:211-222.

Burgess, C.P., Chapman, R., Singleton, P.L., Thom, E.R. 2000. Shallow mechanical loosening of a soil under dairy cattle grazing: Effects on soil and pasture. *New Zealand Journal of Agricultural Research*, 43:279-290.

Buttle, J.M., Turcotte, D.S. 1999. Runoff processes on a forested slope on the Canadian shield. *Nordic Hydrology*, 30:1-20.

Carter, J.A., Deurer, M., Müller, K., van den Dijssel, C., Mason, K. 2010. Soil water repellency — economic and environmental consequences of an emerging issue of soil degradation. Second progress report to AGMARDT. The New Zealand Plant & Food Research Institute Ltd., Auckland, New Zealand. 46p.

Caruso, B.S. 2000. Integrated assessment of phosphorus in the Lake Hayes catchment, South Island, New Zealand. *Journal of Hydrology*, 229:168-189.

Caruso, B.S. 2001. Risk-based targeting of diffuse contaminant sources at variable spatial scales in a New Zealand high country catchment. *Journal of Environmental Management* 63:249-268.

Climo, W.J., Richardson, M.A. 1984. Factors affecting the susceptibility of 3 soils in the Manawatu to stock treading. *New Zealand Journal of Agricultural Research*, 27:247-253

Collins, R., Elliott, S., Adams, R. 2005. Overland flow delivery of faecal bacteria to a headwater pastoral stream. *Journal of Applied Microbiology*, 99:126-132.

Collins, R., McLeod, M., Hedley, M., Donnison, A., Close, M., Hanly, J., Horne, D., Ross, C., Davies-Colley, R., Bagshaw, C., Matthews, L. 2007. Best management practices to mitigate faecal contamination by livestock of New Zealand waters. *New Zealand Journal of Agricultural Research*, 50:267-278.

Cooke, J.G. 1988. Sources and sinks of nutrients in a New Zealand hill pasture catchment II. Phosphorus. *Hydrological Processes*, 2:123-133.

Cooke, J.G., Petch, R.A. 2007. The uncertain search for the diffuse silver bullet: science, policy and prospects. *Water Science & Technology*, 56:199-205.

Cooke, J.G., Dons, T. 1988. Source and sinks of nutrients in a New Zealand hill pasture catchment: I. Stormflow generation. *Hydrological Processes*, 2:109-122.

Cooke, J.G., Cooper, A.B. 1988. Sources and sinks of nutrients in a New Zealand hill pasture catchment III. Nitrogen. *Hydrological Processes*, 2:135-149.

Cooper, A.B., Smith, C.M., Bottcher, A.B. 1992. Predicting runoff of water, sediment, and nutrients from a New Zealand grazed pasture using CREAMS. *Transactions of the ASAE*, 35:105-112.

Davies-Colley, R.J., Nagels, J.W. 2002. Effects of dairying on water quality of lowland streams in Westland and Waikato. *In:* Proceedings of the New Zealand Grassland Association, 64:107-114.

Davies-Colley, R., Nagels, J., Lydiard, E. 2008. Stormflow-dominated loads of faecal pollution from an intensively dairy-farmed catchment. *Water Science & Technology*, 57:1519-1523.

deAlwis, D.A., Easton, Z.M., Dahlke, H.E., Philpot, W.D., Steenhuis, T.S. 2007. Unsupervised classification of saturated areas using a time series of remotely sensed images. *Hydrology and Earth System Sciences*, 11:1609-1620.

DeBano, L.F. 1971. The effect of hydrophobic substances on water movement in soil during infiltration. *Soil Science Society of America Journal*, 35:340-343.

Dekker, L.W., Ritsema, C.J. 1994. How water moves in a water repellent sandy soil 1. Potential and actual water repellency. *Water Resources Research*, 30:2507-2517.

Deurer, M., Bachmann, J. 2007. Modelling water movement in heterogeneous water-repellent soil: 2. A conceptual numerical simulation. *Vadose Zone Journal*, 6:446-457.

Deurer, M., Sivakumaran, S., Müller, K., Clothier, B., 2007. The 'Dry Patch Syndrome' in Hawke's Bay pastures – Is it caused by soil hydrophobicity? HortResearch Client Report No. 21678, Palmerston North, New Zealand. 26p.

Doerr, S.H., Shakesby, R.A., Walsh, R.P.D. 2000. Soil water repellency: its causes, characteristics and hydro-geomorphological significance. *Earth-Science Reviews*, 51:33-65.

Doerr, S.H., Shakesby, R.A., Dekker, L.W., Ritsema, C.J. 2006. Occurrence, prediction and hydrological effects of water repellency amongst major soil and land-use types in a humid temperate climate. *European Journal of Soil Science*, 57:741-754.

Doerr, S.H., Ritsema, C.J., Dekker, L.W., Scott, D.F., Carter, D. 2007. Water repellence of soils: new insights and emerging research needs. *Hydrological Processes*, 21:2223-2228.

Donnison, A., Ross, C., Thorrold, B. 2004. Impact of land use on the faecal microbial quality of hill-country streams. *New Zealand Journal of Marine and Freshwater Research*, 38:845-855.

Dormaar, J. F., Smoliak, A., Willms, W. 1989. Vegetation and soil responses to short-duration grazing on fescue grasslands. *Journal of Range Management*, 42:252-256.

Drewry, J.J., Paton, R.J. 2000. Effects of cattle treading and natural amelioration on soil physical properties and pasture under dairy farming in Southland, New Zealand. *New Zealand Journal of Agricultural Research*, 43:377-386.

Dunne, T. 1978. Field studies of hillslope flow processes. *In:* Ed. Kirkby, M.J. Hillslope Hydrology, Wiley, London. pp. 227-294.

Dunne, T., Black, R. 1970. Partial area contributing to storm run off in a small New England watershed. *Water Resources Research*, 6:1296-1311.

Dunne, T., Moore, T.R., Taylor, C.H. 1975. Recognition and prediction of runoff-producing zones in humid regions. *Hydrological Sciences Bulletin*, 20:305-327.

Edmond, D.B. 1974. Effects of sheep treading on measured pasture yield and physical conditions of four soils. *New Zealand Journal of Experimental Research*, 2:39-43.

Edwards, A.C., Withers, P.J.A. 2008. Transport and delivery of suspended solids, nitrogen and phosphorus from various sources to freshwaters in the UK. *Journal of Hydrology*, 350:144-153.

Elliott, A.H., Ibbitt, R.P. 2000a. Prediction of surface runoff at Pukemanga using TOPMODEL. NIWA Internal Report No. 86, Hamilton, New Zealand. 36p.

Elliott, E., Ibbitt, R.P. 2000b. Application of TOPMODEL to prediction of surface runoff at Waipawa. NIWA Client Report 00AGR206, Hamilton, New Zealand. 48p.

Engman, E.T. 1974. Partial area hydrology and its application to water resources. *Water Resources Bulletin*, 10:512-521

Environment Southland. 2000. State of the environment report—water. Environment Southland, Invercargill, New Zealand. 48 p.

Finlayson, J.D., Betteridge, K., MacKay, A., Thorrold, B., Singleton, P., Costall, D.A. 2002. A simulation model of the effects of cattle treading on pasture production on North Island, New Zealand, hill land, *New Zealand Journal of Agricultural Research*, 45:255-272.

Frasier, G.W., Trlica, M.J., Leininger, W.C., Pearce, R.A., Fernald, A. 1998. Runoff from simulated rainfall in 2 montain riparian communities. *Journal of Range Management*, 51:315-322.

Freeze, A.R. 1972. Role of subsurface flow in generating surface runoff 1. Base flow contributions to channel flow. *Water Resources Research*, 8:609-623.

Freeze, A.R. 1974. Streamflow generation. *Reviews of Geophysics and Space Physics*, 12:627-647

Gao, B.C. 1996. A normalized difference water index for remote sensing of vegetation liquid water from space. *Remote Sensing of Environment*, 58:257-266.

Gbruek, W.J. 1990. Initial contributing area of a small watershed. *Journal of Hydrology*, 118:387-403.

Gburek, W.J., Srinivasan, M.S., Needelman, B.A. 2001. Phosphorus transport in upland watersheds – Connectivity of field to stream. *In:* Changing Wetlands - New Developments in Wetland Sciences, University of Sheffield, Sheffield, UK.

Gburek, W.J., Sharpley, A.N., Heatwaite, L., Folmar, G. 2000. Phosphorus management at the watershed scale: A modification of the Phosphorus Index. *Journal of Environmental Quality*, 29:130-144.

Gburek, W.J., Drungil, C.C., Srinivasan, M.S., Needelman, B.A., Woodward, D.E. 2002. Variable-source-area controls on phosphorus transport: Bridging the gap between research and design. *Journal of Soil and Water Conservation*, 57:534-543.

Gerke, H.H., Hangen, E., Schaaf, W., Hüttl, R.F. 2001. Spatial variability of potential water repellency in a lignitic mine soil afforested with Pinusnigra. *Geoderma*, 102:255-274.

Gillham, R.W. 1984. The capillary fringe and its effect on water-table response. *Journal of Hydrology*, 67:307-324.

Gillingham, A.G., Gray, M.H. 2006. Measurement and modelling of runoff and phosphate movement from seasonally dry hill-country pastures. *New Zealand Journal of Agricultural Research*, 49:233-245.

Gomi, T., Sidle, R.C., Ueno, M., Miyata, S., Kosugi, K. 2008. Characteristics of overland flow generation on steep forested hillslopes of central Japan. *Journal of Hydrology*, 361:275-290.

Gradwell, M.W. 1968. Compaction of pasture top soils under winter grazing. *In:* Proceedings of the Ninth International Congress of Soil Science, Volume 4. Adelaide, South Australia. pp. 429-435.

Guntenspergen, G.R., Peterson, S.A., Leibowitz, S.G., Cowardin, L.M. 2002. Indicators of wetland condition for the Prairie Pothole Region of the United States. *Environmental Monitoring and Assessment*, 78:229-252.

Hallett, P.D. 2008. A brief overview of the causes, impacts and amelioration of soil water repellency - a review. *Soil & Water Research (Special Issue 1),* S21-S29.

Hayward, J.A. 1976. Hydrology of mountain catchments and its implications for management. *In:* Proceedings of soil and plant water symposium. Palmerston North, New Zealand. pp. 126-136.

Heathwaite, A.L., Dils, R.M. 2000. Characterisingphosporus loss in surface and subsurface hydrological pathways. *Science of Total Environment,* 251/252:523-538.

Hedley, M., Lonas, G., Horne, D.J., Hanly, J.A., Collins, R., Donnison, A., Ross, C. 2005. Pathogen transmission routes: farm animals to water bodies. Objectives 9 and 10: Contamination of surface runoff under dairying and mitigation strategies for pathogen contamination of runoff and artificial drainage water. Massey University Report to MAF, July 2005. 19 p.

Hewlett J.D. 1961. Soil moisture as a source of baseflow from steep mountain watersheds. U.S. Forest Research Paper - Southeastern Forest Experimentation Station. Paper # 132. 11 p.

Hewlett, J.D., Hibbert, A.R. 1967. Factors affecting the response of small watersheds to precipitation in humid areas. *In:*Proceedings of 1st International Symposium on Forest Hydrology, Pergamon, Oxford, UK. pp. 275-290.

Hewlett, J.D. Nutter, W. 1970. The varying source area of streamflow from upland basins. *In:*Proceedings of the Symposium on Interdisciplinary Aspects of Watershed Management, American Society of Civil Engineers. pp. 65-83.

Hively, W.D., Bryant, R.B., Fahey, T.J. 2005. Phosphorus concentrations in overland flow from diverse locations on a New York dairy farm. *Journal of Environmental Quality*, 34:1224-1233.

Horne, D.J., McIntosh, J.C. 2000. Hydrophobic compounds in sands in New Zealand - extraction, characterisation and proposed mechanisms for repellency expression. *Journal of Hydrology*, 231-232:35-46.

Houlbrooke, D.J., Horne, D.J., Hedley, M.J., Hanly, J.A. 2004a. Irrigator performance: assessment, modification and implications for nutrient loss in drainage water. *New Zealand Journal of Agricultural Research*, 47:587-596.

Houlbrooke D.J., Morton, J.D., Paton, R.J., Littlejohn, R.P. 2006. The impact of land-use intensification on soil physical quality and plant yield response in the North Otago rolling downlands. *In:* Proceedings of the New Zealand Grassland Association, 68:165-172.

Houlbrooke, D.J., Drewry, J.J., Monaghan, R.M., Paton, R.J., Smith, L.C., Littlejohn, R. P. 2009. Grazing strategies to protect soil physical properties and maximise pasture yield on a Southland dairy farm. *New Zealand Journal of Agricultural Research*, 52:323-336.

Houlbrooke, D.J., Horne, D.J., Hedley, M.J., Hanly, J.A., Snow, V.O. 2004b. A review of the literature on the land treatment of farm dairy effluent in New Zealand and its impact on water quality. *New Zealand Journal of Agricultural Research*, 47:499-511.

Houlbrooke, D.J., Horne, D.J., Hedley, M.J., Hanly, J.A., Scotter, D.R., Snow, V.O. 2004c. Minimising surface water pollution resulting from farm-dairy effluent application to mole-pipe drained soils. I. An evaluation of the deferred irrigation system for sustainable land treatment in the Manawatu. *New Zealand Journal of Agricultural Research*, 47:405-415.

Hutchinson, J.M.S. 2003. Estimating near-surface soil moisture using active microwave satellite imagery and optical sensor inputs. *Transactions of the ASAE*, 46:225-236.

Karunarathna, A.K., Kawamoto, K., Moldrup, P., Wollesen de Jonge, L., Komatsu, T. 2010a. A simple beta-function for soil-water repellency as a function of water and organic carbon contents. *Soil Science*, 175:461-468.

Karunarathna, A.K., Moldrup, P., Kawamoto, K., de Jonge, L.W., Komatsu, T. 2010b. Two-region model for soil water repellency as a function of matric potential and water content. *Vadose Zone Journal*, 9:719-730.

Knisel, W.G. 1980. CREAMS: A field-scale model for chemicals, runoff and erosion from agricultural management systems. United StatesDepatment. of Agriculture, Science and Education Administration, Conservation Report No. 26. p. 643.

Laaha, G., Blochel, G. 2006. A comparison of low flow regionalisation methods—catchment grouping. *Journal of Hydrology*, 323:193-214.

Larned, S.T., Scarsbrook, M.R., Snelder, T.H., Norton, N.J., Biggs, B.J.F. 2004. Water quality in low-elevation streams and rivers of New Zealand: recent state and trends in contrasting low-cover classes. *New Zealand Journal of Marine and Freshwater Research*, 38:347-366.

Leighton-Boyce, G., Doerr, S.H., Shakesby, R.A., Walsh, R.P.D. 2007. Quantifying the impact of soil water repellency on overland flow generation and erosion: a new approach using rainfall simulation and wetting agent on in situ soil. *Hydrological Processes,* 21:2337-2345.

Lemmnitz, C., Kuhnert, M., Bens, O., Güntner, A., Merz, B., Hüttl, R.F. 2008. Spatial and temporal variations of actual soil water repellency and their influence on surface runoff. *Hydrological Processes*, 22:1976-1984.

Leonard, R.A., Knisel, W.G., Davis, F.M. 1990. The GLEAMS model - A tool for evaluating agrichemical ground-water loading as affected by chemistry, soils, climate and management. *In:* American Water Resources Association Conference. *Transferring Models to Users.* pp. 187-197.

Lyon, S.W., Walter, M.T., Gérard-Marchant, P., Steenhuis, T.S. 2004. Using a topographic index to distribute variable source area runoff predicted with the SCS curve-number equation. *Hydrological Processes*, 18:2757-2771.

Martin A, Williams H. 2006. State of the Canterbury region water resource 2006. Report No. U06/68. Environment Canterbury, Christchurch, New Zealand. 62 p.

McColl, R.H.S., Gibson, A.R. 1979. Downslope movement of nutrients in hill pasture, Taita, New Zealand. III. Amounts involved and management implications. *New Zealand Journal of Agricultural Research*, 22:279-286.

McColl, R.H.S., McQueen, D.J., Gibson, A.R., Heine, J.C. 1985. Source areas of storm runoff in a pasture catchment. *Journal of Hydrology(New Zealand),* 24:1-19.

McColl, R.H.S., White, E., Gibson, A.R. 1977. Phosphorus and nitrate run-off in hill pasture and forest catchments, Taita, New Zealand. *New Zealand Journal of Marine and Freshwater Research*, 11:729-744.

McDowell, R.W. 2009. The use of safe wallows to improve water quality in deer farmed catchments. *New Zealand Journal of Agricultural Research,* 52:81-90.

McDowell, R.W. 2008a. Water quality of a stream recently fenced-off from deer. *New Zealand Journal of Agricultural Research*, 51:291-298.

McDowell, R.W. 2008b. Phosphorus in humped and hollowed soils of the Inchbonnie catchment, West Coast, New Zealand: II. Accounting for losses by different pathways. *New Zealand Journal of Agricultural Research*, 51:307-316.

McDowell, R.W., Wilcock, R.J. 2007. Sources of sediment and phosphorus in streamflow of a highly productive dairy farmed catchment. *Journal of Environmental Quality*, 36:540-548.

McDowell, R.W., Wilcock, R.J., 2004. Particulate phosporus transport within streamflow of an agricultural catchment. *Journal of Environmental Quality*, 33:2111-2121.

McDowell, R.W, Condron, L.M. 2004. Estimating phosphorus loss from New Zealand grassland soils. *New Zealand Journal of Agricultural Research*, 47:137-145.

McDowell, R.W., Drewry, J.J., Paton, J.R. 2004. Effects of deer grazing and fence line pacing on water and soil quality. *Soil Use & Management*, 20:302-307.

McDowell, R.W., Houlbrooke, D.J., Muirhead, R.W., Müller, K., Shepherd, M., Cuttle, S.P. 2008. Grazed pastures and surface water quality. Nova Sciences Publishers, Inc., New York, USA.

McDowell, R.W., Drewry, J.J., Paton, R.J., Carey, P.L., Monaghan, R.M., Condron, L.M. 2003. Influence of soil treading on sediment and phosphorus transport in overland flow. *Australian Journal of Soil Research*, 41:949-961.

McDowell, R.W., Nash, D., George, A., Wang, Q.J., Duncan, R. 2009. Approaches for quantifying and managing diffuse phosphorus exports at the farm/small catchment scale.*Journal of Environmental Quality*, 38:1968-1980.

McDowell, R.W., Srinivasan, M.S. 2009. Identifying critical source areas for water quality: 2. Validating the approach for phosphorus and sediment losses in grazed headwater catchments. *Journal of Hydrology*, 379:68-80. doi:10.1016/j.jhydrol.2009.09.045

McKergow, L., Tanner, C. 2011. Reading the landscape: selecting suitable diffuse poulltion tools. *In:* Eds. L.D. Currie and C.L. Christensen. Adding to the knowledge base for the nutrient manager. http://flrc.massey.ac.nz/publications/html. Occasional report no. 24.Fertilizer and Lime Research Centre, Massey University, Palmerston North, New Zealand.11 p.

McKerchar, A.I., Pearson, C.P. 1989. Flood frequency in New Zealand. Publication No. 20. Hydrology Centre. Department of Science and Industrial Research, Christchurch, New Zealand.87p.

McLeod, M., Close, M., Collins, R. 2005. Relative risk indices for microbial transport from land to water bodies. Landcare Research Contract Report: LCR0405/165, Hamilton, New Zealand. 40p.

Meriano, M., Srinivasan, M.S., Zarnetske, J., Thomas, D. 2011. A reach-scale perspective of surface-groundwater exchanges in a lowland stream, Canterbury Plains, South Island, New Zealand *(in preparation)*.

Ministry for the Environment (MfE). 2010. Environmental Snapshot. Land: Land use. January 2010. http://www.mfe.govt.nz/environmental-reporting/report-cards/land-use-environmental-snapshot/2010/land-use.pdf. Last accessed in January 2011.

Miyata, S., Kosugi, K., Gomi, T., Mizuyama, T. 2009. Effects of forest floor coverage on overland flow and soil erosion on hillslopes in Japanese cypress plantation forests. *Water Resources Research*, 45:W06402.

Monaghan, R.M., Smith, L.C. 2004. Minimising surface water pollution resulting from farm-dairy effluent application to mole-pipe drained soils. II. The contribution of preferential flow of effluent to whole-farm pollutant losses in subsurface drainage from a West Otago dairy farm. *New Zealand Journal of Agricultural Research*, 47:417-428.

Mosley, M.P. 1979. Streamflow generation in a forested watershed, New Zealand. *Water Resources Research*, 15:795-806.

Mosley, M.P. 1981. Delimitation of New Zealand Hydrologic regions. *Journal of Hydrology*, 49:173-192.

Muirhead, R.W. 2009. Soil and faecal material reservoirs of Escherichia coli in a grazed pasture. *New Zealand Journal of Agricultural Research*, 52:1-8.

Müller, K., Srinivasan, M.S., Trolove, M., McDowell, R. 2010a. Identifying and linking source areas of flow and P transport in dairy-grazed headwater catchments, North Island, New Zealand. *Hydrological Processes,*doi: 10.1002/hyp.7809

Müller, K., Deurer, M., Slay, M., Aslam, T., Carter, J.A., Clothier, B.E. 2010b. Environmental and economic consequences of soil water repellency under pasture. *In:*Proceedings of the New Zealand Grassland Association, 72:151-154.

Nash, D.M., Hannah, M., Halliwell, D.J., Murdoch, C. 2000. Factors affecting phosphorus export from a pasture-based grazing system. *Journal of Environmental Quality*, 29:1160-1166.

Naz, B.S., Bowling, L.C. 2009. Automated identification of tile lines from remotely sensed data. *Transactions of the ASAE*, 51:1937-1950.

Needelman, B.A., Gburek, W.J., Petersen, G.W., Sharpley, A.N., Kleinman, P.J.A. 2004. Surface runoff along two agricultural hillslopes with contrasting soils. *Soil Science Society of America Journal*, 68:914-923.

Nguyen, M.L., Sheath, G.W., Smith, C.M., Cooper, A.B. 1998. Impact of cattle treading on hill land: 2. Soil physical properties and contaminant runoff. *New Zealand Journal of Agricultural Research*, 41:279-290.

Ocampo, C.J., Sivapalan, M., Oldham, C. 2006. Hydrological connectivity of upland-riparian zones in agricultural catchments: Implications for runoff generation and nitrate transport. *Journal of Hydrology*, 331:643-658.

Parliamentary Commissioner for the Environment. 2004. Growing for good: intensive farming, sustainability and New Zealand's environment. Parliamentary Commissioner for the Environment, Wellington, New Zealand. 236 p.

Pearce, A.J. 1990. Streamflow generation processes: An austral view. *Water Resources Research*, 26:3037-3047.

Pearce, A.J., McKerchar, A.I. 1979. Upstream generation of storm runoff. *In:*Eds, D.L. Murray and P. Ackroyd. Physical hydrology: A New Zealand Experience. New Zealand Hydrological Society Publication. pp. 165-192.

Pearce, A.J., Stewart, M.J., Sklash, M.G. 1986. Storm runoff generation in humid headwater catchments 1. Where does the water come from? *Water Resources Research*, 22:1263-1272.

Pearce, A.J., Rowe, L.K., O'Loughlin, C.L. 1984. Hydrology of mid-altitude tussock grasslands, Upper Waipori catchment: II-Water balance, flow duration and storm runoff. *Journal of Hydrology(New Zealand)* 23:60-72.

Petch, R.A. 1988. Soil saturation patterns in steep, convergent hillslope under forested and pasture vegetation. *Hydrological Processes*, 2:93-103.

Petropoulos, G., Carlson, T.N., Wooster, M. J., Islam, S. 2009. A review of T-s/VI remote sensing based methods for the retrieval of land surface energy fluxes and soil surface moisture. *Progress in Physical Geography*, 33:224-250.

Pierson, F.B., Moffet, C.A., Williams, C.J., Hardegree, S.P., Clark, P.E. 2009. Prescribed-fire effects on rill and interrill runoff and erosion in a mountainous sagebrush landscape. *Earth Surface Processes and Landforms*, 34:193-203.

Pilgrim, D.H., Huff, D.D., Steele, T.D. 1978. A field evaluation of subsurface and surface runoff, II Runoff Processes. *Journal of Hydrology*, 38:319-341.

Pionke, H.B., Gburek, W.J., Sharpley, A.N., Schnabel, R.R. 1996. Flow and nutrient export patterns for an agricultural hill-land watershed. *Water Resources Research*, 32:1795-1804.

Pollard, J.C., Drewry, J.J. 2002. Calving environments for farmed red deer: a review of current knowledge and pilot study on soil quality. *In:* Proceedings of a Deer Course for

Veterinarians, Deer Branch of New Zealand and Veterinary Association, Nelson, New Zealand. pp. 97-104.

Proffitt, A.P.B., Bendotti, S., Howell, M.R., Eastham, J. 1993. The effects of sheep trampling and grazing on soil physical properties and pasture growth for a red-brown earth. *Australian Journal of Agricultural Research*, 44:317-331.

Qiu, Z., Walter, M.T., Hall, C. 2007. Managing variable source pollution in agricultural watersheds. *Journal of Soil & Water Conservation*, 62:115-122.

Quinn, J.M., Stroud, M.J. 2002. Water quality and sediment and nutrient export from New Zealand hill-land catchments of contrasting land use. *New Zealand Journal of Marine and Freshwater Research*, 36:409-429.

Quinn, J.M., Cooper, A.B., Davies-Colley, R.J., Rutherford, J.C., Bruce-Williamson, R. 1994. Land use effects on New Zealand hill country streams and implications for riparian management. *In:* Proceedings of International Workshop on the Ecology and Management of AquaticterrestrialEcotones. February 14, 1994. University of Washington, Seattle, USA. pp. 19-27.

Regalado, C.M., Ritter, A. 2005. Characterizing water dependent soil repellency with minimal parameter requirement.*Soil Science Society of America Journal*, 69:1955-1966.

Rijkse, W.C., Bell, J.L. 1975. Soils of Purukohukohu IHD experimental basin, Rotorua County, North Island, New Zealand. New Zealand Soil Survey Report 18. 12 p.

Rodda, H., Thorrold, B.S., Nguyen, L. 1998. Modelling soluble phosphorus losses in surface runoff from a small hillslope basin in New Zealand. *In:* Eds. Currie, L.D., Loganathan, P. Long-term Nutrient Needs for New Zealand's Primary Industries: Global Supply, Production Requirements and Environmental Constraints, Fertiliser and Lime Research Centre, Massey University, Palmerston North, New Zealand. pp. 215-229.

Rodda, H.J.E., Shankar, U., Demuth, S. 1999. The application of geographical information systems to water quality modeling in New Zealand. *In:* Diekkrüger, B., Kirkby, M.J., Schröder, U. (Eds.), Regionalization in hydrology, IAHS. pp. 243-251.

Rodda, H.J.E., Stroud, M.J., Shankar, U., Thorrold, B.S. 2001. A GIS based approach to modeling the effects of land-use change on soil erosion in New Zealand. *Soil Use & Management*, 17:30-40.

Scott, D.F., Van Wyk, D.B. 1990. The effects of wildfire on soil wettability and hydrological behaviour of an afforested catchment.*Journal of Hydrology*, 121:239-256.

Sharpley, A.N., Daniel, T.C., Wright, B., Kleinman, P.J.A., Sobecki, T., Parry, R., Joern, B. 1999. National research project to identify sources of agricultural phosporus loss. *Better Crops*, 83:12-14.

Sharpley, A.N., Syers, J.K. 1979. Phosphorus inputs into a stream draining an agricultural watershed. *Water, Air and Soil Pollution*, 11:417-428.

Sharpley, A.N., Weld, J.L., Kleinman, P.J.A., Gburek, W.J. 2004. The Phosphorus Index. *In:* Managing manure nutrients at concentrated animal feeding operations. U.S. Environmental Protection Agency USEPA 821-B-04-006. 13 p. http://www.epa.gov/npdes/pubs/cafo_manure_guidance_appendixh.pdf. Last accessed March 2011.

Singleton, P.L., Boyes, M., Addison, B. 2000. Effect of treading by dairy cattle on topsoil physical conditions for six contrasting soil types in Waikato and Northland, New Zealand, with implications for monitoring. *New Zealand Journal of Agricultural Research,*43:559-567.

Sklash, M.G., Stewart, M.K., Pearce, A.J. 1986. Storm runoff generation in humid headwater catchments: A case study of hillslope and lower-order stream response. *Water Resources Research*, 22:1273-1282.

Smith, C.M. 1989. Riparian pasture retirement effects on sediment, phosphorus, and nitrogen in channelised surface run-off from pastures. *New Zealand Journal of Marine and Freshwater Research*, 23:139-146.

Smith, C.M. 1987. Sediment, phosphorus, and nitrogen in channelised surface run-off from a New Zealand pastoral catchment. *New Zealand Journal of Marine and Freshwater Research*, 21:627-639.

Smith, C.M., Wilcock, R.J., Vant, W.N., Smith, D.G., Cooper, A.B. 1993: Towards sustainable agriculture: freshwater quality in New Zealand and the influence of agriculture. MAF Policy technical paper 93/10. Wellington, New Zealand Ministry of Agriculture and Fisheries. 208 p.

Snelder, T., Biggs, B. 2002. Multi scale river environment classification for water resources management. *Journal of American Water Resources Association*, 38:1225-1239.

Soil Conservation Service. 1972. Hydrology. *In:* National Engineering Handbook, Section 4. USDA Soil Conservation Service, Washington DC, USA.

Soil and Water Laboratory. 2002. The Soil Moisture Distribution and Routing (SMDR) model: Documentation. Biological and Environmental Engineering Department, Cornell University, Ithaca, New York, USA.

Srinivasan, M.S. 2000. Dynamics of runoff generation process: A hillslope scale field study. Doctoral thesis. The Pennsylvania State University, University Park, Pennsylvania, USA.220p.

Srinivasan, M.S., McDowell, R. 2009. Identifying critical source areas for water quality: 1. Mapping and validating transport areas in three headwater catchments in Otago, New Zealand. *Journal of Hydrology*, 379:54-67. doi:10.1016/j.jhydrol.2009.09.044

Srinivasan, M.S., McDowell, R.W. 2007. Hydrological approaches to the delineation of critical-source areas of runoff. *New Zealand Journal of Agricultural Research*, 50:249-265.

Srinivasan, M.S., Gburek, W.J., Hamlett, J.M. 2002. Dynamics of stormwater generation - A hillslope-scale field study in east-central Pennsylvania, USA. *Hydrological Processes*, 16:649-665.

Srinivasan, M.S., W.J. Gburek, P. Gerard-Marchant, T.S. Steenhuis. 2003. Identification and modelling of critical source areas of runoff generation and phosphorus transport at a watershed scale. *In:* Spring Speciality Conference on Hydrology and Water Quality, American Water Resources Association, Kansas City, Missouri, USA. 4p.

Srinivasan, M.S., Gerard-Marchant, P., Veith, T.L., Gburek, W.J., Steenhuis, T.S. 2005. Watershed scale modeling of critical source areas of runoff generation and phosphorus transport. *Journal of American Water Resources Association*, 41:361-375.

Stenger, R., Barkle, G.F., Burgess, C., Wall, A., Clague, J. 2008. Low nitrate contamination of shallow groundwater in spite of intensive dairy farming: the effect of reducing conditions in the vadose zone - aquifer continuum. *Journal of Hydrology(New Zealand)*, 47:1-24.

Stenger, R., Barkle, G.F., Bidwell, V., Burgess, C., Wall, A., Haas, M., Mertens, J. 2005. Unravelling the nitrogen flowpaths in a New Zealand catchment. *In:* Moglen, G.E. (Ed.)

2005 Watershed Management Conference, 19-22 July 2005, American Society of Civil Engineers. 11p.

Taylor, C.H., Pearce, A.J. 1982. Storm runoff processes and subcatchment characteristics in a New Zealand hill country catchment. *Earth Surface Processes and Landforms*, 7:439-447.

Tian, Y.Q., Gong, P., Radke, J.D., Scarborough, J. 2002. Spatial and temporal modeling of microbial contaminants on grazing farmlands. *Journal of Environmental Quality*, 31:860-869.

Valeron, B., Meixner, T. 2010. Overland flow generation in chaparral ecosystems: Temporal and spatial variability. *Hydrological Processes*, 24:65-75.

Van den Griend, A.A, Engman, E.T. 1985. Partial area hydrology and remote sensing. *Journal of Hydrology*, 81:211-251.

Vant, W.N., Smith, P. 2004. Trends in river water quality in the Waikato Region, 1987-2002. Environment Waikato Technical Report 2004/02. Environment Waikato, Hamilton, New Zealand. 42p.

Wallis, M., Horne, D.J., McAuliffe, K. 1990. A study of water repellency and its amelioration in a yellow-brown sand. Use of wetting agents and their interaction with some aspects of irrigation. *New Zealand Journal of Agricultural Research*, 33:145-150.

Wallis, M.G., Horne, D.J. 1992. Soil water repellency. *Advances in Soil Science*, 20:91-146.

Walsh, R.P.D., Boakes, D., Coelho, C.O.A., Goncalves, A.J.B., Shakesby, R.A, Thomas, A.D. 1994. Impact of fire-induced water repellency and post-fire forest litter and overland flow in southern and central Portugal. *In:*Second International Conference on Forest Fire Research, Coimbra, Portugal. pp. 1149-1159.

Warren, S.D., Nevil, M.B., Blackburn, W.H., Garza, N.E. 1986a. Soil response to trampling under intensive rotation grazing. *Soil Science Society of America Journal*, 50: 1336-1341.

Warren, S.D., Thurrow, T.L., Blackburn, W.H., Garza, N.E. 1986b: The influence of livestock trampling under intensive rotation grazing on soil hydrologic characteristics. *Journal of Range Management*, 40:491-495.

Whipkey, R.Z. 1965. Subsurface stormflow from forested slopes. *International Association of Hydrological Sciences Bulletin*, 10:74-85.

Wilcock, R.J., Nagels, J.W., O'Connor, M.B., Thorrold, B.S., Barnett, J.W. 1999. Water quality of a lowland stream in a New Zealand dairy farm catchment. New Zealand *Journal of Marine and Freshwater Research*, 33:683-696.

Wilcock R.J., Monaghan R.M., Thorrold B.S., Meredith A.S., Duncan M.J., Betteridge K. 2006a. Dairy farming and sustainability: a review of water quality monitoring in five contrasting regions of New Zealand. *In:* Proceedings of the Water 2006 International Conference, 1-4 August 2006, Auckland, New Zealand.

Wilcock, R.J., Monaghan, R.M., Quinn, J.M., Campbell, A.M., Thorrold, B.S., Duncan, M.J., McGowan, A.W., Betteridge, K., 2006b. Land-use impacts and water quality targets in the intensive dairying catchment of the Toenepi Stream, New Zealand. *New Zealand Journal of Marine and Freshwater Research*, 40:123-140.

Willat, S.T., Pullar, D.M. 1983. Changes in soil physical properties under grazed pastures. *Australian Journal of Soil Research*, 22:343-348.

Wilson, D.J., Western, A.W., Grayson, R.B., Berg, A.A., Lear, M.S., Rodell, M., Famiglietti, J.S., Woods, R.A., McMahon, T.A. 2007. Spatial distribution of soil moisture over 6 and

30 cm depth, Mahurangi river catchment, New Zealand. *Journal of Hydrology*, 276:254-274.

Woods, R.A., Rowe, L. 1996. The changing spatial variability of subsurface flow across a hillside. *Journal of Hydrology(New Zealand)*, 35:51–86.

Zollweg, J.A., Gburek, W.J., Steenhuis, T.S. 1996. SmoRMod - A GIS-integrated rainfall-runoff model applied to a small northeastern U.S. watershed. *Transactions of the ASAE*, 39:1299-1307.

In: Agricultural Research Updates. Volume 2
Editor: Barbara P. Hendriks

ISBN: 978-1-61470-191-0
© 2012 Nova Science Publishers, Inc.

Chapter 3

THE EFFECTIVENESS OF RIPARIAN CONSERVATION PRACTICES IN REDUCING SEDIMENT IN IOWA STREAMS

George N. Zaimes[1], Kye-Han Lee[2], Mustafa Tufekcioglu[3], Leigh Ann Long[4], Richard C. Schultz[5] and Thomas M. Isenhart[6]

[1] Lecturer, Laboratory of Management of Mountainous Waters, Department of Forestry and Natural Environment Management, University of Kavala Institute of Technology, Drama Annex, 1st km Drama-Microhoriou, Drama, Greece

[2] Associate Professor, Department of Forestry, College of Agriculture & Life Sciences, Chonnam National University, YoungBong-ro, Buk-gu, Gwangju, Korea

[3] Assistant Professor, Watershed Management, Faculty of Forestry, Artvin Coruh University, Artvin, Turkey.

[4] Research Associate, Forest Ecology and Hydrology, Department of Natural Resource Ecology and Management, College of Agriculture and Life Sciences, Iowa State University, Ames, Iowa

[5] Professor, Forest Ecology and Hydrology, Department of Natural Resource Ecology and Management, College of Agriculture and Life Sciences, Iowa State University, Ames, Iowa

[6] Associate Professor, Stream Ecosystem Ecology, Department of Natural Resource Ecology and Management, College of Agriculture and Life Sciences, Iowa State University, Ames, Iowa

[1] Email: zaimesgeorge@gmail.com
[2] Email: khl@chonnam.ac.kr
[3] Email: mtufekcioglu61@gmail.com
[4] Email: lalong@iastate.edu
[5] Email: rschultz@iastate.edu
[6] Email: isenhart@iastate.edu

ABSTRACT

In the last 100-150 years, 90 % of Iowa's tall-grass prairies containing many wetlands and some riparian forests has been converted to annual row crops and cool-season grass pastures. Bare soil between crop rows erodes easily and can be transported to streams by overland flow, while grazing in riparian areas can reduce stream bank vegetation, making banks susceptible to erosion. The USDA-NRCS estimated that approximately 12,000 kg ha^{-1} yr^{-1} of soil are eroded from 1/3 of Iowa's land with much of these materials entering streams. As a result, sediment is the most common nonpoint source pollutant and a major contributing factor for stream degradation in agricultural watersheds.

In the agricultural states of the United States including Iowa, conservation practices such as riparian forest buffers and grass filters are being established to improve water quality by reducing sediment delivery to stream channels. These conservation practices provide some financial incentives to farmers because they are part of the United Sstates Department of Agriculture Conservation Reserve Program that began as part of the 1996 Farm Bill. This program subsidizes part of the lost income for establishing perennial plant communities on previous managed row crop or grazed pasture land. When these conservation practices are established in the riparian areas of a watershed, stream water quality can improve while maintaining the largest area of the watershed in agricultural production.

In this chapter results are presented on the effectiveness of riparian forest buffers and grass filters in reducing sediment from surface runoff and streambank erosion that are the major sources of sediment to the stream load. Results are from studies on Bear Creek of central Iowa, a National Restoration Demonstration Watershed as designated by the Interagency Team implementing the Clean Water Action Plan (1999) and a Riparian Buffer National Research and Demonstration Area as designated by the United Sstates Department of Agriculture (1998) and other Iowa streams.

1. CHANGES IN THE IOWA LANDSCAPE

Iowa's landscape has been changed more by agriculture than that of any other state in the United States (Dinsmore, 1994). Since the 1850's, when European settlers first arrived in Iowa, 99.9 % of the tallgrass prairies (Figure 1) have been plowed, 95 % of wetlands have been drained, and 70 % of forests have been cleared (Whitney, 1994). Between 1875-1900, the number of 'improved' acres in Iowa increased from 12,658 to 34,574,337 (Andersen and Bishop, 1996). Most of this 'improved' land has been converted to row crop agriculture and continuously grazed cool-season grass pastures (Burkart et al., 1994) (Figure 2). Extensive artificial drainage has occurred within the state to make soils more favorable for growing row crops (Figure 3); about two-thirds of Iowa's 14.6 million ha is currently in row crops (9.3 million ha), and much of that (about 3.6 million ha) has been artificially drained in the last 120 years, most in the last 95 years (Baker et al., 2004). An additional change in agricultural practices occurred after the 1940's and up to the present day, as the diverse patchwork of mixed crop and livestock farming on smaller fields was replaced by large-scale, intensive row crop farming, primarily of corn (*Zea mays* L.) and soybean [*Glycine max* (L.) Merr.] This was facilitated by larger machinery, improved crop hybrids, chemical pest control, and increased availability of inorganic fertilizer. The end result is that land-use change from perennial vegetation to agricultural uses has had major impacts on hydrologic cycling and soil erosion.

Photo credit Jon Sandor.

Figure 1. A reconstructed prairie in Moore Park, Ames, Iowa.

(a)

(b)

(c)

Photo credit a) Jon Sandor, b) Richard Schultz and c) Ashley Wendt.

Figure 2. Row crop agriculture and continuously grazed cool-season grass pastures are common land-uses in Iowa: a) young soybean plants, b) fully grown corn and soybean plants and c) grazing is primarily confined in the riparian area.

Photo credit Leigh Ann Long.

Figure 3. A drainage ditch and a large drainage tile, Hamilton County, central Iowa. This tile is approximately 3.5-4.5 m in diameter.

The perennial plant community with deep, dense root systems that used to cover the soil year-round was replaced with shallow-rooted row crops present for only part of the year. In the case of introduced cool-season pasture grasses, root distribution is shallower especially when overgrazed. Annual plant communities and perennial ones with shallower roots do not provide as much annual evapotranspiration.

Soil disturbance, notably tillage, physically breaks apart soil macroaggregates, leading to reduced soil porosity and decreased water infiltration (Bharati et al., 2002; Marquez et al., 1999). Tillage can also indirectly lead to physical soil aggregation destruction by exposing new soil to wet-dry and freeze-thaw cycles at the soil surface (Six et al., 1998) and by removing plant residue on the soil surface that protects the soil from raindrop compaction and the surface sealing forces of wind and rainfall. Tillage also allows organic matter protected within soil aggregates and communities of organisms leading to the oxidation of organic matter that helps bind soil aggregates together to be exposed to new environments (i.e. temperature, moisture, and aeration) (Six et al., 1998). Macropore space in the soil can be also be lost by compaction from equipment or grazing animals, leading to reduced rates of water infiltration. Infiltration rates in crop fields can be as much as five times lower than in undisturbed prairie or forest soils (Bharati et al., 2002).

The decreased infiltration rates and increased, more rapid surface runoff, leading to accelerated surface erosion is an issue in agricultural watersheds. Bare soil between crop rows erodes easily and can be transported to streams by surface runoff (Waters, 1995) (Figure 2a). Grazing in riparian areas can reduce stream bank vegetation, making the banks more susceptible to streambank erosion (Belsky et al., 1999) (Figure 2c). Erosion due to land clearing in the early part of the settlement period in Iowa led to the deposition of post-settlement alluvium in the floodplain. Yan et al. (2010) determined that post-settlement alluvium may have reduced the water holding capacity of the floodplain soils by as much as 5.09×10^6 m^3 in the 78,000 ha along the South Fork of the Iowa River watershed in central Iowa, which has low topographical relief. Other physiographic regions in Iowa with higher topographical relief may be expected to have much more significant accumulations of post-settlement alluvium. Rates of erosion and sediment transport to streams have decreased since the 1960's due to the implementation of conservation practices such as reduced tillage, sediment storage basins, and other upland and riparian conservation practices (as will be discussed later in Section 2), but rates of erosion and sediment concentrations in streams today are still significantly elevated above pre-European settlement levels (Trimble and Crosson, 2000; Rakovan and Renwick, 2011). The USDA-NRCS (2000) estimated that approximately 12 tons ha^{-1} yr^{-1} of soil are eroded from 1/3 of Iowa's land; much of these material will enter streams.

Because of increased erosion following agricultural settlement (but before the adoption of modern conservation practices) stream channelization was practiced to address flooding occurring from sediment buildup in stream channels. In addition, channelization allowed fields to be more rectangular and easier to farm. Channelization efforts that began in the late 1800's in Iowa have decreased the original stream length of larger streams (greater than 129 km^2 drainage area) by 45 percent (Bulkley, 1975). However, artificial drainage efforts, including subsurface tile and the digging of agricultural ditches, have increased the total length, drainage density, and channel frequency of intermittent streams by more than 50 percent in most Iowa headwater watersheds (Andersen, 2000) (Figure 4). Today, Iowa farmers continue to install tiles or replace old ones with larger diameter tiles (Figure 3).

Channelization reduces stream length (Yan et al., 2010), which increases stream power. Artificial drainage and reduced infiltration from land-use change increase stream discharge. In a study by Schilling and Libra (2003), historical trends in discharge were made for 11 streams in Iowa for the 1940 to 2000 period. For nearly all streams, annual baseflow, annual minimum flow, and annual baseflow percentage increased with time as a result of increased drainage efforts.

Increases in both stream power and discharge lead to stream instability and greater stream bank erosion, which is a significant portion of modern sediment loads in streams (Wilson et al., 2008). As a result of land-use change, sediment derived from upland and streambank erosion is the most common nonpoint source pollutant and a major factor responsible for stream degradation in agricultural watersheds (USEPA, 1995).

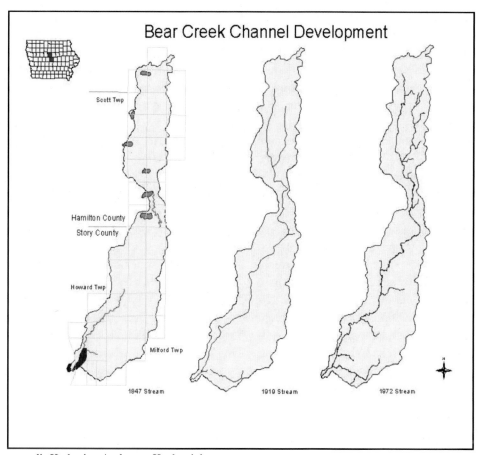

Figure credit Katherine Anderson Koskovich.

Figure 4. Development of the Bear Creek channel network from 1847-1972. In the left-most graphic (1847 stream), red areas are timber, green areas are wetlands or sloughs, and yellow areas are prairie grasses, as noted by the surveyors.

2. CONSERVATION PRACTICES FOR NONPOINT SOURCES POLLUTANTS

2.1. Introduction

The significant land-use changes in Iowa (as discussed earlier in Section 1) have increased surface runoff, peak stream channel discharge, streambank erosion and stream bed incision increasing nonpoint source sediment input to streams and the sediment loads they carry. Sediment input is a natural process, but too much or too little can be detrimental to the functioning of a stream or river (Nerbonne and Vondracek, 2001). Iowa streams have excessive amounts of sediment transported through their channels, especially during high peak flow events. Excessive sediment inputs increase sedimentation in channels, reservoirs and harbours (Owens et al., 2005). Sedimentation alters the natural stream bed substrate composition by increasing primarily the fines. Many invertebrates and fish require stream beds relatively free of fines (Minshall, 1984). In addition, stream beds with excessive amounts of fines can destabilize stream channels (Wilcock, 1997) and are more susceptible to re-suspension, a major source of stream water sediment (Evans et al., 1997) and nutrients (Owens and Walling, 2002).

Sediment and nutrient concentrations in agricultural streams are much higher than in streams of undisturbed forests (Allan, 2004). As the percentage of row crop agriculture (Dodds and Oakes, 2006) or pasture land (Smart et al., 1985) in a watershed increases, so does the sediment and nutrient concentration of its streams. In contrast, increasing the forested land of a watershed is negatively correlated with degraded stream water quality (Sliva and Williams, 2001).

Improving stream water quality is a priority in most agricultural watersheds. To improve the degraded water quality of their streams, Iowa and other agricultural states in the United States, are promoting conservation practices. The Natural Resources Conservation Service (NRCS) has developed numerous Conservation Practice Standards that can help decrease nonpoint source pollutant input to streams in agricultural watersheds. In the following sections we briefly described many of these conservation practices with an emphasis on the riparian forest buffers (USDA-NRCS, 1999a) and filter strips (USDA-NRCS, 1999b). Extensive information on conservation practices is given by the United States Department of Agriculture Natural Resource Conservation Service Field Office Technical Guide at the following website http://efotg.sc.egov.usda.gov/treemenuFS.aspx. In order to best describe these conservation practices we have separated them in three main categories: i) upland, ii) in-stream and iii) riparian areas.

2.2. Upland Conservation Practices

One of the main environmental problems in agricultural watersheds is limiting surface runoff volumes and sheet and rill erosion. The continuous erosion has caused significant degradation of soils and concerns about the sustainability of these agricultural lands. In order to maintain their sustainability a number of different conservation practices have been developed. In the following the most significant and those that have direct impacts on mitigating nonpoint source pollutants are briefly described.

The first area of the upland conservation practices deals with the crops that are planted. In the *Conservation Crop Rotation* (Code 328) practice different crops are grown in a sequence to reduce erosion by including rotations of annual crops as well as perennial grasses and improve soil quality by increasing crop system diversity. The crops alternated include cool- and warm-season grasses and cool-warm-season broadleaf's. In *Stripcropping* (Code 585) rotations of row crops, forages, small grains, or fallow are grown in a systematic arrangement of equal width strips across a field. Finally, narrow strips of perennial, herbaceous vegetative cover can be established across the slope and alternated down the slope with wider cropped strips, and this known as *Contour Buffer Strips* (Code 332).

Changing the soil surface of landform can also reduce overland flow and surface soil erosion. Specifically by plowing, blading, or otherwise elevating the surface of the flat land into a series of broad, low ridges separated by shallow, parallel channels with positive drainage (*Bedding*; Code 310), or by reshaping the surface of the land to planned grades to improve surface drainage and control erosion (*Precision Land Forming*; Code 462), or by farming the sloping land in such a way that the land preparation, planting, and cultivation are done on the contour (*Contour Farming*; Code 330). With the above in many cases a *Diversion* (Code 362) can be constructed, a channel constructed across the slope with a supporting ridge on the lower side to help break up concentrations of water on long slopes or divert water away from certain structures or improvements, or a *Terrace* (Code 600), that is an earth embankment, or a combination of a ridge and channel across the field slope.

Another alternative for conservation practices is proper and effective management of residue and tillage. These include a number of different practices: i) *Residue Management Seasonal* (Code 344); the amount, orientation and distribution of crops or other plant residues are managed on the soil surface for a specific period of time while annual crops are planted or biennial or perennial seed crops are grown when the tilled bed is clean. ii) *Residue and Tillage Management Mulch Till* (Code 345); the amount, orientation and distribution of crop and other plant residue on the soil surface are managed year round while limiting the soil-disturbing activities used to grow crops in systems where the entire field surface is tilled prior to planting. iii) *Residue and Tillage Management No Till/Strip Till/Direct Seed* (Code 329); the amount, orientation and distribution of crop and other plant residue on the soil surface is managed year round while limiting soil-disturbing activities to only those necessary to place nutrients, condition residue and plant crops. iv) *Residue and Tillage Management Ridge Till* (Code 346); the amount, orientation, and distribution of crop and other plant residues is managed on the soil surface year round, while growing crops on pre-formed ridges alternated with furrows, are protected by crop residue.

In many cases measures are taken in specific areas of interest by establishing and maintaining permanent vegetative cover (*Conservation Cover;* Code 327) (Figure 5), establishing a strip of permanent vegetation at the edge or around the perimeter of a field (*Field Border*; Code 386) and/or temporary or permanent exclusion of animals, people, vehicles, and/or equipment from an area (*Access Control*; code 472). In addition, permanent vegetation is established on sites that have or are expected to have high erosion rates and sites that have physical, chemical or biological conditions that prevent the establishment of vegetation with normal practices (*Critical Area Planting*; Code 342). Similarly in areas that are frequently and intensively used by people, animals or vehicles, stabilization is achieved by establishing vegetative cover, surfacing the site with suitable materials, and/or installing needed structures (*Heavy Area Protection*; Code 561).

Another way to mitigate erosion is by focusing on the removal of the surface runoff quickly from the fields with the least possible erosion. The *Drainage Water Management* (Code 554) practice develops a management plan to process water discharges from surface and/or subsurface agricultural drainage systems. One way is by *Surface Drains* that collect and convey excess surface or subsurface water. There are two types of surface drains, the *Main or Lateral* (Code 608) that are an open drainage ditch constructed to a designed cross section, alignment and grade and the *Field Ditch* (Code 607) a graded ditch for collecting excess water in a field. Another way to manage surface runoff and reduce erosion is by placing erosion resistant lining of concrete, stone, synthetic turf reinforcement fabrics, or other permanent material in a waterway or outlet (*Lined Waterway or Outlet*; Code 468A) or by shaping or grading a channel with suitable vegetation to carry surface water at a non-erosive velocity to a stable outlet (*Grassed Waterway*; Code 412). These practices are commonly used to reduce gully erosion. When areas are irrigated, the volume, frequency, and application rate of the irrigation water are planned in an efficient manner (*Irrigation Water*

Management; Code 449) or a lining of impervious material or a chemical treatment, in an irrigation ditch, canal, or lateral is installed (*Irrigation Ditch Lining*; Code 428). When these practices are implemented the land manager needs to keep in mind that they can lead to increased peak flows in streams and rivers (their consequences have been described in Section 1).

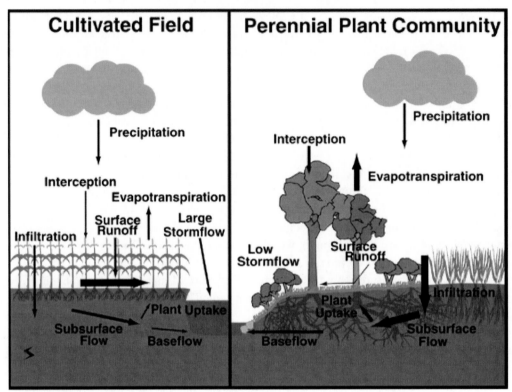

Source Schultz et al., 2000.

Figure 5. Non-tilled row cropped fields have more overland flow and less total evapotranspiration resulting in larger flow compared to areas with perennial vegetation that reduce stormflow and increase baseflow due to higher infiltration and evapotranspiration .

To reduce sediment or other pollutants from reaching the streams in many cases you can trap them before they reach the stream. In the *Sediment Basin* (Code 350), a basin is constructed to collect and store debris or sediment while in the *Water and Sediment Control Basin* (Code 638), an earth embankment or a combination of a ridge and channel is constructed across the slope of minor watercourses to form a sediment trap and water detention basin with a stable outlet.

As discusses in Section 1, in the Iowa landscape a significant number of *subsurface drains* (Code 606) have been placed throughout the state in order to improve drainage and remove the excess water from surface depressions of agricultural fields. The subsurface drain is a conduit, such as a corrugated plastic tubing, tile, or pipe, and installed beneath the ground surface. While this has been effective in removing the drainage water it has also had serious repercussions to the stream water quality, primarily by increasing the nutrients (nitrogen-nitrate) that reach the stream and increasing drainage density and the number of perennial streams being fed by tile baseflow. To minimize these nonpoint source pollutants the *Tile Intake Replacement* is recommended (Interim Code IA-980) that is a replacement of the surface intake which uses an agricultural drainage well as the outlet with subsurface drains. This eliminates surface water intakes and drainage water does not go directly into the ground water but through an agricultural drainage well. Another option is the *Underground Outlet* (Code 620) that is a conduit that is installed beneath the surface of the ground to collect surface water and convey it to a suitable outlet. Finally, a *Denitrifying Bioreactor* (Interim Code 747) could also be used that is a structure containing a carbon source installed to intercept subsurface drain (tile) flow or ground water, and reduce the concentration of nitrate-nitrogen.

2.3. In-Stream Conservation Practices

The conservation practices of this category are located in the stream channel. Typically, they trap sediment, reduce floodwaters and stream bed re-suspension and/or protect streambanks. In stream beds covered with substantial amounts of sediments, re-suspension can provide a major source of stream water sediment (Evans et al., 1997) and nutrients (Owens and Walling, 2002). Streambank erosion is a natural function of streams and rivers (Henderson, 1986) but in many regions of the world the anthropogenic changes have accelerated streambank erosion. In these regions, streambank erosion is a major contributor of sediment and nutrients to the stream load. Streambank erosion can typically contribute 25-60 % of the stream load (Wilkin and Hebel, 1982; Hamlett et al., 1983; Odgaard, 1987; Lawler et al., 1999; Schilling and Wolter 2000; Sekely et al., 2002; Amiri-Tokaldany *et al.,* 2003) and as much as 80-90 % (Simon et al., 1996; Kronvang et al., 1997) in incised streams.

Part of these conservation practices deal with the improvement of the stream bed and the others with the stabilization of the streambanks. In the *Channel Stabilization* (Code 584), measures are used to stabilize the bed of a channel. The channel bed or grade may be maintained or altered, the sediment transport or deposition is modified or the surface and ground water levels of the floodplain riparian areas and wetlands are managed. In other cases, a structure is used to control the channel grade in natural or constructed watercourses (*Grade Stabilization Structure*; Code 410). With this structure the grade can be stabilized, gully erosion reduced and/or water quality improved. Another conservation practice is *Clearing and Snagging* (Code 326) where snags, drifts, or other obstructions are removed from a channel or drainage way that helps restore flow capacity and prevent bank erosion by eddies. *Stream Crossing* (Code 578) is when a stabilized area or structure is constructed across a stream for people, livestock, equipment, or vehicles to travel. This crossing helps improve water quality by reducing the sediment, nutrient, organic, and inorganic loading of the stream and by reducing streambank and stream bed erosion. Dams are also part of the in-stream conservation practices. There are three types: a) *Dam, Floodwater Retarding* (Code 402), b) *Dam, Diversion* (Code 348) and c) *Dam, Multiple Purpose* (Code 349). The purpose of dams is to either retard or divert stream flow that will reduce floods and streambank erosion. These structures can also tarp sediment and nutrients.

For the direct stabilization and protection of the banks of streams or constructed channels, there is the *Streambank and Shoreline Protection* (Code 580) practice. These can include shaping of the slope of the bank, hard structures (e.g. riprap, gabbions), bioengineering

engineering (vegetative material e.g. brushmatresses, live stakes, live fascines) or a combination of the above. Finally, a more holistic approach is taken in the *Stream Habitat Improvement and Management* (Code 395) that maintains, improves, or restores the physical, chemical and biological functions of a stream. To accomplish these goals in-stream practices for the stream bed and bank can be used (described earlier in this Section) in coordination with riparian area conservation practices (described immediately afterwards).

2.4. Riparian Area Conservation Practices

Riparian areas are adjacent to a freshwater body such as a stream, river, lake or pond and dependent on perennial and intermittent water. They are ecotones, transitional zones between aquatic and terrestrial ecosystems. Typically, riparian areas are linear in nature, but difficult to clearly define their boundaries because of their high spatial and temporal variability. The National Research Council (2002) defines them as "Riparian areas - Transitional between terrestrial and aquatic ecosystems and are distinguished by gradients in biophysical conditions, ecological processes, and biota. They are areas through which surface and subsurface hydrology connects waterbodies with their adjacent uplands. They include those portions of terrestrial ecosystems that significantly influence exchanges of energy and matter with aquatic ecosystems (i.e., a zone of influence). Riparian areas are adjacent to perennial, intermittent, and ephemeral streams, lakes, and estuarine–marine shorelines."

To cost-effectively control sediment and nutrient export from a watershed, specific areas need to be targeted (Gburek et al., 2000). These areas are called *critical sources areas* and while these areas account for only a small percentage of the watershed they are often responsible for the majority of the sediment and/or nutrient export of the watershed (Pionke et al., 1997). Such areas are also the riparian areas of the stream since riparian areas can be sinks or sources of sediment and nutrients (Cooper et al., 1995). This is why Gilliam (1994) considers riparian areas "the most important factor influencing nonpoint source pollutants entering surface waters in many areas in the U.S.A." Another advantage of focusing on the riparian areas of the watershed is that you can reduce significantly nonpoint source pollutants while maintaining the largest area of the watershed in agricultural production.

The main conservation practices for riparian areas are some that have been previously described such as *Access Control* (Code 472) when the area that is temporary or permanent excluded from animals, people, vehicles, and/or equipment is the riparian and the *Stream Habitat Improvement and Management* (Code 395). In addition, there are some that focus on riparian areas such as the *Filter Strip* (Code 393), *Fence* (Code 382), *Riparian Forest Buffer* (Code 391) although they could also be used in other areas of the watershed to reduce pollutants. In this section the wetlands will also be described although they are not always located in the riparian areas.

The idea behind the wetland conservation practices is to develop, restore or maintain functional wetlands. This way the wetlands can provide their full ranges of functions and values that help mitigate floodwaters and nonpoint sources pollutants. There are four conservation practices dealing with wetlands: a) *Constructed Wetland* (Code 656), that is intended to treat point and nonpoint sources of water pollution; b) *Wetland Restoration* (Code 657) that rehabilitates a degraded wetland or reestablishes a wetland so that the soil, hydrology, vegetative community and habitat are a close approximation of the original natural condition that existed prior to modification; c) *Wetland Creation* (Code 658) that creates a wetland on a site that was historically non-wetland; and d) *Wetland Enhancement* (Code 659) that deals with the rehabilitation or reestablishment of a degraded wetland, and/or the modification of an existing wetland, which augments specific site conditions for specific species or purposes, possibly at the expense of other functions and other species.

The *Fence* (Code 382) is a constructed barrier that is applied to facilitate the application of conservation practices by providing a means to control movement of animals and people from specific areas. For example a Fence can be placed around a wetland or a riparian area for its protection.

The *Filter Strip* (Code 393) is a strip or area of herbaceous vegetation that removes pollutants from overland flow. Perennial herbaceous vegetation is required in this practice. The correct placement of a Filter Strip can reduce suspended sediment and associated pollutants in surface runoff and irrigation tailwater and dissolved pollutants loadings in surface runoff. It is important to note that Filter Strips are effective when surface runoff is entering the filter strip as uniform sheet flow. When concentrated flow is frequent it should be dispersed before it enters the filter strip in order to be effective.

The *Riparian Forest Buffer* (Code 391) also is a narrow linear area but with trees and/or shrubs located adjacent to and up-gradient from watercourses or water bodies. With the establishment of the riparian forest buffer: a) the shade lowers or maintains water temperatures to improve habitat for aquatic organisms, b) the stream habitat is improved because it provides a source of detritus and large woody debris, c) the excess amounts of sediment, organic material, nutrients and pesticides in surface runoff and the excess nutrients and other chemicals in shallow ground water flow are reduced, d) the riparian plant communities are restored that creates habitat and corridors for wildlife, e) carbon storage in plant biomass and soils is increased, f) flood waters are slowed, g) flood peaks are lowered, h) harvestable crops of timber, fiber, wildlife forage, fruit or other crops consistent are created, and i) space is provided for the water courses to establish geomorphic stability. The riparian forest buffers are applied on areas adjacent to permanent or intermittent streams, lakes, ponds, and wetlands.

The effectiveness of riparian buffers (e.g. filters strips and riparian forest buffers) in reducing nonpoint source pollutants is due to three mechanisms (Schultz et al., 2000) (Figure 6). The first mechanism is reducing overland flow and streambank erosion. Buffers have vegetation that covers the soil year round and increases surface roughness and infiltration thereby reducing overland flow and streambank erosion. The second mechanism is plant assimilation and immobilization of nutrients as long as they are actively accumulating biomass. The third mechanism is the increased soil organic matter that leads to larger microbial populations that are a nutrient sink and improve soil aggregation (infiltration) and denitrification activities.

Conservation practices are more effective when used as a component of a total resource management system including nutrient management, pest management, and erosion runoff and sediment control practices. This leads to combinations of the various conservation practices (discussed in this section). An example is shown in Figure 7 where both riparian area and in-stream conservation practices are utilized. A Riparian Forest Buffer and Filter Strip are established to intercept overland flow and nonpoint source pollutants for the adjacent upland agricultural fields. On the other side of the stream a Fence is established to exclude cattle from the stream in order to reduce grazing impacts. In the stream channel, rock toe control, willow live stakes and cedar bundles are used to stabilize streambanks (Streambank and Shoreline Protection) and boulder weirs to control stream bed incision (Channel Stabilization). Finally, a Constructed Wetland is established to intercept nonpoint source pollutants for tile drainage.

Riparian Buffers Trap Sediment
And Attached Phosphorus From Runoff

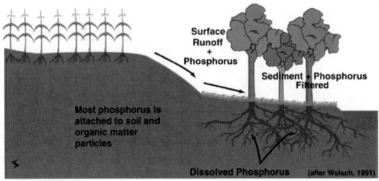

Source Schultz et al., 2000.

Figure 6. Sediment and phosphorus are filtered from overland flow and phosphorus can be taken up by biota of the living filter.

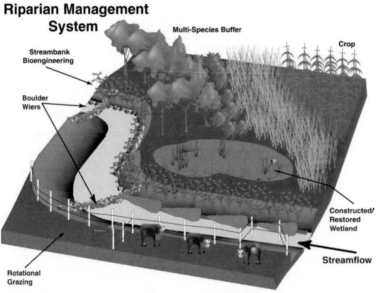

Source Schultz et al., 2000.

Figure 7. Six different conservation practices are combined in order to mitigate nonpoint source pollutants from agricultural actvities: i) Riparian Forest Buffer (Code 391), ii) Filter Strips (Code 393), iii) Constructed Wetland (Code 656), iv) Fence (Code 382), v) Streambank and Shoreline Protection (Code 580) and vi) Channel Stabilization (Code 584).

3. SURFACE RUNOFF AND RIPARIAN CONSERVATION PRACTICES

3.1. Introduction

Runoff and soil erosion from agricultural fields can lead to the loss of productivity and degradation of surface water quality. Sediment and nutrient in runoff from agriculture are leading causes of poor surface water quality in the United States (USEPA, 1995). Riparian buffers are being used as conservation practices to reduce the transport of nonpoint source pollutants in agricultural runoff before they enter surface waters (Lee et al., 2003). Riparian vegetation facilitates the removal of suspended sediments and associated nutrient content from surface runoff (PeterJohn and Correll, 1984; Lowrance et al., 1988). The friction of the soil surfaces can reduce the velocity of runoff that consequently results in the sedimentation of particles, but riparian buffer vegetation and the layer of organic litter on the soil surface are much more effective in slowing the velocity of the surface runoff (Correll, 1997). Whereas the exact role and effectiveness of the various types of buffer vegetation are uncertain, dense stiff grasses are generally considered more effective in trapping particles in surface runoff than most other types of vegetation (Dabney et al., 1993; Meyer et al., 1995). Dabney et al. (1993) demonstrated that as little as a 12 cm wide strip of switchgrass (*Panicum virgatum* L.) can dam water as high as 10 cm deep. Buffer designs that use these stiff-stemmed grasses at their edge can slow surface runoff enough to cause large particles to settle out before entering the buffer. Further filtration by buffer vegetation and surface litter is significant only with large particles and aggregates (Dillaha and Inamdar, 1997). Buffer vegetation, including woody plants, may be effective in removing soluble nutrients from surface runoff by improved infiltration into the buffer soil (Vought et al., 1994). Infiltration is one of the most significant mechanisms influencing buffer performance (Bharati et al., 2002). Not only does it provide the pathway for water and soluble chemicals to enter the profile but suspended fine soil particles with adsorbed chemicals also enter the profile thus decreasing not only surface runoff; but also sediment transport capacity. While the potential exists for some soil pores to plug, the high infiltration rates in buffer soils (Lee et al., 1999) and the dynamic nature of soil structure (Marquez et al., 1999) assures continued high infiltration. This assumes that the large particles in surface runoff are deposited before entering the buffer or shortly thereafter.

The beneficial environmental effects of riparian buffers have led to development of two national standards by the United States Department of Agriculture (USDA) Natural Resources Conservation Service (NRCS) for reduction of agricultural nonpoint source pollution. One of these is the filter strip conservation standard (Code 393) and the other the riparian forest buffer standard (Code 391) (USDA-NRCS, 1999a and b). The national standards can be modified by each state NRCS office to fit local conditions. The riparian forest buffer conservation standard in Iowa consists of two distinct functional zones. Zone 1 begins at the upper edge of the active channel and extends a minimum distance of 12 m or at least one-third the total width of the buffer, with trees and/or shrubs suited to the site and the intended purpose. Zone 2 begins at the up-gradient edge of Zone 1 and extends 6 m to 36.6 m perpendicular to Zone 1. Native warm-season grasses, with or without native forbs, are recommended for vegetation of Zone 2 (USDA-NRCS, 1999a). While the riparian forest buffer standard is believed to provide effective reduction of nonpoint source pollution, there is little quantitative information on its effectiveness for reducing runoff, sediment, and nutrient movement. Such information is needed to modify buffer design and create credibility to improve landowner adoption.

The objective of this study was to determine the effectiveness of an established multispecies riparian forest buffer (RF) in reducing nonpoint source sediment, nitrogen (N), and phosphorus (P) transported from cropland during simulated and natural rainfall events. These buffers were sized to conform to the Code 393 grass filter and the Code 391 riparian forest buffer standards (USDA-NRCS, 1999a and b).

3.2. Methods and Materials

3.2.1. Study Site

This study was conducted in a RF established on a private farm along Bear Creek, in Story County, Iowa, USA (42°11´N, 93°30´W) (Figure 8). The RF consisted of a 7 m wide zone of switchgrass (*Panicum virgatum* L. cv. Cave-in-Rock) (Zone 2) adjacent to the cropland (Figure 9). This RF was used as a model for the development of the NRCS riparian forest buffer conservation standard. The 13 m wide woody plant zone consisted of shrubs and trees. Two rows of shrubs at 1.8 m spacing between rows and 0.9 m between plants within rows were installed. The four rows of trees were installed downslope of the shrubs with 2.4 m spacing between rows and 1.8 m between plants within rows. Tree species included silver maple (*Acer saccharinum* L.), green ash (*Fraxinus pennsylvanica* Marsh.), black walnut (*Juglans nigra* L.), willow (*Salix* spp), cottonwood hybrids (*Populus* spp., e.g., *Populus* clone NC-5326, a designated clone of the North Central Forest Experiment Station), red oak (*Quercus rubra* L.), bur oak (*Quercus macrocarpa* Michx.), and swamp white oak (*Quercus bicolor* Wild.). Details of the RF design, placement, and plant species are given in Schultz et al. (1995). The soil under the multi-species riparian buffer was a Coland (fine-loamy, mixed, mesic cumulic Haplaquoll), with an average natural slope of 5 %. Soil of the adjacent crop field source area was a Clarion (fine-loamy, mixed, mesic typic Hapludoll) with an average slope of 8 % (Dewitt, 1984). The cropland source area was managed under a soybean [*Glycine max* (L.) Merr.] and corn (*Zea mays* L.) rotation.

3.2.2. Field Methodology-Rainfall Simulations

Triplicate plots used were installed in the RF system in April 1997, with a 4.1 by 22.1 m bare cropland source area paired with either no buffer, a 7.1 m wide switchgrass buffer, or a 16.3 m wide switchgrass/woody buffer located at the lower end of each plot. The plots were isolated with sheet metal borders driven into the ground with metal gutters and double-split runoff dividers installed at the downslope end for manual sample collection and flow measurement (Figure 10). The double-split runoff divider separated total runoff volume from each plot to the storage tank at a 25:1 ratio.

Photo credit Thomas Schultz.

Figure 8. Farm in Story County, Iowa where the experiment was conducted.

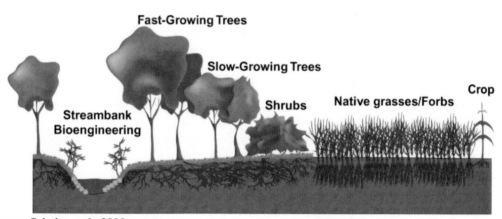

Source Schultz et al., 2000.

Figure 9. Model of the multi-species riparian forest buffer (RF) planted at the study site.

A rainfall simulator was designed and constructed with a 36 m long by 5.1 cm diameter PVC pipe with sprinklers (T40-6 nozzles, RainBird Sales, Inc., Azusa, CA) spaced 4 m apart. The first rainfall simulation was conducted with 2.5 cm h^{-1} of rainfall intensity. The second rainfall simulation, performed 2 d after the first simulation, had 6.9 cm h^{-1} of simulated rainfall intensity. Each rainfall simulation performed once on each of the three replicates. Three rain gauges were placed in each plot. Before each rainfall simulation, soil samples from the upper 1 cm were taken from three random locations within the source area of each plot for particle size determination and nutrient analysis. Runoff samples were collected in 1 L plastic bottles for particle size analysis and 0.5 L bottles for sediment and nutrient analysis at 5-min intervals from the beginning of runoff to its completion. Runoff rates were measured at 5-min intervals by measuring the depth of water in the water tank. To estimate the RF effectiveness, it was assumed that the amount of inflow to the plots with buffers was the same as the discharge amount from non-buffered plots in each set of plots. The total amount of inflow plus rainfall that infiltrated into the buffers was determined from an inflow-rainfall-discharge mass balance.

3.2.3. Field Methodology-Natural Rainfall Events

For the natural rainfall events precipitation was measured by a tipping bucket rain gauge with a CR10 data logger (Campbell Scientific, Logan UT) located in the riparian study area. Triplicate plots used in this study were same as the rainfall simulation study (Figure 10). The water was collected in a tank and amounted to 4 % of the total runoff generated in each plot during each rainfall event. Runoff samples were collected on the day of the rainfall event or on the next day following rainfall events. Multiple events occurring in a day were collected as one runoff sample. Water samples were pulled from the tank after 1 minute of agitation of the collected runoff water. The total runoff volume was determined by measuring the depth of water in the tanks. Runoff samples were collected in 1 L plastic bottles for particle size analysis and 0.5 L bottles for sediment and nutrient analysis. After sampling, the water tanks were cleaned out for the next runoff collection. To estimate the effectiveness of the RF, it was assumed that the amount of input to the plots with buffers was the same as the discharge amount from non-buffered plots in each set of plots.

Total annual precipitation was 738 mm in 1997, and 872 mm in 1998. The total annual precipitation in 1997 was 12 % below, and the total annual precipitation in 1998 was 4 % above the long-term average of 841 mm for the study area. The number of rainfall events that resulted in at least 0.02 mm of runoff at the study plots ranged from six in 1997 to 13 in 1998.

3.2.4. Laboratory Methodology

All runoff samples were analyzed for sediment, total-N, NO_3-N, total-P, and PO_4-P content by using standard procedures (Clesceri et al., 1989) and sediments in runoff samples were separated into particle sizes of 50, 20, 8, and 2 μm by using standard pipette procedures without chemical dispersion (Gee and Bauder, 1986). Particulate-P in runoff samples was calculated as the difference between total-P and PO_4-P, and was used as the basis for calculating the enrichment of soil P in runoff. The amount of runoff, particle size distribution, sediment and nutrient concentration data were used to compute mass transport of each constituent occurring at the end of each plot.

3.2.5. Statistical Analysis

General linear model tests were performed to determine the effects of the switchgrass and the switchgrass/woody buffer in runoff and the concentration and mass transport of the measured variables (SAS Institute, 1999). Least significant difference (LSD) tests were performed to determine differences at p less than 0.05 in buffer treatments for all measured variables.

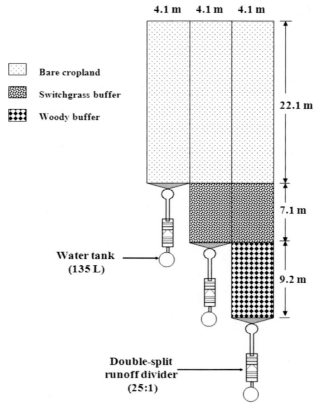

Source Lee et al., 2000.

Figure 10. Schematic diagram of the plots and runoff collector (double-split runoff divider).

3.3. Results

3.3..1 Rainfall Simulation

Buffers reduced mass transport of particle fractions and sediment in both rainfall intensities with greater mass transport of particle fractions and sediment in the higher rainfall intensity (Table 1). The switchgrass buffer removed more than 82 % of sand, 71 % of silt, and 15 % of clay for both rainfall simulations. The switchgrass/woody buffer removed more than 98% of sand, 93% of silt, and 52% of clay for both rainfall simulations. The switchgrass/woody buffer trapped more sediment than the switchgrass buffer. As particle size decreased, the difference in reduction between the switchgrass buffer and the switchgrass/woody buffer increased (Table 2).

Buffers reduced mass transport of nutrients, whereas the higher rainfall intensity increased mass transport of nutrients (Table 3). During a 2-h rainfall simulation, mass transport of total-N and total-P was reduced by switchgrass and switchgrass/woody buffers, whereas NO_3-N and PO_4-P mass transport was reduced by the switchgrass/woody buffer. For the 1-h rainfall simulation with an intensity of 6.9 cm h^{-1}, reduction of mass transport for total-N, NO_3-N, and total-P occurred in both the switchgrass and the switchgrass/woody buffers (p less than 0.05). However, a significant reduction in PO_4-P mass did not occur in the switchgrass/woody buffer for the higher rainfall intensity. As rainfall intensity was increased, percentage mass reduction was decreased. Buffers increased the percentage mass reduction of all measured nutrients and sediment (Table 3). There was no interaction of buffer and rainfall intensity treatments for percentage mass reduction for all nutrients measured.

Table 1. Mass of particle fraction and sediment transported from triplicate plots with a 4.1 by 22.1 m bare cropland source area paired with either no buffer, a 7.1 m wide switchgrass buffer, or a 16.3 m wide switchgrass/woody buffer during rainfall simulations. A pipette method without chemical dispersion was used to analyze particle size for surface runoff samples

Rainfall intensity/ Buffer type	Particle size (μm)			
	Sand (> 50)[†]	Silt (50 - 2)[†]	Clay (< 2)	Sediment[†]
2-h rainfall at 2.5 cm h⁻¹	kg ha⁻¹			
No buffer	5.6 c[‡] (16)[§]	17.9 c (52)	10.8 bc (32)	34.3 c
Switchgrass	0.6 d (6)	4.3 d (41)	5.5 c (53)	10.4 d
Switchgrass/woody	0.1 d (5)	0.8 d (38)	1.2 c (57)	2.1 e
1-h rainfall at 6.9 cm h⁻¹				
No buffer	134.2 a (28)	310.1 a (64)	39.5 a (8)	483.8 a
Switchgrass	23.8 b (16)	88.4 b (61)	33.7 a (23)	145.9 b
Switchgrass/woody	0.9 d (2)	19.1 c (49)	18.8 b (49)	38.8 c

[†]Variables for which there was an interaction of buffer and rainfall intensity treatments.
[‡]Within-column means followed by the same letter are not significantly ($p < 0.05$) different.
[§]Values in parentheses are percentage mass of the particle fraction.
Source Lee et al., 2000.

Table 2. Particle size distribution and phosphorus concentration from the upper 1 cm soil sample from the source area of plots, and of sediment in runoff from triplicate plots with a 4.1 by 22.1 m bare cropland source area paired with either no buffer, a 7.1 m wide switchgrass buffer, or a 16.3 m wide switchgrass/woody buffer during rainfall simulations. The pipette method without chemical dispersion was used to analyze particle size for soil and runoff samples

Rainfall intensity/ Buffer type	Particle size distribution (%)					D50[†] μg	Phosphorus[‡] μg g-1
	Size (μm)						
	2000 - 50	50 - 20	20 - 8	8 – 2	< 2		
Source area soil	78.5	12.8	5.2	2.5	1.0	135	552
2-h rainfall at 2.5 cm h-1							
No buffer	15.3 b[§]	20.2 bc	17.1 a	14.1 a	33.3 ab	10	1732 a
Switchgrass	5.1 c	8.6 d	21.0 a	14.3 a	51.0 a	<2	1321 a
Switchgrass/woody	3.3 c	9.1 d	16.6 a	14.9 a	56.3 a	<2	1698 a
1-h rainfall at 6.9 cm h-1							
No buffer	26.6 a	30.9 a	22.4 a	11.9 a	8.1 c	28	649 b
Switchgrass	14.0 b	22.7 b	25.8 a	13.4 a	24.1 ab	11	798 b
Switchgrass/woody	2.4 c	13.2 cd	16.3 a	16.2 a	51.8 a	<2	1446 a

[†]Size for which 50 % of particles were smaller by weight.
[‡]Total-P content in the upper 1 cm soil from the source area of plots; other phosphorus values are calculated as the difference between total-P and soluble P concentration in runoff.
[§]Within-column means followed by the same letter are not significantly ($p < 0.05$) different.
Source Lee et al., 2000.

Table 3. Percentage mass reduction of total-nitrogen (N), nitrate-nitrogen (NO$_3$-N), total-phosphorus (P), and phosphate-phosphorus (PO$_4$-P), and sediment by 7.1 m wide switchgrass buffers, and 16.3 m wide switchgrass/woody buffers during rainfall simulations. Mass transport from the buffers was compared with the mass transport from non-buffered plots for each rainfall simulation.
Each value is a mean of three replications

Rainfall intensity/ Buffer type	Reduction (%)[†]				
	Total-N	NO$_3$-N	Total-P	PO$_4$-P	Sediment
2-h rainfall at 2.5 cm h^{-1}					
Switchgrass	64.3 bc[‡]	61.1 b	67.6 b	43.7 b	70.2 b
Switchgrass/woody	89.7 a	87.8 a	93.1 a	85.3 a	94.4 a
1-h rainfall at 6.9 cm h^{-1}					
Switchgrass	49.7 c	40.5 c	46.2 c	27.6 b	70.1 b
Switchgrass/woody	72.8 ab	67.5 b	80.7 a	34.7 b	92.2 a

[†]Total input (runoff input from bare source area plus rainfall input) minus runoff output divided by total input, multiplied by 100.
[‡]Within-column means followed by the same letter are not significantly ($p < 0.05$) different.
Source Lee et al., 2000.

3.3.2. Natural Rainfall Events

The switchgrass and switchgrass/woody buffers reduced surface discharge of runoff, sediment and nutrients from the crop fields to the stream. The sediment reduction occurred primarily in the switchgrass buffer. The switchgrass/woody buffer had three times less sediment transported through it than the switchgrass buffer, which had 13 times less sediment transported through it than the non-buffered plots (Table 4).

Table 4. Runoff and mass transport of sediment, total-nitrogen (N), nitrate-nitrogen (NO$_3$-N), total-phosphorus (P), and phosphate-phosphorus (PO$_4$-P), from non-buffered and buffered plots during 1997-1998. Each value is the mean of 19 precipitation events from three replicated plots. The cropland source area for each plot was 4.1 by 22.1 m

Buffer	Runoff mm	Sediment kg ha^{-1}	Total-N g ha^{-1}	NO$_3$-N	Total-P	PO$_4$-P
None	9.7 a[†]	587.1 a	551.0 a	92.2 a	198.8 a	48.4 a
Switchgrass (7 m wide)	4.0 b	45.2 b	119.7 b	34.4 b	40.4 b	20.5 b
Switchgrass/woody (16.3 m wide)	1.8 c	16.4 c	51.3 c	16.6 c	19.4 c	10.2 c

† Values in the same column followed by a different letter are significantly different ($p < 0.05$).
Source Lee et al., 2003.

The buffers reduced the mass transport of total-N and NO$_3$-N, total-P, and PO$_4$-P in surface runoff from cropland. The average mass transport of total-N and NO$_3$-N, total-P, and PO$_4$-P were different (p less than 0.05) among buffer treatments (Table 4). Mean percentage mass reductions in N and P in surface runoff occurred through the buffers, with the greater mass reductions measured in the switchgrass/woody buffer (Table 3). The added width of the switchgrass/woody buffer reduced 23 % more runoff, 2 % more sediment, 14 % more total-N, 23 % more NO_3-N 13 % more total-P, and 22 % more PO_4-P than the switchgrass buffer alone (Table 5).

3.4. Discussion

The RF system reduces sediment and nutrient surface runoff from crop fields during rainfall simulations (Lee et al., 2000) and natural rain events (Lee et al., 2003). Particle size has a dominant effect on trapping potential of the RF. Under simulated rainfall conditions, the 7.1 m wide switchgrass buffer removed 82-89 % of the sand and 72-76 % of the silt. The switchgrass buffers removed only 15-49 % of the clay, whereas the 16.3 m wide switchgrass/woody buffer removed 52-89 % of the clay. This result may be due to the greater width of the switchgrass/woody buffer (16.3 m) compared to the switchgrass buffer (7.1 m). This reduction occurs because sediment deposition is a selective process in which large particles are deposited preferentially to clay particles (Alberts et al., 1981). Particle size distribution in the surface runoff changed through the buffers indicating that there was a selective process in which large particles are deposited prior to small particles and more than 90 % of the sediment in the surface runoff from the buffered plots was in the 0.05 mm, and 85 % of the sediment leaving the residue strip was in the size fractions less than 0.035 mm, which increased the nutrient concentrations of the surface runoff.

The greater sediment reduction by the switchgrass buffer could be due to the differences in growth pattern between the cool-season grasses and the warm-season switchgrass, and the biomass harvesting in the coastal plain buffer may have reduced the effectiveness of sediment removal. The uniform distribution of the plants and large production of litter on the switchgrass buffer may be responsible for the high removal of sediment (Lee et al., 1999).

Table 5. Reduction of runoff and mass transport of sediment, total-nitrogen (N), nitrate-nitrogen (NO_3-N), total-phosphorus (P), and phosphate-phosphorus (PO_4-P), from precipitation events from three replicated plots. The row cropland source area for each plot was 4.1 by 22.1 m

Buffer	Runoff	Sediment	Total-N	NO_3-N	Total-P	PO_4-P
			%			
Switchgrass (7 m wide)	58.3 a[†]	95.3 a	80.3 a	62.4 a	78.0 a	57.5 a
Switchgrass/woody (16.3 m wide)	81.5 b	97.2 b	93.9 b	84.9 b	91.3 b	79.8 b

[†] Values in the same column followed by a different letter are significantly different ($p < 0.05$).
Source Lee et al. 2003.

As soil erosion is a selective process with respect to sediment size, eroded soil is usually richer in fine particles and P than the source area surface soil (Sharpley, 1985). The greater enrichment of clay particles and P from the 2-h rainfall simulation at 2.5 cm h^{-1} indicates that the kinetic energy of rainfall impact on the soil was lower than that of the 1-h rainfall simulation at 6.9 cm h^{-1} (Table 2). The fact that a lower amount of sediment was discharged from the lower intensity rainfall simulation compared with the higher intensity rainfall simulation is consistent with differences in rainfall impact on the surface soil (Table 1). Increased enrichment of clay particles and P by buffers indicates that the selective deposition process of sediment results in a high percentage distribution of clay particles in runoff. Because lowering enrichment through the case of buffers is not feasible, maximizing infiltration capacity of buffers may facilitate retention of clay particles and P by widening the buffers.

Mass transport data for total-N, NO$_3$-N, total-P, PO$_4$-P, and sediment indicate that mass is reduced as the runoff flows through the buffers (Table 2 and 5). As rainfall intensity was increased, percentage mass reduction of nutrients was decreased (Table 3). The 7.1 m wide switchgrass buffer had a low percentage mass reduction of nutrients for the high intensity rainfall simulation compared with the low intensity rainfall simulation. This result may be due to the low infiltration brought about by high soil moisture from the previous rainfall simulation and increased runoff depth and velocity. The wider buffers increased the percentage of mass reduction of nutrients and sediment. Buffers were more effective in removing sediment from runoff than in removing nutrients (Table 3). The 9.1 and 4.6 m grass filters with shallow uniform flow removed an average of 70 and 84 % of the incoming sediment, 61 and 79 % of the incoming P, and 54 and 73 % of the incoming N, respectively. Soluble nutrients in the runoff from the filters were sometimes greater than the incoming soluble nutrient transport. This was attributed to the low removal efficiencies for soluble nutrients and the release of nutrients that had been trapped in the buffers during previous rainfall simulations, and mineralized P from organic residues on the buffer plots. Relatively low removal effectiveness of PO$_4$-P indicates that the buffers should be maintained by harvesting grasses and trees to remove accumulated P or to reduce P saturation. Nutrient removal appears to be maximized when buffers are composed of a combination of dense herbaceous and woody vegetation.

The removal of clay particles and dissolved nutrients like NO$_3$-N was mainly dependent on infiltration, indicating that improving infiltration capacity will increase the removal effectiveness for soluble nutrients. These results have practical implications for the design of riparian buffers. For maximum removal efficiency, the buffer soil should have high porosity to enable infiltration of a large amount of runoff. Infiltration is important because the finer particles and soluble nutrients enter the soil profile along with infiltrating water and because it decreases runoff, thus reducing sediment transport capacity of the runoff (Phillips, 1989; Dillaha and Inamdar, 1997). Moreover, a dense, native grass vegetation with dense surface litter would resist the surface runoff flow and decrease the velocity of surface runoff immediately upslope and within the buffers, causing significant reductions in sediment (Meyer et al., 1995). The switchgrass buffer alone is effective in trapping coarse sediments and sediment-bound nutrients, but to trap the clay particles and soluble nutrients, additional buffer width with high infiltration capacity was provided by the deep-rooted woody plant zone.

During the natural rainfall events, the buffers reduced the mass transport of total-N and NO$_3$-N, total-P, and PO$_4$-P in surface runoff from cropland. Mean percentage mass reduction in N and P in surface runoff occurred through the buffers, with the greater mass reductions measured in the switchgrass/woody buffer (Table 4). The added width of the switchgrass/woody buffer reduced 23 % more runoff, 2 % more sediment, 14 % more total-N,

23 % more NO_3-N, 13 % more total-P, and 22 % more PO_4-P than the switchgrass buffer alone (Total 5). The results indicate that the 7 m wide switchgrass buffer alone was effective in removing sediment and sediment-bound nutrients, and the added woody buffer was effective in removing runoff and soluble nutrients. The importance of vegetation in buffers in improving soil structure and permeability is well known. Vegetation changes the soil structure by adding organic matter that improves aggregations and by creating root channels and thereby increasing the infiltration capacity (Marquez et al., 1999). Infiltration of runoff in buffers may facilitate reduction of both sediment-bound nutrients with small particles and of soluble nutrients. During infiltration, sediment-bound nutrients may be sieved from the water through the soil profile (Dillaha et al., 1988). Furthermore, infiltration into the buffer soil decreases surface runoff, which in turn reduces the ability of the runoff to transport soil particles and particulate P (Dillaha et al., 1988). The lower runoff volume from the buffered plots is attributed to increased infiltration by the vegetation. Lee (1999) reported that total-N and total-P transport was associated with sediment in runoff. The removal of N and P from the runoff was nearly as effective as the sediment removal, and this was expected because 65 and 66 % of the N, and 92 and 90 % of the P leaving the 4.6 m and 9.1 m filter strips, respectively, was sediment-bound.

Mass reductions in sediment, nutrients, and runoff occurred in both the wet and dry year. The degree of runoff depends primarily on antecedent moisture conditions and the return period of rainfall events (Clinnick, 1985). For all storms the trapping effectiveness for sediment did not fall below 71 % for the switchgrass buffer or 86 % for the wider switchgrass/woody buffer.

The switchgrass and switchgrass/woody buffers reduced surface discharge of runoff and mass transport of sediment and nutrients from the crop field to the stream. Under natural rainfall events that generated surface runoff at the end of the buffers, the 7 m wide switchgrass buffer removed more than 92 % of the sediment, and the 16.3 m wide switchgrass/woody buffer removed more than 97 % of the sediment. During one rainfall event in 1997 the difference in removal of sediment was 24 % less in the narrower buffer. The wider switchgrass/woody buffer increased the removal efficiency of soluble nutrients by 20 %. These results would suggest that the narrower switchgrass buffer alone is effective in removing sediment and sediment-bound nutrients but that the wider switchgrass/woody buffer adds a significant ability to also remove soluble nutrients in all but the most intense storm events (more than 75 mm hr^{-1}). Infiltration of runoff water into the soil profile and filtration of sediment by vegetation and organic litter on the buffers were the main mechanisms of nutrient removal from the runoff. Since native grass stands increase in density during the first 4 to 5 years and soil quality changes under the restored buffers is slow, further improvement of soluble nutrient removal can be expected in the future. These results suggest that there are major functional difference between narrow grass filters and wider mixed grass and woody plant buffers. The selection of one over the other should be based on the problems of each particular site.

4. STREAMBANK EROSION AND RIPARIAN CONSERVATION PRACTICES

4.1. Introduction

Streambank erosion along with surface runoff and stream bed re-suspension have been identified as major contributors of sediment and nutrients to streams. Sekely et al. (2002) estimated that streambank erosion in a Minnesota stream contributed 30-45 % of the sediment load to streams, while Odgaard (1984) and Schilling and Wolter (2000) estimated a higher contribution of 45-50 % in several Iowa streams. In other regions of the United States (Simon et al., 1996) and other countries (Kronvang et al., 1997), the contribution was estimated to be up to 80-90 %. Typically incised channels have large erosion rates and contribute significant amounts of sediment to the stream. Streambank widening rates can vary several orders of magnitude from 1.5 m/yr to more than 100 m/yr (Simon et al., 1999). In Iowa and other agriculturally dominated states, streams are deeply incised as a result of human alterations (Figure 11).

Phosphorus (P), moves to surface waters predominantly attached to sediment as particulate P (Sharpley et al., 1987) and has been identified as a limiting nutrient for the eutrophication of many lakes and streams (Correll, 1998). Increased P concentration in streams often promotes algal blooms and excess growth of other aquatic nuisance plants. Aerobic decomposition of the enhanced organic matter production may lead to hypoxic conditions and reduce stream integrity (Carpenter et al., 1998). In Minnesota, Sekely et al. (2002) estimated that only 7-10 % of the total-P in the stream was from streambank erosion, while in Illinois, Roseboom (1987) estimated it to be more than 55 %. In Denmark, Kronvang et al. (1997) estimated streambank erosion to contribute more than 90 % of the stream total-P load.

Photo credit Richard Schultz.

Figure 11. Incised stream channels have accelerated stream bank erosion.

Decreased streambank stability in many cases is the result of reduced vegetation cover on the bank that decreases root length and mass in the soil (Dunaway et al., 1994). Row crop agriculture leaves banks bare for long periods of time with root system that are quite shallow. Livestock overgrazing can also impact vegetation cover substantially. Belsky et al. (1999) reported many studies that have shown livestock grazing reducing streambank stability in the western United States.

Controlling streambank erosion is an expensive task that takes considerable time and effort, especially in heavily disturbed watersheds. Control is difficult because so many interacting factors influence streambank erosion. These factors include watershed and riparian vegetation cover and land-use, topography, bank material, river morphology, weather cycles (especially precipitation patterns), watershed area, and channel stage (Hooke, 1980; Hagerty et al., 1981; Schumm et al., 1984; Geyer et al., 2002). Of major interest are the riparian land-uses that have a direct impact on the vegetation cover and the type and amount of disturbance on the streambanks. This is a factor that humans can manipulate easier compared to the other influential factors.

A relatively inexpensive method for increasing streambank stability is the establishment of riparian forest buffers or grass filters conservation practices. These practices of course would generally take longer to stabilize streambanks than traditional hard and/or soft streambank engineering methods such as rip-rap, live fascines, and brushmattresses (USDA-NRCS, 1996) or in-stream methods such as grade control structures.

The objective of the two studies discussed below was to compare streambank erosion along reaches with different riparian land-uses. The first study was conducted along one single stream reach in central Iowa with three different riparian land-uses, riparian forest buffers (RF), continuous pastures (CP), and annual row cropped fields (RC). The second study was conducted along several streams in three Iowa regions. The riparian land-uses in the second study were RC, CP, rotational (RP), intensive rotational (IP) grazed pastures, pastures where the cattle were fenced out of the stream (FP), grass filters (GF), and RF.

4.2. Methods and Materials

4.2.1. Study Site - Bear Creek in central Iowa

An 11 km reach of Bear Creek was selected that has been designated as a National Restoration Demonstration Watershed by the interagency team implementing the Clean Water Action Plan (1999) and as the Bear Creek Riparian Buffer National Research and Demonstration Area by the United States Department of Agriculture (1998). The 11 km study reach is classified as a second order stream (Strahler, 1957) and lies on the most recently glaciated landform of Iowa (12,000 to 14,000 years), the Des Moines Lobe in central Iowa (latitude 42°11′N, longitude 93°30′W). The area is flat with a poorly integrated natural drainage system that consists of incised stream channels, dredged ditches, and field drainage tiles. The riparian zones of the study reach consist of a RF, two RC, and two CP sub-reaches.

The 20 m wide RF (Figure 9) consists of three vegetation zones parallel to the stream (Isenhart et al., 1998; Schultz et al., 2000). The first zone (10 m wide), nearest the stream, is composed of trees that stabilize the streambank and provide long term nutrient storage. The second zone (3.6 m wide) includes shrubs that increase habitat diversity, reduce floodwater velocities, and trap flood debris. The third zone (6.4 m wide) consists of warm-season grasses and forbs that trap sediment, nutrients, and agricultural chemicals in overland flow. The RF had been established for four or more years.

Corn (*Zea mays* L.) and soybeans (*Glycine max* (L.) Merr.) were grown in alternating years on the RC adjacent to the stream, and both sub-reaches were cropped up to edge but also had narrow strips (less than 4 m) of grasses and annual weeds along the streambank. The two RC sub-reaches were 6.8 km apart. In year 2 a RF was established on the RC 1 sub-reach but was still considered a RC sub-reach because of the relatively young age of the RF.

The riparian CP consisting of cool season grasses, primarily Kentucky bluegrass (*Poa pratensis* L.), were continuously grazed from the beginning of May to the end of October, a practice typical of this region of Iowa. The CP were confined to the riparian zone and back slope of the narrow stream valley, allowing livestock access to the whole pasture and stream channel during the entire grazing season. One of the CP was grazed by beef cattle (CP 1) with much higher animal unit numbers (kg/ha) than the other CP, which was grazed by horses (CP 2).

The dominant soil series for the study reach were Spillville (fine loamy, mixed, mesic Cumulic Haplaquolls) and Coland (fine loamy, mixed, mesic Cumulic Hapludolls) (DeWitt, 1984). Both soils are alluvial, moderately permeable, on 0-2 % slopes, and overall similar topography. Particle analysis for soil samples taken at 30 cm intervals on 13 different streambank faces across all three riparian land-uses classified their texture as clay loam or sandy clay loam.

4.2.2. Study Sites - Iowa Regional Study

Stream reaches selected for this study were located in northeast, central and southeast Iowa. The Iowan Surface and the Paleozoic Plateau are the major landforms in northeast Iowa (Prior, 1991). The Paleozoic Plateau has narrow valleys in sedimentary rock with almost no glacial deposits, and because of the shallow limestone near the surface, there are numerous caves, springs, and sinkholes. The Iowan Surface is dominated by gently rolling terrain created by material loosened and moved by many weathering events caused by conditions during the last glaciation. Central Iowa lies on the Des Moines Lobe landform that has subtly rolling terrain with some broad curved bands or ridges, knobby hills, and irregular ponds and wetlands resulting from the most recent glaciation in Iowa (Prior, 1991). The Southern Iowa Drift Plains landform, in southeast Iowa has many gullies, creeks, and rivers, with steeply rolling hills and valleys (Prior, 1991). Streambank erosion has deepened channels into glacial material deposited 500,000 years ago while a mantle of loess covers the slopes and hills.

Besides riparian land-use, the major criteria for selecting study reaches were as follows: 1) having lengths greater than 300 m with the same land-use on both streambanks, 2) located along first to third order streams (Strahler, 1957), 3) channels in the widening stage (Stage III) of the channel evolution model (Schumm et al., 1984), and 4) owned by private farmers. The focus was on low-order streams because they are in closest contact with their adjacent hillslopes, and therefore can contribute a significant portion of the sediment to larger streams. Low-order streams contribute 30-50 % of the sediment to the Illinois River (Johnson, 2003).

It was not possible to find suitable reaches for all the riparian land-uses in every region. The riparian land-uses and the number of reaches of each riparian land-use in each region are presented in Table 6. In the northeast and southeast region, the watershed area above each reach was less than 52 km^2 while in the central region the area was less than 78 km^2. The hillslopes above all of the riparian areas were dominated by agricultural RC with some pastures and homesteads and occasional small pockets of forests.

Suitable reaches with RC adjacent to the streambanks were found only in the central region. Corn and soybeans were the annual row crops, grown in alternating years. These reaches typically had a narrow strip (less than 4 m) of grasses and/or annual weeds along the streambanks, although many of the crops were grown right up to the streambank edge.

All pastures of this study were grazed by beef cattle and were dominated by vegetation consisting of cool-season grasses and forbs. Each landowner started and ended grazing on different dates for all pasture practices, which led to different numbers of total grazing days. The CP were not divided into paddocks, and the cattle had full access to the stream during the entire grazing period. In the northeast and central region, grazing started in early May and ended in early November. In the southeast one of the CP reaches followed similar dates, while in the other two the cattle remained on the pastures throughout the year. Supplemental feed (like hay) were supplied to cattle that grazed year-around. The grazing period for the IP and RP also ran from early May to early November in all regions. In the RP, the pastures were divided into two to three paddocks. Each paddock was grazed 15-30 d and rested for 30 d. In the IP, the pastures were divided into more than six paddocks and each paddock was grazed 1-7 d and rested for 30-45 d. Because the RP and IP practices had only recently been adopted by farmers with beef cattle in Iowa, study reaches were selected only if they had been converted from continuously grazed or row crop agriculture for more than three years.

RP and IP are slowly replacing traditional CP in Iowa because they maintain more vegetative cover providing better utilization of pasture forages, increase profitability and could be considered more environmentally friendly (USDA-NRCS, 1997). In the RP and IP, the pastures are divided in small sections (paddocks) and livestock are moved from one paddock to the next providing short intensive grazing pressure in a paddock followed by long periods of rest and recovery. The result is more complete utilization of the available forage, with time for that forage to regrow and maintain healthy and strong root systems. While many studies on the influence of intensive rotational and rotational grazing on stream ecosystems have been conducted in the Western United States, very few have been conducted in the Midwest (Lyons et al., 2000).

In the FP reaches, cattle had no access to the channel for at least three years. While this is a practice that might have great potential for decreasing streambank erosion, many cattle farmers in Iowa are reluctant to adopt it because the stream is the main water source for the cattle and because of the extensive and costly fence maintenance that may be required after flashy floods, which often occur in low order streams of Iowa.

The selected GF reaches were vegetated by introduced cool-season grasses (USDA-NRCS, 1999b). The RF reaches were vegetated by zones of trees, shrubs, and warm-season grasses (USDA-NRCS, 1999a). Reaches for both of these land-uses were selected only if they had been established for at least five years.

Table 6. General characteristics of the riparian land-use reaches

Riparian land-use	Reach	Soil series[†]	Average bank height	Width/ Depth	Soil texture	Stocking Rate	Precipitation			
							Year 1[§]	Year 2[§]	Year 3[§]	Year 1-3
	#		m			AUM[‡]	cm			
CENTRAL										
Row cropped fields (RC)	2	Spillville-Coland complex	1.7	3.3	Clay loam, Loam	N/A*	74	75	87	241
Continuous pastures (CP)	2	Coland, Colo, Spillville-Coland complex	1.7	2.7	Silt loam, Clay loam, Loam	16-24	74-91	63-75	79-87	241-246
Rotational Pastures (RP)	2	Coland, Coland-Terrill complex	1.5	2.1	Clay loam	15-34	91	75	79	246
Grass filters (GF)	2	Spillville, Spillville-Coland complex	1.7	2.5	Clay loam, Loam	N/A	74	75	87	241
Riparian forest buffers (RF)	2[x]	Coland, Hanlon-Spillville and Spillville-Coland complexes	1.5	3.8	Clay loam, Loam	N/A	74	75	87	241
NORTHEAST										
Continuous pastures (CP)	3	Dorchester, Radford, Otter-Ossian complex	1.5	3.6	Silt loam	16-23	62-92	52-64	106-107	233-250
Intensive rotational Pastures (IP)	3	Dorchester, Dorchester-Chaeseburge-Viney and Dorchester-Chaeseburge complexes	1.2	3.6	Silt loam	10-20	92-134	52-59	95-108	242-300
Cattle fenced out of streams (FP)	2	Radford, Spillville	1.4	2.3	Silt loam	N/A	91-92	52-55	104-106	249-250
Riparian forest buffers (RF)	2	Colo-Otter-Ossian complex, Spillville	1.0	2.3	Silt loam, Loam	N/A	91-92	52-55	104-106	249-250
SOUTHEAST										
Continuous pastures (CP)	3	Nodaway, Nodaway-Cantril complex	1.8	1.4	Silt loam, Loam	15-23	56-86	54-64	74-86	185-235
Rotational pastures (RP)	2	Nodaway	1.9	2.0	Silt loam	13-29	76-86	59-64	81-86	216-235
Intensive rotational Pastures (IP)	2	Nodaway, Nodaway-Cantril complex	2.1	1.5	Silt loam, Loam	8-13	86-93	54-78	74-100	215-271

Table 6. (Continued)

Riparian land-use	Reach	Soil series[†]	Average bank height	Width/ Depth	Soil texture	Stocking Rate AUM[‡]	Precipitation			
	#		m				Year 1[§] cm	Year 2[§]	Year 3[§]	Year 1-3
Cattle fenced out of stream (FP)	1	Nodaway	1.9	1.4	Silt loam	N/A	86	78	100	271
Grass filters (GF)	2	Amana, Nodaway	1.8	2.1	Silt loam	N/A	76-86	55-59	74-81	215-216

[†] Soil Survey Geographic (SSURGO) Database, 2004.

[‡] AUM = Animal Unit Month.

[§] Year 1: August 2001 to August 2002; Year 2: August 2002 to August 2003; Year 3: August 2003 to August 2004.

[*] N/A=not applicable.

[×] In this region, a natural forest along the stream was used as a riparian forest buffer reach.

Source Zaimes et al., 2008a.

4.2.3. Streambank Erosion Pins

Steel rods, called erosion pins, were inserted perpendicularly into the streambank (Wolman, 1959) (Figure 12). A pin length of 762 mm was used in these studies because erosion rates of up to 500 mm per erosion event had been witnessed in similar size streams. A pin diameter of 6.4 mm was selected because it was small enough to cause minimum disturbance to the banks but large enough to not bend under most high discharge events (Lawler, 1993). Erosion pins are well suited for measuring bank erosion rates for short-time scales and when high resolution is needed (Lawler, 1993). Resolution can be as high as 5 mm (Simon et al., 1999). Accuracy, in these studies, was increased even more because all pin measurements were collected by one operator (Couper et al., 2002).

Only severely eroding banks (Figure 12a) were selected for pin plot placement because these banks are the major source of sediment in streams (Beeson and Doyle, 1995). These banks are bare with slumps, vegetative overhang and / or exposed tree roots (USDA-NRCS, 1998).

In the Bear Creek study the number of pin network plots placed in each riparian land-use was based on its total bank length and number of severely eroded bank sites. The RF had six plots, the RC had eight plots and the CP had eight plots that were randomly selected. The horizontal and vertical distances between the pins in the plots were 1 m and 0.3 m, respectively. Vertical distances were smaller because stream bank erosion is more variable vertically than horizontally (Lawler, 1993).The lengths of the exposed portions of the pins were measured approximately every month from June 1998 to June 2000 and bimonthly from June 2000 to July 2002, except during the winter months. During the winter, most pin plots were typically not easily accessible and / or covered with snow and ice. A much milder winter in 2001 allowed the measurement of the pins in January.

In the Iowa Regional study, a preliminary field survey identified and recorded all the severely eroding banks in each reach. Afterwards, randomly five of these severely eroding banks in each reach were selected and erosion pin plots were established. The erosion pin plot included two horizontal rows with five pins. Pins within these rows were placed 1 m apart for a total horizontal length of 4 m (Figure 12a). The horizontal rows were placed at 1/3 and 2/3 of the height of the bank. Exposed pin lengths were measured once in the spring, summer, and fall of each year from August 2001 to August 2004.

The rate of streambank erosion was estimated by subtracting each measurement on the erosion pin from the previous measurement. Positive differences indicated erosion, while negative differences indicated deposition.

4.2.4. Severely Eroding Bank Areas Survey

Another survey to measure the length and height (at certain intervals) for every severely eroded bank in each riparian land-use reach (or sub-reach) was also conducted. The height of the eroding portion of the streambank was estimated with a scaled height pole. The percentage of total severely eroding bank length for each reach (or sub-reach) was calculated by dividing the total length of its severely eroding banks by its total bank length (USDA-NRCS, 1998). Eroding streambank areas were calculated as the product of the average height and total length of each severely eroded bank within a reach (or sub-reach). The total severely eroding bank area for each reach (or sub-reach) was determined as the sum of all its severely eroding bank areas. This survey was conducted in summer 1998 and 2002 for the Bear Creek study and only in the summer of 2002 for the Three Iowa Regions study.

(a)

(b)

Photo credit a) Tom Schultz and b) Mustafa Tufeckioglu.

Figure 12. a) A severely eroding bank with a network of erosion pins and b) a close-up of an eroding pin.

4.2.5. Total Soil and Phosphorus Losses

The product of streambank erosion rate, streambank soil bulk density, and severely eroding bank area for each riparian land-use was used to estimate its total soil loss from the streambanks. Streambank erosion rate was the average rate of all the pin plots in the riparian land-use. By multiplying the total soil loss by the average streambank soil P concentration in each riparian land-use, total-P loss from streambanks was estimated. For the Bear Creek study only total soil losses were estimated.

Streambank soil and P loss per unit of stream length were estimated by dividing the streambank soil and P loss for each riparian land-use by its total stream reach (or sub-reach) length. This was necessary because each riparian land-use had a different total stream reach (or sub-reach) length.

4.2.6. Rainfall Data

Rainfall data were used from the National Oceanic and Atmospheric Administration (NOAA) weather station closest to each study reach. Yearly rainfall data were correlated to yearly streambank erosion. For these studies, precipitation was the best available variable to correlate to streambank erosion because discharge data were not available for any of the low-order streams. The flashiness of most first to third order streams in Iowa allows good seasonal correlation between precipitation and discharge, especially in spring and early summer (Zaimes et al., 2006)

4.2.7. Statistical Analysis

An analysis of variance (ANOVA) in the Statistical Analysis System (SAS) (SAS Institute, 1999) was conducted for the Bear Creek study. Comparisons for the streambank erosion rate were made among riparian land-uses for each measurement period, for each year, for the sum of the last three years, and for the sum of all four years of the study. For the Iowa Regional study the analysis of covariance in the Statistical Analysis System (SAS) was used to examine the impacts of riparian land-use on streambank erosion rate for each year and for all three years measurements were taken (SAS Institute, 1999). Rainfall was used as a covariate in the above model because even in the same region, some riparian land-use reaches received different amounts of rainfall. In addition an ANOVA in SAS was used to compare the percentage of severely eroding bank lengths. Differences were considered significant when the p-values were less than 0.10.

4.3. Results

4.3.1. Bear Creek in Central Iowa

Statistically, the bank erosion rate for Year 1 was significantly greater than the mean of Years 2, 3, or 4 for all three riparian land-uses combined, and for each riparian land-use. Year 1 had above average yearly precipitation (992 mm) and twice the average for the months in which major bank erosion was measured (302 and 192 mm, in June 1998 and April 1999, respectively), compared to the 51-year averages (833 mm, 128 mm, and 88 mm yearly, June and April, respectively). Years 2, 3, and 4 had annual precipitation amounts below the 51-year average, with 459 mm, 807 mm, and 622 mm, respectively.

In Year 1, streambank erosion rate in the RF sub-reach was significantly less than the rates in the RC and CP reaches (Figure 13). In Year 2, the erosion rate in the RF sub-reach was only significantly less than that in the RC sub-reaches. There were no statistically significant differences for Years 3 and 4 in the erosion rates among the three riparian land-uses. The erosion rates for the entire period (Years 1 through 4) was significantly less in the RF subreach than in the RC and CP sub-reaches. The erosion rate for the entire study period increased in the following order: RF (198 mm), CP (594 mm), and RC (643 mm). Finally, the erosion rate among the two CP sub-reaches differed significantly. The CP 1 sub-reach with higher stocking rates was significantly higher than that in the CP2 sub-reach with lower stocking rates for the entire study period and in Year 3. For the CP 1 sub-each the erosion rates were 278 and 903 mm for Year 3 and the entire period, respectively, while for CP 2 52 and 284 mm for Year 3 and the entire study period, respectively.

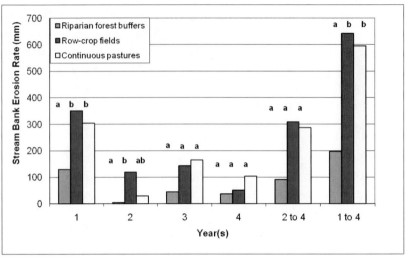

Source Zaimes et al., 2006.

Figure 13. The streambank erosion rate for the three riparian land-uses along Bear Creek in central Iowa. Different letters indicate significant differences ($p < 0.10$). Year 1, June 1998 - June 1999; Year 2, June 1999 - June 2000; Year 3, June, 2000 - July 7, 2001; Year 4, July 2001 - July 2002.

The RF sub-reach had significantly less severely eroded bank length and area than the RC and CP sub-reaches in both 1998 and 2002 (Figures 14 and 15). Lengths and areas of severely eroded banks decreased or remained similar for almost all riparian land-uses between the 1998 survey and the 2002 survey. Eroding lengths and areas in the RF sub-reach decreased both by 56 %. These decreases were much greater than those of the two RC sub-reaches, 11 and 20 %, respectively, and the two CP reaches, both by 2 %. The decreases in the eroding lengths and/or areas between the two RC and two CP sub-reaches were not always consistent. In 1998, RC 1 had larger bank eroding areas than RC 2, but this was reversed in 2002 (Figure 5). The eroding areas were reduced by 41 percent for RC 1 sub-reach but only by 7 percent in RC 2 during that period. The CP 2 sub-reach had smaller eroding lengths and areas than the CP 1. The CP 1 sub-reach was the only riparian land-use to show an increase in severely eroded bank lengths and areas between 1998 and 2002.

Soil loss per unit of bank length was always least for the RF sub-reach (Figure 16). In Year 1 the RC and CP soil losses were similar. In Year 2 the RC had the highest soil loss per unit of bank length, while in Years 3 and 4 the CP had the highest soil loss. Over the entire period, soil loss per unit of bank length for the three riparian land-uses increased in the following order: RF 75 tons km^{-1}; RC 484 tons km^{-1}; and CP 557 tons km^{-1}. Finally, Year 1 had the highest soil loss per unit of bank length for all riparian land-uses. Both the RF and the RC had higher soil losses in this period than in the other three periods combined.

Applying the soil loss per unit length to the entire stream length of the study reach, streambank erosion contributed 2,652 tons soil in Year 1, 554 tons in Year 2, 1,086 tons in Year 3, 582 tons in Year 4 for a total of 4,873 tons over the entire period.

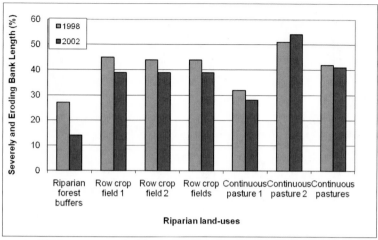

Source Zaimes et al., 2006.

Figure 14. Severely eroded bank lengths for the three riparian land-uses and their sub-reaches along Bear creek in central Iowa, based on streambank erosion surveys taken in 1998 and 2002.

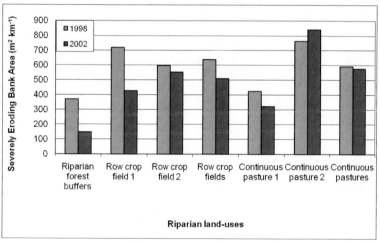

Source Zaimes et al., 2006.

Figure 15. Severely eroded bank areas for the three riparian land-uses and their sub-reaches along Bear creek in central Iowa, based on the streambank erosion surveys taken in 1998 and 2002.

4.3.2. Iowa Regional Study

Over the entire three-year period, average erosion rates among land uses ranked as follows: in the central region, RC > CP > GF > RP > RF; in the northeast region, CP > IP > FP > RF; in the southeast region, RP > CP > IP > FP > GF (Table 7). The differences among the riparian land uses based on the above rankings were not always significant. Specifically, in the southeast region there were no significant differences in the erosion rate among any of the riparian land uses.

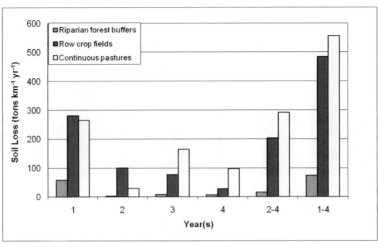

Source Zaimes et al., 2006.

Figure 16. Soil losses per unit lengths for the three riparian land-uses along bear creek in central Iowa. Year 1, June 1998 - June 1999; Year 2, June 1999 - June 2000; Year 3, June, 2000 - July 7, 2001; Year 4, July 2001 - July 2002.

In the central region, the reaches along RC had significantly higher annual and three-year average erosion rates than those along the RF, GF, and RP reaches (Tables 7). The three-year average erosion rates for the CP reaches were significantly higher than those of the RF reaches. The CP reaches also had significantly higher erosion rates than the RF reaches during the last two years of the study, and the GF reaches in Year 3. The differences in this region were the hypothesized, although even more significant differences among the riparian land-uses were expected.

In the northeast region, the CP and IP reaches had significantly higher three-year average erosion rates than the RF and FP reaches (Tables 7). In Year 1, RF and FP reaches had net deposition, while the reaches along the CP and IP had low erosion rates. Because of this fact, the FP reaches had significantly lower erosion rates than the CP reaches. Deposition was probably experienced because of frequent freeze-thaw activities during the winter period that caused material to fall from the top to the bottom of the streambanks while the low streamflows during this period were not able to remove the deposited material. In Year 2 all riparian land-use reaches experienced erosion, even though precipitation total amounts were lower than in Year 1. During Year 2, the erosion rates on the CP reaches were significantly higher than those on the RF reaches. In Year 3, CP and IP reaches had significantly higher erosion rates than the RF and FP banks. In this year IP reaches also had significantly higher erosion rates than the CP reaches. This was something not expected, although it must be noted that over the three-year period CP and IP had very similar erosion rates.

Lyons et al. (2000) found that 1-66 % of the streambank lengths of streams surveyed in Wisconsin were severely eroding, similar to the 10-54 % found in this study (Table 8). The lowest percentage (10 %) was found along RF reaches in northeast Iowa while the highest (54 %) was found along the CP and RP reaches in the southeast (Table 8). In many cases riparian land-uses that had perennial vegetation and fenced out livestock (RF, GF, and FP) had significantly lower percentages than the riparian agricultural land-uses (RC, CP, RP, and IP). Specifically, the RC reaches in the central region had significantly larger severely eroding lengths than the GF and RF reaches. In the northeast region the CP and IP reaches had significantly larger severely eroding lengths than the FP and RF reaches. Finally, in the southeast region all three pasture systems (CP, RP and IP) had significantly larger severely eroding lengths than the GF reaches. Among the grazing riparian land-uses the only significant difference was found in the southeast, where IP severely eroding lengths were lower than the other two grazing practices. In southwestern Wisconsin, Simonson et al. (1994) suggested that streams of high quality should have less than 20 % of their streambank lengths severely eroding. In this study, severely eroding streambank lengths along RF, GF, and FP in all regions were always below this percentage. In contrast, the CP, RP, IP, and RC had 25 % or more of their streambank lengths severely eroding across all regions.

Total soil and P losses among riparian land-uses were strongly correlated to the lengths of severely eroding streambanks (Table 5). Again, RF, GF, and FP streambanks had the lowest losses regardless of region.

Streambanks along RC in the central region had the highest soil losses with 304 tons km^{-1} year^{-1} (Table 8). Across all regions, streambank soil losses along CP ranged from 197 to 264 tons km^{-1} year^{-1}, while those along RP ranged from 94 to 266 tons km^{-1} year^{-1}, and those along IP ranged from 124 to 153 tons km^{-1} year^{-1}. Streambanks along FP and GF had soil loses that ranged from 6 to 61 tons km^{-1} year^{-1} and 22-47 tons km^{-1} year^{-1} respectively, while those along RF had losses ranging 5-18 tons km^{-1} year^{-1}.

Total-P concentration differences in streambank soils among riparian land-uses (Table 8) were not significant (Zaimes et al., 2008b). Total-P losses from streambanks along RC in central Iowa were 108 kg km^{-1} year^{-1}. Across all regions total-P losses along CP ranged from 71 to 123 kg km^{-1} year^{-1}, while along RP losses ranged from 37 to 122 kg km^{-1} year^{-1}, and along IP losses were 66 kg km^{-1} year^{-1}. Streambanks along FP, GF, and RF had the smallest total-P losses ranging from 3 to 34, 9-14, and 2-6 kg km^{-1} year^{-1}, respectively.

4.4. Discussion

Along Bear Creek the RF had the lowest values for all four bank erosion variables (rate, severely eroded bank lengths, severely eroded bank areas, and soil loss) compared to the RC and CP sub-reaches. When streambanks become covered with perennial vegetation, their roughness coefficient should increase, which in turn should decrease stream water velocity and stream power along the bank. Trees, especially willows (*Salix* spp.) and silver maple (*Acer saccharinum* L.), protect streambanks because of their high stem density and deep and fibrous rooting habits that provide support to soil to resist fluid entrainment (Shields et al., 1995; Abernethy and Rutherfurd, 2000). The increase in shear resistance of streambanks from trees depends on the rooting depth of the trees and streambank height (Simon et al., 1999). If roots cover only part of the streambank face (where banks are very tall), then bank shear strength will not increase for the entire bank face, leaving the uncovered parts susceptible to undercutting and streambank erosion. In contrast to trees, annual row crops have very shallow annual roots that are present for only part of the year. Heavy stocking rates in CP also reduce the number and length of roots on perennial pasture grasses that anchor the soil, while heavy trampling along and on the streambank increases streambank instability (Belsky et al., 1999).

Table 7. Streambank erosion rates under different riparian land-uses in three Iowa regions[†]

Riparian land-use	Stream bank erosion rates									
	YEAR 1[‡]	SD[§]	YEAR 2[‡]	SD[3]	YEAR 3[‡]	SD	SUM YEAR 1-3	SD	AVERAGE YEAR 1-3	SD
	mm		mm		mm		mm		mm yr[-1]	
CENTRAL										
Row cropped fields (RC)	225 (74)	a	223 (59)	a	271 (37)	ab	717 (137)	a	239 (46)	a
Continuous pastures (CP)	79 (71)	a	128 (64)	ab	298 (40)	a	499 (133)	ab	166 (44)	ab
Rotational pastures (RP)	70 (73)	a	54 (75)	b	198 (44)	bc	313 (135)	bc	104 (45)	bc
Grass filters (GF)	87 (74)	a	66 (59)	b	168 (37)	c	319 (137)	bc	106 (46)	bc
Riparian forest buffers (RF)	54 (74)	a	4 (59)	b	83 (37)	c	139 (137)	c	46 (46)	c
NORTHEAST										
Continuous pastures (CP)	151 (63)	a	184 (48)	a	137 (45)	b	512 (109)	a	171 (36)	a
Intensive rotational pastures (IP)	114 (65)	ab	98 (53)	ab	313 (39)	a	511 (130)	a	170 (43)	a
Cattle fenced out of streams (FP)	-25 (73)*	b	51 (65)	ab	24 (47)	c	67 (137)	b	22 (46)	b
Riparian forest buffers (RF)	-10 (73)*	ab	36 (65)	b	1 (47)	c	45 (137)	b	15 (46)	b
SOUTHEAST										
Continuous pastures (CP)	127 (61)	a	23 (50)	a	182 (40)	a	302 (125)	a	101 (42)	a
Rotational pastures (RP)	166 (72)	a	16 (59)	a	199 (39)	a	366 (136)	a	122 (45)	a
Intensive rotational pastures (IP)	59 (72)	a	55 (62)	a	169 (37)	a	281 (134)	a	94 (45)	a
Cattle fenced out of stream (FP)	42 (102)	a	-6 (102)	a	95 (55)	a	173 (209)	a	58 (70)	a
Grass filters (GF)	37 (72)	a	12 (61)	a	109 (46)	a	123 (143)	a	41 (48)	a

[†] The mean rainfall that each riparian land-use reach received was used as a covariate to estimate stream bank erosion rate. In parentheses is the standard error.

[‡] Year 1: August 2001 to August 2002; Year 2: August 2002 to August 2003; Year 3: August 2003 to August 2004.

[3] Different letters indicate significant differences (p < 0.10) among riparian land-uses.

* Negative numbers indicate deposition

Source Zaimes et al., 2008a.

Table 8. Total and per unit stream reach length of soil and total phosphorus losses from streambank erosion under different riparian land-uses in three Iowa regions

Riparian land-use	Stream reach length	Severely eroding streambank		Area	Bulk density	Streambank soil loss		Streambank soil phosphorus concentrations	Streambank phosphorus loss	
	total	Length[†]	SD[‡]			total	unit length		total	unit length
	km	%		m^2	tons m^{-3}	tons yr^{-1}	tons km^{-1} yr^{-1}	kg tons^{-1}	kg yr^{-1}	kg km^{-1} yr^{-1}
CENTRAL										
Row cropped fields (RC)	1.6	44 (6)	a	1657	1.23	487	304	354	172	108
Continuous pastures (CP)	1.7	39 (6)	ab	1999	1.35	448	264	349	156	92
Rotational pastures (RP)	1.3	25 (6)	b	899	1.31	122	94	398	49	37
Grass filters (GF)	1.6	16 (6)	bc	615	1.16	76	47	303	23	14
Riparian forest buffers (RF)	1.4	14 (6)	bc	430	1.24	25	18	350	9	6
NORTHEAST										
Continuous pastures (CP)	1.6	38 (5)	a	1935	1.15	381	238	518	197	123
Intensive rotational pastures (IP)	1.5	27 (5)	a	1125	1.20	230	153	432	99	66
Cattle fenced out of streams (FP)	0.8	11 (6)	b	203	1.16	5	6	464	2	3
Riparian forest buffers (RF)	0.8	10 (6)	b	244	1.10	4	5	479	2	2
SOUTHEAST										
Continuous pastures (CP)	1.8	54 (5)	a	2661	1.32	355	197	360	128	71
Rotational pastures (RP)	1.5	54 (6)	a	2403	1.36	399	266	459	183	122
Intensive rotational pastures (IP)	0.7	32 (6)	b	371	1.28	87	124	531	46	66
Cattle fenced out of stream (FP)	0.3	16 (6)	bc	239	1.32	18	61	555	10	34
Grass filters (GF)	0.7	16 (6)	c	289	1.29	15	22	406	6	9

[†] In parentheses is the standard error.

[‡] Different letters indicate significant differences ($p < 0.10$) among riparian land-uses.

Source Zaimes et al., 2008a.

The impact of RF was also evident in the two RC sub-reaches. While the severely eroded bank length for both RC decreased by 5-6 % (Figure 14) from 1998 to 2002, the severely eroded bank area of RC 1 decreased much more (41 %) than the RC 2 (7 %) (Figure 15). The difference between the two RC was the establishment of a RF in spring of Year 2 along the RC 1. Even though newly planted perennial plants in the buffer provided only minimal cover, cultivation and harvest disturbances associated with RC culture were eliminated, minimizing disturbances along the streambanks of RC 1 compared to RC 2.

The impact of different grazing management is evident in the two CP sub-reaches. The CP 2 sub-reach with cattle and higher stocking rates had significantly higher streambank erosion rates than those of CP 1 sub-reach with horses for the entire study period and for Year 3. The CP 2 was also the only sub-reach to have an increase in the severely eroding stream length (3 %) and eroding area (10 %) from 1998 to 2002 (Figure 14 and 15). The reasons for the increase in its severely eroded bank lengths and areas was probably correlated to the increase in cattle stocking rates from 1999 through 2001 compared to 1998 and the smaller avaialability of water in the channel during these drier years that led cattle to enter the stream in more places. In contrast, the CP 1 sub-reach showed a decrease in eroding length (4 %) and an even larger decrease in eroding area (24 %). The main difference among the two adjacent sub-reaches was that the CP 2 had 3-3.5 times more animal units (kg ha^{-1}) than the CP 1, depending on the year. In addition, cattle tend to spend more time in and close to streams than other livestock (Platts, 1981). Finally, the stocking rate for the cattle pasture was particularly high with 2.3-3.1 cow/calf pairs ha^{-1}. For Kentucky bluegrass (*Poa pratensis* L.) pastures in Iowa, the recommended stocking rates are 0.5-1.3 cow/calf pairs ha^{-1} (Barnhart et al., 1998).

In the Iowa Regional study with several streams results were similar to the Bear Creek study. Based on the responses of the four variables (erosion rate, severely eroding bank lengths, soil losses and P losses) of this study (Tables 7 and 8), RF was the land-use that stabilized streambanks and minimized soil and total-P losses the most. These responses are especially encouraging because most of the RF had only recently been established following the abandonment of past riparian management practices such as RC and CP. The GF riparian land-use followed but was not as efficient. This could have been because the GF were even younger than the RF in some cases. In addition, tree root systems probably provide more protection to streambanks than grass roots along the deeply incised channels with nearly vertical banks that were found along our study reaches. There has been a lot of debate about the role of roots in bank stabilization, with some indicating tree roots as more effective (Gregory et al., 1991), while others suggest grass roots are more effective (Lyons et al., 2000). Recent studies indicate that trees stabilize streambanks better because of the greater quantity of larger diameter roots (Wynn et al., 2004; Wynn and Mostaghimi, 2006). In general, when selecting riparian vegetation for streambank stability, it is very important to not only consider the hydrologic channel processes but also the mechanical and ecological processes that control streambank stability (Simon and Collison, 2002). When bank height exceeds the rooting depth of the vegetation other stream stabilization techniques might be necessary.

In the FP reaches, streambank stability was greater than in the other grazing systems that allowed cattle full access to the stream (Tables 7 and 8). Cattle are attracted to riparian areas and tend to spend a lot of time in and around the stream (Trimble and Mendel, 1995). Improvements in streambank stability in FP reaches have also been found in other studies (Laubel et al., 2003), but this practice is not socially and economically acceptable to many farmers in Iowa. Where off-stream water is provided as an alternative to fencing, streambank erosion also has been dramatically reduced (Sheffield et al., 1997) and in some cases cattle weight gains have even been seen (Porath et al., 2002). In Iowa, off-stream water without fencing would not be as effective as it is in some other states because many pastures are confined to the narrow riparian corridors along low order streams.

There were minimal significant differences when comparing RP and IP to CP reaches (Tables 7 and 8). The differences in individual farmer interpretation of each of these grazing practices and the fact that some of the RP and IP systems had been established for no longer than three years may have contributed to the inconsistencies that were found in this study. Work by Lyons et al. (2000) suggested that IP can improve streambank stability and decrease soil losses. Decreased erosion and increased stability could be attributed to the shorter time cattle spend in the stream and the adjacent riparian areas thereby reducing streambank disturbance. Bank stabilization could probably increase more if the number of paddocks along the stream decreased and the number of paddocks in the uplands increased. This would decrease the time that cattle spend in the riparian areas. Even with decreased numbers of riparian paddocks there may not be enough rest to allow plants to get reestablished on heavily disturbed streambanks. It appears that bank healing seems to require more time than regrowth of the forage in the paddocks. So in many cases, the keys to successful recovery of streambank stability in pastures will include decreasing animal stocking rates, controlling the timing of grazing in the riparian paddocks, especially under wet conditions and when the least damage to the plants can be done or by eliminating cattle from the streambanks completely until plants are reestablished (Clary and Kinney, 2002).

The RC reaches had the highest erosion rates, largest severely eroding areas and soil and P losses in the region they were present. Leaving the streambanks bare for long periods of time, the shallow root systems of the crops during the rest of the period and the disturbance by the cultivation machinery appear to destabilize them and make them heavily susceptible to erosion.

The percentage of severely eroding bank lengths showed more significant differences among riparian land-uses (Table 8) than the erosion rates (Tables 7). As the erosion pins were placed on severely eroding banks, high erosion rates were expected. The data from this study suggest that the percentage of severely eroding bank lengths provides a better indicator of the impacts of the adjacent riparian areas.

Overall in both studies the large soil and P losses per unit stream length from streambanks along pastures with full livestock access (CP, RP and IP) to the stream and RC indicate that streambank erosion can be a significant contributor to the stream water sediment and P load. In contrast, regardless of whether it is trees or grasses, perennial plant communities with vigorous root systems increase streambank stability. Vegetation is an integral part of the riparian landscape, and the amount of streambank vegetation, especially in low order streams, is important because of the stabilizing support the roots can provide. Riparian conservation practices can provide a feasible solution to improve stream water quality in agricultural watersheds by cost-effectively stabilizing streambanks.

CONCLUSION

Major land-uses changes in the Iowa landscape have altered the hydrologic cycle and accelerated surface and streambank erosion. These accelerated erosion rates must be reduced to restore stream and riparian area health and maintain sustainable production in agriculture.

Vigorous perennial plant communities in riparian areas are efficient at reducing nonpoint source pollutants from reaching the stream. In the Bear Creek surface runoff study mass reductions in sediment, nutrients, and runoff occurred in both the wet and dry years. Regardless of the type of the buffer, sediment trapping efficiency did not fall below 71 %. Based on this estimate, if buffers were established along all streams in Iowa they would have the potential of capturing 12,417,304 tons yr^{-1} of soil from entering streams across the entire state. This is based on the USDA-NRCS (2000) estimate that 12 ton ha^{-1} yr^{-1} of soil eroded from 1/3 of Iowa's land (Iowa total area is 14,574,300 ha) and that approximately 30 % of the eroded soil reaches the stream. This trapping efficiency for conservation buffers is expected when field surface runoff is uniformly distributed across the entire buffer area. If concentrated flow paths transect the riparian zone and deliver the sediment and associated nutrients and agricultural chemicals directly to a stream, buffer efficiency would be reduced (Knight et al., 2010).

Similarly streambank erosion adjacent to riparian forest buffers, grass filters or riparian areas where cattle were fenced out of the stream, was significantly lower than on banks adjacent to agricultural practices (row crop fields, continuous, rotational and intensive rotational pastures). Specifically, in the Bear Creek study if all non-buffered sub-reaches (7.8 km) had established riparian forest buffers soil losses from stream bank erosion would decrease by a total of 3,709 tons or an 84 % reduction over a four year period (this percentage ranged from 77-97 depending on the year). In the Iowa Regional study soil and phosphorus losses per unit length of the conservation practices were in the range of 2-48 times and 2-62 times less, respectively than the agricultural land-uses (Table 8). However, if stream power is not reduced along with the increased bank protection, the channel will erode sediment from other non-buffered reaches to maintain the channel-equilibrium as suggested by Lane's (1955) model.

The density and type of riparian vegetation also influence surface and stream bank erosion. The switchgrass (warm-season grass) buffer had greater sediment reduction than the cool-season grass buffer. Warm-season grasses have a uniform distribution of their plants with stiff stems and large production of litter. Major functional differences were also evident between the narrower switchgrass buffer and wider switchgrass/woody buffers. The wider switchgrass/woody buffer had higher sediment trapping effectiveness and nutrient removal compared to the narrower switchgrass. Still the narrower switchgrass buffer alone is effective in removing sediment and sediment-bound nutrients but if removal of soluble nutrients is also wanted a wider switchgrass/woody buffer should be established. The additional buffer width with high infiltration capacity provided by the deep-rooted woody plant zone appears to enhance the trapping of clay particles and soluble nutrients. Finally, the relatively low removal of orthophosphate indicates that the buffers should be maintained by harvesting grasses and trees to remove accumulated phosphorus or to reduce phosphorus saturation.

In the Iowa Regional study that included both grass filters and riparian forest buffers no significant streambank erosion differences were found among these practices. Still there are indications that riparian forest buffers stabilized streambanks and minimized soil and total phosphorus losses the most because there were more significant differences between the riparian forest buffers and agricultural practices than between the grass filters and agricultural practices. In the central Iowa region grass filters reaches had more than double the erosion rate and soil and phosphorus losses than the riparian forest buffer reaches. Trees appear to stabilize streambanks better than grasses alone because of the greater quantity of larger diameter roots (Wynn et al., 2004; Wynn and Mostaghimi, 2006). In incised channels, riparian forest buffers can help increase streambank stability because of their deeper, thicker and denser rooting system. Of course when bank height exceeds the rooting depth of the vegetation, undercutting can result in large trees falling into the channel. Under such conditions additional stream stabilization techniques might be necessary (Simon et al., 1999).

Narrow corridors of riparian forest buffers and grass filters effectively reduce sediment and sediment-bound nutrients from streambank erosion and surface runoff. Use of these conservation practices allows the largest part of the watershed to remain in agriculture production. At the same time, riparian forest buffers and grass filters provide some financial incentives to farmers because the Conservation Reserve Program subsidizes some of the lost income for agricultural land planted in this riparian conservation practices. Riparian conservation practices coupled with the appropriate upland and in-stream conservation practices in target sources areas can help achieve the goals of agriculture sustainability and protection of stream waters.

REFERENCES

Abernethy, B. and I.A. Rutherfurd, 2000. Does the weight of riparian trees destabilize riverbanks? Regulated rivers: *Res. Manage.* 16:565-576.

Alberts, E.E., W.H. Neibling, and W.C. Moldenhauer. 1981. Transport of sediment and phosphorus in runoff through cornstalk residue strips. *Soil Sci. Soc. Am. J.* 45:1177-1184.

Allan, J.D. 2004. Landscapes and riverscapes: The influence of land use on stream ecosystems. *Annu. Rev. Ecol. Evol.* S. 35:257-284.

Amiri-Tokaldany, E., S.E. Darby, and P. Tosswell, 2003. Bank stability analysis for predicting reach scale land loss and sediment yield. *J. Am. Water Resourc. Assoc.* 39:897-909.

Andersen, K.L. 2000. Historical alterations of surface hydrology in Iowa's small agricultural watersheds. M.S. Thesis, Iowa State University, Ames, IA.

Andersen, K.L. and T.R. Bishop. 1996. Historical analysis of surface waters: Storm Lake and Bear Creek watersheds. In: Schultz R.C. and T.M. Isenhart, (eds.) Progress report and renewal request – Technical support document – Riparian management systems (RiMS), design, function and location. Agroecology Issue Team, Leopold Center for Sustainable Agriculture, Iowa State University, Ames, IA.

Baker, J.L, S.W. Melvin, D.W. Lemke, P.A. Lawlor, W.G. Crumpton, and M.J. Helmers. 2004. Subsurface drainage in Iowa and the water quality benefits and problem. In: Cooke R. (ed.) Drainage VIII, Proceedings of the Eighth International Symposium, 21-24 March 2004, Sacramento, CA. p. 39-50.

Barnhart, S., D. Morrical, J. Russell, K. Moore, P. Miller, and C. Brummer, 1998. Pasture management guides for livestock producers. Iowa State University Extension PM-1713, Iowa State University. Ames, IA.

Beeson, C.E. and P.F. Doyle, 1995. Comparison of bank erosion at vegetated and non-vegetated channel bends. *Water Resourc. Bul.* 31:983-990.

Belsky, A.J., A. Matzke, and S. Uselman. 1999. Survey of livestock influences on stream and riparian ecosystems in the western United States. *J. Soil Water Conserv.* 54:419-431.

Bharati, L., K-H. Lee, T.M. Isenhart and R.C. Schultz. 2002. Riparian zone soil-water infiltration under crops, pasture, and established buffers. *Agrofor. Syst.* 56:249-257.

Bulkley, R.V. 1975. A study of the effects of stream channelization and bank stabilization on warm water sport fish in Iowa. Subproject No 1, Inventory of major stream alterations in Iowa. U.S. Fish and Wildlife Service FWS/OBS-76-11, Ames, IA.

Burkart, M.R., S.L. Oberle, M.J. Hewitt, and J. Pickus. 1994. A framework for regional agroecosystems characterization using the National Resources Inventory. *J. Environ. Qual.* 23, 866-874.

Carpenter, S.R., N.F. Caraco, D.L. Correll, R.W. Howarth, A.N. Sharpley, and V.H. Smith. 1998. Nonpoint pollution of surface waters with phosphorus and nitrogen. *Ecol. Appl.* 8:559–568.

Clary, W.P. and J.W. Kinney. 2002. Streambank and vegetation response to simulated cattle grazing. *Wetlands* 22:139-148.

Clesceri, L.S., A.E. Greenberg, and R.R. Trussell. 1989. Standard methods for the examination of water and wastewater. 17th ed. American public Health Association, Washington, D.C.

Clinnick, P.F. 1985. Buffer strip management in fire operation: a review. *Aust. For.* 48:33-45.

Cooper, A.B. C.M. Smith, and M.J. Smith. 1995. Effects of riparian set-a side on soil characteristics in an agricultural landscape: Implications for nutrient transport and retention. *Agric. Ecosys. Environ.* 55:61-67.

Correll, D.L. 1997. Buffer zones and water quality protection: General principles. In: Haycock, N.E., T.P. Burt, K.W.T. Goulding, and G. Pinay (ed.) Buffer zone: Their processes and potential in water protection. Quest Environmental, Harpenden, Herts, UK. p. 7-20.

Correll., D.L. 1998. The role of phosphorus in the eutrophication of receiving waters: A review. *J. Environ. Qual.* 27:261-266.

Couper, P., T. Stott, and I. Maddock, 2002. Insights into river bank erosion processes derived from analysis of negative erosion- pin recordings: Observations from three recent UK studies. *Earth Surf. Processes Landforms* 27:59-79.

Dabney, S.M., K.C. Mcgregor, L.D. Meyer, E.H. Grissinger, and G.R. Foster. 1993. Vegetative barriers for runoff and sediment control. In: Mitchell, J.K. (ed.) Integrated Resource Management and Landscape Modification for Environmental Protection. ASAE. St. Joseph, MI. p. 60-70.

Dewitt, T.A. 1984. Soil survey of Story County, Iowa. USDA Soil Conservation Service, Washington, D.C.

Dillaha, T.A. and S.P. Inamdar. 1997. Buffer zones as sediment traps or sources in buffer zones. In: Haycock, N.E., T.P. Burt, K.W.T. Goulding, and G. Pinay (ed.) Buffer zones: Their processes and potential in water protection. Quest Environmental, Harpenden, Herts, UK. p. 33-42.

Dillaha, T.A., J.H. Sherrard, D. Lee, S. Mostaghimi, and V.O. Shanholtz. 1988. Evaluation of vegetative filter strips as a best management practice for feed lots. *J. Water Pol. Control Fed.* 60:1231-1238.

Dodds, W.K. and R.M. Oakes. 2006. Controls on nutrients across a prairie stream watershed: Land use and riparian cover effects. *Environ. Manage.* 37:634-646.

Dinsmore, J.J. 1994. A country so full of game: A story of the wildlife in Iowa. University of Iowa Press, Iowa City, IA.

Dunaway, D., S.R. Swanson, J.Wendel, and W. Clary, 1994. The effect of herbaceous plant communities and soil textures on particle erosion of alluvial streambanks. *Geomorphology* 9:47-56.

Evans, R.D., A. Proving, J. Mattie, B. Hart, and J. Wisniewski. 1997. Interactions between sediments and water summary of the 7th international symposium. *Water Air Soil Pol.* 99:1-7.

Gburek, W.J., A.N. Sharpley, A.L. Heathwaite, and G.J. Foldor. 2000. Phosphorus management at the watershed scale: a modification of the phosphorus index. *J. Environ. Qual.* 29:130-144.

Gee, G.W. and J.W. Bauder. 1986. Particle size analysis. Pp.383-411. In A. Klute (ed.) Methods of soil analysis, Part I. American Society of Agronomy Madison, WI.

Geyer, W.A., T. Neppl, K. Brooks, and J. Carlisle, 2002. Woody vegetation protects streambank stability during the 1993 flood in central Kansas. *J. Soil Water Conserv.* 55:483- 486.

Gilliam, J.W. 1994. Riparian Wetlands and water quality. *J. Environ. Qual.* 23:896-900.

Gregory, S.V., F.J. Swanson, W.A. McKee, and K.W. Cummins, 1991. An ecosystem perspective of riparian zones. *BioSciences* 41:540-551.

Hagerty, D.J., M.F. Spoor, and C.R. Ullrich, 1981. Bank failure and erosion on the ohio river. *Eng. Geol.* 17:141-158.

Hamlett, J.M., J.L. Baker, and H.P. Johnson, 1983. Channel morphology changes and sediment yield for a small agricultural watershed in Iowa. *Trans ASAE* 26:1390-1396.

Henderson, J.E. 1986. Environmental designs for streambank protection projects. *Water Resourc. Bul.* 22:549-558.

Hooke, J.M. 1980. Magnitude and distribution of rates of river bank erosion. *Earth Surf. Processes* 5:143-157.

Isenhart, T.M., R.C. Schultz, and J.P. Colletti, 1998. Watershed restoration and agricultural practices in the Midwest: Bear Creek of Iowa. In: Williams, J.E., M.P. Dombeck, and C.A. Woods (eds.),Watershed restoration: Principles and practices. American Fisheries Society, Bethesda, MD. p. 318-334.

Johnson, C. 2003. 5 Low-cost methods for slowing streambank erosion. *J. Soil Water Conserv.* 58:13-17.

Knight, K.W., R.C. Schultz, C.M. Mabry, and T.M. Isenhart, 2010. Ability of remnant riparian forests, with and without grass filters, to buffer concentrated surface runoff. *J. Am. Water Resourc. Assoc.* 46:311-322.

Kronvang, B., R. Grant, and A.L. Laubel, 1997. Sediment and phosphorus export from a lowland catchment: Quantification of sources. *Water Air Soil Pol.* 99:465-476.

Lane, E.W. 1955. The importance of fluvial morphology in hydraulic engineering. *Proc. Am. Soc. Civ. Eng.* 81:1-17.

Laubel, A., B. Kronvang, A.B. Hald, and C. Jensen, 2003. Hydromorphological and biological factors influencing sediment and phosphorus loss via bank erosion in small lowland rural streams in Denmark. *Hydrol. Process.* 17:3443-3463.

Lawler, D.M., 1993. The measurement of river bank erosion and lateral channel change: A review. *Earth Surf. Processes Landforms* 18:777-821.

Lawler, D.M. J.R., Grove, J.S. Couperwaite, and G.J.L. Leeks, 1999. Downstream change in river bank erosion rates in the Swale-Ouse system, Northern England. *Hydrol. Process.* 13:977-992.

Lee, K-H, T.M. Isenhart, R.C. Schultz, and S.K. Mickelson. 1999. Nutrient and sediment removal by switchgrass and cool-season grass filter strips in central Iowa, USA. *Agrofor. Syst.* 44:121-132.

Lee, K-H., T.M. Isenhart, R.C. Schultz, and S.K. Mickelson. 2000. Multispecies riparian buffers trap sediment and nutrients during rainfall simulations. *J. Environ. Qual.* 29:1200-1205.

Lee, K-H., T.M. Isenhart, R.C. Schultz. 2003. Sediment and nutrient removal in an established multi-species riparian buffer. *J. Soil Water Conserv.* 58:1-8.

Lowrance, R.R., S. McIntyre, and C.L. Lance. 1988. Erosion and deposition in a field/forest system estimated using csium-137 activity. *J. Soil Water Conserv.* 43:195-199.

Lyons, J., B.M. Weasel, L.K. Paine, and D. Undersander. 2000. Influence of intensive rotational grazing on bank erosion, fish habitat quality, and fish communities in southwestern Wisconsin trout streams. *J. Soil Water Conserv.* 55:271-276.

Marquez, C.O., C.A. Cambardella, T.M. Isenhart, and R.C. Schultz. 1999. Assessing soil quality in a riparian buffer by testing organic matter fractions in central Iowa, USA. *Agrofor. Syst.* 44:133-140.

Meyer, L.D., S.M. Dabney, and W.C. Harmon. 1995. Sediment-trapping effectiveness of stiff grass hedges. *Trans ASAE* 38:809-815.

Minshall, G.W. 1984. Aquatic insect-substratum relationships. In: Resh, V.H. and D.M. Rosenberg (Eds.), The ecology of aquatic insects, Praeger, New York, NY. p. 356-400.

National Research Council. 2002. Riparian areas: functions and strategies for management. National Academy of Science. Washington, DC.

Nerbonne, B.A. and B. Vondracek, B. 2001. Effects of local land use on physical habitat, benthic macroinvertebrates, and fish in the Whitewater River, Minnesota, USA. *Environ. Manage.* 28:87-99.

Odgaard, A.J., 1984. Bank Erosion Contribution to Stream Sediment Load. Iowa Institute of Hydraulic Research Report 280. University of Iowa, Iowa City, IA.

Odgaard, A.J., 1987. Streambank erosion along two rivers in Iowa. *Water Resour. Res.* 23:1225-1236.

Owens, P.N., R.J. Batalla, A.J. Collins, B. Gomez, D.M. Hicks, A.J. Horowitz, G.M. Kondolf, M. Marden, M.J. Page, D.H. Peacock, E.L. Petticrew, W. Salomons, and N.A. Trustrum. 2005. Fine-grained sediment in river systems: Environmental significance and management issues. *River Res. Applic.* 21:693–717

Owens, P.N. and D.E. Walling. 2002. The phosphorus content of fluvial sediment in rural and industrialized river basins. *Water Res.* 36: 685-701.

PeterJohn, W.T. and D.L. Correll. 1984. Nutrient dynamics in and agricultural watershed: Observations on the role of a riparian forest. *Ecology* 65:1466-1475.

Phillips, J.D. 1989. An evaluation of the factors determining the effectiveness of water quality buffer strips. *J. Hydrol.* 107:133-145.

Pionke, H.B., W.J. Gburek, A.N. Sharpley, and J.A. Zollweg. 1997. Hydrologic and chemical controls on phosphorus loss from catchments. In Tunney, H., (ed.), Phosphorus loss to water from agriculture. CAB International Press, Cambridge, England. p. 225-242.

Platts, W.S. 1981. Effects of sheep grazing on a riparian-stream environment. Research. Note INT-307, USDA, Forest Service, Intermountain Forest and Range Experiment Station, Ogden, UT.

Prior, J.C. 1991. Landforms of Iowa. Iowa Department of Natural Resources. University of Iowa Press, Iowa City, IA.

Porath, M.L., P.A. Momont, T. DelCurto, N.R. Rimbey, J.A. Tanaka, and M. McInnis, 2002. Offstream water and trace mineral salt as management strategies for improved cattle distribution. *J. Anim. Sci.* 80:346-356.

Rakovan, M.T. and W.H. Renwick. 2011. The role of sediment supply in channel instability and stream restoration. *J. Soil Water Conserv.* 66:40-50.

Roseboom, D.P., 1987. Case Studies of Stream and River Restoration. In: Management of the Illinois River System: The 1990's and Beyond. Illinois River Resource Management, A Governor's Conference, Peoria, IL. p. 184-194.

SAS (Statistical Analysis System) Institute. 1999. SAS Release 8.1 ed. SAS Inst., Cary, NC.

Schilling, K.E. and R.D. Libra. 2003. Increased baseflow in Iowa over the second half of the 20th century. *J. Am. Water Res. Assoc.* 39:851-860.

Schilling, K.E. and C.F. Wolter, 2000. Applications of GPS and GIS to map channel features in Walnut Creek, Iowa. *J. Am. Water Resour. Assoc.* 36:1423-1434.

Schultz R.C., J.P. Colletti, T.M. Isenhart, C.O. Marquez, W.W. Simpkins, and C.J. Ball. 2000. Riparian forest buffer practices. In: Garrett, H.E. (ed.), North America agroforestry: An integrated science and practice. American Society of Agronomy, Madison, WI. p. 189-281.

Schultz, R.C., J.P. Colletti, T.M. Isenhart, W.W. Simpkins, C.W. Mize, and M.L. Thompson. 1995. Design and placement of a multi-species riparian buffer strip system. *Agrofor. Syst.* 29:201-226.

Schumm, S.A., M.D. Harvey, and C.C. Watson 1984. Incised channels: morphology, dynamics and control. Water Resources Publication, Littleton, CO.

Sekely, A.C., D.J. Mulla, and D.W. Bauer, 2002. Streambank slumping and its contribution to the phosphorus and suspended sediment loads of the Blue Earth River, Minnesota. *J. Soil Water Conserv.* 57:243-250.

Sharpley, A. N. 1985. The selective erosion of plant nutrients in runoff. *Soil Sci. Soc. Am. J.* 49:1527-1534.

Sharpley, A.N., S.J. Smith and J.W. Naney. 1987. Environmental impact of agricultural nitrogen and phosphorus use. *J. Agric. Food Chem.* 35: 812-817.

Sheffield, R.E., S. Mostaghimi, D.H. Vaughan, E.R. Collins, Jr. and V.G. Allen, 1997. Off-stream water sources for grazing as a stream bank stabilization and water quality BMP. *Trans ASAE*, 40:595-604.

Shields, Jr., F.D., A.J. Bowie, and C.M. Cooper, 1995. Control of streambank erosion due to bed degradation with vegetation and structure. *Water Res. Bul.* 31:475-489.

Sliva, L. and D.D. Williams. 2001. Buffer zone versus whole catchment approaches to studying land use impact on river water quality. *Water Res.* 35:3462-3472.

Simon, A. and A. Collison. 2002. Quantifying the mechanical and hydrologic effects of riparian vegetation on stream bank stability. *Earth Surf. Processes Landforms* 27:527-546.

Simon, A., M. Rinaldi, and G. Hadish. 1996. Channel evolution in the loess area of the Midwestern United States. In: Proceedings of the sixth federal interagency sedimentation conference, Las Vegas, Nevada. U.S. Government Printing Office, Washington, D.C. pp. III.86-III.93.

Simon, A., A. Curini, S. Darby, and E.J. Langendoen. 1999. Streambank mechanics and the role of bank and near-bank processes in incised channels. In: Darby S.E. and A. Simon (eds.), Incised rivers channels: processes forms, engineering and management. John Wiley and Sons Press, Chichester, England. p. 123-152.

Simonson, T.D., J. Lyons, and P.D. Kanehl 1994. Guidelines for evaluating fish habitat in Wisconsin streams. General Technical Report NC-164. United States Department of Agriculture, Forest Service, North Central Forest Experiment Station, St. Paul, MN.

Six, J., E.T. Elliott, K. Paustian, and J. Doran. 1998. Aggregation and soil organic matter accumulation in cultivated and native grassland soils. *Soil Sci. Soc. Am. J.* 62:1367-1377.

Smart, M.M., J.R. Jones, and J.L. Sebaugh. 1985. Stream-watershed relations in the Missouri Ozark Plateau province. *J. Environ. Qual.* 14:77-82.

Strahler, A.N. 1957. Quantitative Analysis of Watershed Geomorphology. *Trans. Am. Geophys. Un.* 38:913-920.

Trimble, S.W. and P. Crosson. 2000. U.S. soil erosion rates – myth and reality. *Science* 289:248-250.

Trimble, S.W. and A.C. Mendel, 1995. The cow as a geomorphic agent – A critical review. *Geomorphology* 13:133-153.

USDA-NRCS. 1996. Streambank and Shoreline Protection. Engineering Field Handbook, Chapter 16, USDA-NRCS, Washington, D.C.

USDA-NRCS .1997. Profitable Pastures: A Guide to Grass, Grazing and Good Management. USDA-NRCS, Des Moines, IA.

USDA-NRCS 1998. Erosion and Sediment Delivery. Field Office Technical Guide Notice no. IA-198. USDA NRCS, Des Moines, IA.

USDA-NRCS. 1999a. Riparian forest buffer. Conservation practice standard, Code 391. USDA-NRCS, Des Moines, IA.

USDA-NRCS. 1999b. Grass filters. Conservation practice standard, Code 393. USDA-NRCS, Des Moines, IA.

USDA-NRCS. 2000. Natural resources inventory. 1997 Summary Report. USDA-NRCS, Des Moines, IA.

USEPA. 1995. Water quality conditions in the United States: A profile from the 1994 national water quality inventory report to congress. Publication EPA 841-F-95-010. Office of Water, Washington, DC.

Vought, L.B.-M., J. Dahl, C.L. Pedersen, and J.O. Lacoursière. 1994. Nutrient retention in riparian ecotones. *Ambio* 23:342-348.

Waters, T.F. 1995. Sediment in streams: Sources, biological effects, and control. American Fisheries Society, Bethesda, MD.

Whitney, G.G. 1994. From coastal wilderness to fruited plains: A history of environmental change in temperate North America, 1500 to present. Cambridge University Press, Cambridge, England.

Wilcock, P.R. 1997. The components of fractional transport rate. *Water Resour. Res.* 33:247-258.

Wilkin, D.C. and S.J. Hebel. 1982. Erosion, Redeposition, and Delivery of Sediment to Midwestern Streams. *Water Resour. Res.* 18:1278-1282.

Wilson, C.G., R.A. Kuhnle, D.D. Bosch, J.L. Steiner, P.J. Starks, M.D. Tomer, and G.V. Wilson. 2008. Quantifying relative contributions from sediment sources in Conservation Effects Assessment Projects watersheds. *J. Soil Water Conserv.* 63:523-532.

Wolman, M.G. 1959. Factors influencing erosion of a cohesive river bank. *American J. Sci.* 257:204-216.

Wynn, T.M. and S. Mostaghimi. 2006. The effects of vegetation and soil type on streambank erosion, Southwestern Virginia, USA. *J. Am. Water Resourc. Assoc.* 42:69-82.

Wynn, T., S. Mostaghimi, J. Burger, A. Harpold, M. Henderson, and L.-A. Henry. 2004. Variation in root density along stream banks. *J. Environ. Qual.* 33:2030-2039.

Yan, B., M.D. Tomer, and D.E. James. 2010. Historical channel movement and sediment accretion along the South Fork of the Iowa River. *J. Soil Water Conserv.* 65:1-8.

Zaimes, G.N., R.C. Schultz, and T.M. Isenhart. 2006. Riparian land-uses and precipitation influences on stream bank erosion in central Iowa. *J. Am. Water Resour. Assoc.* 42:83-97.

Zaimes, G.N., R.C. Schultz, and T.M. Isenhart. 2008a. Streambank soil and phosphorus losses under different riparian land-uses in Iowa. *J. Am. Water Resour. Assoc.* 44:935-947.

Zaimes, G.N., R.C. Schultz, and T.M. Isenhart. 2008b. Total phosphorus concentrations and compaction in riparian areas under different riparian land-uses of Iowa. *Agric. Ecosyst. Environ.* 127:22-30.

In: Agricultural Research Updates. Volume 2
Editor: Barbara P. Hendriks

ISBN: 978-1-61470-191-0
© 2012 Nova Science Publishers, Inc.

Chapter 4

DIET SELECTION OF HERBIVORES ON SPECIES-RICH PASTURES

P. Hejcmanová[1] and J. Mládek[2]*

[1] Faculty of Forestry and Wood Sciences, Czech University of Life Sciences,
Kamýcká, Suchdol, Czech Republic
[2] Department of Botany, Faculty of Science, Palacký University,
Šlechtitelů, Olomouc, Czech Republic

ABSTRACT

Understanding grazing by domesticated ruminants for animal production is of high economic importance throughout the world and therefore, is paramount in designing management strategies for livestock production. The plant-animal interface is the central feature of these systems. Food quantity and quality are major determinants of animal production. Both food quantity and quality herbivores maintain through selective foraging which alters sward structure, modifies plant species composition and thus produces new patterns of plant biomass production. Therefore, we focus our chapter on the mechanisms of foraging selection which may enable us to have insight into grazing decision-making and processes. The central question is: what are principle drivers in grazing decision processes leading to high selectivity on species rich grasslands and this at different spatial and temporal levels? However, before we answer this question we have to be precise in what we mean under selective grazing and to distinguish it from other terms used in the domain of foraging strategies. We will also summarize known methods and quantifications of grazing selectivity. Only then may we bring a complex view on various factors affecting diet selection strategies of herbivores in species-rich pastures. Finally, we propose management rules in order to use herbivore foraging selectivity to utilize food resources in semi-natural grasslands most efficiently and simultaneously keeping forage production and quality of grasslands from a long-term perspective.

---

* Corresponding author

1. INTRODUCTION

Pastures and grazing livestock managed by man are an inseparable part of animal husbandry. Pastures form our landscape and give it its high economic value. They provide important regulating ecosystem services and have high intrinsic value for future maintenance as a cultural heritage and as biodiversity hotspots (e.g. Austrheim *et al.* 1999, Hart 2001, Pavlů *et al.* 2006, Smit *et al.* 2008). Species-rich grasslands originate predominantly from large herbivores' grazing at moderate grazing pressure which induce high species richness and enhance occurrence of vulnerable species (Collins *et al.* 1998, Pavlů *et al.* 2007). At present, the management of these grasslands fulfill economic goals. The role of grazing systems has been widely reassessed and multiple goals have been assigned to current grassland management with a strong focus on ecosystem function and biodiversity conservation. Within the terms of animal husbandry, the ultimate goal of livestock producers is to achieve an environmentally and economically sustainable livestock production on pastures and to maintain the forage production. This requires a fundamental understanding of the processes at the plant-animal interface. The knowledge of diet selection is therefore vital for agronomists regarding animal nutrition and also for grassland specialists managing the development of sward structure and species composition.

The plant species composition and sward structure on species-rich pastures is substantially formed by selective defoliation which changes micro-site conditions, such as light, moisture or temperature and consequently altered competitiveness among plant species in the sward (Bullock & Marriott 2000). Thus, grazing offers an important tool for conservation management. The success of maintaining high biodiversity levels lies, however, on grazing intensity, grazing distribution pattern and specific exploitation of available forage resources by animals. Grazing may both, either increase, or decrease plant species composition and sward heterogeneity (Adler *et al.* 2001). Hence, the pastoral management needs to predict animals' foraging decisions at various temporal and spatial scales. Given the constraints imposed through management practices, animals are continually faced with a series of short-term decisions about what to forage and where to forage (Taylor 1993). This is fundamental to animal decisions on the trade-off between forage quality and quantity ingested. In intensive production systems with few highly productive plant species, foraging is mainly a function of sward characteristics such as height or structure (homogeneous *versus* heterogeneous sward) integrated to particular patches and their spatial distribution (e.g. WallisDeVries *et al.* 1999, Griffiths *et al.* 2003). On species-rich pastures, the animals' decision-making becomes more complex because animals have not only spatially heterogeneous sward structure, but also more variability in rich food resources of divers quantity and quality (Figure 1). The diversity of forage supply enables large herbivores to select their forage in relation to available plant species and their particular quality. Thus, differences among large herbivore species in forage selectivity offer the potential for efficient utilization of pastures with diverse arrays of plant species. The knowledge of mechanisms of foraging selection may help us to link the grazing processes with management leading to both effective livestock production and biodiversity conservation.

2009, photo Pavla Hejcmanová.

Figure 1. Species-rich pastures under low grazing intensity evolve in highly heterogeneous swards with vegetation patches of diverse quantity and quality. The heifers on the Betlém experimental pasture in the Jizerské Mountains in the northern Czech Republic selectively graze on short, more nutritious sward.

Therefore, we focus our chapter on the mechanisms of diet selection of large herbivores which may enable us to have insight into grazing decision-making and processes. The central question is: what are principle drivers in grazing decision processes leading to high selectivity of animals on species-rich grasslands? However, before we answer this question we have to be precise on what we mean by grazing selectivity and distinguish it from other terms used in the domain of foraging strategies. We will also review methods of evaluating foraging selectivity. Only then we may bring a complex view on various factors affecting diet selection on herbivores with particular attention paying to species richness of the sward. And finally, we emphasize the importance of management of grazing systems for animal performance.

2. What Do We Mean by Grazing Selectivity?

Grazing selectivity presents an important mechanism whereby grazing herbivores search and intake individual food sources. This mechanism helps the animals to respond to changes in actual stages of sward (caused either naturally or by management) and *vice versa*, selective grazing affects plant-species composition and vegetation structure of pasture swards.

First, the terms selective grazing or diet selection can be easily replaced by diet preferences. There is, however, a clear distinction between selection and preference which consists of the difference between what the animals eat and what they would eat if given complete freedom of choice (Hodgson 1979). Such a situation is almost impossible on the pastures. Herbivores face various constraints caused by environmental and management factors while foraging on pastures. Then, the animals start to forage selectively, adjusting their choice to the available food resources. Selection is thus a function of preference, but it is clearly affected by the abundance or the availability of both palatable and unpalatable plant species, and by their spatial distribution (Wang *et al.* 2010a). Selection is further influenced by some of the animals' foraging abilities; for example, their ability to sort one food from the other, to walk long distances and to learn and remember the location of food patches (Dumont 1997, Dumont *et al.* 2002).

On species-rich pastures, the vegetation is heterogeneous and the grazing pattern interacts with the spatial distribution of vegetation. If the spatial pattern of grazing does not follow the pattern of vegetation, we refer to 'patch grazing' or 'homogeneous grazing'. When a grazing pattern closely tracks a vegetation pattern, we refer to 'selective grazing' (Adler *et al.* 2001).

Selective grazing has been described in a conceptual hierarchical model across different spatial and temporal scales. Senft *et al.* (1987) and Bailey *et al.* (1996) defined six basic spatial scales for large herbivores in a foraging hierarchy which span from plant level to regional scale and each scale is functionally defined based on characteristic behaviors that occur at different rates: 1. Bite is the smallest spatial as well as temporal scale and is defined as the complete herbage prehension, jaw and tongue movements, and severance by head movements (Laca *et al.* 1994). 2. Feeding station is a set of plants available to a herbivore without moving their front feet (Novellie 1978). 3. Patch represents a cluster of feeding stations separated from others by a break in the foraging sequence when animals re-orient to a new location. 4. Feeding site refers to a set of patches in a contiguous spatial area that animals graze within a foraging bout. It may contain one plant community or even extend beyond. Foraging bout is the time spent continuously foraging without interruption until a complete change of behavior, for instance, to resting, ruminating or other. 5. Camp is a set of feeding sites which involve places where animals drink, rest or seek shelter. 6. Home range represents a large scale involving several camps, but usually limited by fencing on pastures or other barriers in the landscape. Temporal scales include short-term (instantaneous or daily scale), medium-term (vegetation season) and long-term (seasonal cycles and inter-annual variability) (O'Reagain & Swartz 1995). At each scale, the animals confront a series of decisions. Free ranging herbivores have a free choice within a variety of habitats and food resources. They can independently migrate in an open landscape and to feed selectively to satisfy their energy, nutrient, and minerals requirements (McNaughton 1984, WallisdeVries & Schippers 1994). At pastures in animal production systems, different mechanisms operate because pastures are usually delineated by natural landscape barriers or by a fence imposed by management. In order to obtain an adequate and balanced diet, the animals must adopt a strategy coping with constraints of limited area affording resources limited in quantity, nutrient and mineral content. Pasture constitutes of one or more camps where animals together drink and rest between feeding bouts (Bailey *et al.* 1996). Species-rich pastures create implicitly heterogeneous sward with a number of patches and feeding sites. Regarding that the area of pasture is usually small and readily accessible, the initial decision where to start grazing at the beginning of each bout has little importance. On the other hand, during the grazing, the animals search and select individual vegetation patches. At this level, only bites are aggregated, which animals select only by the head movement in terms of feeding station (Bailey *et al.* 1996, Bailey *et al.* 1998). However, whatever the spatial scale, the animals' selectivity seems to be based on the maximization of daily energy intake rather than instantaneous/momentary maximization (WallisDeVries & Daleboudt 1994).

3. MEASURING GRAZING SELECTIVITY IN HETEROGENEOUS SWARDS

3.1. Electivity Indices

Although forage value is generally considered to be determined by organic matter digestibility (e.g. Marinas *et al.* 2003), palatability for herbivores does not strictly follow this plant characteristic and is also influenced by odor and taste of biomass (Provenza *et al.* 2003). Preference of herbivores differ between (i) cafeteria experiments (e.g. Thomas *et al.* 2010), in which all food types are equally available, and (ii) field studies, where food types vary in availability, accessibility and divergent spatial distributions (Pérez-Harguindeguy *et al.* 2003, Wang *et al.* 2010b). From a practical point of view, measurements of palatability from cafeteria trials are mainly useful for the planning of animal feeding whereas measurements of grazing selectivity on species-rich pastures enable us to predict under what conditions particular food types will be exploited by herbivores.

Measuring diet selection requires a comparison of the relative abundance of food type available to an herbivore with the relative abundance of food type utilized by an herbivore. For this purpose, many electivity indices (e.g. Ivlev 1961, Jacobs 1974, Vanderploeg & Scavia 1979, Johnson 1980) were developed and their suitability has been much debated. But due to their computational simplicity, electivity indices have not been superseded by more complex statistical techniques, for instance, as resource selection functions (Boyce & McDonald 1999). Lechowicz (1982) analyzed strengths and weaknesses of seven most commonly used electivity indices and identified Vanderploeg and Scavia's relativized electivity as the single best, but not perfect, electivity index. Its main advantage lies in the fact that maximum attainable preference is an increasing function of the number of food types, but on the contrary, index is vulnerable to sampling errors. In the last decade, diet selection studies on species-rich pastures (e.g. Dumont *et al.* 2005a, Farrugia *et al.* 2006, Boulanger *et al.* 2009, Fraser *et al.* 2009) most often used Jacobs' modified electivity index (Jacobs 1974) due to its low sensitivity to variations in the relative abundance of food types. However, comparisons of diet selections among studies which use different electivity index are needed very often. As every index reacts differently to changes in availability and use of food type, food type may appear preferred according to one index but avoided according to another. Finally, Lechowicz (1982) and Tanentzap *et al.* (2009) brought empirical evidence that choice of electivity index is unimportant when preference is derived from rank order of species selectivity which is consistent across all electivity indices.

3.2. Measuring Forage Availability

Relative availability of forage (food types) might differ significantly between measurement techniques. Frequency of occurrence represents the presence/absence of food types within a sampling unit. It provides a rapid measure of the spatial patchiness of food types, but Norbury and Sanson (1992) did not recommend it as this measure overemphasizes rare food types. Availability is sometimes expressed as density of food types within a sampling unit (e.g. Van der Wal *et al.* 2000). However, this measure is unsuitable for food types whose individuals cannot be readily separated, such as grass tussocks. Cover, the perpendicular projection of plants onto the ground, was often used in diet selection studies (e.g. Grant *et al.* 1985); it may be measured accurately with point quadrat or visually estimated. As neither frequency nor density and cover account for differences in plant height, more appropriate are techniques based on biomass. Direct assessment of plant biomass in sampling units by harvesting is the most accurate but precludes the evaluation of the amount of food type utilized by herbivores. In relatively homogeneous pastures, this might be surpassed by the use of paired neighbor plots – one for the determination of food type availability before grazing and one for determination of food type utilization after grazing (Lepš *et al.* 1995). Such destructive techniques with harvesting, separating and weighing of each food type can be very labor-intensive and thus, the number of replications would be substantially limited. Alternatively, biomass of food types in the sampling unit may be obtained much faster using a calibrated weight-estimate method (Tadmor *et al.* 1975) as follows: visual estimates of the biomass of food types are calibrated by clipping and weighing in several training plots, when consistent estimates are attained, direct estimations of food type biomass of the studied plots are undertaken.

In the case when food type availability is sufficient to be determined at the scale of sward patches, researchers commonly use double-sampling techniques based on calibration of some easily measured patch quantity (i.e. estimators) by clipping and weighing of biomass in several training patches (e.g. Reese *et al.* 1980). Many instruments for forage estimation have been developed and adapted for day-to-day on farm management. Among the most commonly used biomass estimators, there are the Robel pole (Robel *et al.* 1970), an electronic capacitance meter (Currie *et al.* 1973), rising plate meter (Castle 1976) and sward stick (Stewart *et al.* 2001). The visual obstruction method measures the lowest point on the Robel pole which is not visually obstructed by vegetation. The capacitance meter measures the capacitance of the air–herbage mixture (Curie *et al.* 1987) and responds mainly to the surface area of the foliage. The rising plate meter (Figure 2) integrates sward height and density into one measure, often called bulk density (Michalk & Herbert 1977). Sward stick relies on a positive relationship between biomass yield and canopy height. Commercial products using abovementioned methods are usually calibrated and enable quick assessment of grassland biomass which is needed for whole farm simulation models in order to ensure economic management of grasslands (Sanderson *et al.* 2001). Nevertheless, these calibrated, commercial instruments operate reliably only in relatively homogeneous grasslands (e.g. Currie *et al.* 1987, Murphy *et al.* 1995) while species-rich pastures consist of patches of distinct structure and phenology. Harmoney *et al.* (1997) quantified effectiveness of four indirect methods across heterogeneous pastures and found the Robel pole and rising plate meter to be most efficient biomass estimators but with fairly low coefficients of determination, 0.63 and 0.59, respectively. On that account, most researchers (e.g. Virkajärvi 1999, Sanderson *et al.* 2001) recommend calibrating all of these devices separately for each measurement occasion. Therefore, Martin *et al.* (2005) conducted measurements on naturalized cattle pastures and used local and time-specific calibrations; however, they were not able to identify a method that was consistently accurate to approximate standard quadrant harvesting. Thus, estimating forage biomass in mixed-species pastures is still a challenge. The effectiveness of indirect methods in heterogeneous pastures may be improved by the construction of separate calibration curves for distinct patch types in the pasture.

Czech Republic 2009, photo Pavla Hejcmanová.

Figure 2. Rising plate meter represents a very useful tool for the estimation of available biomass on pastures.

3.3. Measuring Diet Composition

Quantification of diet composition can be generally implemented by three approaches: (i) analyzing plant parts consumed by herbivores either with the help of oesophageal fistulation (e.g. Woji and Iji 1996), by disassembling of stomach contents (e.g. Homolka and Heroldová 1992, Bee *et al.* 2009) or by analyzing animal feces (e.g. Fraser *et al.* 2009, Hejcmanová *et al.* 2010); (ii) recording grazing behavior of foraging animals (Dumont *et al.* 2007a, Hejcmanová *et al.* 2009) and (iii) recording utilization of vegetation (e.g. Lepš *et al.* 1995, Hejcman *et al.* 2008, Mládek *et al.* 2011a). Only the last two approaches appear suitable for quantification of diet selection in species-rich swards composed of many plant species because small masticated or digested particles are almost impossible to identify to a species level (Norbury & Sanson 1992). Moreover, exact feeding trials made by McInnis *et al.* (1983) showed that diets determined by stomach content analysis and fecal analysis significantly overestimate proportions of grasses on the account of forbs. Cell walls of forbs are more readily digested, thus forbs gradually disappear as they pass through the digestive tract. Hence, the least proportions of forbs are usually detected in feces.

Direct observation of individual animals and recording their grazing behavior is the most common technique (e.g. Bailey 1995, WallisDeVries *et al.* 1998, Dumont *et al.* 2007a, Farrugia *et al.* 2008, Hejcmanová *et al.* 2009). Sampling grazing time, biting rate and bite size (weight) gives complex information on grazing behavior patterns. On species-rich swards marked by their heterogeneity and patchy character, directly observing the animals enables the embracing of the spatial pattern of grazing and movements on pastures at different scales. Another advantage of direct animal behavior sampling consists of recording social interactions among animals which may mutually compete for available forage resources or in turn, they may cooperate in terms of vigilance against potential predators. Measuring diet composition on species-rich pastures requires the advanced knowledge of the sward and its components (plant species). In order to precisely identify the plant parts, plant species or defined sward types/patches, the observers have to approach the animals at a rather close distance. For such type of behavior sampling, the animals have to be trained, although the risk of bias by the observer's presence remains relatively high. A flat well-arranged pasture render the investigation possible using a binocular with appropriate parameters (Figure 3). The biggest advantage of direct observation however reposes in the possibility to determine motivation of a concrete animal for the choice of concrete food items at a given moment (and for how long time); and this in relation to their actual physiological state (age, reproduction), abilities or experience. Consequently, the method enables to reveal proper mechanisms of diet selection.

Vegetation utilization techniques provide a clear picture where and to what degree a pasture is being used (Laycock *et al.* 1972); therefore, these methods render information on grazing intensity which may substantially contribute for clarification of possible different patterns of diet composition. Yet, these techniques have been denominated unsuitable for quantification of diet composition due to the possible bias caused by losses of plant parts by trampling, weathering or grazing by other animals than those of interest (e.g. Holechek *et al.* 1982). However, only these techniques enable accurate determination of neighborhood effects on diet selection (e.g. Arnold 1987). Moreover, the knowledge of plant species utilization in permanent plots is extremely important for the assessment of impact of grazing selectivity on community functioning (e.g. Cingolani *et al.* 2005). Principally, there are two approaches for the quantification of vegetation utilization. The first approach relies on evaluating biomass differences between grazed and ungrazed (caged) plots (e.g. Lepš *et al.* 1995), however, this method is dependent on the assumption that species biomass in both grazed and ungrazed plots is the same and this might be difficult to ensure in heterogeneous pastures.

The second approach is based on the comparison of species biomass or height in the same plots before and after grazing (Arnold 1987). The repeated observation of the same plots should exclude omitting utilization of plants grazed totally (without leaving aboveground residues) which may occur when the researcher relies on a single shot after grazing observation of the plot (Laycock *et al.* 1972). Another advantage of observing foraging traces on vegetation consists of the evaluation of selection of diet of the flock of grazing animals as a whole. The method can be practiced in variable terrain where the visibility of individual animals is restricted. Plant species can be easily classified according to selection by herbivores even in less accessible pastures in mountain areas (Hejcman *et al.* 2008).

Czech Republic 2007, photo Pavla Hejcmanová.

Figure 3. Direct observation of diet selection on a pasture using binoculars to identify selected items by an animal at patch- or plant-species level.

4. DRIVERS OF GRAZING SELECTIVITY

Foraging comprises four key phases that can be considered as approach, appraisal, defoliation and ingestion (Griffiths *et al.* 2003). Each of these phases is subject to key cues and rules controlling an animal's decision-making. The animal behavior is driven by an array of factors of different nature and result in a variety of grazing response patterns. Foraging response mechanisms issue essentially from animal characteristics and abilities, from sward characteristics, environmental conditions, the management intervention of man, and interactions among them.

4.1. Animal as a Predictor of Diet Selectivity

Foraging processes are basically predetermined by intrinsic morphological and physiological characteristics of the animal. Then, they are constrained by animal cognitive abilities, capacity of adaptation to concrete conditions and social intra- and inter-species environment.

4.1.1. Morpho-Physiological Adaptation and Body Size

Herbivore ungulates represent a group highly diverse in body size (ranging from very small Royal antelope of 25 cm at shoulder to 5 m tall giraffe). They fill a wide variety of ecological niches and play a key role in ecosystems all over the world, from tropical rain forests to the tundra beyond the polar circle. Their dietary preferences are equally varied. We recognize main foraging types according to their adaptations for consuming (i) a roughage diet composed primarily of grasses - grazers, and (ii) a concentrate diet of browse or forbs – browsers. Transient types consuming various diets are generally known as mixed or intermediate feeders. Varied forage of the animal is reflected in morpho-physiological adaptations in the structure and function of individual parts of the digestive system (Hofmann & Stewart 1972, Hofmann 1989, Gagnon & Chew 2002). Hofmann's nutritional and physiological interpretations of anatomical differences among ruminants were, however, subjected to a series of tests and appear non-supportable. For instance, Pérez-Barbería and Gordon (1999) examined 21 morphological traits of the jaw and skull of 94 species of ungulates to test the differences in of the jaw and skull morphology among feeding types (browsers, grazers, mixed feeders, frugivorous, omnivorous). Results of this study showed that phylogeny has a stronger influence in explaining the differences in jaw and skull morphology which exert in mechanics of chewing than the feeding-type classification. After excluding omnivorous species, there were no differences among the rest of animal feeding types.

On the other hand, the animal feeding types differ in body size and morphological traits functionally related to the ability of forage selection (muzzle width, incisor-arcade shape, incisor shape), prehension of food (incisor protrusion), food comminution (molar occlusal surface area), hypsodonty (high-crowned molar), and intake rate (incisor breadth) (Pérez-Barbería & Gordon 2001). The grazers are usually ranked among large and heavier species and browsers among smaller ones. However, the organs related to food intake and digestion are positively related to body weight and body size rather than to feeding. Gordon and Illius (1994) found that African ruminants with different morphological adaptations of the digestive tract display comparable digestive strategies. Their wet and dry mass of the rumen and hindgut contents, fermentation rate in the rumen and retention time of digesta within digestive tracts do not differ between feeding types and are positively related with the body weight. Similar relations of ruminant digestion to body weight were shown to Robbins et al. (1995). They found no difference in the efficiency of fiber digestion between feeding type and proved that fiber digestion increases as body weight increases. Salivary glands size is approximately four times larger in browsers than grazers and their weight increased linearly with body weight (Robbins et al. 1995). Therefore, body size (and weight) of individual herbivore species cannot be considered a good predictor of the differences between feeding types (Figure 4). Body size, however, regardless of feeding type, represents an important driver of diet selection. Body size is the main variable in the determination of differences in the oral traits related to food selection and processing of food in the individual herbivore's species. So food selection process is similar for species with similar body size and various feeding styles (smaller species are more selective regardless of feeding style). This emphasizes the importance of food structural characteristics in the definition of oral morphology.

Small-sized herbivore species are generally more selective, spending more time searching for high-quality forage than larger-body-sized species. The larger herbivores are forced to feed less-quality forage in order to maintain a certain level of forage intake (Bailey et al. 1996). However, both small and large animals adopt relatively selective strategies if there are adequate available food resources. For instance, Schwartz and Ellis (1981) compared foraging behavior, diet selection and diet overlap of two native herbivores (American bison (*Bison bison*) and pronghorn (*Antilocapra americana*)) and two domestic herbivores (sheep and cattle). They revealed that both small and large animals carry out relatively selective strategies when the forage conditions allow it. They found out that sheep, the smallest of the investigated species, were always less selective than pronghorn and even they were less selective than large cattle in season with most abundant forage. Schwartz and Ellis (1981) relate this behavior by human selection which makes sheep to be food and habitat generalists despite their relatively small size. Therefore, the generally-respected fact that smaller-body size animals are more selective than larger may not always be valid.

Senegal 2008, Michal Hejcman.

Figure 4. The Western Derby eland (*Taurotragus derbianus derbianus*), the critically endangered antelope, belongs to browsers. Despite its large size, the antelope is selective, able to move its muzzle and tongue in order to select their forage at plant part level on an array of woody plants and forbs.

Another important driver of foraging patterns and selectivity represent the physiological state of individual animals, namely age, sex, and reproductive state. The animals adjust their forage intake according to physiological requirements related to growth rate. Growing animals have higher demands on diet quality and have higher energetic demands. For instance, suckling calves and yearling heifers have generally higher crude protein content in their diet than mature steers (Grings *et al.* 2001). In sexually dimorphic animals such as Bighorn sheep (*Ovis canadensis*), sub-adult males have higher forage intake than adult ones, while females do not show any difference between age classes (Ruckstuhl *et al.* 2003). For females, the most important driver in foraging pattern is reproduction. Lactating females have to compensate the energy invested in lactation and face the tradeoffs between foraging and vigilance over predation risk of their offspring. Consequently, they increase the food intake by increasing the bite rate, daily grazing time or by increased selectivity for high quality and energy forage (Ruckstuhl *et al.* 2003, Lamoot *et al.* 2005, Farrugia *et al.* 2006). However, differences in foraging patterns and selectivity between animals of various physiological states are reduced by changes in forage availability and quality during the course of a season (Grings *et al.* 2001). Scarce food sources can create diet overlap which occurs among native and domestic or introduced animals (Schwartz & Ellis 1981, Fritz *et al.* 1996, Baldi *et al.* 2004, Campos-Arceiz *et al.* 2004), among animals with similar body size (Schwartz and Ellis, 1981) and among animals with the same feeding ecology (Fritz *et al.* 1996).

4.1.2. *Animal's Cognitive Abilities*

Cognitive abilities facilitate animals to effectively use their environment and to detect and remember the distribution of high and poor quality food resources. Cognitive mechanisms of diet selection are comprised of an animal's individual ability of learning and spatial memory that both rely on visual, olfactory, and auditory senses. Learning provides animals with the flexibility not only to satisfy nutrient requirements, but most of all, to maintain homeostasis in environments where nutrient content and toxicity of potential foods is variable (Bryant *et al.* 1991).

Mechanisms of diet selection associated with learning processes evolve during an individual's ontogeny from maternal observation, peer interaction and nutritional post-ingestive consequences (Provenza 1995). Social learning, and learning from dams in particular, enables young animals to get critical information about their specific foraging environment, such as the location of food, water and cover resources (Provenza & Balph 1988). For instance, lambs avoid the food which causes post-ingestive distress much sooner than those without a mother example (Provenza 1994). Ganskopp and Cruz (1999) found that naive cattle graze on a broader array of forage and harvest fewer bites than their experienced congeners. Social learning therefore increases not only diet selection, but the efficiency of learning about nutritious foods and reduces the risk of over-ingesting toxic foods (Bryant *et al.* 1991, Provenza 1995). The early experience in life affects animals' distribution on pasture, for instance, as cattle return to areas where they foraged with their dams early in life (Howery *et al.* 1998).

Herbivores associate food items with their nutritional consequences. Animals learn about post-ingestive consequences of foods through two inter-related systems, affective and cognitive. The affective system integrates the taste of food and its post-ingestive consequences, and changes in the amount of ingested food, depending on whether the post-ingestive consequences are aversive (toxicity) or positive (nutritious food). Thus, the affective system provides feedback so animals can learn to ingest nutritious and avoid intoxication. The cognitive system integrates the taste and sight to select or to avoid particular foods. For instance, livestock associate visual cues with feeding sites and if the animals are trained to associate high quality food with some visual cue, they are able to generalize the cue to selecting initial patches regardless of their quality (Renken *et al.* 2008). The post-ingestive effects of nutrients and toxins from food are therefore integrated with the plant's odor, taste, and texture and results in the palatability of that food resource. Species-rich pastures provide a variety of food resources of different quality, thus offering a mixed diet. Such diversity stimulates food intake (see Wang *et al.* 2010c) and encourages the animal to maximize and balance intake of (macro- and micro-) nutrients and to regulate intake of different toxins (Provenza *et al.* 2003).

As another cognitive ability, spatial memory allows animals to remember where they have foraged and to use that information to determine where they will travel and forage. Spatial memory operates as a two-part code: working (short-term) memory and reference (long-term) memory. The functional value of working and reference memory is determined according to temporal and spatial scales of forage and habitat availability (Laca & Ortega 1996).Working memory serves for remembering and avoiding recently grazed areas within grazing bouts where food resources are depleted. The spatial information is retained only long enough to complete a particular task and then is discarded as no longer necessary. Working memory is used at the scale of the feeding station, patch or feeding site, which were visited during the preceding grazing bout, and in cattle, lasts for eight hours at least (Bailey *et al.* 1989). On the contrary, reference memory is retained for longer periods and works from patch level deciding on a daily basis, through feeding sites with their particular abiotic and biotic characteristics to the level of home range, remembering the relative value of the habitats for months or even years (Howery *et al.* 1999).

Nevertheless, grazing herbivores are faced with changes in patch quality and resource distribution in the natural, heterogeneous environment. In the face of such unpredictability, remembering the exact location and the previous quality of patches may be of little informative value (Illius & Gordon 1990). Instead, investing time in exploring patches to determine their quality may reduce the cost of feeding on a low-quality patch where a high-quality alternative is available. Such sampling behavior helps the forager overcome the problem of 'incomplete information' (Stephens & Krebs 1986) by tracking environmental fluctuations and thereby increasing foraging efficiency. An ability to switch between strategies would allow greater foraging success under unpredictable conditions (Bateson & Kacelnik 1998, Inglis *et al.* 2001). So in stable and predictable environments, grazing herbivores can use spatial memory to increase foraging efficiency, while in a more variable environment, spatial memory can be replaced by sampling behavior (Hewitson *et al.* 2005).

4.1.3. Inter-species Interactions

Diet composition and selectivity of herbivores may be influenced by the presence of other herbivore species and their trophic interactions. On natural pastures, where different species of free ranging herbivores share a long evolutionary history, the animals adopted strategies to cope with inter-species competition and to share available common food resources. Competition is considered to be a major selective force leading to resource partitioning (Schoener 1983). The resource partitioning among sympatric herbivores implies the differential temporal and spatial use of available resources, also called niche segregation and resource partitioning which enable species to coexistence despite overlaps of ecological, namely dietary and nutritional, requirements. Trophic inter-species interactions are mediated largely through their grazing and browsing impacts on vegetation. Generally, at high density and limited food resources, the animals with high diet similarity compete, whereas removal of only part of the biomass by one species may facilitate another species access to forage of an adequate height or quality (Vesey-Fitzgerald 1960, Bell 1970, McNaughton 1979). Facilitation occurs in two different manners (i) when grazing by one species makes more grass accessible to another species, e.g. by reducing grass height and removing stems or (ii) when grazing by one species stimulates grass regrowth, thereby enhancing the nutritional quality of forage for another species (Arsenault & Owen-Smith 2002). The resource partitioning has been widely described in temperate zones (Jenkins & Wright 1988, Putman 1996, Johnson *et al.* 2000, Mysterud 2000) as well as tropical native large herbivore assemblages in Africa (Vesey-Fitzgerald 1960, Bell 1970, McNaughton 1979, Jarman & Sinclair 1979, Voeten & Prins 1999, Woolnough & du Toit 2001, Cromsight & Olff 2006) and Asia (Dinerstein 1980, Martin 1982, Johnsingh & Sankar 1991; Bagchi *et al.* 2003; Steinheim *et al.* 2005, Wegge *et al.* 2006).

However, the patterns of resource partitioning may be disrupted by introducing an exotic species into a natural system such as cattle or sheep inducing strong competitive interactions. For instance, in Tanzania, the cattle introduced in an area with wildebeest and zebra selected feeding sites with forage characteristics similar to those of both wild native herbivores, while their selection for feeding sites did not overlap (Voeten & Prins 1999). Similar potential for competition resulting in spatial displacement was observed among mule deer, elk and introduced free-ranging cattle in the Blue Mountains in Oregon, USA. Two native cervids avoided areas used by cattle and their mutual overlap of used habitats resulted in partitioned use of vegetation communities within habitats (Stewart *et al.* 2002).

Still another situation occurs on managed species-rich pastures limited by fencing. Domestic animals such as cattle, sheep or goat differ in dietary and nutritional requirements and intrinsic abilities of diet selection (Van Soest 1994, Gordon *et al.* 1996, Fraser & Gordon 1997, Fraser *et al.* 2009). The species-rich sward enables them to select different items and mitigate potential competition or increase efficient utilization of available herbage on pasture. Sheep and goats consume more forbs than cattle (Rodriguez-Iglesias & Kothman 1998). Cattle, in comparison with sheep, graze more grasses with proportionally more stems and less leaves, or even dead material, and with high content of cellulose. Whilst sheep co-grazing with cattle is able to select sward components low in the profile, selecting grass parts or forbs with higher content of total protein, carbohydrates and phosphorus (Cook *et al.* 1967, Grant *et al.* 1985). Goats consume more forbs and browse than sheep (Bartolome *et al.* 1998), the co-grazing of these two species implies the resource partitioning at pasture. However, the diet selectivity may change in response to different grazing pressure. Animut *et al.* (2005) found that grazing pressure had no effect on selectivity of goats and sheep neither for, nor against grasses, but influenced selectivity for some forbs, namely ragweed (*Ambrosia* spp.).

4.2. Vegetation as a Determinant of Diet Selection

Species-rich pastures present exceptional swards with high diversity of plant species for animals which create a mosaic of patches of different vegetation structure varying in forage availability and nutritive value (Dumont *et al.* 2005b). Heterogeneity of the vegetation is the fundamental factor in the mechanisms and dynamics of selectivity and grazing behavior on pastures (Adler *et al.* 2001). The heterogeneous species-rich swards generate high spatial and temporal variability which frequently mutually interacts (Rychnovská 1993, Mládek *et al.* 2011b). Herbivores, on the contrary, require a relatively constant intake of nutrients to satisfy metabolism, growth, and reproduction (Prins & Langevalde 2008). They are therefore faced with the problem of obtaining a relatively constant supply of nutrients in a relatively variable and fluctuating environment (Frank 2006). Therefore, the basic trade off between quality of forage and ingested quantity differs according to actual conditions at each scale of spatial and temporal variability (Griffiths *et al.* 2003).

4.2.1. Diet Selection from Bite to Landscape

The animals on species-rich swards respond to heterogeneity in food resources by adopting divers foraging strategies. At a small scale, such as feeding stations (*sensu* Senft *et al.* 1987), the animals make short-term decisions to maximize instantaneous nutrient intake, whereas at a larger temporal scale, the animals decide on a daily intake basis (Fryxell 1991, WallisdeVries *et al.* 1999). At plant part level, the animals select the most accessible plant parts offering the largest bite size or rate of nutrient intake (Figure 5). Arnold (1960) conducted a series of experiments where the rate of stocking was such that consumption exceeded growth. Sheep continuously selected leaf in preference to stem where this was physically possible. Within both leaf and stem fractions of the plants, the animals selected material of the highest available nitrogen content. Furthermore, plant level is highly variable within plant species and the animals respond to it by selecting individual plants which maximize their intake of digestible nutrients. For instance, Ganskopp *et al.* (1992) experimentally proved that cattle prefer to graze individual grass bunches without stems. At the level of feeding station, selectivity is limited by the degree of resource depletion. Grazing animals simultaneously evaluate remaining available forage in reach and costs of travelling to another feeding station and then takes a decision to stay or to move on (Bailey *et al.* 1998). Therefore, differential defoliation of feeding stations within a paddock is ruled by the marginal forage value (Charnov 1976) which needs to be determined first by sampling (Dumont *et al.* 2005a). When the best remaining item at the feeding station is below a certain threshold, or when the rate of forage acquisition at that station falls below that threshold, the animal moves forward, establishing a new feeding station at which diet selection proceeds again (Figure 6). Diet selection is further improved by using spatial memory (e.g. Edwards *et al.* 1996). As grazing progresses and the resources are gradually depleted, herbivores are forced to get back to sampling and searching a new threshold value (Hewitson *et al.* 2005). Heterogeneous sward on species-rich pastures creates typical patchy arrangements of vegetation (Correll *et al.* 2003, Parsons & Dumont 2003, Pavlů *et al.* 2007). The animals consistently select high quality and productive patches, rejecting those of low quality (Garcia *et al.* 2003). Large pastures may involve a mosaic of different plant communities which we find at landscape level (Figure 7). The utilization of these communities and particular, landscape units, strongly depends upon animal species (larger herbivores assemble their diet from more units than smaller herbivores, Prins & Langevalde 2008) and the particular constraints within which it operates. For instance, Frank (2006) identified a production threshold (34 g m^{-2}) below which plant communities were not grazed in his study system and further showed that ungulates behave in landscape units by increasing the intake as aboveground production of a unit area increases. The selection of units within a landscape is a complex process involving a trade-off between nutrient requirements, distance to water and predation risk (O'Reagain & Schwartz 1995). The temporal variability in the sward quality may be either arising naturally through normal changes in plant physiology, phenology, and growth associated with seasonal changes in environmental conditions (Albon & Langvatn 1992, Mládek *et al.* 2011b) or may be supported by selective grazing. The defoliation by grazing determines the pattern of organic matter digestibility, biomass growth, and accumulation of senescent material (Illius 1986). Both natural and grazing-induced temporal variability in forage quality and availability occur at short term scales over a few seconds to several hours, at medium term scale over few days to weeks within a vegetation season, and at long term, including seasonal cycles (O'Reagain & Schwartz 1995) and inter-annual variability due to rainfall **variability (Pavlů** *et al.* 2006).

Photo Pavla Hejcmanová.

Figure 5. Cattle may bite off swards of 4 cm of height. Albeit their large muzzle does not generally allow them to be too selective at plant part level, the cows are able to eliminate mosses from a bite and spit them out (dry small bunch of sward in front of cow).

Jizerské Mountains 2009, photo Pavla Hejcmanová.

Figure 6. Pasture with an array of plant species offer a heterogeneous forage in patches. The heifer moves to a new feeding station as forage on the present one is considered to be depleted or insufficient.

Czech Republic 2010, photo Michal Hejcman.

Figure 7. The sheep are herded by a shepherd through the landscape in the Czech Middle Mountains. The animals may spread over the rangeland and seek for forage of their choice.

4.2.2. Sward Characteristics and Plant Functional Traits

A basic parameter characterizing the sward on species-rich pasture is the high diversity of plant species composition. Species-rich swards stimulate large herbivore generalists to make decisions of which plant species to consume. On one hand, the animals have more opportunity to choose palatable plants and optimize their nutrient intake, and on the other hand, higher diversity of food resources make foraging decisions more complex and may make it difficult to determine from which one the animal can obtain the highest nutritional benefit (see experimental study by Wang *et al.* 2010a). Comparing different livestock species, sheep in contrast to cattle, substantially distinguish plant species and select their diet at plant species and even at plant part level with preference for forbs and avoiding grass stems (Grant *et al.* 1985). Sheep on the species-rich pasture (Figure 8) graze on a wide range of present plant species. When meeting a new, unknown plant species, they consume a small quantity and wait for post-ingestive response. Only then, the animals start to graze it in higher quantity (Provenza *et al.* 2003). Sheep is able to learn to select species within several days and recognize favoured species again after six month (Hejcman *et al.* 2008). Mixed species diet in contrast to a single diet positively stimulates voluntary daily intake (it means quantity) and enhances nutrient intake (Wang *et al.* 2010c) and both are positively reflected in animal performance (Atwood *et al.* 2001).

Diet selection, however, reposes mainly on the interactions of quantity and quality parameters of the sward (Mládek *et al.* 2011a). The main indicators of forage quantity are biomass yield, sward height or plant density, and indicators of forage quality consist in nutrient content, organic matter digestibility, and/or maturity of the sward (phenological stage of plants). Animals respond to changes in sward characteristics by changes in their grazing behavior such as total grazing duration, frequency and the number of grazing bouts or by changes in the mechanisms of diet selection and intake rates (number of ingested bites, number of steps in searching diet sources or bite - step ratio).

Forage availability, reflected for instance by sward height, determines bite size (weight) and biting rate (Forbes 1988, WallisDeVries *et al.* 1994, WallisDeVries *et al.* 1998, Griffiths *et al.* 2003). Generally, on short swards, bite size is small and animals display a high biting rate (WallisDeVries & Daleboudt 1994, Barrett *et al.* 2001), increasing approximately by two bites per minute for each 1 cm decrease in sward height (Hejcmanová *et al.* 2009). The animals compensate decreasing forage by increasing daily grazing time or biting rate or by both. If there is abundant forage, the animals selectively feed on high quality patches (Bailey 1995, Wilmshurst *et al.* 1995, Coppedge & Shaw 1998, Ginane *et al.* 2003, Dumont *et al.* 2007a, Dumont *et al.* 2007b). When better quality forage becomes restricted below a certain threshold, for instance, due to seasonal dynamics and reproductive state of the sward (Ginane *et al.* 2003), the animals adjust their grazing time and/or biting rate in order to maintain daily intake (Forbes 1988, Funston *et al.* 1991, WallisDeVries *et al.* 1994, Hejcmanová *et al.* 2009). In other words, when forage is abundant and offers a choice of highly nutritious species, the strategy of maximizing forage quality appears to be the most favorable. On the contrary, if sward consists mainly of species of low forage value, the strategy of maximizing forage quantity seems to be more efficient in maximizing energy gain (Schwartz & Ellis 1981, Dumont *et al.* 2005a, Mládek *et al.* 2011a).

In the eastern part of the Czech Republic 2010, photo Michal Hejcman.

Figure 8. Typical species-rich pasture grazed by sheep in the White Carpathian Mountains.

Species-rich pastures represent for animals a complex, heterogeneous and continually changing environment where various types of patches may occur. The animals continually assess potential costs and benefits and are flexible in adopting their foraging strategy. For instance, Ginane *et al.* (2003) offered to the heifers an alternative between a tall (14 cm) abundant reproductive sward with high biomass gradually decreasing in nutritive value and a vegetative sward (of two heights: tall − 14 cm and short − 8 cm) with lower biomass and higher nutritional value. They revealed that the choice between quality and quantity of offered sward depended on the magnitude of difference between nutritive values. When the difference was moderate, the animals increased their grazing time on more available reproductive sward patches of lower quality and higher quantity. As the difference in quality widened, the animals switched their preference for vegetative, better quality sward. The animals responded to seasonal changes by changes in foraging behavior in order to maintain their total intake as well as the diet digestibility (Ginane *et al.* 2003). However, when animals face a grazing time constraint, they show considerable flexibility in their grazing behavior compensating for the restricted time by grazing, for instance, for fewer and longer foraging bouts (Iason *et al.* 1999). The animals seem to prioritize quality over intake (Ginane & Petit 2005).

In species-rich grasslands, diet selection strategies of domesticated herbivores have rarely been compared between plant communities, because direct assessment of quantity (biomass yield) and quality (digestibility) in fine-grained heterogeneous environment is costly and laborious. Therefore, indirect approaches for quantification of both quality and quantity of herbivore's diet are needed. Most papers from the last 20 years based assessment of foraging strategies in semi-natural grasslands on the following assumptions (i) most legumes have higher forage quality than other plant species (Bruinenberg *et al.* 2002), (ii) leaves provide more digestible biomass than stems including inflorescence in both monocotyledons and dicotyledons (Duru 1997). For instance, Fraser and Gordon (1997) made a comparison of the diet of goats, red deer and guanacos in three contrasting Scottish upland vegetation communities. They classified diet composition of oesophageal-fistulated animals into several forage groups as broad- and fine- leaved grasses, clover and dicotyledons, all further divided to green leaf parts and stem/flower parts. Dumont *et al.* (2007a), in their multi-site study inspecting diet selection of commercial vs traditional livestock breeds, categorized bite types according to their broad botanical classification (grass, legumes, forbs), height (tall or short) and vegetation stage (vegetative, reproductive or dead). Both studies used Jacobs selectivity index (Jacobs 1974) and confronted selectivity patterns between different domesticated ruminants. These studies were important for the recognition of feeding overlap or possible complementary resource use, and facilitated vegetation management using multi-species grazing systems. But these studies did not provide a clear picture how herbivore's selectivity is related with the plant's ability to tolerate or to resist herbivory, and therefore, they did not infer how selectivity affects resource use. All of these links might be investigated with the help of plant functional classification (Cingolani *et al.* 2005, Evju *et al.* 2009, Rusch *et al.* 2009) and freely accessible trait values in databases, for common European species in BIOLFLOR (Klotz *et al.* 2002) and LEDA (Kleyer *et al.* 2008). Plant trait approach enables considerably easier comparison of diet selection strategies between communities of distinct floral composition; on the condition we are able to determine forage availability and diet composition to the plant species level. The quantity-quality dilemma in herbivore's selection may be addressed at the plant species level or at the community level, for which community-weighted means of a particular trait are calculated by abundance weighing of species trait values (Cingolani *et al.* 2005). At the plant species level, forage quantity was recognized to be correlated with canopy height (Cornelissen *et al.* 2003). However, forage quality in species-rich grasslands is difficult to assess with a single functional trait. Organic matter digestibility, at least for grass species, negatively correlates with leaf dry matter content (LDMC) and positively with specific leaf area (SLA) (Al Haj Khaled *et al.* 2006). An herbivore's selectivity for these leaf traits, which are considered the best indicators of resource exploitative vs conservative strategies, remains largely untested. To our knowledge, only one cafeteria experiment evaluated preference of ungulates with respect to leaf traits, Lloyd *et al.* (2010) ranked palatability of 44 New Zealand native grass species for deer and sheep. They concluded that deer had a greater tendency to select grasses with high SLA but this trait was not a good predictor of sheep preference. Similarly, two studies of sheep diet selection in semi-natural grasslands (Cingolani *et al.* 2005, Mládek *et al.* 2011a) consistently reported that SLA was not related to sheep selectivity at the plant species level. But surprisingly, both studies showed that sheep selected species with tougher leaves or with higher LDMC. This could be partially explained by higher amounts of chemical defenses in the softer leaves of dicotyledons compared to tougher leaves of grasses (Long *et al.* 1999). Forage quality is also modified by the stage of species maturity (Thomas *et al.* 2010), which is principally ruled by flowering period. Indeed, in alpine grasslands, Evju *et al.* (2009) demonstrated that sheep selected species with a later onset of flowering. Probably the most informative measure of forage quality would be the forage indicator value (Klapp *et al.* 1953). This expert-based ordinal classification of grassland species, which is included in BIOLFLOR database (Klotz *et al.* 2002), is based on information of protein and mineral biomass concentrations, leaf/stem ratio, palatability, accessibility and seasonal duration of the plant's forage value for livestock. Mládek *et al.* (2011a) showed that sheep grazing mesic semi-natural grasslands selected sward patches and even plant species within patches with higher forage indicator value.

4.2.3. Plant Defense against Herbivory and Neighborhood Effects

Grasslands with long evolutionary history of grazing host a variety of species which survive herbivore pressure due to defense strategy (Díaz *et al.* 2007). A basic idea of plant defenses is that a plant should gain protection from the investment it allocates to its own physical or biochemical defense (Milchunas & Noy-Meir 2002). A plant's physical defense is represented by structural characteristics which can cause injury (spines, thorns, awns), substantially reduce digestion (silica) or can influence searching time (plant crypticity), cropping time (plant fibrousness, tensile and shear strength) and bite size (plant canopy structure); all of these effects depend on the morphology of the herbivore (Laca *et al.* 2001). For instance, Canada thistle (*Cirsium arvense* (L.) Scop.) represents highly defended weed of cattle and sheep pastures with prickly leaves, but C. thistle is preferentially grazed by goats and these are used for its elimination (see review by Popay & Field 1996). Plant's chemical defense constitutes secondary compounds – toxins, which can cause poisoning or reduce digestion (Figure 9). Some secondary metabolites are toxic in very small quantities such as alkaloid colchicine in meadow saffron (*Colchicum autumnale* L.), but most compounds such as tannins become noxious for livestock only after considerable plant consumption (Robbins *et al.* 1987, Frohne *et al.* 2005). Virtually every species in pastures contain at least low concentrations of secondary compounds, even forages sown to agriculturally improved grasslands: saponins in alfalfa (*Medicago sativa* L.), alkaloids in tall fescue (*Festuca arundinacea* Schreb.) and reed canarygrass (*Phalaris arundinacea* L.), tannins in the trefoils (*Lotus* spp.), and cyanogenic glycosides in the clovers (*Trifolium* spp.) (Provenza *et al.* 2007). But grazing herbivores seldom consume enough toxins to result in poisoning because they regulate their intake through post-ingestive feedback and quickly learn to eat mixes of plants that mitigate toxicity (Villalba *et al.* 2004). For instance, tannins contained in many wild plants may interact in rumen with highly toxic alkaloids from the other plants, thus neutralizing their negative effects (Lisonbee *et al.* 2009). Therefore, no plant species is protected absolutely against herbivory (Provenza *et al.* 2003).

White Carpathian Mountains, Czech Republic 2010, photo Jan Mládek.

Figure 9. Cypress spurge (*Euphorbia cyparissias* L.) belongs to plant species avoided by livestock due to its milky irritating sap. This species in a plant community under grazing has therefore competitive advantage and may spread and become a pasture weed.

Some species have been long reported as very poisonous for livestock, such as white hellebore (*Veratrum album* L.) due to the high content of toxic alkaloids. This species achieves local dominance in cattle pastures in the Alps (Kleijn & Steinger 2002), but Hejcman *et al.* (2008) surprisingly found that white hellebore was highly selected by sheep after an introduction to abandoned mountain grasslands despite the general belief that sheep are particularly susceptible to its poison (Frohne *et al.* 2005). Similarly, the abovementioned high toxicity of meadow saffron was not an obstacle for sheep which readily grazed its soft spring leaves in a species-rich pasture with dominance of unpalatable tor-grass (*Brachypodium pinnatum* (L.) P. B.) (Mládek unpublished). However, field consumption of plant species by herbivores depends not only on its physical and chemical characteristics, abundance and spatial distribution, but is greatly influenced by the characteristics of neighboring vegetation (e.g. Atsatt and O'Dowd 1976). In other words, species susceptibility to grazing is altered by the association with alternative forage species. This phenomenon has been debated in vast amounts of literature under a variety of names – plant defense guilds, associational avoidance, a. defense, a. refuge, a. resistance, a. susceptibility, neighbor contrast defense, n. c. susceptibility, shared defense or s. doom. In the last few years, researchers studied neighborhood effects (e.g. Bergvall *et al.* 2006, Miller *et al.* 2009, Bee *et al.* 2009) and agreed that these effects rely on different underlying mechanisms according to spatial scale, and suggested to unify terminology. Principally, herbivores can be selective or unselective between or within patches of plants and this can rise to four different scenarios (Bergvall *et al.* 2006, Rautio *et al.* 2008), which are involved in two alternative hypotheses first explicitly introduced by Hjältén *et al.* (1993). *Repellent-plant hypothesis* operates for the selection between patches in the absence of within patch selectivity and predicts: (i) if there are some less defended, palatable plants within the avoided patch, they would then gain protection from the defended plants (referred as associational defense) or (ii) highly defended, unpalatable plant within patch of palatable plants would be eaten more than if it occurred in a patch of mainly unpalatable plants (referred as associational susceptibility). *Attractant –decoy hypothesis* was proposed for situation when a herbivore is unselective between patches and selective within a patch and asserts: (i) a defended, unpalatable plant in a patch of mainly palatable plants would be less eaten than in a patch with plants of its kind (referred as neighbor contrast defense – opposite to associational susceptibility) or (ii) a palatable plant would be eaten more when growing in a patch of mainly less palatable plants than in a patch with plants of its kind (referred as neighbor contrast susceptibility – opposite to associational defense).

However, foraging of herbivore in natural grasslands has been described as a nested hierarchy of decisions, from landscape to plant level (Senft *et al.* 1987). Moreover, large herbivores continually sample and evaluate food instead of either consuming or rejecting food items (Hewitson *et al.* 2005). Therefore, it is not easily predictable which scenarios from the opposite pairs will take place in a given species assemblage. The foraging selectivity of herbivores is determined not only by species composition within a patch and patch characteristics, but also by the contrast between patches within a community (WallisDeVries *et al.* 1999). When the patches are clearly apparent, large herbivores should graze optimally according to 'patch-use theory' (e.g. Stephens and Krebs 1986) and select food at the stand level exclusively. Indeed, when herbivores have an opportunity to choose both patch and plant levels, individual species are utilized in relation to the quality of the stand; this was evidenced in several bioassay studies (Rautio *et al.* 2008, Miller *et al.* 2009). Also, a few field studies described directly decreased plant defoliation in its natural environment due to associational defense; for instance, the exceptional study by McNaughton (1978) in East African grasslands provided a clear picture that unpalatable plants protected a palatable grass (*Themeda triandra* Forsk.) from grazing by unselective herbivores. Furthermore, Palmer *et al.* (2003) performed a field study in mountain pastures in Scotland and brought an example of associational susceptibility; unpalatable heather (*Calluna vulgaris* L.) was much more utilized by sheep and deer in patches of forage grasses than growing alone. On the contrary, opposite scenarios following attractant–decoy hypothesis might be expected according to bioassay studies (Bergvall *et al.* 2006, Rautio *et al.* 2008) when differences between patches are hardly discernable. In real plant communities, nevertheless, large herbivores usually discern between patches, therefore, exclusive within patch selectivity is seldom found. Hjältén *et al.* (1993) suggested that food selection at the plant level becomes more important where the scale of patchiness is much larger than the home range size of the herbivore, forcing animals to live within just one vegetation patch. Correspondingly, Bee *et al.* (2009) proposed food selection at the plant level might strengthen at high population densities when herbivores are not distributed across patches in the proportion to the food available, that those herbivores in the poorer quality patch may browse the few palatable plants particularly heavily. On the other hand, food selection at the plant level (within patches) applied when domesticated animals grazed poor quality pasture with a few palatable plants under very low stocking rate (Provenza 2003). This leads to a neighbor contrast susceptibility scenario: 'cattle ate the best and left the rest' which, after repetition for many years, evoked that toxin-containing woody plants such as sagebrush (*Artemisia* spp.) and juniper (*Juniperus* spp.) have come to dominate over 39 million hectares of land in the Western USA (West 1993). Up to now, we are uncertain what environmental conditions in natural grassland systems facilitate either between or within patch selection. The first step towards elucidation of herbivore's motives were brought by Miranda *et al.* (2011), who illustrated that in spring, abundant forage on pastures enabled large herbivores to select palatable shrubs. In this case, unpalatable plants in the vicinity of browsed palatable shrubs were less consumed, which refers to neighbor contrast defense. Conversely, in scrubland with limited forage resources, herbivore consuming a given palatable shrub continued, using the same feeding station, foraging on unpalatable neighbors, which refers to association susceptibility. This feeding behavior takes place whenever searching and finding a new, optimal, more nutritive patch is more costly than feeding on unpalatable resources. Miranda *et al.* (2011) consequently outlined that opposite associational effects can be explained by optimal foraging theory (McArthur & Pianka 1966). Courrant and Fortin (2010) asked more explicitly and tested if abovementioned alternative associational scenarios may be clarified by quantification of energy gain. They recorded foraging behavior of free-ranging bison and assessed how spatial patterns of occurrence of highly profitable sedge (*Carex atherodes* Spreng.) could control the risk of herbivory for seven other plant species. Bison had higher energy gain in feeding stations where this sedge was consumed while avoiding plant species that experienced neighbor contrast defense (generally shorter species); whereas energy gain in feeding stations was higher by consuming instead of avoiding the plant species for which they detected associational susceptibility (tall species). Moreover, risk of herbivory for short species experiencing on average neighbor contrast defense increased notably in that feeding stations where their difference in height with *C. atherodes* become smaller. Thus, even large generalist herbivores as bison make simultaneously foraging decisions at both stand and plant level, and individual plant species (here *C. atherodes*) can cause opposite neighboring effects for different species and these are dependent on the spatial variation in sward structure. Overall, alternative neighborhood scenarios can be predicted by simple foraging rules of energy gain maximization.

Another issue is what spatial patterns of neighbor relationship contribute most to defense against herbivory, i.e. provide support for both palatable and unpalatable species which is a prerequisite for stable coexistence between herbivores and plants in grasslands ecosystems (Provenza *et al.* 2003). Wang *et al.* (2010) conducted a manipulative experiment using sheep and three native plant species with different palatability. Their principal finding lies in fact that palatable plants cannot effectively defend themselves against herbivory if the constituted spatial pattern between plant species only compels herbivores to make foraging selections at one scale, either between or within patches. They concluded that the particularly high complexity of spatial neighborhoods reduce herbivore's selectivity, thereby the vulnerability of palatable plants decrease. Remarkably, experimental studies from the last five years showed that particularly high complexity of grassland sward satisfies both nutrient and total food requirements of herbivores best (Wiggins *et al.* 2006, Wang *et al.* 2010b,c).

5. ANIMAL PERFORMANCE ON SPECIES-RICH PASTURES

The ultimate goal of livestock producers is to achieve an environmentally and economically sustainable livestock production and to maintain the forage production on grasslands. Pastures rich in plant species evolve mostly, if not sown *a priori* with a mixture of plant species (e.g. Hofmann & Ries 1989), in traditional grazing systems under low and moderate grazing intensity. Livestock producers may manage grazing systems favoring diversity of plant species by applying both the rotational or continuous grazing (Pavlů *et al.* 2003). The management does not have any remarkable effect on grazing behavior patterns and forage intake (Sharrow 1983, Hejcmanová unpublished data on heifers). If the grazing intensity between the systems is similar, both rotational and continuous grazing allows attaining similar outputs in terms of livestock daily live-weight gain (Hepworth *et al.* 1991, Hart *et al.* 1993, Manley *et al.* 1997, Kitessa & Nikol 2001). The key management variable appears to be the grazing intensity. Grazing intensity represents the intensity of grassland exploitation in relation to productivity of the sward. Comparing pastures with plant communities of similar biomass production, the intensity may be expressed as stocking rate, and this in terms of number of animals per hectare (Gillen & Sims 2002) or as livestock unit (LU) per hectare: 1 LU = 600 kg live weight in studies from Western Europe, e.g. Dumont *et al.* (2007), or 1 LU= 500 kg live weight in Central European studies, e.g. Pavlů *et al.* (2006).

Under high grazing intensity, many plant species are not able to tolerate the grazing pressure and the amount of available forage becomes restricted (Bullock & Marriott 2000, Hofmann *et al.* 2001, Pavlů *et al.* 2007). Low and moderate grazing intensities enable the sward to attain high structural heterogeneity and plant species to coexist. On the other hand, at very low intensity, the plant species diversity declines as a result of competition among plants, namely for light (Marriott *et al.* 2004, Pavlů *et al.* 2007).

2009, photo Pavla Hejcmanová.

Figure 10. Cattle on pasture under intensive grazing pressure generally synchronize their activities and mutually influence their behavior. While grazing, the heifers stay very close one to another, jostle and compete for restricted food sources.

One of the most important and widely used parameters which reflect grazing pressure is the sward height (Figure 10). Tall sward offers more available forage and despite its lower nutritive value, the livestock displays positive linear response of daily live weight gains to the sward height (Manley *et al.* 1997, Realini *et al.* 1999, Barthram *et al.* 2002). Most of these studies, however, have investigated the animal performance on improved pastures with few highly productive plant species, mostly mixtures of perennial ryegrass (*Lolium perenne* L.) and white clover (*Trifolium repens* L.). Fraser *et al.* (2009) compared the performance of steers grazing in ryegrass/white clover-dominated improved and semi-natural pastures in three experiments and found higher liveweight gains and meat quality on the improved one. On the other hand, other studies from Finland revealed no difference in the daily gain of beef calves grown on intensively managed fertilized and natural multi-species pastures (Niemelä *et al.* 2008). Furthermore, in Central Europe, animal performance on species-rich pastures was studied by Pavlů *et al.* (2006). They conducted a long-term grazing experiment with two grazing intensities (intensive and extensive for target sward height of 5 and 10 cm, respectively), and revealed that daily live weight gains per animal were relatively similar under both extensive and intensive management, but slightly higher under extensive grazing with higher sward height of lower nutritive value (Figure 11). This was in concordance with cow and calf performance on *Nardus*-dominated semi-natural grassland in Scotland as reported by Common *et al.* (1998). However, if the animal performance was recalculated to unit land area, seasonal live weight output per hectare was approximately 1.5 times higher under more intensive than under extensive grazing management (Pavlů *et al.* 2006). A series of grazing experiments with moderate and low grazing intensities on pastures across Europe brought very similar results; livestock performance per hectare was lower on low than moderate grazing intensity (Isselstein *et al.* 2007). Similarly, daily live weight gains per animal do not differ between given grazing intensities (Isselstein *et al.* 2007, Dumont *et al.* 2007a). Different situations may, however, occur under different climatic and ecosystem conditions. On sand sagebrush (*Artemisia filifolia* Torr.) rangelands (with a mixture of grasses and forbs), only calf birth weight and weaning weight was a negative function of stocking rate, while cow live weights remained similar among stocking rates. In years of drought, the situation changed and cow live weight declined as stocking rate increased (Gillen & Sims 2002). In semi-arid conditions, for instance, in Mediterranean ecosystems, an early-season deferment of grazing may be applied as an effective management measure. This allows the forage to grow sufficiently and consequently, to support moderate stocking rate with satisfactory daily live weight gains of calves and cows per hectare. The deferment presents an economically sound option there, despite the additional costs of supplementary feed during the deferment period (Gutman *et al.* 1999).

As already stated, multi-species pastures, namely at lower grazing intensities, stand out by high heterogeneity of swards. If the area available for animals is too large, some places may remain underutilized and decrease the efficiency of the grazing system. To ensure an even livestock utilization of pasture, additional management measures may be adopted. These measures may include, for instance, reducing the distance to water on larger pastures (Hart *et al.* 1993), providing shade and shelter for resting sites (Senft *et al.* 1985), placing mineral licks or feed supplements or improving forage quality by burning (all measures reviewed by Bailey *et al.* 1998).

Czech Republic 2010, photo Pavla Hejcmanová.

Figure 11. The vegetation under grazing is liable to changes in the sward structure. On the Betlém experimental pasture in the Jizerské Mountains, the different grazing intensity causes visible differences not only in the structure, but principally in plant species composition (paddock of intensive grazing on the left side, paddock of extensive grazing on the right side, situation in spring before the start of grazing season).

CONCLUSION

Experimental studies in the last decade have pointed out that domesticated ruminants perform better when offered a variety of food types (e.g. Villalba *et al.* 2010). Thus, livestock can better meet their needs for energy, intake of nutrients and regulate their intake of toxins in species-rich than in species-poor swards (Provenza *et al.* 2003, Wang *et al.* 2010c). It has been shown that in the absence of high quality species, the nutrient intake is maintained under high functional group richness, e.g. a combination of legumes, grasses and forbs (Wang *et al.* 2010c). As natural grasslands are generally low productive and mainly support species of lower nutritional quality than standard forage plants in agriculturally improved grasslands, plant species richness will critically influence herbivore food intake and nutrition. Overall, production per animal in species-rich pastures could benefit by low grazing intensity which allows animals to manifest their feeding preferences. However, absolute feeding preferences for plant species, which may be obtained from cafeteria trials, may not be applicable on species-rich pastures. On pastures, herbivore's diet selection is substantially modified by species availability, spatial distribution and neighborhood effects as described in the chapters above.

Pattern of selection is generally subjected to the body-size of the herbivore. Small herbivores such as sheep and goats are generally more selective than large ones such as cattle. The larger herbivores are forced to feed on less-quality forage in order to maintain a certain level of intake. However, both small and large animals adopt relatively selective strategies if there are adequate available food resources. It has been suggested that the coarser the grain of vegetation, the more likely animals select food at the stand level, i.e. highly exploit forage at high quality/quantity patches and omit other less profitable patches. In the case of fine-grained vegetation heterogeneity, selective herbivores choose exclusively palatable plant species since these provide sufficient forage intake. However, even large generalist herbivores make simultaneously foraging decisions at both stand and plant level, and the risk of herbivory for every individual plant in the feeding station can be predicted by simple foraging rules of energy gain maximization (Courrant & Fortin 2010).

Let us consider that if structure of the sward and grazing intensity enable high selectivity of foraging for palatable species, and these plant species are not able to compensate defoliation by fast regrowth (which occur in most unproductive ecosystems − Harrison & Bardgett 2008), pastures become gradually dominated by unpalatable well-defended species such as sagebrush (*Artemisia* spp.) (Provenza 2003). Thus, although selectivity enhances animal performance, i.e livestock production per animal, high selectivity often cause adverse effects on forage quality and livestock production per unit land from long-term perspective. Provenza *et al.* (2003) suggested that higher grazing pressure reduces selectivity, whereby providing support for both palatable and unpalatable species. Furthermore, it has been recently shown that herbivores passively reduce selectivity due to the high complexity of spatial neighborhoods between palatable and unpalatable plants (Wang *et al.* 2010a). In summary, we propose fine-grained swards may be grazed under low grazing pressure because these plant communities compel herbivores to reduce selectivity, whereas coarse-grained swards should be utilized by less selective animals under high grazing intensity in order to preclude expansion of well-defended plants on account of palatable ones. Coexistence of palatable and unpalatable plants is a prerequisite for sustainable utilization of grassland ecosystems by herbivores, i.e. long-term maintenance of livestock production.

REFERENCES

Adler, P. B., Raff, D. A. & Lauenroth, W. K. (2001). The effect of grazing on the spatial heterogeneity of vegetation. *Oecologia*, 128, 465–479.

Al Haj Khaled, R., Duru, M., Decruyenaere, V., Jouany, C. & Cruz, P. (2006). Using leaf traits to rank native grasses according to their nutritive value. *Rangeland Ecology & Management*, 59, 648–654.

Albon, S. D. & Langvatn, R. (1992). Plant phenology and the benefits of migration in a temperate ungulate. *Oikos*, 65, 502–513.

Animut, G., Goetsch, A. L., Aiken, G. E., Puchala, R., Detweiler, G., Krehbiel, C. R., Merkel, R. C., Sahlu, T., Dawson, L. J., Johnson, Z. B. & Gipson, T. A. (2005). Performance and forage selectivity of sheep and goats co-grazing grass/forb pastures at three stocking rates. *Small Ruminant Research*, 59, 203–215.

Arnold, G. W. (1960). Selective grazing by sheep of two forage species at different stages of growth. *Australian Journal of Agricultural Research*, 11, 1026–1033.

Arnold, G. W. (1987). Influence of the biomass, botanical composition and sward height of annual pastures on foraging behaviour by sheep. *Journal of Applied Ecology*, 24, 759–772.

Arsenault, R. & Owen-Smith, N. (2002). Facilitation versus competition in grazing herbivore assemblages. *Oikos*, 97, 313–318.

Atsatt, P. R. & O'Dowd, D. J. (1976). Plant defense guilds. *Science*, 193, 24–29.

Atwood, S. B., Provenza, F. D., Wiedmeier, R. D. & Banner, R. E. (2001). Influence of free-choice versus mixed-ration diets on food intake and performance of fattening calves. *Journal of Animal Science*, 79, 3034–3040.

Austrheim, G., Gunilla, E., Olsson, A. & Grøntvedt, E. (1999). Land-use impact on plant communities in semi-natural sub-alpine grasslands of Budalen, central Norway. *Biological Conservation*, 87, 369–379.

Bagchi, S., Gopal, S. P. & Sankar, K. (2003). Niche relationships of an ungulate assemblage in a dry tropical forest. *Journal of Mammalogy*, 84, 981–988.

Bailey, D. W. (1995). Daily selection of feeding areas by cattle in homogeneous and heterogeneous environments. *Applied Animal Behaviour Science*, 45, 183–200.

Bailey, D. W., Dumont, B. & WallisDeVries, M. F. (1998). Utilization of heterogeneous grasslands by domestic herbivores: Theory to management. *Annales de Zootechnie, 47*, 321–333.

Bailey, D. W., Gross, J. E., Laca, E. A., Rittenhouse, L. R., Coughenour, M .B., Swift, D. M., & Sims, P. L. (1996). Mechanisms that result in large herbivore grazing distribution patterns. *Journal of Range Management*, 49, 386–400.

Bailey, D. W., Rittenhouse, L. R., Hart, R. H. & Richards, R. W. (1989). Characteristics of spatial memory in cattle. *Applied Animal Behaviour Science*, 23, 331–340.

Baldi, R., Pelliza-Sbriller, A., Elston, D. & Albon, S. (2004). High potential for competition between guanacos and sheep in Patagonia. *Journal of Wildlife Management* 68, 924–938.

Barrett, P. D., Laidlaw, A. S., Mayne, C. S. & Christie, H. (2001). Pattern of herbage intake rate and bite dimensions of rotationally grazed dairy cows as sward height declines. *Grass and Forage Science*, 56, 362–373.

Barthram, G. T., Marriott, C. A., Common, T. G. & Bolton, G. R. (2002). The long-term effects on upland sheep production in the UK of a change to extensive management. *Grass and Forage Science*, 57, 124–136.

Bartolome, J., Franch, J., Plaixats, J. & Seligman, N. G. (1998). Diet Selection by Sheep and Goats on Mediterranean heath-woodland range. *Journal of Range Management*, 51, 383–391

Bateson, M. & Kacelnik, A. (1998). Risk-sensitive foraging: decision making in variable environments. In: Dukas, R. (ed), *Cognitive Ecology*, Chicago University Press, Chicago, pp. 297–341.

Bee, J. N., Tanentzap, A. J., Lee, W. G., Lavers, R. B., Mark, A. F., Mills, J. A. & Coomes, D. A. (2009). The benefits of being in a bad neighbourhood: plant community composition influences red deer foraging decisions. *Oikos*, 118, 18–24.

Bell, R. H. V. (1970). The use of the herb layer by grazing ungulates in the Serengeti. In: Watson, A. (ed), *Animal populations in relation to hein food resources*, Blackwell, Oxford, pp. 111–124.

Bergvall, U. A., Rautio, P., Kesti, K., Tuomi, J. & Leimar, O. (2006). Associational effects of plant defences in relation to within and between-patch food choice by a mammalian herbivore: neighbour contrast susceptibility and defence. *Oecologia*, 147, 253–260.

Boulanger, V., Baltzinger, C., Saïd, S., Ballon, P., Picard, J-F. & Dupouey, J-L. (2009). Ranking temperate woody species along a gradient of browsing by deer. *Forest Ecology and Management*, 258, 1397–1406.

Boyce, M. S. & McDonald, L. L. (1999). Relating populations to habitats using resource selection functions. *Trends in Ecology and Evolution*, 14, 268–272.

Bruinenberg, M. H., Valk, H., Korevaar, H. & Struik, P. C. (2002). Factors affecting digestibility of temperate forages from seminatural grasslands: a review. *Grass and Forage Science*, 57, 292–301.

Bryant, J. P., Provenza, F. D., Pastor, J., Reichardt. P. B., Clausen, T. P. & du Toit, J. T. (1991). Interactions between woody plants and browsing mammals mediated by secondary metabolites. *Annual Review of Ecology and Systematics*, 22, 431–446.

Bullock, J. M. & Marriott, C. A. (2000). Plant responses to grazing, and opportunities for manipulation. In: Rook, A. J. & Penning, P. D. (ed), *Grazing management: The principles and practice of grazing, for profit and environmental gain, within temperate grassland systems*, British Grassland Society, Reading, pp. 27–32.

Bullock, J. M., Franklin, J., Stevenson, M. K., Silvertown, J., Coulson, S. J., Gregory, S. J. & Tofts, R. (2001). A plant trait analysis of responses to grazing in a long-term experiment. *Journal of Applied Ecology*, 38, 253–267.

Campos-Arceiz, A., Takatsuki, S. & Lhagvasuren, B. (2004). Food overlap between Mongolian gazelles and livestock in Omnogobi, southern Mongolia. *Ecological Research*, 19, 455–460.

Castle, M. E. (1976). A simple disc instrument for estimating herbage yield. *Journal of British Grassland Society*, 31, 37–40.

Charnov, E. L. (1976). Optimal foraging: The marginal value theorem. *Theoretical Population Biology*, 9, 129–136.

Cingolani, A. M., Posse, G. & Collantes, M. B. (2005). Plant functional traits, herbivore selectivity and response to sheep grazing in Patagonian steppe grasslands. *Journal of Applied Ecology*, 42, 50–59.

Collins, S. L., Knapp, A. K., Briggs, J. M., Blair, J. M. & Steinauer, E. M. (1998). Modulation of diversity by grazing and mowing in native tallgrass prairie. *Science*, 280, 745–747.

Common, T. G., Wright, I. A. & Grant, S. A. (1998). The effect of grazing by cattle on animal performance and floristic composition in *Nardus*-dominated swards. *Grass and Forage Science*, 53, 260–269.

Cook, C. W., Harris, L. E. & Young, M. C. (1967). Botanical and nutritive content of diets of cattle and sheep under single and common use on mountain range. *Journal of Animal Science*, 26, 1169–1174.

Coppedge, B. R. & Shaw, J. H. (1998). Bison grazing patterns on seasonally burned tallgrass prairie. *Journal of Range Management*, 51, 258–264.

Cornelissen, J. H. C., Lavorel, S., Garnier, E., Díaz, S., Buchmann, N., Gurvich, D. E., Reich, P. B., ter Steege, H., Morgan, H. D., van der Heijden, M. G. A., Pausas, J. G. & Poorter, H. (2003). A handbook of protocols for standardised and easy measurement of plant functional traits worldwide. *Australian Journal of Botany*, 51, 335–380.

Correll, O., Isselstein, J. & Pavlů, V. (2003). Studying spatial and temporal dynamics of sward structure at low stocking densities: the use of an extended rising-plate-meter method. *Grass and Forage Science*, 58, 450–454.

Courant, S. & Fortin, D. (2010). Foraging decisions of bison for rapid energy gains can explain the relative risk to neighboring plants in complex swards. *Ecology*, 91, 1841–1849.

Cromsight, J. P. G. M. & Olff, H. (2006). Resource partitioning among savanna grazers mediated by local heterogeneity: an experimental approach. *Ecology*, 87, 1532–1541.

Currie, P. O., Hilken, T. O. & White, R. S. (1987). Evaluation of a single probe capacitance meter for estimating herbage yield. *Journal of Range Management*, 40, 537–541.

Currie, P. O., Morris, M. J. & Neal, D. L., (1973). Uses and capabilities of electronic capacitance instruments for estimating standing herbage. Part 2. Sown ranges. *Journal of British Grassland Society*, 28, 155–160.

De Bello, F., Lepš, J. & Sebastia, M. T. (2005). Predictive value of plant traits to grazing along a climatic gradient in the Mediterranean. *Journal of Applied Ecology*, 42, 824–833.

Díaz, S., Lavorel, S., McIntyre, S., Falczuk, V., Casanoves, F., Milchunas, D. G., Skarpe, C., Rusch, G., Sternberg, M., Noy-Meir, I., Landsberg, J., Zhang, W., Clark, H. & Campbell, B. D. (2007). Plant trait responses to grazing – a global synthesis. *Global Change Biology*, 13, 313–341.

Dinerstein, E. (1980). An ecological survey of the Royal Karnali – Bardia Wildlife Reserve, Nepal. Part III: ungulate populations. *Biological Conservation*, 18, 5–38.

Distel, R. A., Laca, E. A., Griggs T. C. & Demment, M. W. (1995). Patch selection by cattle: maximization of intake rate in horizontally heterogeneous pastures. *Applied Animal Behaviour Science*, 45, 11–21.

Dumont, B. (1997). Diet preferences of herbivores at pasture. *Annales de Zootechnie*, 46, 105–116.

Dumont, B., Carrère, P. & D'Hour, P. (2002). Foraging in patchy grasslands: diet selection by sheep and cattle is affected by the abundance and spatial distribution of preferred species. *Animal Research*, 51, 367–381.

Dumont, B., Renaud, P-C., Morellet, N., Mallet, C. Anglard, F. & Verheyden-Tixier, H. (2005a). Seasonal variations of red deer selectivity on a mixed forest edge. *Animal Research*, 54, 369–381.

Dumont, B., Prache, S., Carrère, P. & Boissy, A. (2005b). How do sheep exploit pastures? An overview of their grazing behaviour from homogeneous swards to complex grasslands. *Options Méditerranéennes, series A*, 74, 317–328.

Dumont, B., Garel, J. P., Ginane, C., Decuq, F., Farruggia, A., Pradel, P., Rigolot, C. & Petit, M. (2007a). Effect of cattle grazing a species-rich mountain pasture under different stocking rates on the dynamics of diet selection and sward structure. *Animal*, 1, 1042–1052.

Dumont, B., Rook, A. J., Coran, C. & Röver, K. U. (2007b). Effects of livestock breed and grazing intensity on biodiversity and production in grazing systems: 2. Diet selection. *Grass and Forage Science*, 62, 159–171.

Duru, M. (1997). Leaf and stem in vitro digestibility for grasses and dicotyledons of meadow plant communities in spring. *Journal of the Science of Food and Agriculture*, 7, 175–185.

Edwards, G. R., Newman, J. A., Parsons, A. J. & Krebs, J. R. (1996). Use of spatial memory by grazing animals to locate food patches in spatially heterogeneous environments: An example in sheep. *Applied Animal Behaviour Science*, 50, 147–160.

Evju, M., Austrheim, G., Halvorsen, R. & Mysterud, A. (2009). Grazing responses in herbs in relation to herbivore selectivity and plant traits in an alpine ecosystem. *Oecologia*, 161, 77–85.

Farruggia, A., Dumont, B., D'hour, P. & Egal, D. (2008). How does protein supplementation affect the selectivity and performance of Charolais cows on extensively grazed pastures in late autumn? *Grass and Forage Science*, 63, 314–323.

Farruggia, A., Dumont, B., D'hour, P., Egal, D. & Petit, M. (2006). Diet selection of dry and lactating beef cows grazing extensive pastures in late autumn. *Grass and Forage Science*, 61, 347–353.

Forbes, T. D. A. (1988). Researching the plant-animal interface: the investigation of ingestive behaviour in grazing animals. *Journal of Animal Science*, 66, 2369–2379.

Frank, D. A. (2006). Large herbivores in heterogeneous grassland ecosystems. In: Danell, K., Bergström, R., Duncan,. P. & Pastor, J. (eds), *Large herbivor ecology, ecosystem dynamics and conservation*, Cambridge University Press, Cambridge, pp. 326–347.

Fraser, M. D. & Gordon, I. J. (1997). The diet of goats, red deer and South American camelids feeding on three contrasting Scottish upland vegetation communities. *Journal of Applied Ecology*, 34, 668–686.

Fraser, M. D., Theobald, V. J., Griffiths, J. B., Morris, S. M. & Moorby, J. M. (2009). Comparative diet selection by cattle and sheep grazing two contrasting heathland communities. *Agriculture, Ecosystems and Environment*, 129, 182–192.

Fritz, H., De Garine-Wichatitsky, M. & Letessier, G. (1996). Habitat use by sympatric wild and domestic herbivores in an African savanna woodland: the influence of cattle spatial behaviour. *Journal of Applied Ecology*, 33, 589–598.

Frohne, D. & Pfänder, H. J. (2005). *Poisonous plants, 2nd edition., A handbook for doctors, pharmacists, toxicologist, biologists and veterinarians*. Timber Press, Portland.

Fryxell, J. M. (1991). Forage quality and aggregation by large herbivores. *American Naturalist*, 138, 478–498.

Funston, R. N., Kress, D. D., Havstad, K. M. & Doornbos, D. E. (1991). Grazing behaviour of rangeland beef cattle differing in biological type. *Journal of Animal Science*, 69, 1435–1442.

Gagnon, M. & Chew, A. E. (2002). Dietary preferences in extant African Bovidae. *Journal of Mammalogy*, 81, 490–511.

Ganskopp, D. & Cruz, R. (1999). Selective differences between naive and experienced cattle foraging aminy eight grasses. *Applied Animal Behaviour*, 62, 293–303.

Ganskopp, D., Angell, R. & Rose, J. (1992). Response of cattle to cured reproductive stems in a caespitose grass. *Journal of Range Management*, 45, 401–404.

Garcia, F., Carrère, P., Soussana, J. F. & Baumont, R. (2003). The ability of sheep at different stocking rates to maintain the quality and quantity of their diet during the grazing season. *Journal of Agricultural Science*, 140: 113–124.

Gillen, R. L. & Sims, P. L. (2002). Stocking rate and cow-calf production on sand sagebrush rangeland. *Journal of Range Management*, 55, 542–550.

Ginane, C. & Petit, M. (2005). Constraining the time available to graze reinforces heifers' preference for sward of high quality despite low availability. *Applied Animal Behaviour Science*, 94, 1–14.

Ginane, C., Petit, M. & D'Hour, P. (2003). How do heifers choose between maturing reproductive and tall or short vegetative swards? *Applied Animal Behaviour Science*, 83, 15–27.

Gordon, I. J. & Illius, A. W. (1994). The functional-significance of the browser-grazer dichotomy in African ruminants. *Oecologia*, 98,167–175.

Gordon, I. J., Illius, A. W. & Milne, J. D. (1996). Sources of variation in the foraging efficiency of grazing ruminants. *Functional ecology*, 10, 219–226.

Grant, S. A., Suckling, D. E., Smith, H. K., Torvell, L., Forbes, T. D. A. & Hodgson, J. (1985). Comparative studies of diet selection by sheep and cattle: The Hill grasslands. *Journal of Ecology*, 73, 987–1004.

Griffiths, W. M., Hodgson, J. & Arnold, G. C. (2003). The influence of sward canopy structure on foraging decisions by grazing cattle. I. Patch selection. *Grass and Forage Science*, 58, 112–124.

Grings, E. E., Short, R. E., Haferkamp, M. R. & Heitschmidt, R. K. (2001). Animal age and sex effects on diets of grazing cattle. *Journal of Range Management*, 54, 77–81.

Gutman, M., Holzer, Z., Baram, H., Noy-Meir, I. & Seligman, N. G. (1999). Heavy stocking and early-season deferment of grazing on Mediterranean-type grassland. *Journal of Range Management*, 52, 590–599.

Harrison, K. A. & Bardgett, R. D. (2008). Impacts of grazing and browsing by large herbivores on soils and soil biological properties. In: Gordon, I. J. & Prins, H. H. T (eds), *The ecology of browsing and grazing*, Springer, Heidelberg, pp. 201–216.

Harmoney, K. R., Moore, K. J., George, J. R., Brummer, E. C. & Russell, J. R. (1997). Determination of pasture biomass using four indirect methods. *Agronomy Journal*, 89, 665–672.

Hart, R. H. (2001). Plant biodiversity on shortgrass steppe after 55 years of zero, light, moderate, or heavy cattle grazing. *Plant Ecology*, 155, 111–118.

Hart, R. H., Bissio, J., Samuel, M. J. & Waggoner, J. W. (1993). Grazing systems, pasture size, and cattle grazing behavior, distribution and gains. *Journal of Range Management*, 46, 81–87

Hejcman, M., Žáková, I., Bílek, M., Bendová, P., Hejcmanová, P., Pavlů, V. & Stránská, M. (2008). Sward structure and diet selection after sheep introduction on an abandoned grassland in the Giant Mts., Czech Republic. *Biologia*, 63, 506–514.

Hejcmanová, P., Stejskalová, M., Pavlů, V. & Hejcman, M. (2009). Behavioural patterns of heifers under intensive and extensive continuous grazing on species-rich pasture in the Czech Republic. *Applied Animal Behaviour Science*, 117, 137–143.

Hepworth, K. W., Test, P. S., Hart, R. H., Smith, M. A. & Waggoner J. W. (1991). Grazing systems, stocking rates, and cattle behavior in southeastern Wyoming. *Journal of Range Management*, 44, 259–262.

Hewitson, L., Dumont, B. & Gordon, I. J. (2005). Response of foraging sheep to variability in the spatial distribution of resources. *Animal Behaviour*, 69, 1069–1076.

Hjälten, J., Danell, K, & Lundberg, P. (1993). Herbivore avoidance by association: vole and hare utilization of woody plants. *Oikos*, 68, 125–131.

Hodgson, J. (1979). Nomenclature and definitions in grazing studies. *Grass and Forage Science*, 34, 11–18,

Hofmann, L. & Ries, R. E. (1989). Animal performance and plant production from continuously grazed cool-season reclaimed and native pastures. *Journal of Range Management*, 42, 248–251.

Hofmann, R. R. & Stewart, D. R. M. (1972). Grazer or browser: a classification based on the stomach-structure and feeding habitats of East African ruminants. *Mammalia*, 36, 226–240.

Hofmann, R. R. (1989). Evolutionary steps of ecophysiological adaptation and diversification of ruminants - a comparative view of their digestive-system. *Oecologia*, 78, 443–457.

Holechek, J. L., Vavra, M. &. Pieper, R. D. (1982). Botanical composition determination of range herbivore diets: a review. *Journal of Range Management*, 35, 309–315.

Homolka, M. & Heroldová, M. (1992). Similarity of results of stomach and faecal contents analyses in studies of the ungulate diet. *Folia Zoologica*, 41, 193–208.

Howery, L. D., Bailey, D. W. & Laca, E. A. (1999). Impact of spatial memory on habitat use. In: Launchbaugh, K. L., Sanders, K. D. & Mosley, J. C. (eds), *Grazing behavior of livestock and wildlife Idaho forest*, Wildlife & Range Experimental Station, Bulletin University of Idaho, Moscow.

Howery, L. D., Provenza, F. D., Banner, R. E. & Scott, C. B. (1998). Social and environmental factors influence cattle distribution on rangeland. *Applied Animal Behaviour Science*, 55, 231–244.

Howery, L. D., Provenza, F. D., Banner, R. E. & Scott, C. B. (1998). Social and environmental factors influence cattle distribution on rangeland. *Applied Animal Behaviour Science*, 55, 231–244.

Illius, A. W. & Gordon, I. J. (1990). Constraints on diet selection and foraging behaviour in mammalian herbivores. In: Hughes, R. N. (ed), *Behavioural mechanisms of food selection*, Springer-Verlag, Heidelberg, pp. 369–392.

Illius, A. W. (1986). Foraging behaviour and diet selection. In: Gudmundsson, O. (ed), *Grazing research at northern latitudes,* Plenum Press, New York, pp. 227–236.

Inglis, R., Langton, S., Forkman, B. & Lazarus, J. (2001). An information primacy model of exploratory and foraging behaviour. *Animal Behaviour*, 62, 543–557.

Isselstein, J., Griffith, B. A., Pradel, P. & Venerus, S. (2007). Effects of livestock breed and grazing intensity on biodiversity and production in grazing systems. 1. Nutritive value of herbage and livestock performance. *Grass and Forage Science*, 62, 145–158.

Ivlev, V. S. (1961). *Experimental ecology of the feeding of fishes*. Yale University Press, New Haven.

Jacobs, J. (1974). Quantitative measurement of food selection, a modification of the forage ratio and Ivlev's electivity index. *Oecologia*, 14, 413–417.

Jaramillo, V. J. & Detling, J. K. (1992). Small-scale heterogeneity in a semi-arid North American grassland. II. Cattle grazing of simulated urine patches. *Journal of Applied Ecology*, 29, 9–13.

Jarman, P. J. & Sinclair, A. R. E. (1979). Feeding strategy and the pattern of resource partitioning in ungulates. In: Sinclair, A. R. E. & Norton-Griffiths, M. (eds), *Serengeti: dynamics of an ecosystem*, University of Chicago Press, Chicago, pp. 130–163.

Jenkins, K. J. & Wright, R. G. (1988). Resource partitioning and competition among cervids in Northern Rocky Mountains. *Journal of Applied Ecology*, 25, 11–24.

Johnsingh, A. J. T. & Sankar, K. (1991). Food plants of chital, sambar and cattle on Mundanthurai plateau, South India. *Mammalia*, 55, 57–66.

Johnson, B. K., Kern, J. W., Wisdom, M. L., Findholt, S. L. & Kie, J. G. (2000). Resource selection and spatial separation of mule deer and elk during spring. *Journal of Wildlife Management*, 64, 685–697.

Johnson, D. H. (1980). The comparison of usage and availability measurements for evaluating resource preference. *Ecology*, 61, 65–71.

Kitessa, S. M. & Nicol, A. M. (2001). The effect of continuous or rotational stocking on the intake and live-weight gain of cattle co-grazing with sheep on temperate pastures. *Animal Science*, 72, 199–208.

Klapp, E., Boeker, P., König, F. & Stählin, A. (1953). *Wertzahlen der Grünlandpflanzen*. Das Grünland, Hannover.

Kleijn, D. & Steinger, T. (2002). Contrasting effects of grazing and hay cutting on the spatial and genetic population structure of *Veratrum album*, an unpalatable, long-lived, clonal plant species. *Journal of Ecology*, 90, 360–370.

Kleyer, M., Bekker, R. M., Knevel, I. C., Bakker, J. P., Thompson, K., Sonnenschein, M., Poschlod, P., Van Groenendael, J. M., Klimeš, L., Klimešová, J., Klotz, S., Rusch, G. M., Hermy, M., Adriaens, D., Boedeltje, G., Bossuyt, B., Dannemann, A., Endels, P., Götzenberger, L., Hodgson, J. G., Jackel, A-K., Kühn, I., Kunzmann, D., Ozinga, W. A., Römermann, C., Stadler, M., Schlegelmilch, J., Steendam, H. J., Tackenberg, O., Wilmann, B., Cornelissen, J. H. C., Eriksson, O., Garnier, E. & Peco, B. (2008). The LEDA Traitbase. A database of lifehistory traits of Northwest European flora. *Journal of Ecology*, 96, 1266–1274.

Klotz, S., Kühn, I. & Durka, W. (eds) (2002). *BIOLFLOR – Eine Datenbank zu biologisch-ökologischen Merkmalen der Gefäßpflanzen in Deutschland*. Schriftenreihe für Vegetationskunde 38, Bundesamt für Naturschutz, Bonn.

Laca, E. A. & Ortega, I. M. (1996). Integrating foraging mechanisms across temporal and spatial scales. In: West, N. E. (ed) *5th International Rangeland Congress*, Vol. 2, Society for Range Management, Salt Lake City, pp. 129–132

Laca, E. A., Shipley, L. A. & Reid, E. D. (2001). Structural anti-quality characteristics of range and pasture plants. *Journal of Range Management*, 54, 413–419.

Laca, E. A., Ungar, E. D. & Dement, M. W. (1994). Mechanisms of handling time and intake rate of a large mammalian grazer. *Applied Animal Behaviour Science*, 39, 3–19.

Lamoot, I., Vandenberghe, C., Bauwens, D. & Hoffmann, M. (2005). Grazing behaviour of free-ranging donkeys and Shetland ponies in different reproductive states. *Journal of Ethology*, 23, 19–27.

Laycock, W. A., Buchanan, H. & Krueger, W. C. (1972). Three methods of determining diet, utilization, and trampling damage on sheep range. *Journal of Range Management*, 25, 352–356.

Lazo, K. M. & Kazzal, N. T. (1971). Separation of true selective grazing by cattle from effects of the esophageal fistula. *Journal of Animal Science*, 33, 1124–1128.

Lechowitz, M. J. (1982). The sampling characteristics of electivity indices. *Oecologia*, 52, 22–30.

Lepš, J., Michálek, J., Kulíšek, P. & Uhlík, P. (1995). Use of paired plots and multivariate analysis for the determination of goat grazing preference. *Journal of Vegetation Science* 6, 37–42.

Lisonbee, L. D., Villalba, J. J. & Provenza, F. D. (2009). Effects of tannin on selection by sheep of forages containing alkaloids, tannins and saponins. *Journal of the Science of Food and Agriculture*, 89, 2668–2677.

Lloyd, K. M., Pollock, M. L., Mason, N. W. H. & Lee, W. G. (2010). Leaf trait–palatability relationships differ between ungulate species: evidence from cafeteria experiments using naïve tussock grasses. *New Zealand Journal of Ecology*, 34, 219–226.

Long, R. J., Apori, S. O., Castro, F. B. & Ørskov, E. R. (1990). Feed value of native forages of the Tibetan Plateau of China. *Animal Feed Science and Technology*, 80, 101–113.

Manley, W. A., Hart, R. H., Samuel, M. J., Smith, M. A., Waggoner, J. W. & Manley, J. T. (1997). Vegetation, cattle and economic responses to grazing strategies and pressures. *Journal of Range Management*, 50, 638–646.

Marinas, A., García-González, R. & Fondevila, M. (2003). The nutritive value of five pasture species occurring in the summer grazing ranges of the Pyrenees. *Animal Science*, 76, 461–469.

Marriott, C. A., Fothergill, M., Jeangros, B., Scotton, M. & Louault, F. (2004). Long-term impacts of extensification of grassland management on biodiversity and productivity in upland areas. A review. *Agronomie*, 24, 447–461.

Martin, C. (1982). Interspecific relationship between barasingha and axis deer in Kanha MP, India and relevance to management. In: Wemmer, C. W. (ed), *Biology and management of the Cervidae*, Smithsonian Institution Press, Washington, pp. 299–306.

Martin, R. C., Astatkie, T., Cooper, J. M. & Fredeen, A. H. (2005). A comparison of methods used to determine biomass on naturalized swards. *Journal of Agronomy & Crop Science*, 191, 152–160.

McInnis, M. L., Vavra, M. & Krueger, W. C. (1983). A comparison of four methods used to determine the diets of large herbivores. *Journal of Range Management*, 36, 302–306.

McNaughton, S. J. (1978). Serengeti ungulates: feeding selectivity influences the effectiveness of plant defence guilds. *Science*, 199, 806–807.

McNaughton, S. J. (1979). Grazing as an optimization process: grass-ungulate relationships in the Serengeti. *American Naturalist*, 113, 691–703.

McNaughton, S. J. (1984). Grazing lawns: animals in herds, plant form and co-evolution. *American Naturalist*, 124, 863–886.

Michalk, D. L. & Herbert, P. K. (1977). Assessment of four techniques for estimating yield on dryland pastures. *Agronomy Journal*, 69, 864–868.

Milchunas, D. G. & Noy-Meir, I. (2002). Grazing refuges, external avoidance of herbivory and plant diversity. *Oikos*, 99, 113–130.

Miller, A. M., McArthur, C. & Smethurst, P. J. (2009). Spatial scale and opportunities for choice influence browsing and associational refuges of focal plants. *Journal of Animal Ecology*, 78, 1134–1142

Miranda, M., Díaz, L., Sicilia, M., Cristóbal, I. & Cassinello, J. (2011). Seasonality and edge effect determine herbivory risk according to different plant association models. *Plant Biology*, 13, 160–168.

Mládek, J., Hejcmanová, P., Dvorský, M., Mládková, P., Pavlů, V., de Bello, F., Hejcman, M. & Duchoslav, M. (2011a). Sheep trade-off in diet selection: forage quality in mesic vs forage quantity in dry species-rich grasslands. Submitted.

Mládek, J., Hejcman, M., Hejduk, S., Duchoslav, M. & Pavlů, V. (2011b). Community seasonal development enables late defoliation without loss of forage quality in semi-natural grasslands. *Folia Geobotanica*, 46, 17–34.

Murphy, W. M., Silman, J. P. & Mena Barreto, A. D. (1995). A comparison of quadrat, capacitance meter, HFRO sward stick, and rising plate meter for estimating herbage mass in a smooth stalked meadowgrass-dominant white clover sward. *Grass and Forage Science*, 50, 452–455.

Mysterud, A. (2000). Diet overlap among ruminants in Fennoscandia. *Oecologia*, 124, 130–137.

Niemelä, M., Huuskonen, A., Jaakola, S., Joki-Tokola, E. & Hyvärinen, M. (2008). Coastal meadows as pastures for beef cattle. *Agriculture, Ecosystems and Environment*, 124, 179–186.

Norbury, G. L. & Sanson, G. D. (1992). Problems with measuring diet selection of terrestrial, mammalian herbivores. *Australian Journal of Ecology*, 17, 1–7.

Novellie, P. A. (1978). Comparison of the foraging strategies of blesbok and springbok on the Transvalal highveld. *South African Journal of Wildlife Research*, 8, 137–144.

O'Reagain, P. J. & Schwartz, J. (1995). Dietary selection and foraging strategies of animals on rangelands, coping with spatial and temporal variability. In: Journet, M. (ed), *Recent Developments in the Nutrition of Herbivore*, INRA Editions, Paris, pp. 407–423.

Pakeman, R. J. (2004). Consistency of plant species and trait responses to grazing along a productivity gradient: a multi-site analysis. *Journal of Ecology*, 92, 893–905.

Palmer, S. C. F., Hester, A. J., Elston, D. A., Gordon, I. J. & Hartley, S. E. (2003). The perils of having tasty neighbors: grazing impacts of large herbivores at vegetation boundaries. *Ecology*, 84, 2877–2890.

Parsons, A. J. & Dumont, B. (2003). Spatial heterogeneity and grazing processes. *Animal Research*, 52, 161–179.

Pavlů, V., Hejcman, M., Pavlů, L. & Gaisler, J. (2003). Effect of rotational and continuous grazing on vegetation of the upland grassland in the Jizerské hory Mts., Czech Republic. *Folia Geobotanica*, 38, 21–34.

Pavlů, V., Hejcman, M., Pavlů, L. & Gaisler, J. (2007). Restoration of grazing management and its effect on vegetation in an upland grassland. *Applied Vegetation Science*, 10, 375–382.

Pavlů, V., Hejcman, M., Pavlů, L., Gaisler, J. & Nežerková, P. (2006). Effect of continuous grazing on forage quality, quantity and animal performance. *Agriculture, Ecosystems and Environment*, 113, 349–355.

Pérez-Barbería, F. J. & Gordon, I. J. (1999). The functional relationship between feeding type and jaw and cranial morphology in ungulates. *Oecologia*, 118, 157–165.

Pérez-Barbería, F. J. & Gordon, I. J. (2001). Relationships between oral morphology and feeding style in the Ungulata: a phylogenetically controlled evaluation. *Proceedings of Royal Society of London, ser. B*, 268, 1023–1032.

Pérez-Harguindeguy, N., Díaz, S., Vendramini, F., Cornelissen, J. H. C., Gurvich, D. E. & Cabido, M. (2003). Leaf traits and herbivore selection in the field and in cafeteria experiments. *Austral Ecology*, 28, 642–650.

Popay, I. & Field, R. (1996). Grazing animals as weed control agents. *Weed Technology*, 10, 217–231.

Prins, H. H. T. & van Langevalde, F. (2008). Assembling a diet from different places. In: Prins, H. H. T. & van Langevalde, F. (eds), *Resource ecology: spatial and temporal dynamics of foraging*, Springer, Dordrecht, pp. 129–155.

Provenza, F. D. & Balph, D. F. (1988). Development of dietary choice in livestock on rangelands and its implications for management. *Journal Animal Science*, 66, 2356–2368.

Provenza, F. D. (1994). Ontogeny and social transmission of food selection in domesticated ruminants. In: Galef, B. G., Mainardi, M. & Valsecchi, P. (eds), *Behavioral aspects of feeding: basic and applied research in mammals*, Harwood, Singapore, pp. 147–164.

Provenza, F. D. (1995). Postingestive feedback as an elementary determinant of food preference and intake in ruminants. *Journal of Range Management*, 48, 2–17.

Provenza, F. D. (2003). Twenty-five years of paradox in plant-herbivore interactions and sustainable grazing management. *Rangelands*, 25, 4–15.

Provenza, F. D., Villalba, J. J., Dziba, L. E., Atwood, S. B. & Banner, R. E. (2003). Linking herbivore experience, varied diets, and plant biochemical diversity. *Small Ruminant Research*, 49, 57–274.

Provenza, F. D., Villalba, J. J., Haskell, J., MacAdam, J. W., Griggs, T. C. & Wiedmeier, R. D. (2007). The value to herbivores of plant physical and chemical diversity in time and space. *Crop Science*, 47, 382–398.

Putman, R. J. (1996). Ungulates in temperate forest ecosystems: perspectives and recommendations for future research. *Forest Ecology and Management*, 88, 205–214.

Rautio, P., Kesti, K., Bergvall, U. A., Tuomi, J. & Leimar, O. (2008). Spatial scales of foraging in fallow deer: Implications for associational effects in plant defence. *Acta Oecologica*, 34, 12–20.

Realini, C. E., Hodgson, J., Morris, S. T. & Purchas, R. W. (1999). Effect of sward surface height on herbage intake and performance of finishing beef cattle. *New Zealand Journal of Agricultural Research*, 42, 155–164.

Reese, G. A., Bayn, R. L. & West, N. E. (1980). Evaluation of double-sampling estimators of subalpine herbage production. *Journal of Range Management*, 33, 300–306.

Renken, W. J., Howery, L. D., Ruyle, G. B. & Enns, R. M. (2008). Cattle generalise visual cues from the pen to the field to select initial feeding patches. *Applied Animal Behaviour*, 109, 128–140.

Robbins, C. T., Hanley, T. A., Hagerman, A. E., Hjeljord, O., Baker, D. L., Schwartz, C. C. & Mautz, W. W. (1987). Role of tannins in defending plants against ruminants: reduction in protein availability. *Ecology*, 68, 98–107.

Robbins, C. T., Spalinger, D. E. & Vanhoven, W. (1995). Adaptation of ruminants to browse and grass diets ± are anatomical-based browser-grazer interpretations valid. *Oecologia*, 103, 208–213.

Robel, R. J., Briggs, J. N., Dayton, A. D. & Hulbert, L. C. (1970). Relationships between visual obstruction measurements and weight of grasslands vegetation. *Journal of Range Management*, 23, 295–297.

Rodriguez-Iglesias, R. M. & Kothman, M. M. (1998). Best linear unbiased prediction of herbivore preferences. *Journal of Range Management*, 51, 19–28.

Roguet, C., Prache, S. & Petit, M. (1998). Development of a methodology for studying feeding station behaviour of grazing ewes. *Applied Animal Behaviour Science*, 55, 307–316.

Ruckstuhl, K. E., Festa-Bianchet, M. & Jorgenson, J. T. (2003). Bite rates in Rocky Mountain bighorn sheep (*Ovis canadensis*): effects of season, age, sex and reproductive status. *Behavioral Ecology and Sociobiology*, 54, 167–173.

Rusch, G. M., Skarpe, C. & Halley, D. J. (2009). Plant traits link hypothesis about resource-use and response to herbivory. *Basic and Applied Ecology*, 10, 466–474.

Rychnovská, M. (ed) (1993). *Structure and functioning of semi-natural meadows.*Elsevier, Amsterdam.

Sanderson, M. A., Rotz, C. A., Fultz, S. W. & Rayburn, E. B. (2001). Estimating forage mass with a commercial capacitance meter, rising plate meter, and pasture ruler. *Agronomy Journal*, 93, 1281–1286.

Schoener, T. W. (1983). Field experiments on interspecific competition. *American Naturalist*, 122, 240–285.

Schwartz, C. C. & Ellis, J. E. (1981). Feeding ecology and niche separation in some native and domestic ungulates on shortgrass prairie. *Journal of Applied Ecology*, 18, 343–353.

Seman, D. H., Stuedemann, J. A. & Hill, N. S. (1999). Behavior of steers grazing monocultures and binary mixtures of alfalfa and tall fescue. *Journal of Animal Science* 77: 1402–1411.

Senft, R. L., Coughenour, M. B., Bailey, D. W., Rittenhouse, L. R., Sala, O. E. & Swift, D. M. (1987). Large herbivore foraging and ecological hierachies. *BioScience*, 37, 789–799.

Senft, R. L., Rittenhouse, L. R. & Woodmansee, R. G. (1985). Factors influencing selection of resting sites by cattle on shortgrass steppe. *Journal of Range Management*, 38, 295–299.

Sharrow, S. H. (1983). Forage standing crop and animal diets under rotational vs. continuous grazing. *Journal of Range Management*, 36, 447–450.

Smit, H. J., Metzger, M. J. & Ewert, F. (2008). Spatial distribution of grassland productivity and land use in Europe. *Agricultural Systems*, 98, 208–219.

Spalinger, D. E. & Hobbs, N. T. (1992). Mechanisms of foraging in mammalian herbivores: new models of functional response. *American Naturalist*, 140, 325–348.

Steinheim, G., Wegge, P., Fjellstad, J. I., Jnawali, S. R., Weladji, R. B. (2005). Dry season diets and habitat use of sympatric Asian elephants (*Elephas maximus*) and greater one-horned rhinoceros (*Rhinoceros unicornis*) in Nepal. *Journal of Zoology*, 265, 377–385.

Stephens, D. W. & Krebs, J. R. (1986). *Foraging Theory.* Princeton University Press, Princeton, New Jersey.

Stewart, K. M., Bowyer, R. T., Kie, J. G., Cimon, N. J. & Johnson, B. K. (2002). Temporospatial distributions of elk, mule deer, and cattle: resource partitioning and competitive displacement. *Journal of Mammalogy*, 83, 229–244.

Stewart, K. E. J., Bourn, N. A. D. & Thomas, J. A. (2001). An evaluation of three quick methods commonly used to assess sward height in ecology. *Journal of Applied Ecology*, 38, 1148–1154.

Tadmor, N. H., Brieghet, A., Noy-Meir, I,, Benjamin, R. W. & Eyal, E. (1975). An evaluation of the calibrated weight-estimate method for measuring production in annual vegetation. *Journal of Range Management*, 28, 65–69.

Tanentzap, A. J., Bee, J. N., Lee, W. G., Lavers, R. B., Mills, J. A., Mark, A. F. & Coomes, D. A. (2009). The reliability of palatability estimates obtained from rumen contents analysis and a field-based index of diet selection. *Journal of Zoology*, 278, 243–248.

Taylor, J. A. (1993). *Chairperson's summary paper*. Proceedings of the XVII International Grassland Congress, pp. 739–740.

Thomas, D. T., Milton, J. T. B., Revel, C. K., Ewing, M. A., Dynes, R. A., Murray, K. & Lindsay, D. R. (2010). Preference of sheep among annual legumes is more related to plant nutritive characteristics as plants mature. *Animal Production Science*, 50, 114–123.

Van der Wal, R., Madan, N., van Lieshout, S., Dormann, C., Langvatn, R. & Albon, S. D. (2000). Trading forage quality for quantity? Plant phenology and patch choice by Svalbard reindeer. *Oecologia*, 123, 108–115.

Van Soest, P. J. (1994). *Nutritional Ecology of Ruminants, 2nd edition*. Comstock Publishing, Ithaca.

Vanderploeg, H. A. & Scavia, D. (1979). Two electivity indices for feeding with special reference to zooplankton grazing. *Journal of Fisheries Research Board of Canada*, 36, 362–365.

Vesey-Fitzgerald, D. F. (1960). Grazing succession among East African game animals. *Journal of Mammalogy*, 41, 161–172.

Villalba, J. J., Provenza, F. D. & Han, G. (2004). Experience influences diet mixing by herbivores: Implications for plant biochemical diversity. *Oikos*, 107, 100–109.

Villalba, J. J., Provenza, F. D. & Manteca, X. (2010). Links between ruminants' food preference and their welfare. *Animal*, 4, 1240–1247.

Virkajärvi, P. (1999). Comparison of three indirect methods for prediction of herbage mass on timothy-meadow fescue pastures. *Acta Agriculturae Scandinavica, Section B - Soil & Plant Science,* 49, 75–81.

Voeten, M. M. & Prins, H. H. T. (1999). Resource partitioning between sympatric wild and domestic herbivores in the Tarangire region of Tanzania. *Oecologia*, 120, 287–294.

WallisDeVries, M. F. & Daleboudt, C. (1994). Foraging strategy of cattle in patchy grassland. *Oecologia*, 100, 98–106.

WallisDeVries, M. F. & Schippers, P. (1994). Foraging in a landscape mosaic: selection for energy and minerals in free-ranging cattle. *Oecologia*, 100, 107–117.

WallisDeVries, M. F., Laca, E. A. & Demment, M. W. (1998). From feeding station to patch: scaling up food intake measurements in grazing cattle. *Applied Animal Behaviour Science*, 60, 301–315.

WallisDeVries, M. F., Laca, E. A. & Demment, M. W. (1999). The importance of scale of patchiness for selectivity in grazing herbivores. *Oecologia*, 121, 355–363.

Wang, L., Wang, D., Bai, Y., Huang, Y., Fan, M., Liu, J. & Li, Y. (2010a). Spatially complex neighboring relationships among grassland plant species as an effective mechanism of defense against herbivory. *Oecologia*, 164, 193–200.

Wang, L., Wang, D., Bai, Y., Jiang, G., Liu, J., Huang, Y. & Li, Y. (2010b). Spatial distributions of multiple plant species affect herbivore foraging selectivity. *Oikos*, 119, 401–408.

Wang, L., Wang, D., He, Z., Liu, G. & Hodgkinson, K. C. (2010c). Mechanisms linking plant species richness to foraging of a large herbivory. *Journal of Applied Ecology*, 47, 868–875.

Wegge, P., Shrestha, A. K. & Moe, S. R. (2006). Dry season diets of sympatric ungulates in lowland Nepal: competition and facilitation in alluvial tall grasslands. *Ecological Research*, 21, 698–706.

West, N. E. (1993). Biodiversity of rangelands. *Journal of Range Management*, 46, 2–13.

Wiggins, N. L., McArthur, C. & Davies, N. W. (2006). Diet switching in a generalist mammalian folivore: fundamental to maximising intake. *Oecologia*, 147, 650–657.

Wilmshurst, J. F., Fryxell, J. M. & Hudson, R. J. (1995). Forage quality and patch choice by wapiti (*Cervus elaphus*). *Behavioral Ecology*, 6, 209–217.

Woji, A.Y. & Iji, P. A. (1996). Oesophageal fistulation of West African Dwarf sheep and goats for nutritional studies. *Small Ruminant Research*, 21, 133–137.

Woolnough, A. P. & du Toit, J. T. (2001). Vertical zonation of browse quality in tree canopies exposed to size-structured guild of African browsing ungulates. *Oecologia*, 129, 585–590.

In: Agricultural Research Updates. Volume 2
Editor: Barbara P. Hendriks
ISBN: 978-1-61470-191-0
© 2012 Nova Science Publishers, Inc.

Chapter 5

IMPACTS, ECOLOGY AND DISPERSAL OF THE INVASIVE ARGENTINE ANT

Eiriki Sunamura[1], Shun Suzuki[1], Hironori Sakamoto[2], Koji Nishisue[1], Mamoru Terayama[1] and Sadahiro Tatsuki[1]

[1] Graduate School of Agricultural and Life Sciences,
The University of Tokyo, Bunkyo-ku, Tokyo, Japan
[2] Graduate School of Environmental Sciences,
Hokkaido University, Kita-ku,
Sapporo, Japan

ABSTRACT

Introduction of alien organisms is a major risk that follows international trade and globalization. Ants are among the most harmful groups of invasive organisms, with five species including the Argentine ant *Linepithema humile* listed among the world's 100 worst invasive species by the IUCN. We review the impacts, ecology, and dispersal of invasive ants, with the Argentine ant as a representative. Invasive ants attain high population densities in the introduced range, and cause damage to ecosystems, agriculture, and human well-being by the sheer number. The high densities stem partly from their characteristic social structure 'supercolonies', i.e., aggregations of numerous, mutually cooperative nests. In the Argentine ant, high consistency of their supercolony identities makes supercolony an important unit in inferring dispersal history of the species: colonies originating from a common source colony remain mutually compatible even if they are isolated for a long time. We highlight two topics in the dispersal history of the Argentine ant: 1) formation of an unprecedented intercontinental supercolony by the 150 year international trade; and 2) recent successive introductions to Pan-Pacific region seemingly in accordance with globalization.

INTRODUCTION

Introduction of alien organisms is a major risk that accompanies international trade. Global biodiversity is being lost at an unprecedented rate [Pimm et al. 1995], and biological invasion is one of its main causes, along with changes in land use, climate, and biogeochemical cycles [Vitousek 1994; Vitousek et al. 1996; Wilcove et al. 1998; Sala et al. 2000]. Alien species have strong competitive ability against indigenous species, often owing to release from natural enemies [Mitchell and Power 2003; Torchin et al. 2003]. For instance, introduced Nile perch in Lake Victoria may have caused disappearance of hundreds of endemic fish species by predation and competition [Witte et al. 1992]. Introduced species are also economically destructive [Pimentel et al. 2005; Xu et al. 2006; Pejchar and Mooney 2009]. In the Great Lakes region of the U.S.A., maintenance of water intake pipes clogged by introduced zebra mussels costs millions of dollars every year [Pejchar and Mooney 2009]. Moreover, introduced species threaten human health, as represented by the cases of disease outbreak mediated by alien mosquitoes [Juliano and Lounibos 2005]. In order to stop the damage by invasive alien species, prevention of introduction and spread, as well as control of established population, should be addressed [Mack et al. 2000; Simberloff et al. 2005; Hulme 2006]. Because eradication or successful management is often very difficult and costly, and because globalization is accelerating transportation of alien species, many authors put more emphasis on preventive approaches [Ricciarddi and Rasmussen 1998; Mack et al. 2000; Leung et al. 2002; Perrings et al. 2002; Westphal et al. 2008; Hulme 2009]. Prevention requires knowledge on pathways of dispersal, and corresponding quarantine system and surveillance network.

Ants are recognized as one of the most harmful groups of alien species. In fact, 17 invertebrate species are listed among the world's 100 worst invasive species by the IUCN (International Union for Conservation of Nature), and five of them are ants [Lowe et al. 2000]. The five species comprise the Argentine ant *Linepithema humile*, the big-headed ant *Pheidole megacephala*, the little fire ant *Wasmannia auropunctata*, the red imported fire ant *Solenopsis invicta*, and the yellow crazy ant *Anoplolepis gracilipes*. In addition, the garden ant *Lasius neglectus* is now rapidly expanding its distribution and damage in Europe [Ugelvig et al. 2008]. These species are derived from various subfamilies and genera, as well as various native ranges (Table 1), but all of them attain high population densities in introduced ranges and directly or indirectly displace indigenous ants, other invertebrates, vertebrates and plants [Holway et al. 2002]. They also become agricultural pests by tending honey-dew producing homopteran insects (aphids, scale insects and mealybugs). Furthermore, they are significant nuisance pests in urban areas that invade buildings with high frequency and in large numbers. In particular, the red imported fire ant and the little fire ant are pronounced sanitary pests that torment farmers, gardeners and those indoor, with their poison stings [Rhoades et al. 1989; Wetterer and Porter 2003]. The damage to livestock, wildlife, and public health caused by *Solenopsis* fire ants in the U.S.A. can be $1 billion per year [Pimentel et al. 2005]. Despite the enormous efforts to cope with introduced invasive ants, no truly effective control methodology has been established: eradication or successful management is extremely hard with existing methods [Soeprono and Rust 2004; Silverman and Brightwell 2008]. Given the hardship of control, preventive measures should be strengthened to suppress ant invasions. However, knowledge on their invasion history and pathways of dispersal is still limited. For the big-headed ant and yellow crazy ant, even their native ranges have not been clarified [Holway et al. 2002].

Table 1. Representative invasive alien ant species and the distribution. All of them but the garden ant are listed among the world's 100 worst invasive species by the IUCN

Species	Subfamily	Native range	Introduced range
Argentine ant *Linepithema humile*	Dorichoderinae	South America	Africa, Asia, Australia, Europe, North America, Oceanic Islands (Atlantic and Pacific)
Big-headed ant *Pheidole megacephala*	Myrmicinae	Africa or Asia?	Africa, Australia, North America, South America, Oceanic Islands (Indian and Pacific)
Garden ant *Lasius neglectus*	Formicinae	Middle East?	Europe
Little fire ant *Wasmannia auropunctata*	Myrmicinae	Middle and South America	Africa, Australia, North America, Oceanic Islands (Pacific)
Red imported fire ant *Solenopsis invicta*	Myrmicinae	Middle and South America	Asia, Australia, North America
Yellow crazy ant *Anoplolepis gracilipes*	Formicinae	Africa or Asia?	Africa, Asia, Australia, Oceanic islands (Indian and Pacific)

In this chapter we review the pest status (consequences of invasion), ecology (intrinsic characteristics that enhance invasion success), and dispersal (an ultimate cause of invasion) of invasive ants, with the Argentine ant as an exponent. This species is one of the most intensively studied invasive ants. In most sections of this review, brief comparisons are made between Argentine ants and other invasive ants.

THE ARGENTINE ANT *LINEPITHEMA HUMILE*: PEST STATUS

Distribution and Habitat

The Argentine ant (Figure 1) is native to Paraná River drainage of northern Argentina and surrounding countries in South America [Tsutsui et al. 2001; Wild 2007]. During the last 150 years, this species has been unintentionally introduced to all continents except Antarctica, and many oceanic islands with Mediterranean or mild temperate climate [Suarez et al. 2001; Roura-Pascual et al. 2004; Wetterer et al. 2009]. First recorded from Madeira somewhere between 1847 and 1858, Argentine ants landed Europe (Portugal and France), North America (Louisiana and California), and Africa (South Africa) during 1890-1910, Central America (Mexico and Bermuda) and Australia (Victoria, Western Australia, New South Wales, and Tasmania) around 1940-1950, and finally Asia (Japan) in 1993, and have spread in each area. Ecological niche characteristics of Argentine ants are similar between native and introduced ranges: in the introduced range Argentine ants typically become established along coastal areas and major river corridors [Suarez et al. 2001; Espadaler and Gómez 2003; Roura-Pascual et al. 2006, 2009a]. Dry inlands and high-elevation areas are rarely invaded. Access to permanent sources of water or soil moisture may be important abiotic conditions for establishment [Holway 1998a; Menke et al. 2007, 2009]. Argentine ants mostly invade disturbed environments such as urban district and agricultural land, but they sometimes penetrate natural environments [Holway et al. 2002], where impacts on endemic species are of particular concern (e.g., South Africa's fynbos shrubland [Bond and Slingsby 1984] and Hawaii's subalpine shrubland [Cole et al. 1992]).

Photographs by Taku Shimada.

Figure 1. The Argentine ant *Linepithema humile*. Upper: a worker carrying a larvae; lower: a fertile queen. Workers are 2.2-2.6 mm in length and from light to dark brown in color [Newell and Barber 1913]. Queens are 4.5-5 mm long.

Although native range and optimal abiotic conditions (e.g., climate and nesting microhabitat) vary among invasive ant species, predominant establishment in disturbed environment [Holway et al. 2002] and occasional infestation in natural environment (e.g., Galápagos Islands for little fire ants [Lubin 1984]; tropical rain forest of Christmas Island for yellow crazy ants [O'Dowd et al. 2003]) is the common pattern among species.

IMPACTS

There are many similarities among invasive ant species in their effects on local ecosystems, economy, and human well-being, with the *Solenopsis* fire ants and the little fire ant causing additional effects with their stings [Holway et al. 2002]. Here we survey case studies with Argentine ants.

1) Impacts on Ecosystems

Ecological impacts of the introduced Argentine ants range over many taxa and are caused by various means [Holway et al. 2002; Lach 2003; Ness and Bronstein 2004; Krushelnycky et al. 2005; Lach & Thomas 2008]. Among the impacts, effects on invertebrates are the most notable. Argentine ants competitively displace almost all of the native ant species throughout the invaded area [North America: Erickson 1971; Ward 1987, Human and Gordon 1997, Holway 1998b; Australia: Heterick 2000, Walters 2006, Rowles and O'Dowd 2009a Asia: Miyake et al. 2002, Touyama et al. 2003; Europe: Cammell et al. 1996, Carpintero et al. 2005; Africa: Bond and Slingsby 1984] by their large number and aggressiveness [Human and Gordon 1999; Holway 1999; Rowles and O'Dowd 2007; Carpintero and Reyes-López 2008]. Argentine ants can also reduce the abundance or diversity of non-ant ground-dwelling invertebrates or change their composition [Cole et al. 1992; Human and Gordon 1997; Bolger et al. 2000; Krushelnycky and Gillespie 2008; Rowles and O'Dowd 2009a], probably by predation or competition, though in some ecosystems the impacts are little apparent [Holway 1998; Walters 2006]. The affected species are from many orders (e.g., Collembola, Diptera, Lepidoptera, Coleoptera and Araneae) and functional groups (e.g., decomposers, herbivores, predators and scavengers). Wide range of rare, floral, or arboreal invertebrates may also be affected by Argentine ant infestation [Huxel 2000; Altleld and Stiling 2006, 2009; Lach 2007, 2008; Krushelnycky and Gillespie 2008; Nygard et al. 2008].

The Argentine ant reduces the abundance of some vertebrates in the introduced range, by eliminating their prey. One well-studied example is the coastal horned lizard *Phynosoma coronatum* in southern California [Fisher et al. 2002]. This reptile feeds mainly on ants, and Argentine ants are unsuitable nutritional alternatives to native ants they displace [Suarez et al. 2000; Suarez and Case 2002]. Argentine ants also negatively affect the abundance of the grey shrew *Notiosorex crawfordi* in southern California [Laakkonen et al. 2001]. Disruption of native arthropod community by the Argentine ants may affect the prey availability of the shrew. Other than reptiles and mammals, impacts of Argentine ants are concerned for birds [Holway et al. 2002]. In the cork oak forest of the Iberian Peninsula, Argentine ants affect arthropod prey availability for foliage-gleaning birds by reducing order diversity and ant species richness [Estany-Tigerström et al. 2010]. Argentine ants can also harm birds by nest predation. Nest predations by Argentine ants have been documented for many birds [Newell and Barber 1913], including endangered species such as the California gnatcatcher *Polioptila californica californica* [Sockman 1997] and Hawaiian goose Nene *Branta sandvicensis* [Krushelnycky et al. 2005]. However, relative importance of predation pressure by Argentine ants compared to those by other ants, mammals and birds, has been little addressed [Sockman 1997; Suarez et al. 2005a].

Plants may be negatively affected by Argentine ants by several, and often indirect, means. First, plants may suffer disruption of ant-plant mutualism on seed dispersal [Rodriguez-Cabal et al. 2009]. In not a few plant species, ants disperse the seeds in lieu of an elaiosome, a lipid-rich appendage attractive to ants [Giladi 2006]. They transport seeds to their nest, consume only the elaiosome, and then dispose the intact seeds in a nest chamber or a refuse pile outside. This not only enables distant dispersal of the seeds, but also reduces location and consumption of the seeds by predators. Argentine ants do not replace native ants as seed dispersers, especially for large seeds, in many places around the world: they transport seeds for only short distances, leave the seeds above ground (leading to high rates of predation), and thus change the plant community structure [Bond and Slingsby 1984; Christian 2001; Carney et al. 2003; Gómez et al. 2003; Rowles and O'Dowd 2009b] Second, plants can suffer disruption of pollination and consequent reduction of seed-set by Argentine ants [Blancafort and Gómez 2005]. Ants are generally poor pollinators [Hölldobler and Wilson 1990], and consumption of floral nectar by Argentine ants might be costly to insect-pollinated plants [Holway et al. 2002]. In South Africa, Argentine ant can be a strong competitor with the honeybee for nectar, collecting 42% of the nectar of black ironbark before honeybees start foraging in the morning [Buys 1987]. Pollination failure can also be induced by direct displacement of pollinators [Blancafort and Gómez 2005]. Argentine ants can aggressively drive flower-visiting pollinators away [Lach 2007]. They can also displace pollinators by preying on the larvae or disturbing the nests [Cole et al. 1992]. Third, Argentine ants can damage plants by tending homopteran insects [Altfeld and Stiling 2006], as in the agriculture (see below). Homopteran outbreaks caused by Argentine ants impose net fitness costs to the host plants, e.g., reduced seed mass and smaller early leaves [Brightwell and Silverman 2010]. In some cases plants may experience reduced mortality owing to exclusion of herbivores by Argentine ants, despite the negative effect of Homoptera tending [Altfeld and Stiling 2009], but it might not be the general pattern. In Christmas Island, sooty mold induced by mutualism of yellow crazy ant and Homoptera have led to conspicuous deaths of trees [O'Dowd et al. 2003].

2) Impacts on Agriculture

Argentine ants have been widely recognized as agricultural pests, mainly for causing outbreak of phloem-feeding homopteran insects [Newell and Barber 1913; Vega and Rust 2001]. In the introduced range, Argentine ants form mutualistic relationships with various homopteran species opportunistically [Holway et al. 2002; Lester et al. 2003]. Like many other ant species, Argentine ants protect these insects from predators or parasitoids in exchange of carbohydrate-rich honeydew. They displace natural enemies of Homoptera

[Flanders 1945; Bartlett 1961; Daane et al. 2007; Mgocheki and Addison 2009] or selectively remove parasitized homopteran individuals [Frazer and Van den Bosch 1973]. Outbreak of homopteran insects is detrimental to crop growth because of excess consumption of the phloem, injury on the product apparence by galls, transmission of pathogens, and encouragement of the growth of sooty mold over the leaves. Homoptera outbreak via mutualism with Argentine ants is documented from citrus orchards, vineyards [e.g., Phillips and Sherk 1991; Addison and Samways 2000; Daane et al. 2007], and many other crops [Lester et al. 2003]. A few studies compared the intensity of Homoptera tending between Argentine and native ants. These studies suggested moderate or strong ability of Argentine ants to displace parasitoids compared to other ants [Martinez-Ferrer et al. 2003; Mgocheki and Addison 2009; Powell and Silverman 2010]. In these laboratory experiments, colony size of studied ants was controlled. The effects of Homoptera tending by Argentine ants would be increased in the field, considering the high density of this species. According to some scientists in the U.S.A., homopteran outbreak was observed only in association with Argentine ants [Newell and Barber 1913; Phillips and Sherk 1991].

Argentine ants also directly damage crops. In Japan, we often hear complaints from owners of fields and kitchen gardens such as: Argentine ants infest figs before humans harvest; Argentine ants damage and deform root crops (e.g., carrot and white radish) [E. Sunamura, personal communication].

3) Impacts in Urban Area

Argentine ants are nuisance pests that inhabit urban districts and frequently intrude into structures [e.g., Gordon et al. 2001]. Examples of damages in Japan include: swarm on foods; bite humans and pets; crawl into bed and disturb a sleep; these annoyances occur so often that the residents get on the verge of nervous breakdown [E. Sunamura, personal communication]. In the U.S.A., Argentine ants went on a rampage as household pests soon after the initial introduction [Newell and Barber 1913], forcing some people to move to uninfested localities and making real estate values of invaded districts fall [De Ong 1916]. The continuous effort to control Argentine ants since then to the present [Newell and Rouge 1909; Sunamura et al. 2011a] also bears out their significance as urban pest, as well as the hardship of control. Indeed, ants are today ranked among the greatest urban pests by the structural pest control industry in the U.S.A., and Argentine ants account for a considerable portion of the number of ant control by pest management professionals (e.g., 85% in San Diego) [Field et al. 2007].

ECOLOGY

In this section we review the ecology of Argentine ants relevant to their invasiveness. Because of the ecological and economic significance, ecology of Argentine ants has been investigated intensively in the introduced range. Though not yet plenty, knowledge on the ecology of the native South American population has been updated greatly in recent years.

The native habitat of the Argentine ant, Paraná River drainage [Wild 2004, 2007], is an unstable environment with repeated flooding [LeBrun et al. 2007]. In such an environment with frequent disturbance, characteristics such as high reproductive ability (typical of r-strategists), high migration ability, and broad dietary spectrum, may be adaptive. The following characteristics of Argentine ants may have evolved in such context, but those characteristics may also allow their establishment, abundance, and frequent human-mediated dispersal out of the native range.

1) Polygyny [Multi-Queen System]

In many ant species, a nest contains a single reproductive queen [Hölldobler and Wilson 1990]. However, in Argentine ants, a nest contains multiple queens (>10 queens/1000 workers) [Keller et al. 1989]. An individual queen lays up to dozens of eggs per day [Abril et al. 2008].

2) Colony Budding

In many ant species, winged queens undertake mating flight and found new nests at distant sites from their natal nests [Hölldobler and Wilson 1990]. In contrast, Argentine ant queens do not engage in mating flight, and instead they mate within natal nests [Markin 1970a; Passera et al. 1988]. New nests are founded by colony budding, in which queens leave for new nesting sites nearby the natal nests, accompanied by workers, on foot [Ingram and Gordon 2003]. Because queens undertaking colony budding can rely on worker forces from the start, they reach their maximum fecundity earlier and thus have higher intrinsic rate of natural increase, compared to queens undertaking mating flight and founding nests independently [Tsuji and Tsuji 1996]. Simulations showed that when nest mortality is dependent on the nest size [e.g., in local competition: Holway and Case 2001; Walters and MacKay 2005; Sagata and Lester 2009], colony budding is more adaptive than independent founding by mating flight, under frequent disturbances [Nakamaru et al. 2007].

3) Opportunistic Nesting Behavior

Argentine ants readily relocate nests when nest condition becomes unfavorable [Gordon et al. 2001; Heller and Gordon 2006]. Other than digging shallow nests in the soil, Argentine ants make use of various microhabitats for nesting sites such as under mulch, under paving stones, between cracks in stone wall and concrete blocks, within and under flowerpots, inside empty cans, in garbage bags and garbage boxes [Vega and Rust 2001; S. Tatsuki, personal communication]. These traits may increase the chance of human transportation of this species.

4) Supercoloniality

In typical ant species, colonies are composed of one or several cooperative nests, and territorial aggression among conspecifics from neighboring nests is commonly observed

[Hölldobler and Wilson 1990]. In Argentine ants, however, individuals can move freely among many nests without incurring territorial aggression (nests are often interconnected with trails), and the aggregation of mutually non-aggressive nests is referred to as a 'supercolony'. Formation of supercolony reduces the cost associated with territorial defense, and enables Argentine ants to invest more in colony growth [Holway et al. 1998].

Supercoloniality may have arisen from the traits of polygyny and colony budding, but one supercolony can cover an extremely large area. For example, Argentine ants form a vast supercolony for more than 6000 km along the Mediterranean coast, though the supercolony is not perfectly continuous [Giraud et al. 2002]. Such large-scale supercolonies are common in the introduced range [>900 km across coastal California, U.S.A.: Tsutsui et al. 2000; >900 km across New Zealand: Corin et al. 2007a; >2700 km across Australia: Suhr et al. 2011; >400 km across Japan: Sunamura et al. 2009a; >100 km in the Western Cape, South Africa: Mothapo and Wossler 2011].

Territorial aggression occurs among individuals from different supercolonies [Thomas et al. 2006]. When individuals from different supercolonies encounter, they usually run away or attack each other. The attack often escalates to fierce fight, where individuals incur severe injury (lose legs and antennas) or die. In Europe, two populations different from the aforementioned supercolony exist, one expanding over >700 km along the Iberian Mediterranean coast [Giraud et al. 2002], and the other recently found from Corsica [Blight et al., 2010]. Other than the large one in California, the U.S.A. harbors several, much smaller supercolonies in California [Tsutsui et al. 2003a] and southeastern part of the country [Buczkowski et al. 2004]. In Japan, three small supercolonies (<1 km in diameter) were found from Kobe Port, other than the largest one in the country [Sunamura et al. 2007]. In the Western Cape of South Africa, a supercolony different from the dominant one was detected from Elim [Mothapo and Wossler 2011]. These supercolony identities were confirmed by behavioral experiments in which individuals from distant localities were artificially put together, but the behavioral patterns in these laboratory experiments may be consistent with the actual behavioral patterns in the field [Thomas et al. 2006]. In Japan, combats among Argentine ant individuals have been observed in the contact zone of distinct supercolonies in Kobe Port (Figure 2).

Argentine ants in the introduced range are characterized by single massive supercolonies and a few smaller supercolonies. In contrast, Argentine ants in the native range form multiple, mutually incompatible supercolonies that are typically less than hundreds of meters in diameter [Tsutsui et al. 2000; Heller 2004; Pedersen et al. 2006; Vogel et al. 2009]. For example, Vogel et al. [2009] found as many as 11 supercolonies along the 3 km study transect. The difference in supercolony size and number between the introduced and native range may have arisen from introduction of few supercolonies from the native range to new environments [Tsutsui et al. 2000; Giraud et al. 2002; Helanterä et al. 2009]. The release from fierce competition with conspecific supercolonies may enable Argentine ants in the introduced range to attain elevated population density [Suarez et al. 1999; Tsutsui et al. 2000]. High population density may be responsible for the displacement of native ants and other species by numerical advantage, prominent outbreak of homopteran insects in agricultural environment, and terrible nuisance in urban area [Holway et al. 2002; Silverman and Brightwell 2008].

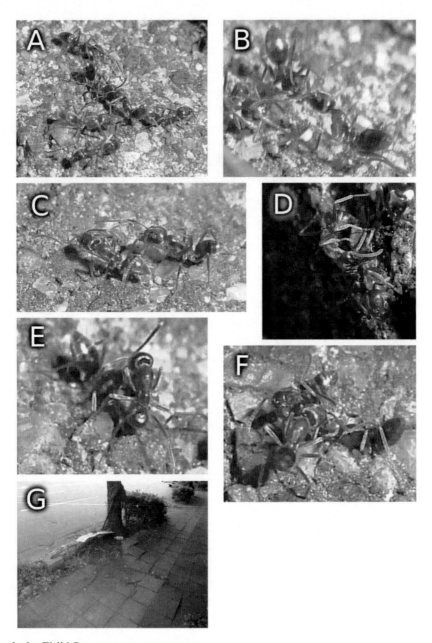

Photographs by Eiriki Sunamura.

Figure 2. Aggressive interactions between Argentine ant supercolonies in the field. In Kobe Port, Japan, by-ship multiple introductions may have led to the distribution of mutually incompatible supercolonies. A) Trails of different supercolonies collide and turn into chaos. Two individuals in the center flex their gasters menacingly. B) Two workers bite each other's mandibles. C) The left worker tries to run away, but the right worker bites it in the leg. D) The upper worker bites the lower individual in the petiole and emits chemical defensive compounds from the gaster. E) Two workers bite each other. The front individual uses chemical weapon. F) Fights often lead to death (right). The winner (lower left) meets another opponent (upper left), and the war continues. G) Pavement of Maya Wharf, Kobe Port, where the ant combat (A-E) was observed. This is the first record of field observation of aggressive interaction between Argentine ant supercolonies in Japan (April 2009).

5) Broad Dietary Spectrum

Argentine ants are omnivorous [LeBrun et al. 2007], and show flexible patterns of food resource use [Tillberg et al. 2007]. They prey on live arthropods, scavenge dead animals, and collect plant remains in the field [Abril et al. 2007]. They swarm on meat, vegetables and sweets in houses [E. Sunamura, personal obseravtion]. Among the broad dietary spectrum, however, Argentine ants feed mainly on liquid food such as the honeydew produced by homopteran insects [Markin 1970b; Abril et al. 2007]. Mass consumption of the carbohydrate-rich honeydew in the introduced range might fuel Argentine ant activity, as well as reduce the trophic level of Argentine ants to maintain high population density [Davidson 1998; Grover et al. 2007; Tillberg et al. 2007].

As described above, preadaptation (e.g., of the colony structure) to disturbed habitat, numerical dominance gained by formation of expansive supercolony, and perhaps dietary flexibility, may contribute to the success of Argentine ants in the introduced range (see also Suarez et al. 2008). Supercoloniality and omnivory are common among the invasive ant species (some species undergo mating flight along with colony budding) listed by the IUCN [Lowe et al. 2000] and other notorious ants (e.g., garden ant *Lasius neglectus*: Cremer et al. 2008) [Passera 1994; Holway et al. 2002]. In addition to these factors, release from natural enemies, a general cause of invader's success, may contribute to the success of these ants. For example, more than 30 natural enemies of *Solenopsis* fire ants have been discovered in their native South America (e.g., social parasitic ants, mites, parasitic flies and entomopathogenic microsporidia) [Williams et al. 2003]. As for the Argentine ants, however, natural enemies in the native range have not been detected [but see Markin and McCoy 1968; Tsutsui et al. 2003b; Reuter et al. 2005; Touyama et al. 2008]. Finally, release from interspecific competition may be also important [LeBrun et al. 2007]. Fire ants and Argentine ants co-occur in a part of their native ranges, and competition with supercolonies of each other appears to be as fierce as that with supercolonies of their own species. Because an ant species which possesses any one of the above characteristics is not necessarily invasive, invasiveness of alien ants might arise from addition of some characteristics or their synergistic effect.

CONTROL

Major control methods currently used for Argentine ants are chemical means such as toxic baits and barrier treatments [Soeprono and Rust 2004; Silverman and Brightwell 2008]. Argentine ants are readily susceptible to the active ingredients, but treatments often fail to achieve successful control. The difficulties in controlling Argentine ants are backed by formation of supercolony [Silverman and Brightwell 2008]. Partial treatments of a supercolony may achieve temporary decline or elimination of Argentine ants, but members of the supercolony soon move into the vacant space from surrounding area by tracking trails, relocating nests, or budding. For example, in a survey of homeowners living in a neighborhood infested with Argentine ants in California, approx. 40% of homeowners who hired professional pest control services did not obtain satisfactory results, with Argentine ants kept coming back [Klotz et al. 2008]. There are a few important cases where feasibility of eradication was shown for incipient small colonies with intense bait applications and follow-up surveys (a quarter of 26 ha residential area of Bunbury, Australia: Davis et al. 1998; 9.3 ha natural environment on Tiritiri Matangi Island, New Zealand: Harris 2002), but it is virtually impossible to eradicate large infestations, in the face of limited resources and numerous stakeholders [Forschler 1997; Silverman and Brightwell 2008].

Control of other invasive ants faces the same problem with Argentine ants, since they also form supercolonies. However, eradication from small infestation or successful large-scale management have been achieved in concerned natural environments, with adequate evaluation of distribution range, sufficient resources, and good cooperation among stakeholders (big-headed ant *Pheidole megacephala* in northern Australia [Hoffmann and O'Connor 2004; Hoffmann, 2010]; little fire ant *Wasmannia auropunctata* in Santa Fe and Marchena Islands, Galápagos [Abedrabbo 1994; Causton et al. 2005]; yellow crazy ant *Anoplolepis gracilipes* in some part of Seychelles and Christmas Island [Haines and Haines 1978; Green and O'Dowd 2009]). For fire ants, release trials of natural enemies moved from their native range have been conducted, and successful establishment of self-sustaining populations has been confirmed for two species of parasitic flies [Graham et al. 2003; Porter et al. 2004] and an entomopathogenic microsporidium [Williams et al. 1999].

DISPERSAL

Means of Dispersal

International dispersal of Argentine ants naturally depends on human activity [Suarez et al. 2001]. The type of commerce involved is obscure. Even for exotic ants as a whole [McGlynn 1999], there have been few systematic analyses. In case of the U.S.A., 94% of the ants intercepted in commerce at its border between 1927 and 1985 (394 records in total) were detected on plants (only one record for Argentine ants and no record for fire ants *Solenopsis invicta* and *Solenopsis richteri*, though they are now widespread in the country: but see Lewis et al. 1992) [Suarez et al. 2005b]. Major means of transportation for ants intercepted at the New Zealand border between 1955-2005 were fresh produce of air cargo and air passenger, as well as container of maritime cargo (The big-headed ant *Pheidole megacephala* was the number 1 intercepted species, and the yellow crazy ant and Argentine ant were among the top 20 species) [Ward et al. 2006]. There have been few records of Argentine ant interception at Japan border: the first record was made recently (2005) from air plants which came from Guatemala through Los Angeles by airway [T. Kishimoto, personal communication]. The air plants were infested by a colony with approx. 150 workers and two queens. Occurrence of Argentine ants has not been recorded from Guatemala [Wetterer et al. 2009], but the country includes areas suitable for their establishment [Roura-Pascual et al. 2004; Hartley et al. 2006].

The primary mode of dispersal of introduced Argentine ants within countries or continents is also human-mediated long distance jump dispersal [Suarez et al. 2001; Ward et al. 2005; Okaue et al. 2007; Blight et al. 2009; Pitt et al. 2009; Roura-Pascual et al. 2009b], which can proceed at a rate of >100 km/year [Suarez et al. 2001]. Unaided dispersal of Argentine ants is made exclusively via colony budding on foot, generally at a rate of only <300 m/year [Suarez et al. 2001]. The dual model of dispersal, human-mediated jump dispersal and colony budding, may be a general rule in invasive ants, but some species (e.g.,

Solenopsis invicta) show more complex patterns of dispersal because they also undertake mating flight [Holway et al. 2002; Espadaler et al. 2007].

Statistics on what vehicles Argentine ants use in jump dispersal events are scarce, but potted plants and nursery stock have often been cited in the literature [Newell and Barber 1913; Smith 1965; Madge and Caon 1987; Costa and Rust 1999]. These may be one of the major means of human-mediated dispersal for invasive ants in general [e.g., Lewis et al. 1992; Ugelvig et al. 2008]. They appear very significant vehicle, because ant colonies can be transported intact. Steamboats and railroads were also major means of Argentine ant dispersal in the U.S.A. at the beginning of the last century [Newell and Barber 1913], but their relative contribution might have declined with the development of other means of transportation. Roura-Pascual et al. [2009b] suggested with simulations relatively low but certain contribution of cork industry in the Argentine ant jump dispersal in the Iberian Peninsula. For *Solenopsis invicta* and *S. richteri* intercepted at California border during 1987-1992 (758 records in total: note that they include both international and domestic transportation), the sources were divided into nursery stock, agricultural shipments, non-agricultural shipments, empty trucks, rental trucks, and autos [Lewis et al. 1992]. Among them, non-agricultural shipments (e.g., pallets, roofing materials and carpets) comprised the most (approx. 300 records), followed by empty trucks and agricultural shipments (approx. 200 and 180 records, respectively). In Florida and probably in other parts of the southern U.S.A. also, road maintenance practice with soil depot may be a major source for dispersal of the polygyne form of *Solenopsis invicta* [King et al. 2009].

RELEVANCE OF SUPERCOLONY IDENTITY TO DISPERSAL HISTORY

In attempts to reconstruct the dispersal history of alien species, genetic analyses and historical records provide vital information. In case of Argentine ants, their social structure is also closely related to dispersal history. Previous studies showed genetic similarity within supercolonies [Tsutsui et al. 2000; Giraud et al. 2002; Corin et al. 2007a; Suhr et al. 2009], but disparity between supercolonies [Jaquiéry et al. 2005; Thomas et al. 2006; Pedersen et al. 2006]. As for mitochondrial DNA sequence, presence of single haplotypes within supercolonies, but different haplotypes between supercolonies, was confirmed [Vogel et al. 2009, 2010]. As for microsatellite loci, differences in the presence/absence or frequencies of alleles were confirmed [Jaquiéry et al. 2005; Thomas et al. 2006; Pedersen et al. 2006]. Furthermore, gene flow is strongly restricted between supercolonies [Jaquiéry et al. 2005; Thomas et al. 2006; Pedersen et al. 2006; Vogel et al. 2009], possibly due to aggression of workers against males from alien supercolonies [Sunamura et al. 2011b]. These findings suggest that distinct supercolonies in the introduced range represent independent sources of introduction. On the other hand, members of a single supercolony in the introduced range are likely to derive from a single introduction, or repeated introductions of a single source or perhaps very similar sources.

Argentine ants use hydrocarbons on their cuticle, the outer layer of the exoskeleton, to discriminate members and non-members of supercolonies [Liang and Silverman 2000; Greene and Gordon 2007; Torres et al. 2007], like many other ant species use their cuticular hydrocarbons to discriminate members and non-members of nests [Howard and Blomquist

2005]. Workers share similar cuticular hydrocarbon profiles (compounds and their relative proportions) within supercolonies, whereas workers from distinct supercolonies have different hydrocarbon profiles [Suarez et al. 2002; Sunamura et al. 2009a]. Many studies have suggested that cuticular hydrocarbon profiles are genetically based in Argentine ants [e.g., Tsutsui et al. 2000; Giraud et al. 2002; Suarez et al. 2002]. Behavioral patterns between Argentine ant colonies remain consistent under usual rearing conditions in laboratories [Giraud et al. 2002; Suarez et al. 2002], but under extreme rearing condition where the ants are fed with only a particular insect prey, prey-derived hydrocarbons are acquired by Argentine ants and inter-colony behavioral patterns are strongly affected [Liang and Silverman 2000]. This shows that cuticular hydrocarbon profiles are influenced by environment. However, such phenomenon has never been observed in the field. Environment-mediated change in hydrocarbon profiles is unlikely to happen in the field, because prey species which can induce the change may be limited [Liang et al. 2001], and Argentine ants usually consume variety of food resources, not specific prey [Markin 1970b; Abril et al. 2007].

Supercolonies expanding for hundreds or thousands of kilometers in the introduced range [Tsutsui et al. 2000; Giraud et al. 2002] may have been formed via numerous human-mediated dispersal events of particular source populations. Evidence supporting this view has been obtained from studies on social structure in recently invaded countries: New Zealand and Japan [Corin et al. 2007a; Sunamura et al. 2009a]. In both countries, Argentine ants were noted in the early 1990s, and now show patchy distribution resulting from human-mediated dispersal [Ward et al. 2005; Okaue et al. 2007]. In New Zealand, behavioral and genetic experiments demonstrated lack of aggression among all of the sampled localities and that the local colonies may originate from a single source population from Australia [Corin et al. 2007a, b]. In Japan, genetic similarity and lack of aggression among all of the localities, except Kobe Port, were revealed [Sunamura et al. 2007, 2009a]. In addition, chemical analyses demonstrated similarity of cuticular hydrocarbon profiles among disjunct members of the supercolony [Sunamura et al. 2009a]. These results suggest that supercolonies expand via human-mediated jump dispersal, without experiencing significant change in hydrocarbon profiles and recognition patterns by founder effect or change in environment concomitant to the dispersal.

The apparent permanence of supercolony identity in Argentine ants makes supercolony an important unit in inferring their dispersal history, although it is necessary to confirm behavioral data with genetic data because the possibilities of a new supercolony arising from mutation in the introduced range and effect of non-genetic factors on supercolony identity cannot be excluded [Liang and Silverman 2000; Blight et al., 2010]. It must also be noted that relationship between supercolony identity and dispersal history varies among ant species. Invasive ants come from various subfamilies in which supercoloniality may have evolved independently [Helanterä et al. 2009] via different mechanisms [Keller and Ross 1998; Fournier et al. 2005]. For example, supercolony identity is less consistent in the garden ant *Lasius neglectus*. In this species, variation in cuticular hydrocarbon profile and the resultant aggression is observed even among populations presumably originated from a common source population [Cremer et al. 2008; Ugelvig et al. 2008]. Effect of environment on cuticular hydrocarbon profile may be relatively strong in this species.

KNOWLEDGE ON DISPERSAL HISTORY

In this section, we gather previous findings on the dispersal history of the Argentine ant.

There have been some large-scale genetic studies examining international dispersal events of Argentine ants, and cases where introduction pathways are suggested have been accumulating. Tsutsui et al. [2001] suggested with microsatellite data that many of the introduced populations around the world originated from southern part of Paraná River drainage. Wetterer and Wetterer [2006] suggested the route from South America through Madeira to Portugal with combination of mitochondrial data, behavioral data and historical records. Tsutsui et al. [2001] suggested with microsatellite data that the population of Maui Island derived from one of the two distinct populations of Hawaii Island. Suhr et al. [2011] suggested with microsatellite data and trade records introduction from Europe to Australia. Corin et al. [2007b] suggested with mitochondrial data and interception records introduction from Australia to New Zealand. Buczkowski et al. [2004] revealed that microsatellite alleles of the large supercolony of California were a subset of the total alleles found from the diminutive supercolonies in southeastern U.S.A., but the alleles of the large supercolony were not a subset of the alleles found from either one of the small supercolonies [van Wilgenburg et al. 2010]. Therefore, the large supercolony may not derive from the small supercolonies.

There have been some other speculations based on historical records, that need to be tested with future research. For example, it was speculated that Argentine ants first reached the U.S.A. in New Orleans from Brazil via coffee ship [Foster 1908]. Also, Argentine ants are thought to have entered South Africa via cattle fodder from Argentina during the Anglo-Boer War [Skaife 1955]. In France, Argentine ants were apparently introduced with orchids and ferns imported from South America to green houses in the Côte d'Azur [Passera 1994]. Argentine ants are thought to have landed Hawaii either with troops in the Second World War [Passera 1994] or with goods from California [Zimmerman 1941].

Though not reached identification of specific introduction pathways, supercolony identities and their genetic analyses have provided information on more general patterns of dispersal. The presence of several, behaviorally and genetically distinct supercolonies in many introduced areas around the world [Giraud et al. 2002; Tsutsui et al. 2003a; Buczkowski et al. 2004; Sunamura et al. 2009a; Mothapo and Wossler 2011] suggests that Argentine ants have come out of their native range at least several times, and that multiple introductions into a country is a common pattern. Intensive investigations of Argentine ant social structure at fine spatial scales, especially in the hubs for international trade, have so far been scarce, but such approach may help identify focal ports of entry. For example, localization of three out of four Japanese supercolonies in Kobe Port may indicate direct, multiple introductions into the port via international trade [Sunamura et al. 2007, 2009a]. Also, Argentine ant colonies in seaport sites in Victoria, Australia, are mutually non-aggressive but exhibit relatively high antennation frequency to recognize each other, possibly due to multiple introductions from slightly different sources [Björkman-Chiswell et al. 2008].

Global-Scale Supercolony: Originated from the Most Ancient Introduction?

With knowledge on the social structure of Argentine ants in various introduced countries accumulating, studies on the relationships among supercolonies around the world are now becoming active. Wetterer and Wetterer [2006] found that the supercolony of Madeira and the largest supercolony in southern Europe are mutually non-aggressive. Sunamura et al. [2009b] examined behavioral relationships among supercolonies from California, Europe, and Japan, and found that the largest supercolonies in these areas are mutually non-aggressive. Their focal populations were selected based on their finding that the large supercolony of Japan has cuticular hydrocarbon profile apparently similar to those of the large supercolonies in California and Europe [Liang et al. 2001; de Biseau et al. 2004; Sunamura et al. 2009a]. During the same period, Brandt et al. [2009] performed comprehensive analyses using supercolonies from various introduced areas of the world (North America, Hawaii, Europe, and Australia), and showed that the largest supercolonies in respective areas are genetically very similar and have very similar cuticular hydrocarbon profiles. The research group then performed behavioral experiments and demonstrated lack of aggression among the large supercolonies [van Wilgenburg et al. 2010]. Furthermore, Vogel et al. [2010] revealed that the large supercolonies shared a single mitochondrial DNA haplotype.

These studies suggest the presence of a transcontinental supercolony. This supercolony is the largest cooperative unit ever known from any social insect species that exhibits intraspecific aggression. Moreover, the extent of this ant society is paralleled by human society only. This supercolony is also a very unique existence in the point that humans created it unwittingly via international trade. Considering that the formation of expansive supercolony is responsible for the invasiveness of Argentine ants, as well as for the difficulties in their control, the global-scale supercolony might be another "inconvenient truth".

The above studies also suggest that the mutually non-aggressive supercolonies around the world arose from a common introduction pathway, or even from a single native supercolony. Wetterer and Wetterer [2006] proposed an interesting hypothesis regarding the dispersal history of the largest supercolony in Europe. Argentine ants were first discovered out of their native range on Madeira between 1847 and 1858, followed by detection in some localities in Portugal in 1890s. At that time, Madeira was an important hub for commerce between Portugal and its colonies in South America. Based on behavioral, genetic, and historical evidence, it was suggested that the European large supercolony originated from the route from South America to Portugal via Madeira, and then throughout southern Europe. The finding that the European supercolony is a part of the intercontinental supercolony gives rise to an intriguing hypothesis that descendants of the most ancient introduced population have spread to many parts of the world, without forgetting their roots, i.e., supercolony identity. Indeed, Argentine ants may have spread from Europe to Australia [Suhr et al. 2011], and then from Australia to New Zealand [Corin et al. 2007b].

There are some points that merit further investigation to verify the 'Madeira single-origin' hypothesis. For example, we do not yet fully understand the distribution pattern of supercolonies in the native range. Are native populations comprised of thousands of mutually distinct supercolonies, or are they mosaic of a countable number of mutually incompatible supercolonies? In other words, are genetic and chemical (in terms of cuticular hydrocarbon profile) variations inexhaustible or limited? In addition, are important ports in the native range dominated by a particular supercolony?

Interestingly, the major supercolony of South Africa and probably the minor one there [Mothapo and Wossler 2011] are behaviorally and genetically different from the dominant supercolonies around the world [van Wilgenburg et al. 2010; ogel et al. 2010]. Argentine ants of South Africa may have derived from source populations that are different from that of the intercontinental supercolony. The genetic and behavioral data do not deny the supposition that Argentine ants were introduced to South Africa directly from Argentina via cattle fodder [Skaife 1955].

In other invasive ant species, relationships among international supercolonies are unknown. Although expansion of very limited propagules in the introduced range might be a common pattern, not all of invasive ants are able to form global-scale supercolonies. For example, in the garden ant, supercolony identity can change during the course of range expansion, as mentioned above [Cremer et al. 2008; Ugelvig et al. 2008].

Successive Introductions to Pan-Pacific Region: Product of Globalization?

Recent successive introductions of invasive ants to Pan-Pacific region is perhaps the most notable trend in ant invasions. In the nearly 100 or more years of invasion histories, the red imported fire ant had been confined in southeastern U.S.A. and Middle America, and the Argentine ant has not penetrated Asia, until 1990s [Callcott and Collins 1996; Suarez et al. 2001]. However, the red imported fire ant was noted in California in 1998 [anonymous 1999], Australia in 2001 [Lach and Thomas 2008], New Zealand on three occasions since 2001 (each time soon eradicated) [Pascoe 2002; Corin et al. 2008], Taiwan in 2003 [Chen et al. 2006], Hong Kong, Macao, and mainland China in 2005 [Zhang et al. 2007]. Australia, New Zealand, Taiwan, and China may have received multiple introductions [Henshaw et al. 2005; Corin et al. 2008; Yang et al. 2008]. Argentine ants were noted in New Zealand in 1990 [Green 1990], Japan in 1993 [Sugiyama 2000], and some other Asian and Oceanian countries in later years [Wetterer et al. 2009]. Sunamura et al. [2009a] concluded from supercolony number that Japan may have received at least four introductions. However, the large supercolony of Japan may actually derive from more than one introduction. The Argentine ant colony in Yanai, the southernmost distribution patch of the large supercolony, possesses unique microsatellite alleles [Hirata et al. 2008], and is likely to represent different invasion history from the other parts of the supercolony. Because this supercolony belongs to the transcontinental supercolony, separate introductions from some of its world domains are possible [Sunamura et al. 2009b].

Rigorous studies to elucidate the cause of the successive introductions to Pan-Pacific countries are yet to be performed, but globalization and liberalization may be at least partly responsible. Global-scale analysis showed that the overall degree of international trade best explains the number of invasive alien species in a country [Westphal et al. 2008]. Growth in the volume and complexity of international trade, combined with the liberalization of regulatory regimes to encourage trade, increase the frequency of introductions along existing pathways and the number of new pathways [Perrings et al. 2002]. Global increase in trade volume might have promoted the recent introductions of invasive ants to previously unexplored territories in Pan-Pacific region.

At the same time, perhaps already invaded areas of the world have also received new introductions, but the new introductions might have been confounded with the local spread of the older invasions. New introductions may be especially hard to be noticed in areas where spread of the ant is close to equilibrium (e.g., Argentine ants in the Iberian Peninsula: Roura-Pascual et al. 2009b). This explanation appears consistent with the more marked introductions of the red imported fire ant compared to Argentine ant in recent years. The red imported fire ant has the potential to invade large portions of all continents, but had not gone out of North and Middle America until recently [Morrison et al. 2004]. In contrast, of the potential geographical range of Argentine ants, the major vulnerable areas left uninvaded until 1990 were only southeastern Asia or tropica coattal Africa [Roura-Pascual et al. 2004; Hartley et al. 2006].

Alternatively, increase in transportations of exotic ants might have been specifically high in Pan-Pacific region. For instance, Asian countries have achieved great economic growth in recent decades [The Global Social Change Research Project 2007], and may have met dramatic rise in invasive species propagule pressure, as shown for China [Ding et al. 2008]. Important hubs for the trade between Asian countries and the trading partners (perhaps other parts of Pan-Pacific region for the type of commerce involved in ant transportation) may suffer similar risk, as was the case for Argentine ants in Madeira [Wetterer and Wetterer 2006]. Presence or absence of new ant introduction events to already widespread areas merits investigation in order to understand current dispersal dynamics of invasive ants.

CONCLUSION

International trade has spread Argentine ants over many parts of the world during the last 150 years, and has constructed an unparalleled ant colony in history. The damage caused by the invasion has been tremendous, and there may be still various impacts to be uncovered by scientists. Now the ant invasion is accelerated, apparently in accordance with the rapid growth in the volume of international trade. Increase in trade volume should be accompanied by advance in preventive measures against increased chances of invasion. Currently, knowledge on the means of transportation (both international and domestic) of invasive ants is especially lacking, and more statistics on transportation events is necessary. In addition, there is still limited amount of information on the pathways of dispersal, and the population and social structure should be studied for more locations, perhaps with special focus on international ports, around the world. Associations between ant invasions and trend in economy also need to be addressed. These studies may constitute vital components to reconstruct detailed dispersal history, as well as predict dispersal dynamics of invasive ants at the global scale.

ACKNOWLEDGMENTS

We would like to thank all of our current and previous collaborators, cooperators, and colleagues for nourishing our view on the subject. Special thanks go to Prof. Yukio Ishikawa of The University of Tokyo for advice, Ms. Ayako Hamada of The University of Tokyo for assistance and encouragement, Mr. Taku Shimada of the Ant Room for providing photographs of Argentine ants, and Dr. Toshio Kishimoto of the Japan Wildlife Research Center for sharing information on Argentine ant interception record in Japan. This work was supported by a Grant-in-Aid for young scientists to Eiriki Sunamura (20-6386) from the Japan Society for the Promotion of Science.

REFERENCES

Abedrabbo, S. (1994). Control of the little fire ant *Wasmannia auropunctata*, on Santa Fe Island in the Galápagos Islands. In Williams, D. F. (Eds.), *Exotic ants: biology, impact, and control introduced species* (pp. 219-227). Westview Press, Boulder, Colorado.

Abril, S., Oliveras, J. & Gómez, C. (2007). Foraging activity and dietary spectrum of the Argentine ant (Hymenoptera: Formicidae) in invaded natural areas of the northeast Iberian Peninsula. *Environmental Entomology, 36*, 1166-1173.

Abril, S., Oliveras, J. & Gómez, C. (2008). Effect of temperature on the oviposition rate of Argentine ant queens (*Linepithema humile* Mayr) under monogynous and polygynous experimental conditions. *Journal of Insect Physiology, 54*, 266-272.

Addison, P. & Samways, M. J. (2000). A survey of ants (Hymenoptera: Formicidae) that forage in vineyards in the Western Cape Province, South Africa. *African Entomology, 8*, 251-260.

Altfeld, L. & Stiling, P. (2006). Argentine ants strongly affect some but not all common insects on *Baccharis halimifolia*. *Environmental Entomology, 35*, 31-36.

Altfeld, L. & Stiling, P. (2009). Effects of aphid-tending Argentine ants, nitrogen enrichment and early-season herbivory on insects hosted by a coastal shrub. *Biological Invasions, 11*, 183–191.

Anonymous. (1999). Fire ant invades southern California. *California Agriculture, 53*, 5.

Bartlett, B. R. (1961). The influence of ants upon parasites, predators, and scale insects. *Annals of the Entomological Society of America, 54*, 543-551.

Björkman-Chiswell, B. T., van Wilgenburg, E., Thomas, M. L., Swearer, S. E. & Elgar, M. A. (2008). Absence of aggression but not nestmate recognition in an Australian population of the Argentine ant *Linepithema humile*. *Insectes Sociaux, 55*, 207-212.

Blancafort, X. & Gómez, C. (2005). Consequences of the Argentine ant, *Linepithema humile* (Mayr), invasion on pollination of *Euphorbia characias* (L.) (Euphorbiaceae). *Acta Oecologica, 28*, 49–55.

Blight, O., Orgeas, J., Renucci, M., Tirard, A. & Provost, E. (2009). Where and how Argentine ant (*Linepithema humile*) spreads in Corsica? *Comptes Rendus Biologies, 332*, 747-751.

Blight, O., Renucci, M., Tirard, A., Orgeas, J. & Provest, E. (2010). A new colony structure of the invasive Argentine ant (*Linepithema humile*) in Southern Europe. *Biological Invasions, 12*, 1491-1497.

Bolger, D. T., Suarez, A. V., Crooks, K. R., Morrison, S. A. & Case, T. J. (2000). Arthropods in urban habitat fragments in southern California: area, age, and edge effects. *Ecological Applications, 10*, 1230–1248.

Bond, W. & Slingsby, P. (1984). Collapse of an ant-plant mutalism: the Argentine ant (*Iridomyrmex humilis*) and myrmecochorous Proteaceae. *Ecology, 65,* 1031-1037.

Brandt, M., van Wilgenburg, E. & Tsutsui, N. D. (2009). Global-scale analyses of chemical ecology and population genetics in the invasive Argentine ant. *Molecular Ecology, 18,* 997-1005.

Brightwell, R. J. & Silverman, J. (2010). Invasive Argentine ants reduce fitness of red maple via a mutualism with an endemic coccid. *Biological Invasions, 12,* 2051-2057.

Buczkowski, G., Vargo, E. L. & Silverman, J. (2004). The diminutive supercolony: the Argentine ants of the southeastern United States. *Molecular Ecology, 13,* 2235-2242.

Buys, B. (1987). Competition for nectar between Aregntine ants (*Iridomyrmex humilis*) and honeybees (*Apis mellifera*) on black ironbark (*Eucalyptus sideroxylon*). *South African Journal of Zoology, 22,* 173-174.

Callcott, A. A. & Collins, H. L. (1996). Invasion and range expansion of imported fire ants (Hymenoptera: Formicidae) in North America from 1918-1995. *Florida Entomologist, 79,* 240-251.

Cammell, M. E., Way, M. J. & Paiva, M. R. (1996). Diversity and structure of ant communities associated with oak, pine, eucalyptus and arable habitats in Portugul. *Insectes Sociaux, 43,* 37-46.

Carney, S. E., Byerley, M. B. & Holway, D. A. (2003). Invasive Argentine ants (*Linepithema humile*) do not replace native ants as seed dispersers of *Dendromecon rigida* (Papaveraceae) in California, USA. *Oecologia, 135,* 576-582.

Carpintero, S., Reyes-López, J. & Arias de Reyna, L. (2005). Impact of Argentine ants (*Linepithema humile*) on an arboreal ant community in Doñana National Park, Spain. *Biodiversity and Conservation, 14,* 151–163.

Carpintero, S. & Reyes-López, J. (2008). The role of competitive dominance in the invasive ability of the Argentine ant (Linepithema humile). *Biological Invasions, 10,* 25-35.

Causton, C. E., Sevilla, C. R. & Porter, S. D. (2005). Eradication of the little fire ant, *Wasmannia auropunctata* (Hymenoptera: Formicidae), from Marchena Island, Galápagos: on the edge of success? *Florida Entomologist, 88,* 159-168.

Chen, J. S. C., Shen, C. & Lee, H. (2006). Monogynous and polygynous red imported fire ants, *Solenopsis invicta* Buren (Hymenoptera: Formicidae), in Taiwan. *Environmental Entomology, 35,* 167-172.

Christian, C. E. (2001). Consequences of a biological invasion reveal the importance of mutualism for plant communities. *Nature, 413,* 635-639.

Cole, F. R., Medeiros, A. C., Loope, L. L. & Zuehlke, W. W. (1992). Effects of the Argentine ant on arthropod fauna of Hawaiian high-elevation shrubland. *Ecology, 73,* 1313-1322.

Corin, S. E., Abbott, K. A., Ritchie, P. A. & Lester, P. J. (2007a). Large scale unicoloniality: the population and colony structure of the invasive Argentine ant (*Linepithema humile*) in New Zealand. *Insectes Sociaux, 54,* 275-282.

Corin, S. E., Lester, P. J., Abbott, K. L. & Ritchie, P. A. (2007b). Inferring historical introduction pathways with mitochondrial DNA: the case of introduced Argentine ants (*Linepithema humile*) into New Zealand. *Diversity and Distributions, 13,* 510-518.

Corin, S. E., Ritchie, P. A. & Lester, P. J. (2008). Introduction pathway analysis into New Zealand highlights a source population 'hotspot' in the native range of the red imported fire ant (*Solenopsis invicta*). *Sociobiology, 52,* 129-143.

Costa, H. S. & Rust, M. K. (1999). Mortality and foraging rates of Argentine ant (Hymenoptera: Formicidae) colonies exposed to potted plants treated with fipronil. *Journal of Agricultural and Urban Entomology, 16,* 37-48.

Cremer, S., Ugelvig, L. V., Drijfhout, F. P., Schlick-Steiner, B., Steiner, F. M., Seifert, B., Hughes, D. P., Schulz, A., Petersen, K. S., Konrad, H., Stauffer, C., Kiran, K., Espadaler, X., d'Ettorre, P., Aktaç, N., Eilenberg, J., Jones, G. R., Nash, D. R., Pedersen, J. S. & Boomsma, J. J. (2008). The evolution of invasiveness in garden ants. *PloS ONE, 3,* e3838.

Daane, K. M., Sime, K. R., Fallon, J. & Cooper, M. L. (2007). Impacts of Argentine ants on mealybugs and their natural enemies in California's coastal vineyards. *Ecological Entomology, 32,* 583-596.

Davidson, D. W. (1998). Resource discovery versus resource domination in ants: a functional mechanism for breaking the trade-off. *Ecological Entomology, 23,* 484-490.

Davis, P. R., van Schagen, J. J., Widmer, M. A. & Craven, T. J. (1998). The trial eradication of Argentine ants in Bunbury, Western Australia. *Internal report, Social Insect Research Section, Agriculture Western Australia.*

de Biseau, J.-C., Passera, L., Daloze, D. & Aron, S. (2004). Ovarian activity correlates with extreme changes in cuticular hydrocarbon profile in the highly polygynous ant, *Linepithema humile. Journal of Insect Physiology, 50,* 585-593.

De Ong, E. R. (1916). Municipal control of the Argentine ant. *Journal of Economic Entomology, 9,* 468-472.

Ding, J., Mack, R. N., Lu, P., Ren, M. & Huang, H. (2008). China's booming economy is sparking and accelerating biological invasions. *BioScience, 58,* 317-324.

Erickson, J. M. (1971). The displacement of native ant species by the introduced Argentine ant *Iridomyrmex humilis* mayr. *Psyche, 78,* 257-266.

Espadaler, X. & Gómez, C. (2003) The Argentine ant, *Linepithema humile,* in the Iberian Peninsula. *Sociobiology, 42,* 187-192.

Espadaler, X., Tartally, A., Schultz, R., Seifert, B. & Nagy, Cs. (2007). Regional trends and preliminary results on the local expansion rate in the invasive garden ant, *Lasius neglectus* (Hymenoptera, Formicidae). *Insectes Sociaux, 54,* 293-301.

Estany-Tigerström, D., Bas, J. M. & Pons, P. (2010). Does Argentine ant invasion affect prey availability for foliage-gleaning birds? *Biological Invasions, 12,* 827-839.

Field, H. C., Evans Sr., W. E., Hartley, R., Hansen, L. D. & J. H. Klotz. (2007). A survey of structural ant pests in the southwestern U.S.A. (Hymenoptera: Formicidae). *Sociobiology, 49,* 151-164.

Fisher, R. N., Suarez, A. V. & Case, T. J. (2002). Spatial patterns in the abundance of the coastal horned lizard. *Conservation Biology, 16,* 205–215.

Flanders, S. E. (1945). Coincident infestations of *Aonidiella citrina* and *Coccus hesperidum,* a result of ant activity. *Journal of Economic Entomology, 38,* 711-712.

Forschler, B. (1997). A prescription for ant control success. *Pest Control, 65,* 34-38.

Foster, E. (1908). The introduction of Iridomyrmex humilis (Mayr) into New Orleans. *Journal of Economic Entomology, 1,* 289-293.

Fournier, D., Estoup, A., Orivel, J., Foucaud, J., Jourdan, H., Le Breton, J. & Keller, L. (2005). Clonal reproduction by males and females in the little fire ant. *Nature, 435,* 1230-1234.

Frazer, B. D. & Van den Bosch, R. (1973). Biological control of the walnut aphid in California: the interrelationship of the aphid and its parasite. *Environmental Entomology, 2,* 561-568.

Giladi, I. (2006). Choosing benefits or partners: a review of the evidence for the evolution of myrmecochory. *Oikos, 112,* 481-/492.

Giraud T., Pedersen, J. S. & Keller, L. (2002). Evolution of supercolonies: the Argentine ants of southern Europe. *Proceedings of the National Academy of Sciences of the United States of America, 99,* 6075-6079.

Gómez, C., Pons, P. & Bas, J. M. (2003). Effects of the Argentine ant *Linepithema humile* on seed dispersal and seedling emergence of *Rhamnus alaternus. Ecography, 26,* 532-538.

Gordon, D. M., Moses, L., Falkovitz-Halpern, M. & Wong, E. H. (2001). Effect of weather on infestation of building by the invasive Argentine ant, *Linepithema humile* (Hymenoptera: Formicidae). *The American Midland Naturalist, 146,* 321-328.

Graham, L. C. F., Porter, S. D., Pereira, R. M., Dorough, H. D. & Kelley, A. T. (2003). Field releases of the decapitating fly *Pseudacteon curvatus* (Diptera: Formicidae) in Alabama, Florida, and Tennessee. *Florida Entomologist, 86,* 334-339.

Green, O. R. (1990). Entomologist sets new record at Mt Smart for *Iridomyrmex humilis* established in New Zealand. *The Weta, 13,* 14-16.

Green, P. T. & O'Dowd, D. J. (2009). Management of invasive invertebrates: lessons from the management of an invasive alien ant. In Clout, M. N. & Williams, P. A. (Eds.), *Invasive Species Management: a Hand Book of Principles and Techniques* (pp. 153-172). Oxford University Press.

Greene, M. J. & Gordon, D. M. (2007). Structural complexity of chemical recognition cues affects the perception of group membership in the ants *Linepithema humile* and *Aphaenogaster cookerelli. The Journal of Experimental Biology, 210,* 897-905.

Grover, C. D., Kay, A. D., Monson, J. A., Marsh, T. C. & Holway, D. A. (2007). Linking nutrition and behavioural dominance: carbohydrate scarcity limits aggression and activity in Argentine ants. *Proceedings of the Royal Society B: Biological Sciences, 274,* 2951-2957.

Haines, I. H. & Haines, J. B. (1978). Pest status of the crazy ant, *Anoplolepis longipes* (Jerdon) (Hymenoptera: Formicidae), in the Seychelles. *Bulletin of Entomological Research, 68,* 627-638.

Harris, R. J. (2002). Potential impact of the Argentine ant (*Linepithema humile*) in New Zealand and options for its control. *Science for Conservation, 196,* 1-36.

Hartley, S., Harris, R. & Lester, P. J. (2006). Quantifying uncertainty in the potential distribution of an invasive species: climate and the Argentine ant. *Ecology Letters, 9,* 1068-1079.

Helanterä, H., Strassmann, J. E., Carrillo, J. & Queller, D. C. (2009). Unicolonial ants: where do they come from, what are they and where are they going? *Trends in Ecology and Evolution, 24,* 341-349.

Heller, N. E. (2004). Colony structure in introduced and native populations of the invasive Argentine ant, *Linepithema humile. Insectes Sociaux, 51,* 378-386.

Heller, N. E. & Gordon, D. M. (2006). Seasonal spatial dynamics and causes of nest movement in colonies of the invasive Argentine ant (*Linepithema humile*). *Ecological Entomology, 31,* 499-510.

Henshaw, M. T., Kunzmann, N., Vanderwoude, C., Sanetra, M. & Crozier, R. H. (2005). Population genetics and history of the introduced fire ant, *Solenopsis invicta* Buren (Hymenoptera: Formicidae), in Australia. *Australian Journal of Entomology, 44,* 37-44.

Heterick, B. E. (2000). Influence of Argentine and coastal brown ant (Hymenoptera: Formicidae) invasions on ant communities in Perth gardens, Western Australia. *Urban Ecosystems, 4,* 277-292.

Hirata, M., Hasegawa, O., Toita, T. & Higashi, S. (2008). Genetic relationships among populations of the Argentine ant *Linepithema humile* introduced into Japan. *Ecological Research, 23,* 883-888.

Hoffmann, B. D. & O'Connor, S. (2004). Eradication of two exotic ants from Kakadu National Park. *Ecological Management and Restoration, 5,* 98-105.

Hoffmann, B. D. (2010). Ecological restoration following the local eradication of an invasive ant in northern Australia. *Biological Invasions, 12,* 959-969.

Hölldobler, B. & Wilson, E. O. (1990). *The Ants.* Cambridge, Mass: The Belknap Press of Harvard University Press.

Holway, D. A. (1998a). Factors governing rate of invasion: a natural experiment using Argentine ants. *Oecologia, 115,* 206-212.

Holway, D. A. (1998b). Effect of Argentine ant invasions on ground-dwelling arthropods in northern California riparian woodlands. *Oecologia, 116,* 252-258.

Holway, D. A., Suarez, A. V. & Case, T. J. (1998). Loss of intraspecific aggression in the success of a widespread invasive social insect. *Science, 282,* 949-952.

Holway, D. A. (1999). Competitive mechanism underlying the displacement of native ants by the invasive Argentine ant. *Ecology, 80,* 238-251.

Holway, D. A. & Case, T. J. (2001). Effects of colony-level variation on competitive ability in the invasive Argentine ant. *Animal Behaviour, 61,* 1181-1192.

Holway, D. A., Lach, L., Suarez, A. V., Tsutsui, N. D. & Case, T. J. (2002). The causes and consequences of ant invasions. *Annual Review of Ecology and Systematics, 33,* 181–233.

Howard, R. W. & Blomquist, G. J. (2005). Ecological, behavioral, andbiochemical aspects of insect hydrocarbons. *Annual Review of Entomology, 50,* 371-393.

Hulme, P. E. (2006). Beyond control: wider implications for the management of biological invasions. *Journal of Applied Ecology, 43,* 835-847.

Hulme, P. E. (2009). Trade, transport and trouble: managing invasive species pathways in an era of globalization. *Journal of Applied Ecology, 46,* 10-18.

Human, K. G. & Gordon, D. M. (1997). Effects of Argentine Ants on Invertebrate Biodiversity in Northern California. *Conservation Biology, 11,* 1242–1248.

Human, K. G. & Gordon, D. M. (1999). Behavioral interactions of the invasive Argentine ant with native ant species. *Insectes Sociaux, 46,* 159-163.

Huxel, G. R. (2000). The effect of the Argentine ant on the threatened valley elderberry longhorn beetle. *Biological Invasions, 2,* 81–85.

Ingram, K. K. & Gordon, D. M. (2003). Genetic analysis of dispersal dynamics in an invading population of Argentine ants. *Ecology, 84,* 2832-2842.

Jaquiéry, J., Vogel, V. & Keller, L. (2005). Multilevel genetic analyses of two supercolonies of the Argentine ant, *Linepithema humile. Molecular Ecology, 14,* 589-598.

Juliano, S. A. & Lounibos, L. P. (2005). Ecology of invasive mosquitoes: effects on resident species and on human health. *Ecology Letters, 8,* 558-574.

Keller, L., Passera, L. & Suzzoni, J.-P. (1989). Queen execution in the Argentine ant, *Iridomyrmex humilis. Physiological Entomology, 14,* 157-163.

Keller, L. & Ross, K. G. (1998). Selfish genes: a green beard in the red fire ant. *Nature, 394,* 573-575.

King, J. R., Tschinkel, W. R. & Ross, K. G. (2009). A case study of human exacerbation of the invasive species problem: transport and establishment of polygyne fire ants in Tallahassee, Florida, USA. *Biological Invasions, 11,* 373-377.

Klotz, J. H., Rust, M. K., Field, H. C., Greenberg, L. & Kupfer, K. (2008). Controlling Argentine ants in residential settings (Hymenoptera: Formicidae). *Sociobiology, 51,* 579-588.

Krushelnycky, P. D., Loope, L. L. & Reimer, N. J. (2005). The ecology, policy, and management of ants in Hawaii. *Proceedings of the Hawaiian Entomological Society, 37,* 1-25.

Krushelnycky, P. D. & Gillespie, R. G. (2008). Compositional and functional stability of arthropod communities in the face of ant invasions. *Ecological Applications, 18,* 1547-1562.

Laakkonen, J., Fisher, R. N. & Case, T. J. (2001). Effect of land cover, habitat fragmentation and ant colonies on the distribution and abundance of shrews in southern California. *Journal of Animal Ecology, 70,* 776–788.

Lach, L. (2003). Invasive ants: unwanted partners in ant–plant interactions? *Annals of the Missouri Botanical Garden, 90,* 91–108.

Lach, L. (2007). A mutualism with a native Membracid facilitates pollinator displacement by Argentine ants. *Ecology, 88,* 1994–2004.

Lach, L. (2008). Argentine ants displace floral arthropods in a biodiversity hotspot. *Diversity and Distributions, 14,* 281–290.

Lach, L. & Thomas, M. L. (2008). Invasive ants in Australia: documented and potential ecological consequences. *Australian Journal of Entomology, 47,* 275-288.

LeBrun, E. G., Tillberg, C. V., Suarez, A. V., Folgarait, P. J., Smith, C. R. & Holway, D. A. (2007). An experimental study of competition between fire ants and Argentine ants in their native range. *Ecology, 88,* 63-75.

Lester, P. J., Baring, C. W., Longson, C. G. & Hartley, S. (2003). Argentine and other ants (Hymenoptera: Formicidae) in New Zealand horticultural ecosystems: distribution, hemipteran hosts, and review. *New Zealand Entomologist, 26,* 79-89.

Leung, B., Lodge, D. M., Finnoff, D., Shogren, J. F., Lewis, M. A. & Lamberti, G. (2002). An ounce of prevention or a pound of cure: bioeconomic risk analysis of invasive species. *Proceedings of the Royal Society of London Series B Biological Sciences, 269,* 2407-2413.

Lewis, V. R., Merrill, L. D., Atkinson, T. H. & Wasbauer, J. S. (1992). Imported fire ants: potential risk to California. *California Agriculture, 46,* 29-31.

Liang, D. & Silverman, J. (2000). "You are what you eat": diet modifies cuticular hydrocarbons and nestmate recognition in the Argentine ant, *Linepithema humile. Naturwissenschaften, 87,* 412-416.

Liang, D., Blomquist, G. J. & Silverman, J. (2001). Hydrocarbon-released nestmate aggression in the Argentine ant, *Linepithema humile,* following encounters with insect prey. *Comparative Biochemistry and Physiology Part B, 129,* 871-882.

Lowe, S, Browne, M. & Boudlejas, S. (2000). 100 of the world's worst invasive alien species. *Aliens, 12*, 1–12.

Lubin, Y. D. (1984). Changes in the native fauna of the Galápagos Islands following invasion by the little red fire ant, *Wasmannia auropunctata. Biological Journal of the Linnean Society, 21*, 229-242.

Mack, R. N., Simberloff, D., Lonsdale, W. M., Evans, H., Clout, M. & Bazzaz, F. A. (2000). Biotic invasions: causes, epidemiology, global consequences, and control. *Ecological Applications, 10*, 689-710.

Madge, P. E. & Caon, G. (1987). Argentine ant: an historical review. *Technical Report, Department of Agriculture South Australia, 111*.

Markin, G. P. & McCoy, C. W. (1968). The occurrence of a nematode, *Diploscapter lycostoma*, in the pharyngeal glands of the Argentine ant, *Iridomyrmex humilis. Annals of the Entomological Society of America, 61*, 505-509.

Markin, G. P. (1970a). The seasonal life cycle of the Argentine ant, *Iridomyrmex humilis* (Hymenoptera: Formicidae), in southern California. *Annals of the Entomological Society of America, 63*, 1238-1242.

Markin, G. P. (1970b). Foraging behavior of the Argentine ant in a California citrus grove. *Journal of Economic Entomology, 63*, 740-744.

Martinez-Ferrer, M. T., Grafton-Cardwell, E. E. & Shorey, H. H. (2003). Disruption of parasitism of the California red scale (Homoptera: Diaspididae) by three ant species (Hymenoptera: Formicidae). *Biological Control, 26*, 279–286.

McGlynn, T. P. (1999). The worldwide transfer of ants: geographical distribution and ecological invasions. *Journal of Biogeography, 26*, 535-548.

Menke, S. B., Fisher, R. N., Jetz, W. & Holway, D. A. (2007). Biotic and abiotic controls of argentine ant invasion success at local and landscape scales. *Ecology, 88*, 3164-3173.

Menke, S. B., Holway, D. A., Fisher, R. N. & Jetz, W. (2009). Characterizing and predicting species distributions across environments and scales: Argentine ant occurrences in the eye of the beholder. *Global Ecology and Biogeography, 18*, 50–63.

Mgocheki, N. & Addison, P. (2009). Interference of ants (Hymenoptera: Formicidae) with biological control of the vine mealybug *Planococcus ficus* (Signoret) (Hemiptera: Pseudococcidae). *Biological Control, 49*, 180–185.

Mitchell, C. E. & Power, A. G. (2003). Release of invasive plants from fungal and viral pathogens. *Nature, 421*, 625-627.

Miyake, K., Kameyama, T., Sugiyama, T., Ito, F. (2002). Effect of Argentine ant invasions on Japanese ant fauna in Hiroshima Prefecture, western Japan: a preliminary report (Hymenoptera : Formicidae). *Sociobiology, 39*, 465-474.

Morrison, L. W., Porter, S. D., Daniels, E. & Korzukhin, M. D. (2004). Potential global range expansion of the invasive fire ant, *Solenopsis invicta. Biological Invasions, 6*, 183-191.

Mothapo, N. P. & Wossler, T. C. (2011). Behavioural and chemical evidence for multiple colonisation of the Argentine ant, *Linepithema humile*, in the Western Cape, South Africa. *BMC Ecology, 11*, 6.

Nakamaru, M., Beppu, Y. & Tsuji, K. (2007). Does disturbance favor dispersal? An analysis of ant migration using the colony-based lattice model. *Journal of Theoretical Biology, 248*, 288-300.

Ness, J. H. & Bronstein, J. L. (2004). The effects of invasive ants on prospective ant mutualists. *Biological Invasions, 6*, 445–461.

Newell, W. & Rouge, B. (1909). Measures suggested against the Argentine ant as a household pest. *Journal of Economic Entomology, 2,* 324-332.

Newell, W. & Barber, T. C. (1913). The Argentine ant. *U. S. Department of Agriculture, Bureau of Entomology Bulletin, 122,* 1-98.

Nygard, J., Sanders, N. J. & Connor, E. F. (2008). The effects of the invasive Argentine ant (*Linepithema humile*) and the native ant *Prenolepis imparis* on the structure of insect herbivore communities on willow trees (*Salix lasiolepis*). *Ecological Entomology , 33,* 789–795.

O'Dowd, D. J., Green, P. T. & Lake, P. S. (2003). Invasional meltdown on an oceanic island. *Ecology Letters, 6,* 812–817.

Okaue, M., Yamamoto, K., Touyama, Y., Kameyama, T., Terayama, M., Sugiyama, T., Murakami, K. & Ito, F. (2007). Distribution of the Argentine ant, *Linepithema humile,* along the Seto Inland Sea, western Japan: result of surveys in 2003-2005. *Entomological Science, 10,* 337-342.

Pascoe, A. (2002). Red imported fire ant response stood down. *Biosecurity, 45,* 7.

Passera, L., Keller, L. & Suzzoni, J.-P. (1988). Queen replacement in dequeened colonies of the Argentine ant *Iridomyrmex humilis* (Mayr). *Psyche, 95,* 59-65.

Passera, L. (1994). Characteristics of tramp species. In Williams, D. F. (Eds.), *Exotic ants: biology, impact, and control introduced species* (pp. 23-43). Westview Press, Boulder, Colorado.

Pedersen, J. S., Krieger, M. J. B., Vogel, V., Giraud, T. & Keller, L. (2006). Native supercolonies of unrelated individuals in the invasive Argentine ant. *Evolution, 60,* 782-791.

Pejchar, L. & Mooney, H. A. (2009). Invasive species, ecosystem services and human well-being. *Trends in Ecology and Evolution, 24,* 497-504.

Perrings, C., Williamson, M., Barbier, E. B., Delfino, D., Dalmazzone, S., Shogren, J., Simmons, P. & Watkinson, A. (2002). Biological invasion risks and the public good: an economic perspective. *Conservation Ecology, 6,* 1.

Phillips, P. A. & Sherk, C. J. (1991). To control mealybugs, stop honeydew-seeking ants. *California Agriculture, 45,* 26-28.

Pimentel, D., Zuniga, R. & Morrison, D. (2005). Update on the environmental and economic costs associated with alien-invasive species in the United States. *Ecological Economics, 52,* 273-288.

Pimm, S. L., Russell, G. J., Gittleman, J. L. & Brooks, T. M. (1995). The future of biodiversity. *Science, 269,* 347-350.

Pitt, J. P. W., Worner, S. P. & Suarez, A. V. (2009). Predicting Argentine ant spread over the heterogeneous landscape using a spatially explicit stochastic model. *Ecological Applications, 19,* 1176-1186.

Porter, S. D., Nogueira de Sá, L. A. & Morrison, L. W. (2004). Establishment and dispersal of the fire ant decapitating fly *Pseudacteon tricuspis* in North Florida. *Biological Control, 29,* 179-188.

Powell, B. E. & Silverman, J. (2010). Population growth of *Aphis gossypii* and *Myzus persicae* (Hemiptera: Aphididae) in the presence of *Linepithema humile* and *Tapinoma sessile* (Hymenoptera: Formicidae). *Environmental entomology, 39,* 1492-1499.

Radchenko, A. (2005). Monographic revision of the ants (Hymenoptera: Formicidae) of North Korea. *Annales Zoologici, 55,* 127-221.

Reuter, M., Pedersen, J. S. & Keller, L. (2005). Loss of *Wolbachia* infection during colonisation in the invasive Argentine ant *Linepithema humile*. *Heredity, 94,* 364-369.

Rhoades, R. B., Stafford, C. T. & James Jr., F. K. (1971). Survey of fatal anaphylactic reactions to imported fire ant stings. *Journal of Allergy and Clinical Immunology, 84,* 159-162.

Ricciardi, A. & Rasmussen, J. B. (1998). Predicting the identity and impact of future biological invaders: a priority for aquatic resource management. *Canadian Journal of Fisheries and Aquatic Sciences, 55,* 1759-1765.

Rodriguez-Cabal, M., Stuble, K. L., Nuñez, M. A. & Sanders, N. (2009). Quantitative analysis of the effects of the exotic Argentine ant on seed-dispersal mutualisms. *Biology Letters, 5,* 499-502.

Roura-Pascual, N., Suarez, A. V., Gómez, C.,Pons, P., Touyama, Y., Wild, A. L. & Peterson, T. (2004). Geographical potential of Argentine ants (*Linepithema humile* Mayr) in the face of global climate change. *Proceedings of the Royal Society of London Series B Biological Sciences, 271,* 2527-2535.

Roura-Pascual, N., Suarez, A. V., McNyset, K., Gómez, C., Pons, P., Touyama, Y., Wild, A. L., Gascon, F. & Peterson, A. T. (2006). Niche differentiation and fine-scale projections for Argentine ants based on remotely sensed data. *Ecological Applications, 16,* 1832-1841.

Roura-Pascual, N., Brotons, L., Peterson, A. T. & Thuiller, W. (2009a). Consensual predictions of potential distributional areas for invasive species: a case study of Argentine ants in the Iberian Peninsula. *Biological Invasions, 11,* 1017-1031.

Roura-Pascual, N., Bas, J. M., Thuiller, W., Hui, Cang, Krug, R. M. & Brotons, L. (2009b). From introduction to equilibrium: reconstructing the invasive pathways of the Argentine ant in a Mediterranean region. *Global Change Biology, 15,* 2101-2115.

Rowles, A. D. & O'Dowd, D. J. (2007). Interference competition by Argentine ants displaces native ants: implications for biotic resistance to invasion. *Biological Invasions., 9,* 73-85.

Rowles, A. D. & O'Dowd, D. J. (2009a). Impacts of the invasive Argentine ant on native ants and other invertebrates in coastal scrub in south-eastern Australia. *Austral Ecology, 34,* 239-248.

Rowles, A. D. & O'Dowd, D. J. (2009b). New mutualism for old: indirect disruption and direct facilitation f seed dispersal following Agentine ant invasion. *Oecologia, 158,* 709-716.

Sagata, S. & Lester, P. J. (2009). Behavioural plasticity associated with propagule size, resources, and the invasion success of the Argentine ant *Linepithema humile*. *Journal of Applied Ecology, 46,* 19-27.

Sala, O. E., Chapin III, F. S., Armesto, J. J., Berlow, E., Bloomfield, J., Dirzo, R., Huber-Sanwald, E., Huenneke, L. F., Jackson, R. B., Kinzig, A., Leemans, R., David M. Lodge, D. M., Mooney, H. A., Oesterheld, M., Poff, N. L., Sykes, M. T., Walker, B. H., Walker, M. & Wall, D. H. (2000). Global Biodiversity Scenarios for the Year 2100. *Science, 287,* 1770-1774.

Silverman, J. & Brightwell, R. J. (2008). The Argentine ant: challenges in managing an invasive unicolonial pest. *Annual Review of Entomology, 53,* 231-252.

Simberloff, D., Parker, I. M. & Windle, P. N. (2005). Introduced species policy, management, and future research needs. *Frontiers in Ecology and the Environment, 3,* 12-20.

Skaife, S. H. (1955). The Argentine ant *Iridomyrmex humilis* Mayr. *Transactions of the Royal Society of South Africa, 34,* 355-377.

Smith, M. R. (1965). House-infesting ants of the eastern United States. *United States Department of Agriculture Technical Bulletin, 1326,* 1-105.

Sockman, K. W. (1997). Variation in life-history traits and nest-site selection affects risk of nest predation in the California gnatcatcher. *The Auk, 114,* 324-332.

Soeprono, A. M. & Rust, M. K. (2004). Strategies for controlling Argentine ants (Hymenoptera: Formicidae). *Sociobiology, 44,* 669-682.

Suarez, A. V., Tsutsui, N. D., Holway, D. A. & Case, T. J. (1999). Behavioral and genetic differentiation between native and introduced populations of the Argentine ant. *Biological Invasions, 1,* 43-53.

Suarez, A. V., Richmond, J. Q. & Case, T. J. (2000). Prey selection in horned lizards following the invasion of Argentine ants in southern California. *Ecological Applications, 10,* 711–725.

Suarez, A. V., Holway, D. A. & Case, T. J. (2001). Patterns of spread in biological invasions dominated by long-distance jump dispersal: Insights from Argentine ants. *Proceedings of the National Academy of Sciences of the United States of America, 98,* 1095-1100.

Suarez, A. V. & Case, T. J. (2002). Bottom-up effects on persistence of a specialist predator: ant invasions and horned lizards. *Ecological Applications, 12,* 291-298.

Suarez, A. V., Holway, D. A., Liang, D., Tsutsui, N. D. & Case, T. J. (2002). Spatiotemporal patterns of intraspecific aggression in the invasive Argentine ant. *Animal Behaviour, 64,* 697-708.

Suarez, A. V., Yeh, P. & Case, T. J. (2005a). Impacts of Argentine ants on avian nesting success. *Insectes Sociaux, 52,* 378-382.

Suarez, A. V., Holway, D. A. & Ward, P. S. (2005b). The role of opportunity in the unintentional introduction of nonnative ants. *Proceedings of the National Academy of Sciences of the United States of America, 102,* 17032–17035.

Suarez, A. V., Holway, D. A. & Tsutsui, N. D. (2008). Genetics and behavior of a colonizing species: the invasive Argentine ant. *The American Naturalist, 172,* 72-84.

Sugiyama, T. (2000). Invasion of Argentine ant, *Linepithema humile,* into Hiroshima Prefecture, Japan. *Japanese Journal of Applied Entomology and Zoology, 44,* 127-129.

Suhr, E. L., McKechnie, S. W. & O'Dowd, D. J. (2009). Genetic and behavioural evidence for a city-wide supercolony of the invasive Argentine ant *Linepithema humile* (Mayr) (Hymenoptera: Formicidae) in southeastern Australia. *Australian Journal of Entomology, 48,* 79-83.

Suhr, E. L., O'Dowd, D. J., McKechnie, S. W. & Mackay, D. A. (2011). Genetic structure, behaviour and invasion history of the Argentine ant supercolony in Australia. *Evolutionary Applications, 4,* 471-484.

Sunamura, E., Nishisue, K., Terayama, M. & Tatsuki, S. (2007). Invasion of four Argentine ant supercolonies into Kobe Port, Japan: their distributions and effects on indigenous ants (Hymenoptera: Formicidae). *Sociobiology, 50,* 659-674.

Sunamura, E., Hatsumi, S., Karino, S., Nishisue, K., Terayama, M., Kitade, O. & Tatsuki, S. (2009a). Four mutually incompatible Argentine ant supercolonies in Japan: inferring invasion history of introduced Argentine ants from their social structure. *Biological Invasions, 11,* 2329-2339.

Sunamura, E., Espadaler, X., Sakamoto, H., Suzuki, S., Terayama, M. & Tatsuki, S. (2009b). Intercontinental union of Argentine ants: behavioral relationships among introduced populations in Europe, North America, and Asia. *Insectes Sociaux, 56,* 143-147.

Sunamura, E., Suzuki, S., Nishisue, K., Sakamoto, H., Otsuka, M., Utsumi, Y., Mochizuki, F., Fukumoto, T., Ishikawa, Y., Terayama, M. & Tatsuki, S. (2011a). Combined use of a synthetic trail pheromone and insecticidal bait provides effective control of an invasive ant. *Pest Management Science,* in press.

Sunamura, E., Hoshizaki, S., Sakamoto, H., Fujii, T., Nishisue, K., Suzuki, S., Terayama, M., Ishikawa, Y. & Tatsuki, S. (2011b). Workers select mates for queens: a possible mechanism of gene flow restriction between supercolonies of the invasive Argentine ant. *Naturwissenschaften 98,* 361-368.

The Global Social Change Research Project. Basic guide to the world economic growth, 1970 to 2007. 2007. Available from: http://gsociology.icaap.org

Thomas, M. L., Payne-Makrisâ, C. M., Suarez, A. V., Tsutsui, N. D. & Holway, D. A. (2006). When supercolonies collide: territorial aggression in an invasive and unicolonial social insect. *Molecular Ecology, 15,* 4303-4315.

Tillberg, C. V., Holway, D. A., LeBrun, E. G. & Suarez, A. V. (2007). Trophic ecology of invasive Argentine ants in their native and introduced ranges. *Proceedings of the National Academy of Sciences of the United States of America, 104,* 20856-20861.

Torchin, M. E., Lafferty, K. D., Dobson, A. P., McKenzie, V. J. & Kuris, A. M. (2003). Introduced species and their missing parasites. *Nature, 421,* 628-630.

Torres, C. W., Brandt, M. & Tsutsui, N. D. (2007). The role of cuticular hydrocarbons as chemical cues for nestmate recognition in the invasive Argentine ant (*Linepithema humile*). *Insectes Sociaux, 54,* 329-333.

Touyama, Y., Ogata, K. & Sugiyama, T. (2003). The Argentine ant, *Linepithema humile*, in Japan: assessment of impact on species diversity of ant communities in urban environments. *Entomological Science, 6,* 57-62.

Touyama, Y., Ihara, Y. & Ito, F. (2008). Argentine ant infestation affects the abundance of the native myrmecophagic jumping spider *Siler cupreus* Simon in Japan. *Insectes Sociaux, 55,* 144-146.

Tsuji, K. & Tsuji, N. (1996). Evolution of life history strategies in ants: variation in queen number and mode of colony founding. *Oikos, 76,* 83-92.

Tsutsui, N. D., Suarez, A. V., Holway, D. A. & Case, T. J. (2000). Reduced genetic variation and the success of an invasive species. *Proceedings of the National Academy of Sciences of the United States of America, 97,* 5948-5953.

Tsutsui, N. D., Suarez, A. V., Holway, D.A. & Case, T. J. (2001). Relationships among native and introduced populations of the Argentine ant (*Linepithema humile*) and the source of introduced populations. *Molecular Ecology, 10,* 2151–2161.

Tsutsui, N. D., Suarez, A. V., Holway, D. A. & Case, T. J. (2003a). Genetic diversity, asymmetrical aggression, and recognition in a widespread invasive species. *Proceedings of the National Academy of Sciences of the United States of America, 100,* 1078-1083.

Tsutsui, N. D., Kauppinen, S. N., Oyafuso, A. F. & Grosberg, R. K. (2003b). The distribution and evolutionary history of *Wolbachia* infection in native and introduced populations of the invasive Argentine ant (*Linepithema humile*). *Molecular Ecology, 12,* 3057-3068.

Ugelvig, L. V., Drijfhout, F. P., Kronauer, D. J. C., Boomsma, J. J., Pedersen, J. S. & Cremer, S. (2008). The introduction history of invasive garden ants in Europe: integrating genetic, chemical and behavioural approaches. *BMC Biology, 6,* 11.

van Wilgenburg, E., Torres, C. W. & Tsutsui, N. D. (2010). The global expansion of a single ant supercolony. *Evolutionary Applications, 3,* 136-143.

Vega, S. J. & Rust, M. K. (2001). The Argentine ant – a significant invasive species in agricultural, urban and natural environments. *Sociobiology, 37,* 3-25.

Vitousek, P. M. (1994). Beyond Global Warming: Ecology and Global Change. *Ecology, 75,* 1861-1876.

Vitouselk, P. M., D'Antonio, C. M., Loope, L. L. & Westbrooks, R. (1996). Biological invasions as global environmental change. *American Scientist, 84,* 468-478.

Vogel, V., Pedersen, J. S., d'Ettorre, P., Lehmann, L. & Keller, L. (2009). Dynamics and genetic structure of Argentine ant supercolonies in their native range. *Evolution, 63,* 1627-1639.

Vogel, V., Pedersen, J. S., Giraud, T., Krieger, M. J. B. & Keller, L. (2010). The worldwide expansion of the Argentine ant. *Diversity and Distributions, 16,* 170-186.

Walters, A. C. & Mackay, D. A. (2005). Importance of large colony size for successful invasion by Argentine ants (Hymenoptera: Formicidae): evidence for biotic resistance by native ants. *Austral Ecology, 30,* 395-06.

Walters, A. C. (2006). Invasion of Argentine ants (Hymenoptera: Formicidae) in south Australia: impacts on community composition and abundance of invertebrates in urban parklands. *Austral Ecology, 31,* 567-576.

Ward, P. S. (1987). Distribution of the introduced Argentine ant (*Iridomyrmex humilis*) in natural habitats of the lower Sacramento Valley and its effects on the indigenous ant fauna. *Hilgardia, 55,* 1-16.

Ward, D. F., Harris, R. J. & Stanley, M. C. (2005). Human-mediated range expansion of Argentine ants *Linepithema humile* (Hymenoptera: Formicidae) in New Zealand. *Sociobiology, 45,* 401-407.

Ward, D. F., Beggs, J. R., Clout, M. N., Harris, R. J. & O'Connor, S. (2006). The diversity and origin of exotic ants arriving in New Zealand via human-mediated dispersal. *Diversity and Distributions, 12,* 601-609.

Westphal, M. I., Browne, M., MacKinnon, K. & Noble, I. (2008). The link between international trade and the global distribution of invasive alien species. *Biological Invasions, 10,* 391-398.

Wetterer, J. K. & Porter, S. D. (2003). The little fire ant, *Wasmannia auropunctata*: distribution, impact and control. *Sociobiology, 42,* 1-41.

Wetterer, J. K. & Wetterer, A. L. (2006). A disjunct Argentine ant metacolony in Macaronesia and southwestern Europe. *Biological Invasions, 8,* 1123–1129.

Wetterer, J. K., Wild, A. L., Suarez, A. V., Roura-Pascual, N. & Espadaler, X. (2009). Worldwide spread of the Argentine ant, *Linepithema humile* (Hymenoptera: Formicidae). *Myrmecological News, 12,* 187-194.

Wilcove, D. S., Rothstein, D., Dubow, J., Phillips, A. & Losos, E. (1998). Quantifying threats to imperiled species in the United States. *BioScience, 48,* 607-615.

Wild, A. L. (2004). Taxonomy and distribution of the Argentine ant *Linepithema humile* (Hymenoptera: Formicidae). *Annals of the Entomological Society of America, 97,* 1204-1215.

Wild, A. L. (2007). Taxonomic revision of the ant genus *Linepithema* (Hymenoptera: Formicidae). *University of California Publications in Entomology, 126,* 1-159.

Williams, D. F., Oi, D. H. & Knue, G. J. (1999). Infection of red imported fire ant (Hymenoptera: Formicidae) colonies with the entomopathogen *Thelohania solenopsae* (Microsporidia: Thenohaniidae). *Journal of Economic Entomology, 92,* 831-836.

Williams, D. F., Oi, D. H., Porter, S. D., Pereira, R. M. & Briano, J. A. (2003). Biological control of imported fire ants. *American Entomologist, 49,* 150-163.

Witte, F., Goldschmidt, T., Wanink, J., van Oijen, M., Goudswaard, K., Witte-Maas, E. & Bouton, N. (1992). The destruction of an endemic species flock: quantitative data on the decline of the haplochromine cichlids of Lake Victoria. *Environmental Biology of Fishes, 34,* 1-28.

Xu, H., Ding, H., Li, M., Qiang, S., Guo, J., Han, Z., Huang, Z., Sun, H., He, S., Wu, H. & Wan, F. (2006). The distribution and economic losses of alien species invasion to China. *Biological Invasions, 8,* 1495-1500.

Yang, C. C., Shoemaker, D. D., Wu, W. J. & Shih, C. J. (2008). Population genetic structure of the red imported fire ant, *Solenopsis invicta*, in Taiwan. *Insectes Sociaux, 55,* 54-65.

Zhang, R., Li, Y., Liu, N. & Porter, S. D. (2007). An overview of the red imported fire ant (Hymenoptera: Formicidae) in mainland China. *Florida Entomologist, 90,* 723-731.

Zimmerman, E. C. (1941). Argentine ant in Hawaii. *Proceedings of the Hawaiian Entomological Society, 11,* 108.

In: Agricultural Research Updates. Volume 2
Editor: Barbara P. Hendriks

ISBN: 978-1-61470-191-0
© 2012 Nova Science Publishers, Inc.

Chapter 6

COMMON BACTERIAL BLIGHT IN *PHASEOLUSVULGARIS*

Gregory Perry and K. Peter Pauls

Department of Plant Agriculture
University of Guelph, Guelph, ON, Canada

ABSTRACT

Common bacterial blight (CBB) is a significant foliar disease of dry bean caused by the pathogen *Xanthomonasaxonopodis* pv. *phaseoli*; a gram-negative bacillus with a genome of approximately 3.9Mb. This disease is endemic to most regions where *P. vulgaris* is cultivated and is annually responsible for millions of dollars of crop loss worldwide. The bacteria are soil residents and initial infections occur predominantly through the stomata of leaves, or through plant wounds. Infected plants generally display symptoms on the leaves first, with the formation of small water soaked lesions appearing 1-2 weeks after infection. The lesions gradually enlarge, and become encircled by a region of yellow tissue. As the disease progresses these lesions become necrotic and extensive defoliation is common in infected plants. The bacteria can migrate throughout the plant, including into the seeds through the vascular system of the pedicle. Treatment options for infected plant tissues are limited and the most effective preventative measure is to grow breeder seeds in regions that are free of the pathogen. Topical application of antibiotics and anti-microbial compounds are used as secondary control measures.

Research efforts to mitigate the damage caused by the pathogen have focused on developing bean germplasm that is resistant or tolerant to the pathogen and integrating these sources of resistance into new varieties. Two main sources of resistance have been discovered, *P. vulgaris*-derived resistance and *P. acutifolius*-derived resistance. CBB resistance in *P. vulgaris* is relatively weak but, interspecific crosses between *P. vulgaris* and *P. acutifolius* have been shown to possess high levels of CBB resistance. Molecular studies have shown that resistance in the interspecific lines is conditioned by several quantitative trait loci (QTL), which interact in various ways. Molecular markers that are associated with the resistance QTL have been used for marker assisted selection and are the starting points for studies to identify resistance genes.

INTRODUCTION

The *Xanthomonas* genus is contained in the family Pseudomonadacae, and is comprised predominantly of obligate plant pathogens; many of which have worldwide distribution and have significant negative effects on crop production and quality. The Pseudomonadacae are gram negative, non-spore forming bacilli; measuring 0.2-0.6µm by 0.8-2.9µm. The cells have flagella and are generally mobile with varying level of xanthan gum production, a hallmark of the genera. Most strains of *Xanthomonas* form yellow to cream coloured colonies when cultured on *Xanthomonas campestris* (XCP) agar, Glucose-Yeast-Calcium Carbonate Agar (GYCA), Levure Peptone Glucose Agar (LPGA) or Milk-Tween (MT) media. However, some variants (fuscans) produce a brown, water soluble, melanin pigment (Sheppard and Remeeus, 2005; Glathe and Singh-Verma, 1968; Godcynska and Serfontein, 1998). In culture, *Xanthomonas* spp. colonies are generally visible 48-96hr after plating when incubated at 28°C. However some strains may require 5-6 days to grow. The colonies are convex, round, and mucoid and have even margins (Starr, 1981; Schaad and Stall, 1990; Swings *et al.*, 1993; Sheppard and Remeeus, 2005).

The Pseudomonadacae are split into five pathogenicity groups (A-E), although there has been some debate over the placements of some members in recent years (Rediers and Vanderleyden, 2004). The host range for these plant pathogens extends across 11 monocot families and at least 70 genera and 57 dicot families representing over 150 genera. Although the genus is capable of infecting a wide range of plant species, the host ranges of individual *Xanthomonas* species are generally quite limited (Gabriel *et al.*, 1989; Vauterin *et al.*, 1991).

XANTHOMONAS AXONOPODIS PV. PHASEOLI

Xanthomonas axonopodis pv. *phaseoli*, also referred to as *Xanthomonas campestris* pv. *phaseoli* and its melanin-producing, fuscan variant, *Xanthomonas fuscans* (Figure 1), are causative agents of common bacterial blight (CBB) in *Phaseolus vulgaris* (dry bean, Hayward and Waterson, 1965). The disease pathologies caused by these two strains are virtually indistinguishable in foliar tissue, but infections with the fuscan variant can lead to more severe disease symptoms in some instances (Chan and Goodwin, 1999). However, the pathogenic strains of *X. axonopodis* pv. *phaseoli* and *X. fuscans* subsp. *fuscans* appear to lack any significant genetic geographical distribution,with no one virulence trait showing any significant geographic prevalence. As a result, host genetic resistance to one strain has been shown to provide at least some broad resistance to the pathogen as a whole (Mutlu *et al.*, 2008). There has been some debate over the taxonomy of *X. axonopodis* pv. *phaseoli* over the last 20 years, and consensus has yet to be reached as to whether *X. axonopodis* pv. *phaseoli* or *X. campestris* pv. *phaseoli* is the proper classification for the pathogen. As a result, the two are used interchangeably in the literature (Vauterin*et al.*, 1995; Mutlu *et al.*, 2008).

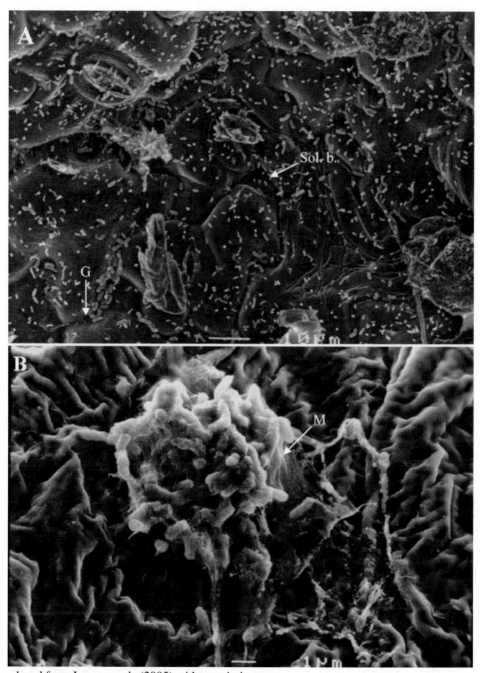

Reproduced from Jaques *et al.,* (2005) with permission.

Figure 1. Scanning electron microscopic micrographs of field-grown bean leaf surfaces colonized by seed-borne *X. axonopodis* pv. phaseoli. (A) Leaf surface showing mostly solitary bacterial populations (Sol. b.). Note the accumulation of bacterial cells in grooves (G) between epidermal cells. Bar, 10 μm. (B) Focus on a bacterial biofilm. Note the matrix (M) embedding bacterial cells constituting a typical biofilm. Bar, 1 μm.

The pathogen has a genome composed primarily of a single circular chromosome of 3938±68Kb (Chan and Goodwin, 1999), which is similar to other members of the Pseudomonadacae that have genomes between 3.3 to 6.6Mb. In addition, many *Xanthomonas* species have small accessory plasmids which may contain a variety of advantageous genes coding for pathogenesis factors (Kim *et al.*, 2008), or antibiotic resistance (Szczepanowski *et al.*, 2004; Bender *et al.*, 1990). DNA sequencing results for several *Xanthomonas* species, including full genome sequence information for *X. axonopodis pv. citri* and *X. campestris pv. campestris* are detailed in the *Xanthomonas* Genome Browser (http://xgb.fli-leibniz.de/cgi/index.pl) (da Silva *et al.*, 2002; Alavi *et al.*, 2008). Only limited sequence information is available for *X. axonopodis* pv. *phaseoli.*Currently, there are 61 entries in Genbank for this species, comprising 10 unique genes from several strains for comparative analyses. Additionally, there are 3 entries for *X. fuscans* subsp. *fuscans*, covering 26,975Kb, including the sequence of the hypersensitive response and pathogenicity (hrp) cluster, which is the location of many of the genes related to infection and host colonization.

CBB is endemic to dry bean producing regions in the world, and is most prevalent in areas where warm, wet or humid conditions facilitate the spread of the pathogen. The bacteria are seed borne, and under field conditions, dissemination can occur through wind-driven rain or mechanical transfer by insect vectors. In addition to natural methods, the bacteria can be spread by overhead sprinkler systems, and the use of infected seeds (Vivader, 1993). The bacteria are capable of surviving in plant debris from previous crops for up to 22 months under field and greenhouse conditions. However *X. axonopodis* pv. *phaseoli* survival in the field is reduced to a maximum of 4 months by tilling. (Gillard *et al.*, 2010)

Although the primary host of *X. axonopodis* pv. *phaseoli* is *P. vulgaris*, the pathogen is capable of infecting other legume species including, *P. lunatus*, *Vigna aconitifolia* and *V. radiata*, *Lablab purpureus* and *Mucuna deeringiana* (Bradbury, 1986; Saettler, 1989). It was previously thought that *X. axonopodis* pv. *phaseoli* could infect *Glycine max*, however these reports can likely be attributed to infections with the related pathovar *Xanthomonas axonopodis* pv. *glycines*. Studies with *X. campestris* pv. *campestris* have indicated that a wider host range may be possible for *Xanthomonas* species and is largely dependent on the number and type of effector molecules inherent in each strain (Alavi *et al.*, 2008; Lin and Martin, 2007; Wei *et al.*, 2007). The environmental ranges of these secondary hosts are predominantly in Asia and the Americas, with only limited cultivation or natural populations in Europe and Africa. In all regions the most significant host for *X. axonopodis* pv. *phaseoli* is *P. vulgaris*.

Studies of the genetic diversity of *X. axonopodis* pv. *phaseoli* have found that the pathogen can be grouped into four genetic lineages. Three of these groups are composed exclusively of *X. axonopodis* pv. *phaseoli*, while the last group contains the *X. fuscans* subsp. *fuscans* strains (Alavi et al, 2008). These estimates of genetic similarity are based on analyses of 21 strains of CBB-causing *Xanthomonas* bacteria with unique, gene-based, DNA fragments for suppression subtractive hybridization analyses to detect differentially expressed genes and fluorescent amplified fragment length polymorphism analyses of the same bacteria. From an initial tester population of 353 subtracted DNA fragments, 75 were found to differentiate among the test strains and 39 unique DNA sequences were characterized.

These results were also supported by Chan and Goodwin (1999) who found that X. *axonopodis* pv. *phaseoli* was highly homologous to X. *campestris* pv. *campestris* as determined by restriction fragment length polymorphism analyses, and Mutlu et al, (2008) who compared the relative pathogenicity of 84 strains of *X. axonopodis* pv. *phaseoli* and *X. fuscans* subsp. *fuscans* and showed that there was some variation in the pathogenicity of different strains of *Xanthomonas* spp., with *X. fuscan* subsp. *fuscan* strains causing the highest level of disease. Interestingly, strain virulence was not related to geographic distribution, indicating that only broad selective pressures appear to be at work between the pathogen and *P. vulgaris*.

DISEASE PATHOLOGY

At a cellular level, *X. axonopodis* pv. *phaseoli* makes use of a type III secretion system (TTSS) to facilitate the insertion of a variety of effector molecules into the host cells (for a review of TTSS see Troisfontaines and Cornelis, 2005). The TTSS is comprised of two main components, the needle structure and the translocation apparatus. The physical structures of these complexes have not been studied in *X. axonopodis* pv. *Phaseoli*. However, the structures of the TTSSs of *Pseudomonas syringae* and *X. fuscans* subsp. *fuscans* have been examined, and they are very similar to TTSSs in other bacteria. The needle is a supramolecular structure responsible for transporting type-III proteins from the bacterial cytosol to the host cell. The structure has two main parts; a multi-ring base which spans and anchors the needle complex to the cytoplasmic and outer membranes, and needle-like filament, which serves to transport the secreted factors. The translocation apparatus is a serves to create a pore in the host cytoplasmic membrane (Gurlebeck *et al.*, 2006; Hauser, 2009).

The genes coding for TTSSs are highly conserved across most pathogenic bacterial species. They occur in clusters and are named the hypersensitive response and pathogenicity (*hrp*) genes. Although the *hrp* region of *X. axonopodis* pv. *phaseoli* strains has not yet been characterized, studies of *X. fuscans* subsp. *fuscans*, *X. axonopodis* pv. *citri* and *X. campestris* pv. *vesicatoria* have shown that the TTSS genes are contained within a genomic island of approximately 35Kb. In *X. campestris* pv. *vesicatoria*, the genomic island contains 37 protein coding regions with 18 encoding *hrp* proteins. Eleven of these proteins are highly conserved between pathogenic plant and animal bacteria (Thieme *et al.*, 2005). These *hrp* conserved (*hrc*) proteins are thought to be part of the core structure of the TTSS and are essential for a compatible pathogen-host interaction. Evidence for this was obtained by Darsonval *et al.* (2008), who compared the sequences of 24 TTSS structural and regulatory genes including *hrpB2*, *hrcJ*, *hrcR*, *hrcT*, *hrcV*, *hrpB2*, *hrpG* and *hrpX* in *X. fuscan* subsp. *fuscan* to the *hrp* genes from different sequenced xanthomonads, including:*X. campestris* pv. *campestris*, *X. axonopodis* pv. *vesicatoria*, *X.citri* subsp. *citri*, *X. axonopodis* pv. *glycines*and *X. oryzae* pv. *oryzae*(Figure 2). They also using targeted gene disruptions to remove one or more of the *hrp* genes from the cluster and found that all of the *hrc* proteins were required for infection, and strains deficient in any one of these genes resulted in bacterial numbers comparable with the incompatible negative control. The disruption of the regulatory proteins yielded a more significant response, with the HrpG and HrpX proteins resulting in the lowest epiphytic bacterial populations in the study.

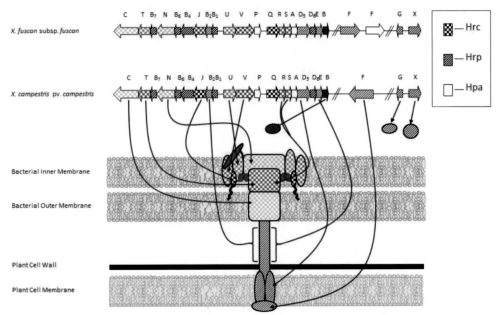

Adapted from Darsonval *et al.* (2008), Büttner and He (2009) and Qian *et al.* (2005).

Figure 2. Overview of the Hrc, Hrp and Hpa genes from the Hrp cluster in X. fuscan pv. fuscans along with a structural model of the TTSS from *X. campestris* pv. campestris. The function identity of the genes in *X. fuscans* subs. fuscans is inferred from sequence homology with other Xanthamonads (Darsonoval *et al.*, 2008). The genes in the Hrp cluster are associated with the type III secretion system which is used to inject effector molecules into the host cell.

The role for the non-conserved *hrp* proteins may vary from species to species, and are generally thought to be responsible for species-specific modifications to the TTSS to facilitate infection. Studies of *X. campestris* pv. *vesicatoria, X. axonopodis* pv. *citri, X. oryzae* pv. *oryzae* and *X. campestris* pv. *campestris* have indicated that the effector arsenal of the *Xanthomonas* genera is composed of approximately 15-20 effector molecules and that small changes in the expressed effector molecules can have large effects on the effective host range (Gurlebeck *et al.*, 2006; Alavi *et al.*, 2008; Lin and Martin, 2007; Wei *et al.*, 2007). No conclusive analysis of the number and type of effector molecules employed by *X. axonopodis* pv. *phaseoli* is currently available.

Xanthomonads have also been shown to possess Type II (T2SS) and Type IV (T4SS) secretions systems, which have been associated with bacterial virulence in other species, although they have not yet been characterized in *X. axonopodis* pv. *phaseoli*. The T2SS is responsible for the secretion of effector molecules from the bacterial cytoplasm to the intercellular space in the host cell. In *X. axonopodis* pv. citri, *X. campestris* pv. *vesicatoria*, and *X. campestris* pv. *campestris*, two sets of T2SSs have been identified (*xps* and *xcs*) (Cianciotto, 2005; Jha *et al.*, 2005). These secretion systems are encoded by two gene clusters composed of 11 (*xpsD-N*) and 12 genes (*xcsC-N*), respectively (Szczesny *et al.*, 2010). During infection, the T2SS secretes cell wall-degrading enzymes, including proteases, cellulases and xylanases across the inner bacterial membrane. However the timing and regulation of enzyme secretion seems to vary, depending on the bacterial pathovar. The formation of the T2SS is partially regulated by the HrpGand HrpX proteins, which also regulate the expression and assembly of the TTSS. In *X. campestris* pv. *campestris*, and *X.*

axonopodis pv. *citri,* the expression of HrpG and HrpX were shown to upregulate the expression of the T2SS genes and were associated with an increase in the secretion of xylanases and proteases (Furutani *et al.,* 2004; Wang *et al.,* 2008; Yamazaki *et al.,* 2008). In *X. campestris* pv. *vecisatoria* HrpG and HrpX expression repress the T2SS genes. This differential expression probably represents an evolutionary adaptation of the pathovars to optimize infection of their respective hosts, by regulating the expression of the two secretions systems to better overcome the host's defenses.

The T4SS is generally associated with the secretion of macromolecules, such as single-stranded DNA for horizontal gene transfer (such as the TDNA in *Agrobacterium tumefaciens*) or protein effectors and toxins into host cells in *Helicobacter pylori and Bordetella pertussis* (For review see Backert and Meyer, 2006). In *A. tumefaciens,* the T4SS is encoded by 11 proteins in the virB group (virB1-11) and the virD4 protein (Cascales and Christie, 2003). Studies of *X. campestris* pv. *campestris* identified analogues to the virB8 protein found in *A. tumefaciens,* and mutational analyses of this gene have indicated that its expression is associated with enhanced pathogenicity (da Silva *et al.,* 2002). However, homologs of other T4SS genes, or possible effector molecules have not been identified (Qian *et al.,* 2005; da Silva *et al.,* 2002; Van Sluys *et al.,* 2002).

Infection by *X. axonopodis* pv. *phaseoli* is highly dependent on permissive environmental conditions. Temperatures higher than 30°C and humidity levels greater that 80% provide are ideal for the pathogen to grow. Frequent rainstorms and sprinkler irrigation can also favor infection and spread of the pathogen (Schwartz *et al.,* 2003).

X. axonopodis pv. *phaseoli*gains entry into plants through existing openings in leaves, including through stomata or pre-existing wounds. Entry into stem tissue occurs through stomata on the hypocotyl, epicotyl, or cotyledons. Once inside the host plant, the bacterium invades the intercellular spaces and causes dissolution of the middle lamella. The latter can lead to the development of a wrinkled appearance to the leaves (Vivader, 1993).

Infected plants generally display symptoms on the leaves first, with the formation of small water soaked lesions appearing on the leaves within 4-10 days after infection. The lesions gradually enlarge, become flaccid and can merge with adjacent lesions. The lesions develop necrotic spots, which may be surrounded by a characteristic yellow boarder (Figure 3). This is caused by localized chlorosis of the leaf tissue at the margin of the nercotic spots. Not all plants will exhibit the yellow zone, as individual pathovars, hosts and environmental conditions can strongly affect lesion characteristics. Over, time the lesions turn brown and become necrotic, with large-scale defoliation common in advanced stages of infection (Coyne and Schuster, 1974). In the later stages, a*Xanthomonas* infection can be distinguished from halo blight infection, caused by *Pseudomonas* syringaepv.*phaseolicola,* because*Pseudomonas*-infected plants take on a brown burned look, which contrasts with CBB's yellow colouration (Zaumeyer and Thomas, 1957).

Infected stems may show reddish streaks, extending longitudinally along the stem. As the disease progresses, the stem surface can split and girdling of the stem can occur above the 1st node. This girdling can severely weaken the plant, and the stem will often break at the node (Coyne and Schuster, 1974).

Infection of the pods normally occurs through wounds in the pods made by insects. Water soaked lesions, surrounded by narrow regions of yellow to reddish brown tissue will form on the surface of the pods, giving them a mottled appearance. In advanced infections, the vascular system may be compromised and symptoms similar to those that develop in stems and leaves may be seen (Hayward and Waterson, 1965).

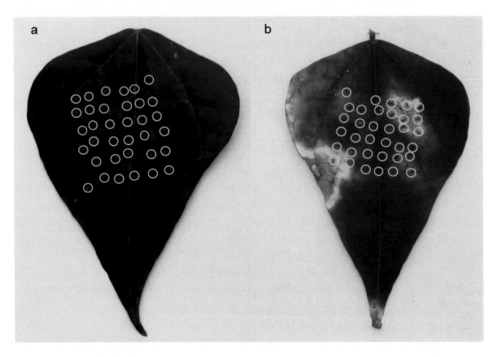

Figure 3. Symptoms of CBB on *P. vulgaris* leaves. CBB-susceptible *P. vulgaris* leaves after a mock inoculation (a) or inoculation with *X. axonopodis* pv. *phaseoli* (b). The infected leaf displays the characteristic yellow chlorotic lesions with brown necrotic centres around the site of infection in addition to diffuse chlorosis around the margins of the leaf. The pin holes from the inoculation are circled in yellow.

Seed infection can occur if the vascular system of the plant has been compromised and the bacterium enters the seed through the pedicel and the funiculus. Additionally, the micropyle of the seed can serve as a potential entry point for the bacteria. The epidemiology of seed infection varies depending on the age of the pods at the time of infection. If the pods are young when infected, the seeds may be completely destroyed by the pathogen. If older pods become infected the seeds can take on a wrinkled and shriveled appearance, and the seed coatmay become discoloured, although the latter can be difficult to detect in coloured varieties. Seed discolouration is more common in infections caused by *fuscans*strains of *X. axonopodis*pv.*phaseoli*, because the bacterium secretes a brown pigment into the tissue surrounding the infected area. In non-fuscan infections, there may be no discolouration of infected seed (Coyne and Schuster, 1974).

During a systemic infection the vascular system of the plant becomes compromised, and a reddish-brown discolouration of the veins can occur. Water soaked lesions occur in adjacent tissues as the disease spreads, and symptoms of infection will present themselves in individual tissues as described above. New symptom development on leavesin a systematically infected plant will often be preceded by the discolouration of the main vein and petiole.

DETECTION

X. axonopodis pv. *phaseoli* is detected in the field predominantly through the recognition of the characteristic symptoms it elicits in leaves. . Visual post-harvest detection is possible in seeds that have been infected by the fuscan variety, when significant seed staining has occurred. But identification of the pathogen in infected seed may require additional procedures to isolate internal bacteria. The most common diagnostic method involves producing a slurry from surface-sterilized seed, followed by plating on selective or semi-selective media such as the XCP and MT medias described earlier. Molecular detection methods exist for pathovars of *X. axonopodis* pv. *phaseoli* that are based on the use of polymerase chain reaction (PCR) primers (X4c: 5'-GGCAACACCCGATCCCTAAACAGG-3', X4e: 5'CGCCGGAAGCACGATCCTCGAAG-3'; Audy *et al.* 1994) for the p7X4 gene (Figure 4). PCR assays with primers for the p7X2 and p7X3 genes in a number of *Xanthomonas* species, including *X. campestris* pv. *campestris* and *X. campestris* pv. *vesicatoria* have also been developed, but a commercial application for this technique is not currently available.

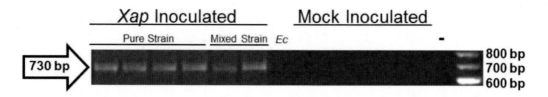

Figure 4. PCR amplification of isolates from *X. axonopodis* pv. *phaseoli* or mock inoculated leaves using the X4c and X4e primers (Alavi *et al.*, 1994*).* Pure strains, along with a mixed *X. axonopodis* culture show the characteristic 730bp amplification product associated with the *phaseoli* pathovar, while the *Escherichia coli (Ec)*, and isolates from the leaves of mock inoculated plants show no amplification.

METHODS OF DISEASE CONTROL

CBB is controlled by: 1) growing foundation and elite seed from seed produced in regions that are certified-free of the disease, 2) treating seeds or plants with bactericidal compounds 3) developing tolerant and the incorporation of resistance through selective breeding and interspecific crossing. The primary method of CBB control in many regions is through the planting of disease free seed. In North America, certified seed is imported from locations such as Idaho, where occurrence of the disease is limited by the dry climate (Cafati and Saettler, 1980; Saettler, 1989). Fields where beans are cultivated should be kept free from weeds and other hosts to minimize the risk of disease carry-over. Crop rotation, with a minimum of two years between bean plantings can also lead to a significant reduction in disease occurrence, with early studies in Romania indicated that a bean/corn rotation resulted in an 85% reduction in pathogen attack (Severin, 1971), although this is not always feasible for growers. Many regions in the developing world do not have any form of seed certification, and control of the pathogen is almost entirely in the hands of the local growers. Under these conditions, it is recommended that infected crops be tilled under the soil to minimize the survivability of the pathogen for future plantings.

Chemical controls, such as copper compounds (blue stone) or antibiotics (such as streptomycin), can be applied over crops by either spraying or dusting (Rosen, 1954). These methods of control were more extensively used during the 1970s-1990s but their use has been reduced in recent years due to limited efficacy and increased cost to the grower. In particular, the treatments are difficult to time, and do not lead to any yield increase (Nuland et al., 1983). Direct treatment of the seeds with streptomycin is still commonly practiced in the United States when there is a risk of contamination in seed stocks. However, countries such as Canada have moved to halt the practice, with domestic seed treatments banned since 2005, and a desired moratorium on the importation of treated seed in the near future (Nuland et al., 1983; Gillard, 2004; AAFC, 2005). The ban on the import of treated seed was to come into effect in 2005 in Canada, however the lack of CBB-resistance in all cultivated bean varieties has led to a delay in its implementation. . The efficacy of these treatments is somewhat questionable, as the ability of these compounds to penetrate under the seed coat and remove bacteria that have invaded the seed tissueis limited.

Development of CBB Resistant *P. vulgaris*

The best method for controlling plant diseases is through the use of resistant crops. Since 2002, CBB-resistant varieties of *P. vulgaris* have been available to growers, with OAC-Rex representing the first commercially available cultivar with this trait in Canada (Tar'an et al., 2002).

There are approximately 50 members of the *Phaseolus* genus, comprising, primary, secondary, tertiary and quaternary gene pools for *P. vulgaris* improvement.The primary germplasm for *P. vulgaris* breeding and variety development is comprised of two geographically isolated pools; named Middle American and Andean (Evans, 1973; Koening and Gepts, 1989; Koening et al., 1990; Singh et al., 1991a; Debouck et al., 1993). Although individuals in these populations are all *P. vulgaris*, with genomes of 11 chromosomes and total genome sizes of approximately 580 million base pairs,they exhibit remarkable differences in seed size and seed coat colour, which form the basis for the different market classes of beans that are used throughout the world (Singh et al., 1991a).

Variability among individuals in the primary gene pool is also observed at the molecular levels including for allozyme markers (Bennet and Leitch, 2005).Singh et al.(1991b),compared nine enzyme systems among 227 landraces of *P. vulgaris*and found a clear separation between the Middle American and Andean gene pools in a cluster analysis of the results (Gepts and Debouck 1991;Gepts 1998). More recent studies using microsatellite markers confirmed the existence of the two populations (Kwak et al., 2009). A recent study of variation in three navy (Middle American) and three white kidney lines (Andean) utilized comparative transcriptomics, proteomics and metabolomics (Mensack et al., 2010). All three approaches clustered the beans according to their centres of origin, suggesting that differences exist between beans from the Middle American and Andean centres of origin in gene transcription, protein expression, and metaboliteaccumulation.

The primary gene pools also exhibit different resistance profiles to diseases affecting common dry bean. Many members of the Middle American pool have resistance to bean common mosaic virus and tolerance to angular leaf spot and bean golden mosaic virus (Sing *et al.*, 1991a). Members of the Andean pool generally show tolerance to halo blight, anthracnose and angular leaf spot.However, immunity to infection by *X. axonopodis* pv. *phaseoli* has not been observed in the primary gene pool of *P. vulgaris.*

The secondary gene pool is comprised of species closely related to *P. vulgaris*, such as *Phaseolus coccineus* and *Phaseolus polyanthus*. These species were originally thought to be members of *P. vulgaris*, but both species exhibit significant genetic divergence from *P. vulgaris*, and crosses between *P. vulgaris* and *P. coccineus*, and *P. vulgaris* and *P. polyanthus* are hampered by improper chromosome pairing(Guo *et al.*, 1991). *P. coccinius* is moderately resistant to CBB, and was used in interspecific crosses with *P. vulgaris* to introduce CBB resistance into cultivated beans However, these attemptswere of limited success, and crosses with species in the secondary gene pool have not yielded commercial disease resistant common bean varieties (Hucl and Scoles, 1985).

The tertiary gene pool contains species such as *P. acutifolius*, *P. filiformis* and *P. augustissmus* (Debouck *et al.*, 1993) that are distantly related to *P. vulgaris*. Commercial interest in these species has focused on their resistance to drought, high temperature, low temperature and common bacterial blight (CBB) resistance. The greatest success has come from the use of*P. acutifolius*in interspecific crosses with *P. vulgaris*as a genetic source of resistance to CBB (Parker, 1985; Singh, 1999; Hucl and Scoles, 1985). The success of interspecific crosses involving *P. vulgaris*species is determined by two incompatibility genes, DL1 and DL2 (Singh and Guiterrez, 1984). Some *P. vulgaris* lines , like ICA Piajo, containalleles for compatibility, thus enabling it to be crossed with other *Phaseolus* species (Singh and Guiterrez, 1984). Due to the genetic divergence between *P. vulgaris* and the tertiary gene pool species, crosses between these species often require embryo rescue in order to yield viable F_1 plants. Recovery of normally reproducing and growing plants and plants with good plant architecture and yieldcan require several generations of backcrossing with the parental *P. vulgaris* line (for review see Gepts *et al.*, 2008).

Hybridization between *P. vulgaris* and *P. acutifolius* followed by embryo rescue, recurrent and congruity backcrossing (i.e., backcrossing alternately to either species) of the interspecific hybrids, gene pyramiding (i.e., combining different sources of CBB resistance genes) and field screening by scientists at CIAT, Palmira, Columbia led to the development of six CBB resistant lines, called VAX lines (Mejia-Jimenez *et al.*, 1994). The 6 lines were all derived from a cross between ICA Piajo (*P. vulgaris*) and G40001(= PI319443), an accession of *P. acutifolius*. The F_1 was crossed with the common bean line A775, followed by a cross with A769 and three cycles of selfing to give VAX1 and VAX 2. Further crossing with Xan263 (RAB73 × (BAT1579 × XAN159) gave rise to the VAX4 and VAX5 lines, while crosses with Xan309 ((RAB39 × XAN90) × XAN263) yielded lines VAX3 and VAX6. The morphology of the individual VAX lines is quite varied, with seed colour ranging from cream (VAX1, VAX2 and VAX4) to red (VAX3, VAX6) to black (VAX5). Due to the varied sources of CBB resistance in their pedigrees, the VAX lines are a valuable resource for breeders, although the propensity of VAX1 and VAX2 towards type III growth habit (indeterminate prostrate) and their low combining ability with elite bean breeding materials limits their use (Singh and Munoz, 1999; Munoz *et al.*, 2004).

CBB Resistant Varieties

CBB-tolerant/resistant dry bean varieties have been developed from germplasm derived from interspecific crosses by various breeding programs in North America. A great northern variety GN#1, released in 1961(Coyne, 1961), was selected for CBB resistance and good agronomic characteristics and was developed from a cross between Montana No. 5 and Tepary #4 (Honma, 1956). Studies of the CBB-resistance in GN#1 (Miklas *et al.* 2003) showed that it contains two major resistance QTL associated with marker molecular markers SU91 and $SAP6_{820}$. Great Northern Nebraska #1 Selection 27 (GN#1 Sel 27) was an off-type selected from GN#1 that had higher CBB-resistance and became the source of CBB resistance in a number of great northern cultivars developed by Coyne and coworkers, including: Jules, Harris, Star, and Starlight, as well as a pinto cultivar Chase (described in Miklas *et al.* 2003).

OAC-Rex is a white bean developed at the University of Guelph. It was derived from the cross HR20-728 x MBE7 made in 1988. MBE7 was a selection from the cross ICA Pijao (*P. vulgaris*)/PI440795 (*P. acutifolius*)// Ex Rico 23 (*P. vulgaris*). PI440795 was the source of CBB resistance in this cross (Parker 1985). HR20-728 is a black bean variety selected from a cross between Ex Rico 23 and Midnight. OAC-Rex was tested in field trials as OAC 95-4, with full registration occurring in 2002 (Tar'an *et al.*, 2001).

HR67 is a CBB resistant white bean line developed from a cross between OAC Rico and XAN 159 with additional crossing with Centrialia-3 and HR13-621 (ref). XAN 159 is a line derived from a cross between ICI Piajo and *P. acutifolius* (PI 319443; McElroy, 1985) and it is the source of CBB resistance in this cross. OAC Rico was a registered white bean variety of in Ontario, although it is no longer widely cultivated. The OAC Rico/XAN 159 line was backcrossed to productive white bean lines Centrailia-3 and HR13-621 at the Harrow research station (Agriculture and Agri-Food Canada). The CBB-resistant line HR45 is a sister line to HR67 and both have been extensively used in the bean breeding program at AAFC (Harrow) and have been valuable for studying the genetic basis of CBB-resistance (Yu *et al.*, 2000; Liu *et al.*, 2010). However, the major complication with using these lines in breeding is the close association between the UBC_{420} CBB resistance QTL and the V gene coding for dark seed coat colour. To make this source of CBB resistance more widely suited for breeding the linkage between the UBC_{420} QTL and the V gene needs to be broken.

Work to develop CBB-resistant white beans outside of North America has yielded breeding lines, but no registered varieties (For review see Gepts *et al.*, 2008). White bean CBB-Teebus germplasm was developed by ARC-Grain Crops Institute in South Africa by backcross breeding using Teebus, a registered white bean variety, Xan159, and Wilkinson-2 (Wilk-2). The pedigree of the Wilk-2 gerplasm, created at Cornell in the mid 1980's, is lost (Singh and Munoz, 1999), but it appears to have *P. acutifolius* and *P. coccineus*-derived CBB resistance. CBB-Teebuslines have moderate resistance to the pathogen, and have not been directly developed into varieties but they are used for breeding (Fourie and Herselman, 2002; Asensio-S.-Manzanera *et al.*, 2005).

Several coloured bean varieties and breeding lines have been registered by the USDA, although none are in significant commercial production. For example the dark red kidney variety, USDK-CBB-15, was derived from a modified $BC_3F_{1:4}$ bulk from K97305/3/SVM-2242//I9566-21-4-2/Montcalm (Miklas *et al.*, 2006). This variety possesses two major resistance QTLs, SU91 and $SAP6_{820}$. SU91 was inherited from SVM-2242, which is a F_3-

derived line developed from Montcalm/Xan159 and SAP6$_{820}$ which was inherited from Montcalmand has Montanna No. 5 in its pedigree. USDK-CBB-15 has moderate CBB-resistance, and has been integrated into the University of South Dakota breeding program.

CBB resistant dark red kidney lines ACUG 10-D1 and ACUG 10-D3 were tested in registration trials in Canada. These lines, have moderate CBB resistance and have the SU91 QTL derived from XAN159. The pedigree of ACUG 10-D1 includes several dark red kidney varieties (AC_Calmont/4/Isle/3/Redhawk//Camelot/Xan159) and the ACUG 10-D3 line has Montcalm and Darkid in its background. Studies of possible *P. vulgaris*-derived CBB resistance QTL from Montcalm have been initiated (Navabi, personal communication).

A CBB-resistant cranberry bean USCR-CBB-20 was recently been registered by the USDA for breeding purposes. This variety was derived from the cross OT9461-281-1/3/USCR-CBB-13*2/I9566-21-5-1 where OT9641-281-1 is an advanced cranberry breeding line developed from a three-way cross including Montcalm, K59-7 and Cardinal,while I9566-21-5-1 is derived from a Montcalm/Xan159 cross. The variety has moderate CBB-resistance and is positive for the SAP6$_{820}$ and SU91 CBB-markers (Miklas *et al.*, 2011). During its development, this germplasm also had the UBC$_{420}$ marker,introducedfrom I9566-21-5-1, which includes Xan159 in its pedigree.However, this marker was screened out of the segregating population because of its tight association with the *V* locus, which codes for dark seed-coat colour not appropriate for a cranberry bean (Mutulu *et al.*, 2005).

MOLECULAR MARKERSASSOCIATED WITH CBB RESISTANCE QTL AND MARKER ASSISTED SELECTION FOR RESISTANCE

Markers Linked to CBB Resistance QTL

Over 30 individual genes and a greater number of QTLs with major contributions to disease resistance have been linked to molecular markers on the BAT93 x Jalo EPP553 core bean linkage map (Freyer *et al.*, 1998); including resistance against CBB (Tar'an, 1998; Tar'an *et al.*, 2001; Larsen and Miklas, 2004, Miklas *et al.*, 2006), rust, anthracnose, angular leaf spot, bean common mosaic virus, beet curly top virus, bean golden yellow mosaic virus and halo blight. The majority of the disease resistance markers that have been developed to date are based upon RAPD (Random Amplified Polymorphic DNA) markers that have been converted to SCAR (Sequence Characterized Amplified Region) markers (For review see Gepts*et al.*, 2008).

CBB resistance in *P. vulgaris* lines obtained from various sources is conditioned by several interacting genes so that inheritance for this trait appears to be quantitative. Although true CBB resistance has been described in *P. acutifolius* and *P. coccineus* (Singh and Munoz, 1999), CBB resistance in *P. vulgaris* is more properly described as varying levels of tolerance of the pathogen, since no typical hypersensitive resistance responses to *X. axonopodis* pv. *phaseoli* infection have been described, to date. The number of genes involved in CBB resistance in any germplasm source is a matter of some debate, with one to five genes hypothesized to be of importance in various lines (Drijfhout and Bok, 1987; Park and Dhanvantari, 1987; Eskridge and Coyne, 1996, Miklas *et al.*, 2006). From mapping studies with several CBB resistant sources at least 22 quantitative trait loci (QTL) associated with CBB resistance have been discovered, located on all 11 bean chromosomes (Figure 5, Table 1; Miklas *et al.*, 2006).

Studies of interactions between CBB resistance QTL have been aided by the identification of molecular markers associated with particular QTL which have shown that they are derived from *P. vulgaris*, P. *coccineus* and *P. acutifolius* backgrounds. For example, in the white bean variety Great Northern Nebraska #1 (GN1) one of its major CBB resistance QTL ($SAP6_{820}$) is hypothesized to be *P. vulgaris*-derived and the other QTL (SU91) is *P. acutifolius*-derived (Miklas et al 2003).

Three QTL were identified for CBB resistance in a population of $F_{2:4}$ lines derived from an OAC Seaforth x OAC 95-4 (OAC Rex) cross (Tar'an *et al.*, 2001). In this material one major and two minor QTL, all derived from OAC 95-4 (OAC Rex), act in an additive fashion. The major CBB resistance QTL (accounting for 42% of the variability) was mapped using the SSR marker PV-ctt001. This marker is located on linkage group B4 or B5 on the core map (Yu *et al.*, 2000; Tar'an *et al.*, 2001; Perry, 2010). The ambiguity regarding the location of the PV-ctt001 QTL is likely due to its distal location on both linkage groups. The two minor QTL for CBB resistance in the same material are associated with the RFLP markers BNG71 and BNG21 and are located on B4 and B3, respectively. These QTLs explained 36% and 10% of the phenotypic variance, respectively (Tar'an *et al.*, 2001).

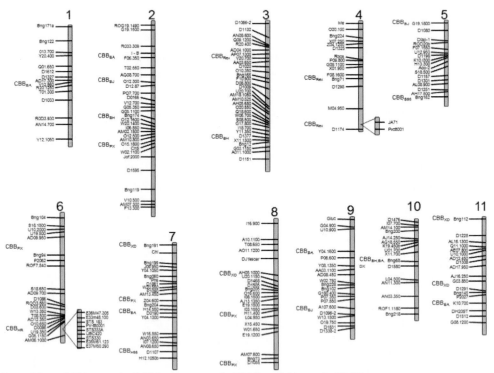

Adapted from Miklas *et al.*, (2006), Gepts *et al.* (2008) and Liu*et al.* (2008).

Figure 5. Linkage map of CBB resistance QTLs from *P. vulgaris* varieties (BA: BelNeb-RR-1/A 55 (Ariyarathne *et al.*, 1999), BJ: Bat 93/Jalo EEP558 (Nodari *et al.*, 1993), BH: Bunsi/Huron, PX-50/Xan 159 (Jung *et al.*, 1997), Rex: OAC Seaforth/OAC Rex (Tar'an *et al.*, 2001), XD: XR235-1-1/DIACOL Calima (BC) (Vallejos *et al.*, 1992), H95: HR67/OAC 95-4 (Yu *et al.*, 2004).

The breeding lines HR67 and HR45derive their CBB resistance from XAN159 and contain one major QTL, accounting for 70% of the phenotypic variation, which was mapped to linkage group B6 and is associated with markers UBC_{420} and PV-tttc001 (Yu *et al.*, 2004). The location of the UBC_{420} marker has also been unclear. It was mapped to B7 by Yu *et al.*2004 but other studies have placed it on linkage group B6, and closely linked to the *V* locus, which codes for seed colour (Jung *et al.*, 1999; Liu *et al.*, 2008; Liu *et al.*, 2010).

Interactionsbetween CBB resistance QTL on the expression of resistance to CBB were studied by using a real-time PCR assay to determine the zygosity of the markersUBC_{420} and SU91 in a BC_6-derived F_2 population from a cross between XAN159 and Teebus (a CBB susceptible *P. vulgaris* variety). (Vandermark *et al.*, 2008).This study found that the QTL associated with the UBC_{420}or SU91 markers conferred only limited CBB resistance on their own. Strong resistance was only observed when plants contained at least one copy of the QTL associated with each marker, and that the interaction was strongly additive. In a similar study to examine the interaction between SAP6 and SU91 on CBB resistance in a pinto bean F_2 population and dark red kidney bean F_2 population, Vandermark *et al.*(2008) found that resistance was primarily conditioned by the presence of at least on copy of the SU91 QTL.

In a cross between OAC-Rex and HR67 the UBC_{420} QTL is responsible for 36% to 47% of the variation in CBB symptoms in field trials, whereas, the PV-ctt001 QTL was responsible for 1% (Durham,personal communication). The effect of the SU91 QTL in this cross could not be studied as both lines have this marker.

Table 1. Phaseolus vulgaris DNA markers associated with CBB resistance QTLs. Markers with known DNA sequence are listed along with PCR primer sequences or Genbank accession numbers

Marker Name	Chromosome	Primer Sequence or Genbank Accession	Size	Reference
Bng165	3	N/A		
Bng21	3	F: ACCCAACTTACAGAGCTGTTTG R: TTCGATTTGGAACATTGGCTG	1430	Murray *et al.*, 2002
Bng71	4	F-GTTGCTGTTAAGATATCAAACG R-GGGACATGTTATTCATATGGTT	1690	
PV-ctt001	4	F-GAGGGTGTTTCACTATTCTCACTGC R-TTCATGGATGGTGGAGGAACAG	152	Tar'an *et al.*, 2001
Bng162	5	N/A		Murray *et al.*, 2002
BC_{420}	6	F-GCAGGGTTCGAAGACACAGTGG R-GCAGGGTTCGCCCAATAACG	900	Yu *et al.*, 2000
STS 183	6	F-CCTATGTACTTCTTGAGGGAGAC R-AGAAGCCCAGGGACTTGGAT	280	Liu *et al.*, 2008
STS333a	6	F-CATAAGATGAATGGTTCTTGAC R-CCATTTGGTGAGATTCACTT	150	Liu *et al.*, 2008
Phs	7	F-AGCATATTCTAGAGGCCTCC R-GCTCAGTTCCTCAATCTGTTC	Multiple	Nodari *et al.*, 1993
PV-tttc001	7	F-TTTACGCACCGCAGCACCAC R-TGGACTCATAGAGGCGCAGAAAG	161	Yu *et al.*, 2004
R4865	8	F-TCCAAAGCCATTGTAGTT R-CAGCTACTTTCAAACTGGG	950	Bai *et al.*, 1997
R7313	8	F-ATTGTTATCGTCGACACG R-AATATTTCTGATCACACGAG	700	Bai *et al.*, 1997

Table 1. (Continued)

Marker Name	Chromosome	Primer Sequence or Genbank Accession	Size	Reference
SU91	8	F-CCACATCGGTTAACATGAGT R-CCACATCGGTGTCAACGTGA	700	Pedraza et al., 1997
BAC6	10	F-TAGGCGGCGGCGCACGTTTTG R-TAGGCGGCGGAAGTGGCGGTG	1250	Jing et al., 1999
SAP6	10	F-GTCACGTCTCCTTAATAGTA R-GTCACGTCTCAATAGGCAAA	820	Miklas et al., 2000

Marker Assisted Selection

Because of the complex nature of resistance to CBB in Phaseolus and the good progress that has been made in identifying molecular markers linked to resistance QTL there is interest in utilizing the markers to select resistant varieties in bean breeding programs. The potential advantages of marker assisted selection (MAS) are that it is easier and faster to screen large numbers of genotypes in a breeding program for molecular markers than screening them for symptoms after inoculating them with the bacteria. In addition, the use of MAS in bean breeding has been shown to be economical. Yu et al.(2000) estimated that PCR-based marker selection for CBB resistant genotypes cost between \$2.25-\$4.24 per data point, as compared to greenhouse or field trials which cost more than \$7.00 per data point, depending on the length of the trials and the local labour costs. Other studies have shown that the cost for an individual PCR-based data point, excluding labour costs, can be as low as \$0.25 (Blair et al., 2007). MAS can be much fasterthan conventional disease screening techniques. For example, greenhouse or field screening requires a period of 4-6 weeks to establish the plants, inoculate them, and score for symptoms. PCR-based screening can be conducted on the seed or seedings lessthan 1 week after establishment. Studies of rice or cassava have indicated that the inclusion of MAS into breeding programs can reduce the time required for the development of a new cultivar from 10-15 years to 7-12 years (Alpuerto et al., 2008; Rudi et al., 2010).

However, MAS is not without difficulties. Since the majority of markers used for disease resistance are not directly associated with the resistance genes themselves, there is a possibility that the linkage between the marker and the trait can be broken due to genetic recombination. This can be mitigated by developing additional markers, closer to the desired QTL or gene, as the more tightly the marker and the QTL are linked, the less likely it will be broken (Robinson and Ruane, 2007). Currently, MAS is used in conjunction with conventional breeding programs as a way of rapidly screening plant lines for advancement, while continuing to test these lines using field and greenhouse screening to validate the results. This has the effect of reducing the number of lines requiring the more intensive field testing, while at the same time, accelerating the propagation of lines containing the markers of interest (for review see Yu et al., 2000; Edwards and McCouch, 2007; Nayak et al., 2010).

Mutlu et al. (2005) used MAS to combine different QTL for CBB resistance into a single pinto bean (P. vulgaris) genotype. They observed that multiple QTL for CBB resistance provided better resistance than either QTL alone. In addition, selection for the QTL linked to the BC_{420} marker yielded plants with a significant CBB resistance, but the use of this marker was not desirable because it is linked to the V locus that results in an undesirable dark seedcoat color (Miklas *et al.*, 2006; Mutlu *et al.*, 2005). O'Boyle *at al.* (2007) examined the potential for using molecular markers linked to CBB loci for MAS of CBB resistant germplasm and pyramiding resistance QTL. They found that the presence of the marker SU91 was significantly correlated with high levels ofCBB leaf resistance and was not associated with any agronomic traits. Interestingly, the presence of both SU91 and BC_{420} resulted in lower levels of CBB resistance than provided by either marker alone. They suggested that this is evidence of epistatic interactions between the independent loci conditioning CBB resistance in common bean.

CONCLUSION

CBB continues to be an important threat for common bean cultivation worldwide. The pathogen *X. axonopodis* pv. *phaseoli* causes significant crop losses on an annual basis(Saettler, 1989), and the ability of the pathogen to colonize the seed makes it highly transmissible from year to year. The development of CBB-resistant lines of *P. vulgaris* in several market classes has represented a significant advancement for the management of this disease, but the limited release of CBB-resistant varieties to growers, particularly of Andean origin indicates that more breeding efforts will be required to fully incorporate CBB-resistance into the germplasm.

The molecular basis of the pathogen-host interaction in CBB is still poorly understood. Limited research on the effector molecules expressed by other members of the *Xanthomonas* genera have indicated that there is considerable variability in the non-conserved *hrp* proteins, and a more detailed examination of these factors would provide valuable information on the virulence factors important to the pathogen.

Genetic resistance to CBB in *P. vulgaris* is a complex trait, controlled by multiple QTLs; however the actual genes involved in this resistance have not been identified. The number and location of these QTLs is largely a question of pedigree, as several breeding programs have independently worked to incorporate CBB-resistance into their local varieties. Current bean genome sequencing projects in Canada and the US are scheduled to produce draft sequences for two *P. vulgaris*genotypes (OAC-Rex, G19833) by 2015 (McClean *et al.*, 2009; Pauls *et al.*, 2010). The examination of these sequences will facilitate the identification of resistance-associated genes, with the long term goal of utilizing these genes to produce CBB resistant varieties.

REFERENCES

Agriculture and Agri-Food Canada (2005). Crop profile for dry bean in Canada. *Pesticide Risk Reduction Program*,01B68-3-0046.

Alavi SM, Sanjari S, Durand F, Brin C, Manceau C and Poussier S (2008). Assesment of Genetic Diversity of *Xanthomonas axonopodis* pv. *phaseoli* and *Xanthomonasfuscans* subsp. fuscans as a basis to identify putative pathogenicity genes and a type III secretion system of the SPI-1 family by multiple suppression subtractive hybridizations. *Applied and Environmental Microbiology*, 74:3295-3301.

Alpuerto V, Norton GW and Alwang J (2008). Economic impact analysis of marker-assisted breeding in rice. Selected paper for the American Agricultural Association Annual Meeting, 1-29.

Ariyarathne HM, Coyne DP, Jung G, Skroch PW Vidaver AK, Steadman JR, Miklas PN and Bassett MJ(1999). Molecular mappingof disease resistance genes for halo blight, common bacterialblight, and bean common mosaic virus in a segregating populationof common bean. Journal of the American Society of Horticultural Science, 124:654–662.

Asensio-S.-Manzanera MC, Asensio C, Singh SP (2006). Gamete selection for resistance to common and halo bacterial blights in dry bean intergene pool populations. *Crop Science*, 46:131–135.

Audy P, Laroche A, Saindon G, Huang HC and Gilbertson RL (1994). Detection of the bean common blight bacteria, *Xanthomonas campestris* pv. *phaseoli* and *Xanthomonas campestris* pv. *phaseoli* var. *fuscans*, using polymerase chain reaction. *Phytopathology*, 84:1185-1192.

Bai Y, Michaels TE and Pauls KP(1997). Identification of RAPD markers linked to common bacterial blight resistance genes in *Phaseolus vulgaris* L. Genome 40:544-551.

Backert S and Meyer TF (2006). Type IV secretion systems and their effectors in bacterial pathogenesis. *Current Opinion in Microbiology*, 9:207-217.

Bender CL, Malvick DK, Conway KE, George S and Pratt P (1990). Characterization of pXV10A, a copper resistance plasmid in*Xanthomonas campestris* pv. *vesicatoria*. Applied and Environmental Microbiology, 56:170–175.

Bennett M, Leitch I (2005). Angiosperm DNA C-Values database. Release 4.0 http://www.rbgkew.org.uk/cval/database1.html.

Blair MW, Diaz JM, Hidalgo R, Diaz LM and Duque MC (2007). Microsatellite characterization of Andean races of common bean(*Phaseolus vulgaris* L.). Theoretical and Applied Genetics, 116:29-43.

Bradbury JF (1986). Guide to plant pathogenic bacteria. CAB International, Farnham House, UK.Buttner D and He SY (2009). Type III protein secretion in plant pathogenicbacteria. Plant Physiology 150:1656-1664.

Cafati CR and Saettler AW (1980). Transmission of *Xanthomonas phaseoli* in seed of resistant and susceptible *Phaseolus* genotypes. *Phytopathology*, 70:638-340.

Cascales E and Christie PJ (2003), The versatile bacterial type IV secretion systems, *Nature Reviews Microbiology*,**1**:137–149.

Chan JWYF and Goodwin PH (1999). A physical map of the chromosome of *Xanthomonas campestris* pv. *phaseoli* var. *fuscans* BXPF65. *FEMS Microbiology Letters*, 180:85-90.

Cianciotto NP (2005). Type II secretion: a protein secretion system for all seasons. *Trends in Microbiology*, 13:581-588.

Coyne DP (1961). Characteristics and performance of the Nebraska 1 dry bean. *Annual Report of the Bean Improvement Cooperative*, 5:50-51.

Coyne DP and Schuster ML (1974). Differential reaction of pods and foliage of beans (*Phaseolus vulgaris*) to *Xanthomonas phaseoli*. *Plant Disease Reporter*,58:278-282.

da Silva AC, Ferro JA, Reinach FC, Farah CS, Furlan LR, Quaggio RB, Monteiro-Vitorello CB, Van Sluys MA, Almeida NF, Alves LM, do Amaral AM, Bertolini MC, Camargo LE, Camarotte G, Cannavan F, Cardozo J, Chambergo F, Ciapina LP, Cicarelli RM,

Coutinho LL, Cursino-Santos JR, El-Dorry H, Faria JB, Ferreira AJ, Ferreira RC, Ferro MI, Formighieri EF, Franco MC, Greggio CC, Gruber A, Katsuyama AM, Kishi LT, Leite RP, Lemos EG, Lemos MV, Locali EC, Machado MA, Madeira AM, Martinez-Rossi NM, Martins EC, Meidanis J, Menck CF, Miyaki CY, Moon DH, Moreira LM, Novo MT, Okura VK, Oliveira MC, Oliveira VR, Pereira HA, Rossi A, Sena JA, Silva C, de Souza RF, Spinola LA, Takita MA, Tamura RE, Teixeira EC, Tezza RI, Trindade dos Santos M, Truffi D, Tsai SM, White FF, Setubal JC and Kitajima JP (2002). Comparisons of the genomes of two Xanthomonas pathogens with differing host specificities. *Nature*, 417(6887):459-463.

da Silva AC, Ferro JA, Reinach FC, Farah CS, Furlan LR, Quaggio RB, Monterio-Vitorello CB, Van Sluys MA, Almeida NF and Alves LM, do Amaral MC, Bertolini LEA, Camargo G, Camarotte F, Cannavan J, Cardozo F, Chambergo LP, Ciapina RMB, Cicarelli LL, Coutinho JR, Cursino-Santos H, El-Dorry JB, Faria AJS, Ferreira RCC, Ferreira MIT, Ferro EF, Formighieri MC, Franco CC, Greggio A, Gruber AM, Katsuyama LT, Kishi RP, Leite EGM, Lemos MVF, Lemos15 EC, Locali MA, Machado AMBN, Madeira NM, Martinez-Rossi EC, Martins J, Meidanis CFM, Menck CY, Miyaki DH, Moon LM, Moreira MTM, Novo VK, Okura MC, Oliveira VR, Oliveira HA, Pereira A, Rossi JAD, Sena C, Silva RF, de Souza LAF, Spinola MA, Takita RE, Tamura EC, Teixeira RID, Tezza M, Trindade dos Santos D, Truffi SM, Tsai FF, White JC, Setubal and Kitajima JP (2002). Comparison of the genomes of two Xanthomonas pathogens with differing host specificities. *Nature*, 417:459-463.

Darsonval A, Darrasse A, Meyer D, Demarty M, Durand K, Bureau C, Manceau C and Jacques MA (2008). The Type III secretion system of *Xanthomonas fuscans* subsp. *fuscans* is involved in the phyllosphere colonization process and in transmission to seeds of susceptible beans. *Applied and Environmental Microbiology*, 74(9):2669-2678.

Debouck DG, Toro O, Paredes OM, Johnson WC and Gepts P (1993). Genetic diversity and ecological distribution of *Phaseolus vulgaris* in north-western South America. *Economic Botany*, 47:408–423.

Drijfout E and Bok WJ (1987). Inheritence of resistance to *Xanthomonas campestris* pv *phaseoli* in tepary bean (*Phaseolus acutifolius*). *Euphytica*, 36:803-808.

Edwards JD and McCouch SR (2007). Molecular markers for use in plant molecular breeding and germplasm evaluation. In Marker-Assisted Selection: Current status and future perspectives in crops, livestock, forestry and fish (eds.) Guimaraes EP, Ruane J, Scherf BD, Sonnino A and Dargie JD. *Food and Agriculture organization of the United Nations*, 29-50.

Eskridge KM and Coyne DP (1996). Estimation and testing hypothesis about the number of genes using inbred-backcross data. *Journal of Heredity*, 87:728-736.

Evans AM (1973). Genetic improvement of *Phaseolus vulgaris*, pp. 107–115. In Nutritional Improvement of Food Legumes by Breeding. Ed. Milner M New York: 389.

Fourie D and Herselman L (2002). Breeding for common blight resistance in dry beans in South Africa. *Annual Report of the Bean Improvement Cooperative*, 45: 50—51.

Freyre R, Skoch PW, Adam-Blondom AF, Geffroy V, Shirmohamadali A, Johnson, WC, Llaca, V, Nodari RO, Pereira PA, Tsai, SM, Tohme J, Dron M, Nienhuis J and Gepts P (1998). Towards on integrated linkage map in common bean IV: Correlations among RFLP maps. *Theoretical and Applied Genetics*, 97:834-846.

Furutani A, Tsuge S, Ohnishi K, Hikichi Y, Oku T, Tsuno K, Inoue Y, Ochiai H, Kaku H and Kubu Y (2004). Evidence for Hrp-Xo-dependent expression of type Ii secretory proteins in *Xanthomonas oryzae* pv. oryzae. *Journal of Bacteriology*, 186:1374-1380.

Gabriel DW, Kingsley MT, Hunter JE and Gottwald T (1989). Reinstatement of *Xanthomonas citri*(ex Hasse) and *X. phaseoli* (ex Smith) to species and reclassification of all *X. campestris* pv. *citri* strains. *International Journal of Systemic and Evolutionary Microbiology*, 39:14-22.

Gepts P (1998). Origin and evolution of common bean: past events and recent trends. HortScience, 33:1124–1130.

Gepts P, Aragão F, Barros E de, Blair MW, Brondani R, Broughton W, Galasso I, Hernández G, Kami J, Lariguet P, McClean P, Melotto M, Miklas P, Pauls P, Pedrosa-Harand A, Porch T, Sánchez F, Sparvoli F and Yu K (2008). Genomics of *Phaseolus* beans, a major source of dietary protein and micronutrient in the Tropics. In: Genomics of Tropical Crop Plants, Edited by Moore PH and Ming R. Springer, Berlin. pp 113-143.

Gepts P and Debouck DG 1991. Origin, domestication,and evolution of the common bean, *Phaseolus uulgans*.In: Common beans: research for crop Improvement (van Schoonhoven A and Voysest O, eds). Wallingford, U.K.: CAB Intl./Cali, Colombia: CIAT; 7-53.

Gillard CL (2004). New strategies for the integrated pest management in dry edible beans to manage risk and to enhance enconomic viability. Agricultural Adaptation Council and the Ontario Coloured Bean Growers Association, London ON.

Gillard CL, Conner RL, Balasubramanian P and Boland G (2010). New IPM strategies to control dry bean anthracnose. 8th Canadian Pulse Research Workshop, Winnepeg AB.

Glathe H and Singh-Verma SB (1968). Glucose peptone agar, a culture medium suited for the isolation and cultivation of bacterial antagonists from soil, and antibacterial action of various soil bacteria against other non-pathogenic and pathogenic bacteria. ZentralblBakteriolParasitenkdInfektionskr Hyg 122(1):3-21.

Goszczynska, T and Serfontein JJ(1998). Milk-Tween agar, a semiselective medium for isolation and differentiation of *Pseudomonas syringae* pv. *syringae*, *Pseudomonas syringae* pv. *phaseolicola* and *Xanthomonas axonopodis* pv. *phaseoli*. *Journal of Microbiological Methods*, 32:65-72.

Guo X, Castillo-Ramirez S, Gonzalez V, Bustos P, Fernandez-Vazquez JL, Santamaria RI, Arellano J, Cevallos MA and Davila G (2007). Rapid evolutionary change of common bean (*Phaseolus vulgaris* L) plastome, and the genomic diversification of legume chloroplasts. *BMC Genomics*, 8:228.

GurlebeckD, ThiemeF and BonasU (2006).Type III effector proteins from the plant pathogen *Xanthomonas* and their role in the interaction with the host plant. *The Journal of Plant Physiology*, 163:233-255.

Hauser AR (2009). The type III secretion system of *Pseudomasaeruginosa*: Infection by injection. *Nature Reviews Microbiology*, 7:654-665.

Hayward AC, Waterston JM (1965). *Xanthomonas phaseoli* var. *fuscans*. *CMI Descriptions of Pathogenic Fungi and Bacteria* No. 49. CAB International, Wallingford, UK.

Honma S(1956). A bean interspecific hybrid. Journal of Heredity 47:217-220.

Hucl P and Scoles GJ (1985). Interspecific Hybridization in the common bean: A Review. HortScience 20:352-357.

Jacques MA, Josi K, Darrasse A, and Samson R (2005). *Xanthomonas axonopodis* pv. *phaseoli* var. fuscans is aggregated in stable biofilm population sizes in the phyllosphere of field-grown beans. Applied and Environmental Microbiology 71(4):2008-2015.

Jha G, Rajeshwari R and Sonti R (2005). Bacterial type two secretion system secreted proteins: double-edged swords for plant pathogens. *Molecular Plant-Microbe Interactions*, 18:891-898.

Jung G, Coyne DP, Skroch P, Nienhuis J, Bokosi JM and Steadman JR (1996). RAPD marker linked to a gene for specific rust resistance in common bean. Annual Report of the Bean Improvement Cooperative, 39:59–60.

Jung G, Skroch P, Coyne DP, Nienhuis J, Ariyarathne H, Kaeppler S and Bassett M(1997). Molecular-marker-based geneticanalysis of tepary-bean-derived common bacterial blight resistancein different developmental stages of common bean. Journal of the American Society of Horticultural Science 122:329–337.

Jung G, Skroch PW, Nienhuis J, Coyne DP, Arnaud-Santana E, Ariyarathne HM and Marita JM (1999). Confirmation of QTL associated with common bacterial blight resistance in four different genetic backgrounds in common bean. *Crop Science*, 39:1448-1455.

Kim, JG, Taylor KW, Hotson A, Keegan M, Schmelz EA and Mudgett MB (2008). XopD SUMO protease affects host transcription, promotes pathogen growth, and delays symptom development in *Xanthomonas*-infected tomato leaves. Plant Cell 20:1915–1929.

Koening R and Gepts P, (1989). Allozyme diversity in wild *Phaseolus vulgaris*: further evidence for two major centers of diversity. *Theoretical and Applied Genetics*, 78: 809-817.

Koening RL, Singh SP and Gepts P (1990). Novel *phaseolin* types in wild and cultivated common bean (*Phaseolus vulgaris*, Fabaceae). *Economic Botany*, 44: 50–60.

Kwak M, Kami JA, Gepts P (2009). The putative Mesoamerican domestication center of *Phaseolus vulgaris* is located in the Lerma-Santiago basin of Mexico. *Crop Science*, 49:554-563.

Larsen RC and Miklas PN (2004). Generation and molecular mapping of a sequence characterized amplified region marker linked with the *Bct*gene for resistance to Beet curly top virus in common bean. *Phytopathology*, 94:320–325.

Lin NC and Martin (2007). *Pto/Prf*-mediated recognition of *AvrPto* and *AvrPtoB* restricts the ability of diverse *Pseudomonas syringae* pathovars to infect tomato. *Molecular Plant Microbe Interactions*, 17:162-174.

Liu S, Yu K and Park SJ (2008). Fine mapping of a major qtl for common bacterial blight resistance from the bean cultivar HR67. Annual reports from the 7th CPRW meeting.

Liu SY, Yu K, Haffner M, Park SJ, Banik M, Pauls KP, and Crosby WL (2010).Construction of a BAC library and a physical map of a major QTL for CBB resistance of common bean (Phaseolus vulgaris L.). *Genetica*, 138(7), pp. 709−716.

McElroy JB (1985). Breeding dry beans, *P. vulgaris* L., forcommon bacterial blight resistance derived from *Phaseolus acutifolius*A. Gray. Cornell University, PhD Thesis.

Mejia-Jimenez A, Munoz C, Jacobson HJ, Roca WM and Singh SP (1994). Interspecific hybridization between common and tepary beans: increased hybrid embryo growth, fertility and efficiency of hybridization through recurrent and congruity backcrossing. *Theoretical and Applied Genetics*, 88:324-331.

Mensack MM, Fitzgerald VK, Ryan EP, Lewis MR, Thompson HJ and Brick MA (2010). Evaluation of diversity among common beans (Phaseolus vulgaris L.) from two centers of domestication using 'omics' technologies. *BMC Genomics*, 11:686.

Miklas PN, Coyne DP, Grafton KF, Mutulu N, Reiser J, Lindgren DT and Singh SP (2003). A major QTL for common bacterial blight resistance derives from the common bean great northern landrace cultivar Montana No. 5. *Euphytica*, 131:137-146.

Miklas PN, Kelly JD, Beebe SE and Blair MW (2006). Common bean breeding for resistance against biotic and abiotic stresses: From classical to MAS breeding. *Euphytica*, 147:105-131.

Miklas PN, Singh SP, Teran H, Kelley JD and Smith JR (2011). Registration of common bacterial blight resistant cranberry dry bean germplasm line USCR-CBB-20. *Journal of Plant Registrations*, 5(1):98-102.

Miklas PN, Smith JR, Riley R, Grafton KF, Singh SP, Jung G and Coyne DP(2000). Marker-assisted breeding for pyramided resistance to common bacterial blight in common bean. Annual Report of the Bean Improvement Cooperative, 43:39-40.

Miklas PN, Smith JR and Singh SP (2006). Registration of common bacterial blight resistant dark red kidney bean germplasm line USDK-CBB-15. *Crop Science*, 46:1005-1007.

Mochida K, Yoshida T, Sakurai T, Yamaguchi-Shinozaki K, Shinozaki K and Tran LSP (2010). Legume TFDB: an integrative database of Glycine max, Lotus japonicas and Medicago truncatula transcription factors. *Bioinformatics*, 26(2):290-291.

Muñoz LC, Blair MW, Duque MC, Roca W andTohme J (2004). Level of introgression in inter-specific (*Phaseolus vulgaris* × *P. acutifolius*) congruity-backcross lines. Crop Science, 44:637–645.

Murray J, Larsen J, Michaels TE, Schaafsma A, Vallejos CE, and Pauls KP (2002). Identification of putative genes in Bean (*Phaseolus vulgaris*) genomic (Bng) RFLP clones and their conversion to STS's. Genome, 45: 1013-1024.

Mutlu N, Miklas PN, Reiser J, and Coyne DP (2005). Backcross breeding for improved resistance to common bacterial blight in pinto bean (*Phaseolus vulgaris* L.). Plant Breeding, 124:282–287.

Nayak SN, Zhu H, Varghese N, Datta S, Choi HK, Horres R, Jungling R, Singh J, Kishor PBK, Sivaramakrishnan S, Hosington DA, Khal G, Winter P, Cook DR and Varshney RK (2010). Integration of novel SSR and gene-based SNP marker loci in the chickpea genetic map and establishment of new anchor points with *Medicago truncatula* genome. *Theoretical and Applied Genetics*, 120:1514-1441.

Nodari RO, Tsai SM, Guzman P, Gilbertson RL and Gepts P (1993). Towards an integrated linkage map of common bean. III.Mapping genetic factors controlling host–bacteria interactions. Genetics, 134:341–350.

Nuland DS, Schwartz HF, and Forster RL (1983). Recognition and management of dry bean production problems. North Central Regional Extension Publication 198.

O'Boyle PD, Kelly JD and Kirk WW (2007). Use of marker assisted selection to breed for resistance to common bacterial blight in common bean. Journal fo the American Soceity of Horticultural Science, 132:381–386.

Park SJ and Dhanvantari BN (1987). Transfer of common bean blight (*Xanthomonas campestris* pv. *phaseoli*) resistance from *Phaseolus coccineus* Lam. to *P. vulgaris* L. through interspecific hybridization. *Canadian Journal of Plant Science*, 67:685-695.

Parker JPK (1985). *Interspecific transfer of common bacterial blight resistance from Phaseolus acutifolius A. Gray to Phaseolus vulgaris L.*University of Guelph, M.Sc. thesis.

Pedraza F, Gallego G, Beebe S and Tohme J(1997). Marcadores SCAR y RAPD para la resistencia a la bacteriosiscomun (CBB). p.130-134. En Singh, S.P. y O. Voysest (eds.). Taller de mejoramiento de frijol para el Siglo XXI: Bases paraunaestrategiapara America Latina. 559 pp. CIAT, Cali, Colombia.

Perry GE (2010). Common bacterial blight resistance in Phaseolus vulgaris: AN examination of the major resistance QTLs of OAC-Rex and HR67. University of Guelph, Ph.D. Thesis.

Qian W, Jia Y, Ren SX, He YQ, Feng JX, Lu LF, Sun Q, Ying G, Tang DJ, Tang H, Wu W, Hao P, Wang L, Jiang BL, Zeng S, Gu WY, Lu G, Rong L, Tian Y, Yao Z, Fu G, Chen B, Fang R, Qiang B, Chen Z, Zhao GP, Tang JL and He C (2005). Comparative and functional genomic analyses of the pathogenicity of phytopathogen*Xanthomonas campestris* pv. *campestris*.

Rademaker JLW, Louws FJ,Schultz MH, Rossbach U, Vauterin L, Swings J, and de Bruijn FJ (2005) A Comprehensive Species to Strain Taxonomic Framework for Xanthomonas *Phytopathology*, 95: 1098.

Rediers H and Vanderleyden J (2004). *Azotobactervinelandii*: a *Pseudomonas* in disguise?. *Microbiology*,150(5):1117–1119.

Robinson J and Ruane J (2007). Marker assisted selection as a potential tool for genetic improvement in developing countries: debating the issues. In Marker-Assisted Selection: Current status and future perspectives in crops, livestock, forestry and fish (eds.) Guimaraes EP, Ruane J, Scherf BD, Sonnino A and Dargie JD. *Food and Agriculture organization of the United Nations*, 427-440.

Rosen WG (1954). Effects of streptomycin on certain green plants. *The Ohio Journal of Science*, 54(2):73-78.

Rudi N, Norton GW, Alwang J and Asumugha G (2010). Economic impact analysis of marker assisted breeding for resistance to pests and post-harvest deterioration in cassava. *African Journal for Agricultural and Resource Economics*, 4(2):110-122.

Saettler AW (1989). Assessment of yield loss caused by common blight of beans in Uganda. *Annual Report of the Bean Improvement Cooperative*, 35: 113-114.

Saettler AW (1989). Common bacterial blight. In: Schwartz HF, Pastor Corrales MA (eds). Bean Production Problems in the Tropics. Cali, Columbio: Contro International de Agricultura Tropical 261-83.

Schaad NW and Stall RE (1988).*Xanthomonas*. In: *Laboratory guide for identification of plant pathogenic bacteria* ed. Schaad NW, 2nd edition, *American Phytopathological Society*, St. Paul: 81-94.

Schwartz HF, Otto KJ and Gent DH (2003). Relation of temperature and rainfall to development of Xanthomonas and Pantoea leaf blights of onion in Colorado. Plant Disease, 87:11-14.

Szczepanowski R, Krahn I, Linke B, Goesmann A, Pühler A, Schlüter A (2004). Antibiotic multiresistance plasmid pRSB101 isolated from a wastewater treatment plant is related to plasmids residing in phytopathogenic bacteria and carries eight different resistance determinants including a multidrug transport system. Microbiology, 150(11):3613-3630.Severin V (1971). Investigations on the prevention of the common blight of beans

(*Xanthomonas phaseoli*). AnaleleInstitutului de CercetaripentruProtectiaPlantelor,7:125-139.

Sheppard JE and Remeeus PM, 2005. Proposal for a new method for detecting *Xanthomonas axonopodis* pv. *phaseoli* on bean seeds. ISTA Method Validation Reports 3.

Singh P and Guiterrez JA (1984). Geographical distribution of the DL1 and DL2 genes causing hybrid dwarfism in *Phaseolus vulgaris* L., their association with seed size and their significance to breeding. *Euphytica*, 33:337-345.

Singh SP (ed) (1999) Common Bean Improvement for the Twenty-First Century. Kluwer Academic Publising, Dordrecht, Germany.

Singh SP and Munoz CG (1999). Resistance to common bacterial blight among *Phaseolus* species and common bean improvement. *Crop Science*, 39:80-89.

Singh SP, Gepts P and Debouck DG (1991a). Races of common bean (*Phaseolus vulgaris* L., Fabaceae). *Economic Botany*, 45:379–396.

Singh SP, Nodari R, Gepts P (1991b). Genetic diversity in cultivated common bean. I. Allozymes. *Crop Science*, 31:19–23.

Starr MP (1981). The genus *Xanthomonas*. In *The Procaryotes*, eds Starr MP, Stolp H, Trüper HGB, Balows A and Schlegel HG, Springer-Verlag Berlin: 742–763.

Swings J, Vauterin L and Kersters K (1993). The bacterium Xanthomonas. In *Xanthomonas* eds. Swings JG and Civerolo EL. Chapman & Hall, London: 121-157.

Szczesny R, Jordan M, Schramm C, Schuls S, Cogez V, Bonas U and Buttner D (2010). Functional characterization of the Xcs and Xps type II secretion systems from the plant pathogenic bacterium *Xanthomonas campestris* pv. *vesicatoria*. *New Phytologist*, 187:983-1002.

Tar'an B, Michaels TE and Pauls KP (1998). Stability of the association of molecular markers with common bacterial blight resistance in common bean (*Phaseolus vulgaris* L.). *Plant Breeding*, 117:553-558.

Tar'an B, Michaels TE and Pauls KP (2001). Mapping genetic factors affecting the reaction to *Xanthomonas axonopodis* pv. *phaseoli* in *Phaseolus vulgaris* L. under field conditions. *Genome*, 44:1046-1056.

Tar'an B, Michaels TE, Pauls KP (2002). Genetic mapping of agronomic traits in common bean. *Crop Science*, 42:544–556.

Thieme F, Koebnik R, Bekel T, Berger C, Boch J, Büttner D, Caldana C, Gaigalat L, Goesmann A, Kay S, Kirchner O, Lanz C, Linke B, McHardy AC, Meyer F, Mittenhuber G, Nies DH, Niesbach-Klösgen U, Patschkowski T, Rückert C, Rupp O, Schneiker S, Schuster SC, Vorhölter FJ, Weber E, Pühler A, Bonas U, Bartels D and Kaiser O (2005). Insights into genome plasticity and pathogenicity of the plant pathogenic bacterium Xanthomonas campestris pv. vesicatoria revealed by the complete genome sequence. *Journal of Bacteriology*, 187:7254-7266.

Troisfontaines P and Cornelis GR (2005). Type III secretion: More systems than you think. *Physiology*, 20(5):326-339.

Vallejos CE, Sakiyama NS and Chase CD (1992). A molecularmarker-based linkage map of *Phaseolus vulgaris* L. Genetics,131:733–740.

Van Sluys MA, Moteiro-Vitorello CB, Camargo LE, Menck CF, da Silva AC, Ferro JA, Oliveira MC, Setubal JC, Kitajima JP and Simpson AJ (2002). Comparative genomic analysis of plant-associated bacteria. *Annual Review of Phytopathology*, 40:169-189.

Vandermark GJ, Fourie D and Miklas PN (2008). Genotyping with real-time PCR reveals recessive epistasis between independent QTL conferring resistance to common bacterial blight in dry bean. *Theoretical and Applied Genetics*, 117(4):531-522.

Vauterin L, Hoste B, Kersters K and Swings J (1995). Reclassification of Xanthomonas. *International Journal of Systemic and Evolutionary Microbiology*, 45:472-489.

Vauterin, L, Yang P,Hoste B, Vancanneyt M, Civerolo EL, Swings J andKersters J (1991).Differentiation of *Xanthomonas campestris* pv. *citri* strains by sodium dodecyl sulfate polyacrylamide gel electrophresis of proteins, fatty acid analysis and DNA-DNA hybridization. *International Journal of Systemic Bacteriology,*41:535–542.

Vivader AK (1993). *Xanthomonas campestris* pv. *phaseoli*: cause of common bacterial blight. In *Xanthomonas* eds. Swings JG and Civerolo EL, Chapman & Hall, London: 121-146.

Wang L, Rong W and He C (2008). Two *Xanthomonas* extracellular polygalacturonases, PghAxc and PhgBxc are regulated by type III secretion regulators HrpX and HrpG and are required for virulence. *Molecular Plant-Microbe Interactions*, 21:555-563.

Wei CF, Kvitko BH, Shimizu R, Crabill E, Alfano JR, Lin NC, Martin GB, Huang HC and Collmer A (2007). A *Pseudomonas syringae* pv. *tomato* DC3000 mutant lacking the type III effector HopQ1-1 is able to cause disease in the model plant *Nicotiana benthamiana*. *Plant Journal*, 51:32-46.

Yamazaki A, Hirata H and Tsuyumu S (2008). HrpG regulates type II secretory proteins in Xanthomonas axonopodis pv. citri. *Journal of General Plant Pathology*, 74:138-150.

Yu K, Park SJ, Poysa V (2000). Marker-assisted selection of common beans for resistance to common bacterial blight: efficiency and economics. *Plant Breeding*, 119:411-416.

Yu K, Park SJ, Zhang B, Haffner M and Poysa V (2004). An SSR marker in the nitrate reductase gene of common bean is tightly linked to a major gene conferring resistance to common bacterial blight. *Euphytica*, 138:89-95.

Zaumeyer, W.J.; Thomas, H.R. (1957) A monographic study of bean diseases and methods for their control. Technical Bulletin US Department of Agriculture No. 868, pp. 65-74. USDA, Washington, USA.

In: Agricultural Research Updates. Volume 2   ISBN: 978-1-61470-191-0
Editor: Barbara P. Hendriks   © 2012 Nova Science Publishers, Inc.

Chapter 7

HEALTHIER FUNCTIONAL BEEF BURGERS

S. C. Andrés[1], S. C. Pennisi Forell[2], N. Ranalli[1,3,4], N. E. Zaritzky[1,4] and A. N. Califano [1]*

[1] CIDCA, CONICET – Facultad de Ciencias Exactas, UNLP,
La Plata, Argentina
[2] INIDEP, Mar del Plata - CONICET
[3] CIC.PBA
[4] Dep. Ingeniería Química, Facultad de Ingeniería, UNLP,
La Plata, Argentina

ABSTRACT

Beef burgers were traditionally associated with nutrients and nutritional profiles that are often considered negative including high levels of saturated fatty acids, cholesterol, sodium, and high fat and caloric content. However, meat is a major source for many bioactive compounds including iron, zinc, conjugated linoleic acid, and B vitamins. By selection of lean meat cuts, removal of adipose fat, including oils of vegetal or marine origin (with high content of polyunsaturated fatty acids), and addition of phytosterols, this type of products could become into healthier functional foods. To obtain a product with similar characteristics to the "classical" beef burgers emulsifiers or binding agents are needed, as well as antioxidants to control lipid oxidation. Low-fat beef burgers with high oleic sunflower, deodorized fish oil, and phytosterols were formulated including whey proteins or egg white and natural antioxidants (tocopherols and/or oregano-rosemary extract). Products were characterized and the effect of frozen storage at -20ªC on the quality of the cooked hamburgers was studied. Cooking yield, moisture and lipid retention, press juiciness, texture profile analysis, microstructure, oxidative stability, color, fatty acid profile, phytosterols contents, microbiological counts, and sensory acceptation were determined.

Cooking yields ranged between 79.4 and 84.5 % for all the formulations, and became slightly lower with frozen storage. Press juiciness also diminished with storage time while hardness increased, for both emulsifiers used. Lipid retention was higher than 95 %

---

* Corresponding author. Tel: +54-221-4254853. FAX: +54-221-4254853
E-mail address: scandres@biol.unlp.edu.ar

while water retention was higher than 70 %, for whey proteins or egg white and these parameters did not change during frozen storage.

Whey proteins protected better from oxidation than egg white, and tocopherols demonstrated an adequate antioxidant effect in formulations with egg white. For all the formulations unsaturated/saturated fatty acids ratio was higher than 5.8, showing a good lipid balance in the products.

Global acceptability for all the formulations presented sensory scores higher than 7 (in a 1 to 9 scale). More than 82.3 % of the panelists liked the taste of the products, 78.7 % liked the texture, and over 86.8 % of the panelists liked the products, considering the overall acceptability of the burgers. These results showed that the presence of the high oleic sunflower and deodorized fish oil did not adversely affect the low-fat beef burgers.

The consumption of 100 g of the cooked product would provide 6% of the recommended daily intake of phytosterols to decrease cholesterol and heart disease risk.

INTRODUCTION

Recommendations of a number of health organizations, including the World Health Organization (WHO), have promoted the integration of health policy with agriculture and food production with a view to improving eating habits and correcting imbalances; in this area, meat industry has an important role in making the necessary modifications of the composition of many processed meats to adapt to this situation (Jiménez-Colmenero, 2000).

Not only the total amount of fat consumed but also the type of fat in the diet has health implications. Beef is one of the widely consumed protein sources in the world. Modern consumers are increasingly concerned about production of safe meat with no undesirable effects on their health. For meat products the energy (fat) level, the sodium level, and fat quality in terms of fatty acids composition are the main priorities.

Beef fat is a significant source of saturated fatty acids in the human diet because red meat has a relatively high proportion of saturated fatty acids in its lipids (Muchenje, Dzama, Chimonyo, Strydom, Hugo, & Raats, 2009). Thus palmitic acid, the predominant saturated fatty acid present in red meat, has a cholesterol-elevating effect (Wolmarans, 2009).

Numerous researchers are endeavoring to optimize the amounts of lipids and the fatty acid profile of various meat products in order to achieve a more convenient composition related to nutrient intake goals. There exist different strategies for the development of healthier lipid meat products, in which reformulation of meat derivatives is one of them. Generally it is based on the replacement (to a greater or lesser extent) of the animal fat normally present in the product with another fat whose characteristics are more in line with health recommendations (Jiménez-Colmenero, 2007). The manufacture of low-fat products generally follows two basic approaches: the use of leaner raw materials (which raises the cost) and/or the reduction of fat and caloric contents by adding water and other ingredients that contribute few or no calories.

As ratios between polyunsaturated (PUFA) to saturated (SFA) fatty acids (PUFA/SFA) and n-6/n-3 PUFA of some meats are naturally somewhat removed from the recommended values (WHO, 2004), changes in the amounts and the lipid profiles of such products could help to improve the nutritional quality of the Western diet. Meat, particularly red meat, is already an important dietary source of long chain n-3 PUFA, in which docosapentaenoic acid

(DPA) predominates (Jiménez-Colmenero, 2007); however, further enrichment of meat with unsaturated fatty acids may be a practical mean of increasing their intake in the population.

Vegetable and marine oils have been used to substitute saturated fatty acids with unsaturated fatty acids, and as these new lipids materials have different physicochemical characteristics, the processing conditions have to be adjusted to induce the desired quality attributes in the reformulated product. Pre-emulsion is generally used to incorporate fats, with an emulsifier, typically a protein of non-meat origin, prior to the manufacture. Also, if the oil added is highly susceptible to lipid oxidation, an antioxidant must be incorporated. Other possibility to improve the different categories of processed meats, especially ground products, is to include different additives with a "healthy perception", such as phytosterols or special oils.

A variety of non-meat fats of plant (olive, high oleic sunflower, linseed, soybean) and marine origin have been added to different meat products as partial substitutes for meat fats (mainly from pork and beef) (Pelser, Linssen, Legger, & Houben, 2007; Lee, Faustman, Djordjevic, Faraji, & Decker, 2006a; Park, Rhee, Keeton, & Rhee, 1989; Park, Rhee, & Ziprin, 1990; Jimenez-Colmenero, 2007; Paneras & Bloukas, 1994; Pappa, Bloukas, & Arvanitoyannis, 2000; Yilmaz, Şimşek, & Işikh, 2002; Andrés, Zaritzky, & Califano, 2009) since there are rich sources of MUFA and PUFA and are cholesterol-free.

These approaches can be supplemented by the use of a number of technological procedures that help offset undesirable side effects produced as a result of changes to the composition and nature of the product (Muchenje et al., 2009). Reduction of fat in finely ground meat products presents a number of difficulties in terms of appearance, flavor, and texture, with less acceptation by the consumers. The use of non-meat ingredients could help to convey desirable texture and, more important, enhance water-holding capacity (Piñero, Parra, Huerta-Leidenz, Arenas de Moreno, Ferrer, Araujo, & Barboza, 2008).

Protein-based fat mimetics, which are considered to be non–meat ingredients, include egg proteins, blood plasma, vegetable protein, non-fat dry milk, casein and whey proteins. Traditional use of such protein ingredients has been as fillers, binders and extenders to improve the flavor, texture, appearance, and nutritional value of comminute products (sausages, patties) as well as providing a greater flexibility of formulations to minimize processing losses (El-Magoli, Laroia, & Hansen, 1996; Serdaroğlu, 2006). Non-meat proteins may become dispersed in the salt-soluble muscle protein gel-matrix formed during processing to bind water, or they may gel, and thus, interact with the muscle proteins. The results may be several types of multi-component gels that can impart different textural properties to meat, depending on the gel formed (El-Magoli et al., 1996).

Among milk proteins, whey protein concentrate (WPC) has been exhibit functional properties proven to be useful in fat replacement, as their gelation characteristics, high water and fat binding abilities, and their ultimate effect on emulsion stability. Results of El-Magoli et al. (1996) showed that WPC at 4% can be effectively used as a functional ingredient in low fat beef patties, and also, that sodium tripolyphosphate (TPP) enhanced the effectiveness of WPC in the low fat meat system. They found that WPC may masks some oxidized volatiles in the meat systems.

Also, there have been in demand by many processors hydrocolloids as binders (Lu & Chen, 1999). Starch is an important biological macromolecule in foods. Various starches have been added in protein gels for reducing cost and improving texture. Starch has been recognized as filler to increase the firmness of products and to enhance the gel strength (Li,

Yeh, & Fan, 2007). The composite reinforcing effect of starch may be due to the swelling of starch granules embedded in the protein gel, which compresses the matrix, and that the protein matrix losses moisture and becomes firmer (Kim & Lee, 1987). Starch added in meat emulsions favors the formation of a more compact and stronger heat induced protein network (Carballo, Fernández, Barreto, Solas, & Jiménez-Colmenero, 1996). When starch is added in meat emulsion, starch results in a decrease in cooking loss and an increase in both storage modulus (G') and loss modulus (G'') (Li & Yeh, 2003). The gelling characteristics of the starch–protein mixture depend upon the thermal behavior of starch and protein.

Khalil (2000) had demonstrated that some of the physical and sensory characteristic problems associated with low-fat beef patties could be eliminated by replacing fat with a starch/water combination, which proved to be more effective than replacing fat with water alone. Incorporation of starch with water resulted in patties that were higher in cooking yield than the control and provided improvements in texture characteristics.

The interaction of starch and protein is important in determining macroscopic properties of food products such as flow, stability, texture and mouth feel. Literature results suggest that starch granule surface components dictate the manner in which starch interacts with its surrounding proteins (Li et al., 2007). In starch/meat system, meat protein denatured at temperature lower than starch gelatinization and formed a continuous matrix. The gelatinized starch was either coating on the peripheral surface of protein network or trapped in the protein matrix (Li & Yeh, 2002).

Texture is the most important factor in deciding overall acceptance of patty products. Sarıçoban, Yılmaz, and Karakaya (2009) had demonstrated that finding optimum levels of textural parameters based on maximum sensory score should be final goal for investigations studying the effect of processing variables on the texture profile analysis (TPA) parameters and sensory properties.

Velioğlu, Velioğlu, Boyacı, Yılmaz, and Kurultay (2010) evaluated the fat, water and textured soy protein (TSP) proportions influence over the shrinkage, fat loss and moisture loss of regular hamburger patties. The results showed that the most significant factor for fat and moisture losses was TSP content. However fat content was much more effective than TSP in determining the shrinkage characteristic of hamburger; although an excessive amount of fat could cause a slight decrease in fat loss.

This chapter deals with healthier beef burgers enriched in unsaturated fatty acids and phytosterols, focusing on the effect of adding different emulsifiers and natural antioxidants to the formulations. Physicochemical characteristics of the products such as cooking yield, lipid and water retention, juiciness, texture, microstructure, microbiological and sensory quality, oxidative stability, color, and fatty acid profile were determined on products cooked after frozen storage.

BURGER INGREDIENTS

Throughout the experiments described in this chapter low-fat burgers were prepared using fresh lean beef meat (*adductor femoris* and *semimembranosus* muscles) obtained from local processors (pH: 5.48±0.01) (Pennisi Forell, Ranalli, Zaritzky, Andrés, & Califano, 2010) grounded with a commercial food processor (Universo, Rowenta, Germany) equipped with a 14 cm blade. Sodium chloride (NaCl) and tripolyphosphate (TPP) were added to the ground meat. Water was incorporated to replace part of the beef fat as oil in water emulsion.

Pre-Emulsified Oil Addition

Oil in water (o/w) emulsions may be easier to disperse into water-based foods (such as muscle foods), than bulk oil that could physically separate from the aqueous phase during storage. Oil pre-emulsion technology with a non-meat protein improves the ability of the system, since the oils can be stabilized or immobilized in a protein matrix. This reduces the chances of bulk oil to physically separate from the structure of the meat product so that it remains stable throughout processing, storage and consumption.

Whey proteins can be used to produce a low-viscosity O/W emulsion at oil concentration ranging from 5 % to 30 %, with excellent physical stability that is not adversely affected by thermal processing (Djordjevic, McClements, & Decker, 2004).

Siegel, Church and Schmidt (1979) demonstrated that egg white is a good binder for meat pieces. Due to the crude protein concentration and the stronger gel strength, dried egg white may be more useful in restructuring muscle products than was once thought for raw egg white (Lu & Chen, 1999).

As emulsifiers, whey protein concentrate (W, 80%, Arla Foods Ingredients S.A., Martínez, Argentina) or dry egg white (E, 80%, Tecnovo SA, Entre Ríos, Argentina) dissolved in cold distilled water were used. As fat sources, high oleic acid sunflower oil (85% n-9, Ecoop, Cooperativa Obrera, Bahía Blanca, Argentina) and deodorized refined fish oil (26% n-3, Omega Sur S.A., Mar del Plata, Argentina) were used.

Mixed phytosterols (Advasterol 90% with 16-24% campesterol, 19-32% stigmasterol and 32-50% β-sitosterol, AOM SA, Buenos Aires, Argentina), mixed tocopherols (TO, Mixed tocopherols 70% with d-γ-/d-β-tocopherol 43.81%, d-δ-tocopherol 19.31% and d-α-tocopherol 7.40%, AOM SA, Buenos Aires, Argentina); besides a mixture of rosemary oleoresin (5%) and distilled fraction of oregano (5%) (OR, Río Arnedo SA, Buenos Aires, Argentina) were incorporated.

The emulsion obtained with a hand-held food processor (Braun, Buenos Aires, Argentina), during 1.5 min was added to the meat. At last, corn starch was added to enhance the water holding capacity of the systems. The batter obtained was stored 1 h at 4 °C.

BURGER MANUFACTURE AND STORAGE

The ingredients of the formulations are listed in Table 1 where letter W indicates those formulations containing 3% whey protein concentrate while E corresponds to 3% added dry egg white. Numbers 1 to 4 designate the type of antioxidant considered: 1 = none, control, 2 = 0.1% TO + 0.05% OR, 3 = 0.1% TO, and 4 = 0.05% OR.

Table 1. Formulation (%) of the different products*

COMPONENTS	W1	W2	W3	W4	E1	E2	E3	E4
Beef	74.65	74.50	74.55	74.60	74.65	74.50	74.55	74.60
Distilled water	10	10	10	10	10	10	10	10
High oleic sunflower oil	9.9	9.9	9.9	9.9	9.9	9.9	9.9	9.9
Fish oil	0.1	0.1	0.1	0.1	0.1	0.1	0.1	0.1
Whey protein concentrate (W)	3	3	3	3	-	-	-	-
Dry egg white (E)	-	-	-	-	3	3	3	3
Tocopherols (TO)	-	0.1	0.1	-	-	0.1	0.1	-
Oregano-rosemary (OR)	-	0.05	-	0.05	-	0.05	-	0.05

* All the samples also contained 1% NaCl, 0.2% TPP, 0.15% mixed phytosterols, and 1% corn starch.

Burgers (40±1g) were formed using a hamburger mould (5 cm diameter and 1.2 cm in high), wrapped separately in polyethylene cling film and sealed in lots of eight in Zip-Lock pouches (C. S. Johnson & Sons, S.A.I.C., Buenos Aires Argentina). They were frozen and stored at –20 °C up to 6 months. At different storage times (0, 2, 4, and 6 months) burgers were removed from the freezer and immediately cooked (without previously defrosting) in a commercial, double-sided electric household grill (3882, Oster, China) preheated during 30 min to reach the maximum temperature (210 °C). The burgers were cooked until a final internal temperature of 73 °C for 15 s was reached, according to the recommendations of the FDA-CFSAN (2003). Then, samples were cooled immediately at room temperature over absorbent paper and further processed as needed according to the different methodologies. The procedure was replicated twice.

PROXIMATE ANALYSIS AND CALORIC CONTENT

Proximate analysis of all the cooked formulations showed no significant differences (P > 0.05) between them. Mean values of the main components were: protein 23.61±0.48%, lipids 13.84±0.15%, ash 2.21±0.09%, and moisture 58.80±0.25%. Due to the formulation, more than 87% of the protein content of the products was from meat, and the rest from milk or egg, all of them, proteins with high nutritional value. The low salt addition to the products (1% NaCl) produced low ash levels.

Caloric contents (kcal) were calculated using the Atwater values corresponding to lipids (9 kcal/g), proteins (4.02 kcal/g) and carbohydrates (3.87 kcal/g) (Cáceres, García, & Selgas, 2006). Caloric content of the burgers was 225.4 kcal/100 g, of which total lipids represented 55.2% (124.6 kcal/100g), and 105.9 kcal/100g come from oils.

Process Yields

Cooking yield determined as the percent of mass kept in weight during the cooking treatment at different storage times (El-Magoli et al., 1996; Candogan & Kolsarici, 2003; Piñero et al., 2008) showed that all formulated burgers had similar yields (P > 0.05), being 83 % at initial of storage, and diminished to 81.14 % after six months of frozen storage. Moisture and lipid retention (amount of moisture or lipids retained in the cooked products per 100 g of raw sample after cooking, El-Magoli et al., 1996) were similar (P > 0.05) and high for all the burgers formulations, with mean values of 71.078 % (SEM = 1.058) and 97.44 % (SEM = 0.78), respectively. These results showed a good retention of oils incorporated to the products by the matrix, mainly of meat proteins, after a frozen storage and grill cook.

Color

Color was measured at room temperature on the internal surface of recently cut cooked burgers (five replicates), using a Chroma Meter CR-400 colorimeter (Minolta Co., Osaka, Japan) and CIE-LAB parameters (L*, a*, and b*) were determined. Within the approximate uniform color space CIELAB, two color coordinates, redness, a*, and yellowness, b*, and lightness, L*, are defined. Coordinate a* takes positive values for reddish colors and negative values for the greenish colors, whereas b* takes positive values for yellowish colors and negative values for the bluish colors. L* is an approximate measurement of luminosity, which is the property according to which each color can be considered as equivalent to a member of the grey scale, between black and white, taking values within the range 0–100.

During cooking of meats, a hemichrome pigment (denatured globin and oxidized heme iron) that is tan in color is formed (Foegeding, Lanier, & Hultin, 1996). Healthier beef burgers samples with no storage presented higher L* values than those measured in commercial "light" hamburgers (52.4 ± 0.26) with less than one months of frozen storage, which could be attributed to a milky appearance that was imparted by the oil emulsion (Table 2). Similar results were obtained by Lee et al. (2006a) and Lee, Hernández, Djordjevic, Faraji, Hollender, Faustman, & Decker (2006b), who developed meat products (patties, sausages, restructured hams), fortified with *n-3* fatty acids incorporated as a pre-emulsion of algal oil stabilized with whey proteins isolate.

For L* parameter, all the factors considered (emulsifier, antioxidant and storage time) and their interactions were significant. Formulations containing egg albumin were lighter than products that contained whey proteins (58.9±0.66 and 56.8±0.77, respectively, P < 0.05). Regardless the storage time, the control group containing whey proteins presented the lowest lightness (53.7±0.75), and the L* highest value (60.7±1.65) corresponded to the tocopherols + OR formulation with egg white (P < 0.05). There were no significant differences between the formulations containing those antioxidants (tocopherols or OR) added individually (P > 0.05). As the third level interaction was highly significant (P < 0.01) and considering that this term explained the highest percentage of total variance (23%), the changes produced by storage time depended on the emulsifier and antioxidant combination (Table 2). Comparing for each formulation, the initial and final mean L* values (at 0 and 6 months storage), non significant differences were found, even though several authors working on meat lamb found that L* increased during refrigerated storage (Linares, Bórnez, & Vergara, 2006, 2008; Linares, Berruga, Bórnez, & Vergara, 2007; Santé´-Lhoutellier, Engel, & Gatellier, 2008; Soldatou, Nerantzaki, Kontominas, & Savvaidis, 2009); in contrast, Rojas and Brewer (2008) reported an increase in L* of vacuum-packaged pork hamburgers after 2 months of frozen storage, remaining constant thereafter.

Table 2. Changes in lightness (L*) and redness (a*) of low-fat cooked beef burgers formulated with whey protein concentrate (W) or dry egg white (E), as a function of the antioxidant added and frozen storage time

Color parameter	Month 0	Month 2	Month 4	Month 6
Lightness (L*)				
W1	53.2^{aD}	56.3^{aAB}	52.8^{aD}	52.6^{aD}
W2	59.6^{aABC}	56.7^{abAB}	52.7^{bD}	60.3^{aAB}
W3	57.2^{abCD}	56.7^{bAB}	61.3^{aA}	60.3^{abAB}
W4	56.6^{abCD}	53.5^{bB}	59.9^{aAB}	59.6^{aB}
E1	61.7^{abAB}	58.3^{bA}	53.2^{cD}	62.9^{aAB}
E2	63.8^{aA}	58.8^{bA}	56.3^{bBCD}	64.0^{aA}
E3	56.1^{aCD}	58.0^{aA}	58.1^{aAC}	56.4^{aCD}
E4	58.0^{aB}	59.2^{aA}	58.1^{aAC}	59.0^{aBC}
Redness (a*)				
W1	7.6^{cABC}	10.1^{bA}	15.4^{aA}	13.4^{aA}
W2	5.7^{cCD}	9.6^{bAB}	12.5^{aB}	7.2^{cB}
W3	9.2^{aA}	7.3^{aC}	7.8^{aC}	7.5^{aB}
W4	8.9^{aAB}	7.0^{aBC}	7.7^{aC}	7.6^{aB}
E1	5.1^{cD}	8.2^{bA}	12.7^{aB}	6.9^{bcB}
E2	6.2^{cCD}	9.5^{bA}	14.2^{aAB}	8.5^{bcB}
E3	6.6^{aBCD}	7.6^{aBC}	7.6^{aC}	7.9^{aB}
E4	7.1^{aABD}	8.3^{aAC}	7.3^{aC}	8.4^{aB}

[a-b] Means with the same superscript within same row do not differ significantly (P>0.05) according to Tukey's test

[A-B] Means with the same superscript within same column do not differ significantly (P>0.05) according to Tukey's test.

Products containing whey proteins as emulsifier presented higher redness (9.03±0.74) than egg white formulations (8.30±0.57, P < 0.05). At the beginning of the storage period individual a* values of the cooked hamburgers ranged between 4.31 and 12.39; formulations with individual antioxidants (TO or OR, W3, W4, E3, and E4) showed no significant differences during storage at -20 °C. Control and TO + OR samples (W1, W2, E1, and E2) showed an increase in redness up to the fourth months of frozen storage, decreasing thereafter (Table 2). The decrease in a* values has frequently been associated with the formation of metmyoglobin and thus with meat discoloration (Jeremiah, 2001).

There were no significant differences of yellowness (b*values, P > 0.05) attributable to the different factors analyzed. Yellowness ranged between 11 and 17 during frozen storage.

Juiciness and Texture

Juiciness of the samples was determined as the liquid extracted (%) by compression of cooked burgers at room temperature at different times of storage. Duplicate pre-waited cooked burgers (cooked weight) were placed between a pair of two pieces of filter paper (Whatman Nro. 5, 10 cm diameter, Whatman International Ltd., Maidstone, U.K.) and a pair of aluminum foils sheets (10 cm diameter) (all this set also pre-waited, w_{before}). Then, the whole set was placed in a TAXT2i Texture Analyzer (Stable Micro Systems, UK) and a 100-N force was applied to the set for 2 min (Zorrilla, Rovedo & Singh, 2000). The filter and aluminum sheets were waited after pressing (w_{after}), and the mass of the extracted juice was determinate as follows:

$$\text{Press juice } (\%) = (w_{after} - w_{before}) \text{ cooked weight } \times 100$$

The different formulations presented similar average juiciness values ($P > 0.05$), being 4.431 % at the beginning of the storage, decreasing to 2.848 % after six months (Figure 1).

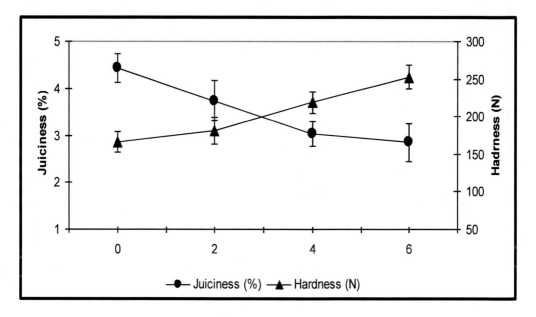

Figure 1. Effect of frozen storage on average values of juiciness (%) and hardness (N) of cooked burgers formulated with oils.

Texture Profile Analysis (TPA) (Bourne, 1978; Brennan & Bourne, 1994) was performed on cooked burgers every two months during storage. Samples were compressed twice to 30 % of their original height between flat plates using a TAXT2i Texture Analyzer (Stable Micro Systems, UK). In these experiments the head was operated at 0.5 mm/sec. Hardness (peak force of first compression cycle, N), springiness (distance of the detected height of the product on the second compression divided by the original compression distance, mm/mm), cohesiveness (ratio of positive areas of second cycle to area of first cycle, J/J), adhesiveness (negative force area of the first byte represented the work necessary to pull the compressing plunger away from the sample, J), chewiness (hardness x cohesiveness x springiness, N), and resilience (area during the withdrawal of the first compression divided by the area of the first compression, J/J) were determined. Figure 2 shows a typical profile obtained from a cooked sample.

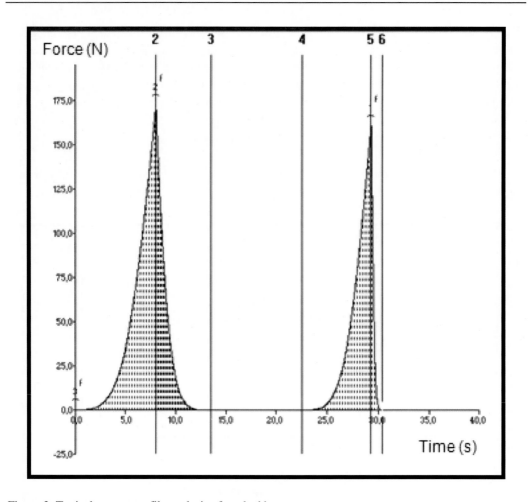

Figure 2. Typical texture profile analysis of cooked burger.

TPA for low-fat beef burgers formulated with high oleic sunflower and fish oils was similar to those obtained for the commercial hamburgers with 10% beef fat (Figure 3). The profiles did not present the peak associated to fracturability and showed very lower adhesiveness (average 0.050 J, SEM = 0.025).

During freezing meat tissue undergoes a series of changes that affect their organoleptic properties. Changes in hardness, loss of juiciness, increase in gumminess and loss of water holding capacity (WHC) have been attributed to protein denaturation, mainly myofibrilar proteins. (Lanari Vila, 1988). Denatured proteins have less WHC thus the products are less juicy. Besides aggregated myofibrilar proteins enhance the hardness of the burgers. Stanley (1983) observed coagulation and firmness increase of the myofibrilar proteins during meat, rising hardness and reducing WHC with fibers rupture and water and lipid losses, decreasing juiciness of the cooked products.

Others TPA parameters, such as chewiness and resilience also increased with storage time, but did not differ between formulations (Figure 4). Cohesiveness of all samples was constant during the frozen storage, with an average value of 0.553 J/J (SEM = 0.004).

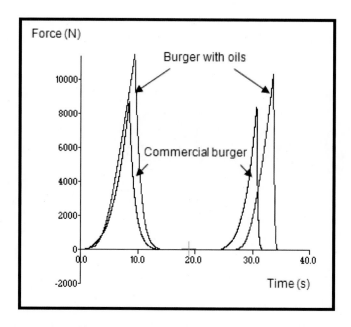

Figure 3. Texture profile analysis of cooked low-fat beef burgers formulated with high oleic sunflower and fish oils and commercial hamburgers with 10% beef fat.

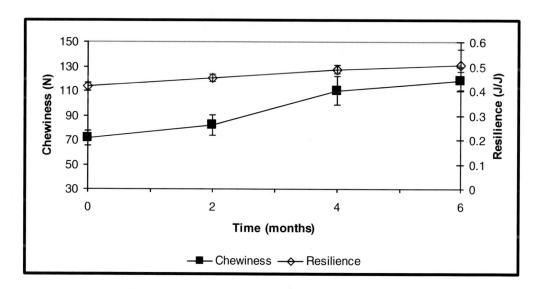

Figure 4. Effect of frozen storage on average values of chewiness (N) and resilience (J/J) of cooked burgers formulated with oils.

Microstructure

Raw burgers formulated with egg white or whey proteins without storage were observed using an environmental scanning electron microscope (ESEM). The micrographs showed a cohesive and smooth matrix, with coated oil drops of 40 μm, with a homogeneous distribution, results of a good emulsion of the oil added by the proteins. Also, well distributed starch granules of about 20 μm in size were observed. In frozen samples, it was observed an increase in droplets size, probable by coalescence process (Figure 5).

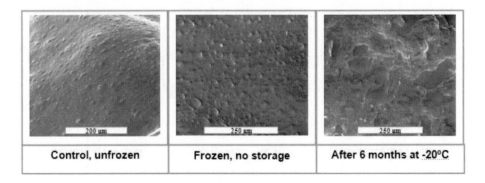

| Control, unfrozen | Frozen, no storage | After 6 months at -20°C |

Figure 5. Micrographs of raw burgers (W4 formulation) obtained by ESEM. Effect of frozen storage.

In cooked burgers, due to protein denaturation and partial dehydration by heat, more granular homogeneous matrixes were observed with fine strands and sheets with gel-like appearance (Figure 6).

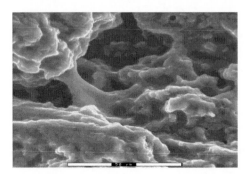

Figure 6. Enviromental scanning electron micrography of cooked burger (W2) showing gel-like structures.

As frozen storage progressed the porosity of cooked matrixes increased which is related to the higher cooking losses and less juiciness (Figure 7).

| Control, unfrozen | Frozen, no storage | 6 meses, congelado |

Figure 7. Micrographs of cooked burgers (W4 formulation) obtained by ESEM. Effect of frozen storage.

LIPID OXIDATION

Incorporation of unsaturated fatty acids into food systems is potentially problematic due to their propensity to readily oxidize. Furthermore, muscle foods are quite susceptible to oxidation, and meat processing operations facilitate particle size reduction and exposure of increased surface area, addition of potential pro-oxidant ingredients, and heat-induced changes that decrease oxidative stability (Lee, Decker, Faustman, & Mancini, 2005). Preventive lipid oxidation during processing and storage of meat is essential in order to maintain the quality and safety.

In the last years, the interest in using natural antioxidants to prevent meat lipid oxidation has increased in order to avoid the possible harmful effects of adding synthetic substances. Many researchers have evaluated the antioxidant properties of extracts from different herbs and spices. Rosemary (*Rosmarinus officianalis*) is a popular *Labitae* herb with a verified potent antioxidant activity (Estévez & Cava, 2006, Estévez, Ramírez, Ventanas, & Cava, 2007) which has been used traditionally to improve the sensory characteristics and extend the shelf-life of foods. Essential oils are regarded as ''natural'' alternatives to chemical preservatives and their use in foods meets the demand of consumers for minimally processed products. Oregano is a characteristic spice of the Mediterranean cuisine, obtained by drying leaves and flowers of *Origanum vulgare subsp. hirtum* plants, well known for its antioxidative and antimicrobial activity (Chouliara, Karatapanis, Savvaidis, & Kontominas, 2007; Hernández-Hernández, Ponce-Alquicira, Jaramillo-Flores, & Guerrero Legarreta, 2009). However, the practical application of several essential oils in foods is limited due to the strong flavour they impart to foods and also to their interaction with some food ingredients (Chouliara et al., 2007).

Another alternative is the use of mixed tocopherol isomers, by-products of the vegetable oil industry, which may be more effective in preventing lipid oxidation in muscle foods than α-tocopherol. The reason for this is that some of the tocopherol isomers have a superior antioxidant activity than α-tocopherol. Tocopherols are usually used at levels up to 500 mg/kg in food product because above this concentration they may act as pro-oxidants (Channon & Trout, 2002).

The oxidation processes are slowed down during frozen storage but not completely hindered. In fact, some lipid soluble radicals may even be more stable at the lower temperatures and thereby propagate oxidation (Kanner, 1994)

Georgantelis, Blekas, Katikou, Ambrosiadis, and Fletouris (2007) reported that rancid flavor is initially detected in meat products with thiobarbituric acid reactive substances (TBARS) values higher than 0.6 mg MDA/kg. This fact was also emphasized by Campo, Nute, Hughers, Enser, Wood, and Richardson (2006) who reported that a TBARS value of around 2 mg MDA/kg could be considered the limiting threshold for the acceptability of oxidized beef.

Table 3 shows the changes in TBARS levels found in cooked burgers containing different antioxidants during storage at -20 °C.

All the considered factors (type of emulsifier, antioxidant, and storage time) and their interactions were highly significant ($P < 0.01$) on lipid oxidation of the products. When the type of emulsifier was compared, whey proteins protected better from the oxidation than egg white, and this fact was in accordance with the antioxidant properties of whey proteins described by other authors (Hu, McClements, & Decker, 2003). All cooked burgers containing whey proteins as emulsifier (even the control W1) were protected against lipid oxidation, maintaining TBARS levels lower than 0.6 mg MDA/kg after 6 months of frozen storage and average values were not significantly different (P > 0.05).

Table 3. Lipid oxidation of low-fat cooked beef burgers expressed as TBARS (mg MDA/kg of cooked product) for samples formulated with whey protein concentrate (W) or dry egg white (E), as a function of the antioxidant added and frozen storage time

	Month 0	Month 2	Month 4	Month 6
W1	0.31^{aA}	0.24^{aB}	0.34^{aB}	0.48^{aABC}
W2	0.35^{aA}	0.37^{aB}	0.52^{aB}	0.43^{aBC}
W3	0.24^{aA}	0.33^{aB}	0.42^{aB}	0.50^{aABC}
W4	0.46^{aA}	0.51^{aAB}	0.43^{aB}	0.47^{aBC}
E1	0.40^{cA}	0.46^{bcB}	1.12^{aA}	0.77^{bA}
E2	0.16^{aA}	0.27^{aB}	0.29^{aB}	0.26^{aC}
E3	0.21^{aA}	0.38^{aB}	0.46^{aB}	0.40^{aBC}
E4	0.22^{cA}	0.82^{abA}	0.98^{aA}	0.65^{bAB}

[a-b] Means with the same superscript within same row do not differ significantly (P>0.05) according to Tukey's test

[A-B] Means with the same superscript within same column do not differ significantly (P>0.05) according to Tukey's test.

Tocopherols basically were effective in the burgers with egg white but brought no additional effect to the whey protein burgers. TBARS content in formulations with dry egg white without tocopherols (control E1 and E4 with OR antioxidant) increased during storage and exceeded the level of lipid oxidation which produce a rancid odor and taste. On the other hand those samples containing tocopherols (E2 and E3) maintained low TBARS levels until the end of frozen storage (P > 0.05).

It can be concluded that when whey proteins were used as emulsifier it was not necessary to incorporate any antioxidant, while on the products containing egg white proteins it was necessary to add tocopherols to control lipid oxidation. This result can be explained through the combination of two different mechanisms of action: the tocopherols may act as free radical scavenging of both antioxidant components while the TPP present in the formulations may act as chelator. Lee et al. (2005) found antioxidant combinations (sodium citrate, sodium erythorbate and rosemary) with beneficial effects working in *n-3* fortified patties.

Fatty Acid Profile

The most often cited criteria for saturated fat recommendations considered that saturated fats increase LDL-cholesterol, but also increase HDL-cholesterol and decrease triglyceride levels (Mensink, Zock, Kester, & Katan, 2003). However limits on saturated fat intake should be considered in the specific context of the replacement nutrient, as replacement with carbohydrates may have little benefit (Smit, Mozaffarian, & Willet, 2009).

Regarding the effect of *n-6* and *n-3* fatty acid consumption, both types of fatty acids have shown to have anti-inflammatory properties that are protective of atherogenic changes in vascular endothelial cells (De Caterina, Liao, & Libby, 2000). Linoleic acid is an essential n-6 fatty acid that favorably affects the blood lipid profile, and is associated with a lower risk of coronary heart disease events and reduced risk of type II diabetes (Smit et al., 2009). Long chain polyunsaturated *n-3* fatty acids have been implicated as critical nutrients for human health, and fortification of foods with these fatty acids is an emerging area of commercial and academic interest (Lee et al., Mancini, 2005; Jiménez-Colmenero, 2007). There is considerable evidence from epidemiological clinical and biochemical studies that eicosapentaenoic acid (EPA) and docosahexaenoic acids (DHA) consumption have demonstrated physiological benefits on blood pressure, heart rate, triglycerides, and a reduced risk of fatal coronary heart disease and sudden cardiac death (Smit et al., 2009).

Total lipids of cooked burgers were extracted (in duplicate) using chloroform-methanol (2:1, v/v) according to the procedure of Folch, Lees and Sloane Stanley (1957), and then methylated with 10% boron-trifluoride methanol complex in methanolic solution for the analysis (Morrison & Smith, 1964); fatty acids methyl esters (FAME) were quantified by gas chromatography on a Hewlett Packard 6890 gas chromatograph (Hewlett Packard, USA) equipped with a flame ionization detector (FID) and a split/splitless injector (Chrompack, Middleburg, The Netherlands) following the protocol of Arici, Tasan, Gecgel, & Ozsoy (2002).

Atherogenicity Index and Thrombogenicity Index

Atherogenicity and thrombogenicity indices (López-López, Cofrades, Ruiz-Capillas, Jiménez-Colmenero, 2009; Subhadra, Lochmann, Rawles, & Chen, 2006; Ulbricht & Southgate, 1991) were calculated using the fatty acid composition of the samples to determine the potential health impact on human consumers. The following equations were used:

Atherogenicity Index:
AI = [C12:0 + (4×C14:0) + (C16:0)]/[(n-6 PUFA + n-3 PUFA)+MUFA]

Thrombogenicity Index:
TI = [C14:0+C16:0+C18:0]/[(0.5×MUFA) + (0.5×n-6 PUFA)+(3×n-3 PUFA) + (n-3 PUFA/n-6 PUFA)]

where PUFA=polyunsaturated fatty acids including 18:2n-6, 18:3n-3 and 20:4n-6; and MUFA=monounsaturated fatty acids including all isomers of 16:1 and 18:1. To perform the calculations concentrations of the fatty acids were expressed as g/100 g.

Fatty acid profiles of the different burgers initially and after six months of frozen storage are presented in Table 4.

No differences were found for elaidic, linoleic and araquidonic acids contents between formulations or storage times (P > 0.05).

Saturated fatty acid content (SFA) of the products was between 9.4 and 14.6% FA, palmitic acid being the major component of SFA, followed by stearic acid. Monounsaturated fatty acids contents (MUFA) were in the range 78.8-85.0% FA, comprising oleic acid more than 97% of MUFA. The difference between unsaturated and MUFA contents corresponded to linoleic and linolenic acids (4.4-6.7% FA). Unsaturated/SFA ratio for all the formulations was higher than 5.8, showing a good lipid balance in the products.

The PUFA content ((18:2 *n-6* + 18:3 *n-3* + 20:4 *n-6*) of the burgers was 5.78±0.42% and was not affected by the time of frozen storage or formulation (P > 0.05). Linoleic acid (18:2 *n-6*) accounts more than 95% of this PUFA content.

When SFA and MUFA contents of the hamburgers formulated with vegetable and marine oils are compared to those informed by USDA (2009) for a commercial product (pan-broiled cooked patty, 90% lean meat, 10% fat, NDB No: 23564) with similar beef fat content (49.8% SFA, 41.8% MUFA and 4% PUFA) it can be seen that SFA were reduced by at least 60% and MUFA content nearly doubles (Table 3).

There is compelling evidence to indicate that the type of fat is more important than the total amount of fat for cardiovascular disease risk. In this context, predictive equations, like atherogenicity index (AI) and thrombogenicity index (TI), have been used to estimate the effects of the foods fatty acid profile on plasma cholesterol and lipoprotein concentrations (Ayo, Carballo, Serrano, Olmedilla-Alonso, Ruiz-Capillas, & Jiménez-Colmenero, 2007).

The AI and TI take the interactions among different fatty acids into account, allowing an integrated assessment of dietary lipid on human coronary health (Ulbricht & Southgate, 1991). Higher values of TI and AI (>1.0) are detrimental to human health (Subhadra et al.; Bobe, Zimmerman, Hammond, Freeman, Lindberg, & Beitz, 2004).

AI values for all burgers formulated ranged between 0.09 and 0.16 (Table 3); similar results have been reported by Ayo et al. (2007) for low fat frankfurters formulated with walnut. López-López et al. (2009) studied frankfurters with olive oil and seaweed and founded higher AI values. Besides, TI values of the burgers were between 0.20 and 0.33 (Table 3), being in agreement with López-López et al (2009).

Phytosterols

Phytosterols/stanols are used as a novel food ingredient with plasma cholesterol lowering activity. There is substantial evidence from a vast number of clinical studies that plant sterols and stanols lower total and LDL-cholesterol, by partly inhibiting cholesterol absorption, and that their effect is additional to that achieved by other strategies (e.g. a low-fat diet and/or the use of cholesterol-lowering drugs like statins). A wide variety of phytosterol structure exists but the phytosterols found most frequently in nature are β-sitosterol, campesterol and stigmasterol. Many pre-market controlled clinical trials have demonstrated that an intake of 2 g/day of phytosterol/stanols reduces serum LDL cholesterol concentration by approximately 10% (Ostlund, 2007). In addition, the safety of these compounds has been studied extensively; the US Food and Drug Administration has accepted that phytosterols/stanols are Generally Recognized as Safe (GRAS) (de Jong, Ros, Ocké, & Verhagen, 2008) and the Scientific Committee on Foods of the Commission of European Communities came to the conclusion that the addition of phytosterols is safe, provided that the daily consumption does not exceed 3 g (Scientific Committee on Food, 2002).

Table 4. Effect of frozen storage time on fatty acid profile (% of total fatty acids) of the cooked formulations (n=2)

Storage (months)	W1		W2		W3		W4		E1		E2		E3		E4	
	0	6	0	6	0	6	0	6	0	6	0	6	0	6	0	6
Miristic 14:0	0.61[cd]	0.58[ab]	0.98[ab]	0.55[cd]	0.55[cd]	0.59[cd]	0.58[cd]	0.56[cd]	0.59[cd]	0.63[bc]	0.38[d]	0.30[d]	0.77[abc]	1.12[a]	0.6[bcd]	1.09[a]
Palmitic 16:0	7.16[bcd]	6.89[cd]	8.13[b]	7.21[bcd]	7.32[bcd]	6.63[cd]	7.84[bc]	5.88[d]	7.25[bcd]	8.11[b]	6.78[cd]	8.9[ab]	8.53[ab]	9.45[a]	6.37[cd]	9.23[a]
Palmitoleic 16:1 n-7	0.76[d]	0.70[d]	0.74[d]	0.70[d]	0.81[cd]	0.86[cd]	0.60[d]	0.66[d]	1.29[bcd]	1.51[bcd]	0.55[d]	0.83[cd]	1.57[abc]	1.77[ab]	1.25[bcd]	1.96[a]
Stearic 18:0	4.89[ab]	4.52[ab]	5.45[a]	5.21[ab]	4.39[b]	3.73[b]	5.24[ab]	2.91[b]	4.30[ab]	3.76[b]	4.25[ab]	3.24[b]	3.64[b]	3.80[b]	3.55[b]	3.71[b]
Oleic 18:1 n-9 c	79.76[bc]	82.09[cd]	77.43[d]	81.28[ab]	80.9[ab]	81.28[ab]	78.95[bcd]	84.19[a]	81.54[cd]	78.76[abc]	82.81[ab]	80.81[b]	78.52[bcd]	78.13[cd]	83.27[a]	78.19[cd]
Total 18:1 trans	N.D.	N.D.	0.64[a]	N.D.	0.01[a]	0.28[a]	0.07[a]	0.1[a]	N.D.	N.D.	0.14[a]	N.D.	0.24[a]	0.59[a]	0.48[a]	N.D.
Linoleic 18:2 n-6	5.61[a]	5.37[a]	5.52[a]	5.10[a]	5.95[a]	5.47[a]	6.69[a]	6.21[a]	5.31[a]	5.72[a]	4.72[a]	4.8[a]	5.9[a]	5.45[a]	4.42[a]	6.02[a]
Linolenic 18:3 n-3	0.33[b]	0.25[b]	1.17[a]	0.21[b]	0.15[b]	N.D.	0.09[b]	N.D.	0.19[b]	N.D.	N.D.	0.26[b]	N.D.	N.D.	N.D.	N.D.
Arachidonic 20:4 n-6	N.D.	N.D.	N.D.	N.D.	N.D.	0.63[a]	N.D.	0.21[c]	N.D.	N.D.	0.46[b]	N.D.	0.2[c]	N.D.	N.D.	N.D.
SFA	12.66	11.99	14.56	12.97	12.26	10.95	13.66	9.35	12.14	12.5	11.41	12.44	12.94	14.37	10.52	14.03
MUFA	80.52	82.79	78.81	81.98	81.72	82.42	79.62	84.95	82.83	80.27	83.5	81.64	80.33	80.49	85.00	80.15
Unsat	86.46	88.41	85.5	87.29	87.82	88.52	86.40	91.37	88.33	85.99	88.68	86.70	86.43	85.94	89.42	86.17
PUFA	5.94	5.62	6.69	5.31	6.1	6.1	6.78	6.42	5.5	5.72	5.18	5.06	6.1	5.45	4.42	6.02
Unsat/SFA	6.83	7.37	5.87	6.73	7.16	8.08	6.33	9.77	7.28	6.88	7.77	6.97	6.68	5.98	8.50	6.14
AI	0.11	0.10	0.14	0.11	0.11	0.10	0.12	0.09	0.11	0.12	0.09	0.12	0.13	0.16	0.10	0.16
TI	0.29	0.27	0.32	0.29	0.28	0.25	0.31	0.20	0.27	0.29	0.26	0.28	0.30	0.33	0.24	0.33

Different superscripts within the same file indicate that average values differ significantly ($P < 0.05$).

N.D.: not detected

SFA: saturated fatty acids (14:0 + 16:0 + 18:0); MUFA: monounsaturated fatty acids (16:1 n-7 + 18:1 n-9 c + 18:1 n-9 t), PUFA: polyunsaturated fatty acids (18:2 n-6 + 18:3 n-3 + 20:4 n-6)

AI: Atherogenicity Index

TI: Thrombogenicity Index

Phytosterols content was determined over the lipid phase extracted of the cooked burgers by Folch´s procedure (Folch et al., 1957) by gas chromatography in an Agilent 6890N capillary gas chromatograph (Agilent Technologies, CA, USA) equipped with an automatic split/splitless injection (1 µl), flame-ionization detector and HP-5 MS column (30 m x 0.25 mm I.D., 0.25 µm film thickness; operating conditions were based on American Oil Chemists' Society (AOCS) recommended practice Ce 7-87 (AOCS, 1998).

The concentrations of individual and total phytosterols were not affected either by formulation or storage time at -20 ºC (P > 0.05). Total phytosterols content of cooked burgers was 0.314±0.020% db, while campesterol, stigmasterol and β-sitosterol levels were 0.071±0.005, 0.103±0.006 and 0.138±0.011% db, respectively. The obtained results were in accordance to the proportions of each type of phytosterol in the mix (22.8% campesterol, 25.3% stigmasterol and 40.6% β-sitosterol).

Total phytosterol content of 100 g of cooked burgers represented 6% of the daily recommended dose necessary to decrease cholesterol content and heart disease risk (de Jong et al., 2008).

MICROBIOLOGICAL ANALYSIS

Total mesophilic and psychrotrophic aerobic count (Plate Count Agar, PCA, Oxoid; 30°C for 2 days or 4°C for 7 days, respectively), Enterobacteriaceae (Violet Red Bile Agar, Merck KGaA, Darmstadt, Germany; 37°C for 24 h) and lactic acid bacteria (Man Rogosa Sharp Agar, MRS, Oxoid; 30°C for 2 days) were performed by pour plate method every two months during frozen storage of raw burgers. Data were expressed as log colony forming units (CFU)/g sample.

Initial total mesophilic aerobic counts in all raw formulations were between 4 - 5 log CFU/g showing a good microbiological quality of them, due to the quality of ingredients used and the adequate methodology and sanitization conditions of elaboration. These levels were kept during the frozen storage at -20°C in all burger formulations.

Total psychrotrophic aerobic and lactic acid bacteria showed similar counts, without variation between formulations or storage time.

Microbial counts of cooked burgers were lower than detection limit of this methodology, showing that thermal treatment reached the microbiological safety of products.

SENSORY EVALUATIONS

Sensory analyses were conducted by 25 panelists from graduate students and faculty members in our Institute who were experienced in sensory evaluation of foods, but received no specific training relevant to these products.

For sensory evaluations only formulations that did not show significant lipid oxidation were selected (W2, W3, W4, E2, and E3). Samples that have been stored for 2 months were cooked as described above and holding on a warming tray in covered plates for no longer than 10 minutes. Warm pieces were distributed in white polystyrene plates and presented to the panelist with three-digit codes and in random order for evaluation. Tap water was supplied to the panelist for rinsing between samples. Experiments were conducted in an appropriately designed and lighted room. Panelists were asked to indicate how much they liked or disliked the products on a 9-point hedonic scale (9= like extremely; 5= indifferent; 1= dislike extremely) according to flavor, texture, color and overall acceptability characteristics.

Global acceptability for all the formulations presented sensory scores higher than 7 (in a 1 to 9 scale) with mean value of 7.21 (SEM = 0.1276). Taste and texture mean scores were 7.16 (SEM = 0.1528) and 7.08 (SEM = 0.157), respectively. More than 82.3 % of the panelists liked the taste of the products, 78.7 % liked the texture, and over 86.8 % of the panelists liked the products, considering the overall acceptability of the burgers. These results showed that the presence of the 9 % high oleic sunflower and 1 % deodorized fish oil did not adversely affect the low-fat beef burgers.

CONCLUSION

Healthier low-fat beef burgers containing 9 % of high oleic sunflower oil and 1 % of refined fish oil, thus, enriched with unsaturated fatty acids, and phytosterols were produced using only natural antioxidants to delay rancidity. Whey proteins protected the burgers against lipid oxidation, maintaining TBARS levels of all the formulations lower than 0.6 mg MDA/kg after 6 months of frozen storage. Formulations containing egg albumin were lighter in color than whey protein products, and were more susceptible to lipid oxidation. On products containing egg white proteins tocopherols demonstrated an adequate antioxidant effect. All products showed very good stability and quality attributes. The incorporation of egg white or whey proteins contributed to obtain high process yields which slightly diminished during storage. All the formulations showed similar juiciness that decreased during frozen storage; opposite result was observed for hardness, chewiness and resilience parameters that increased.

The fatty acids profile of these low-fat meat burgers showed a high monounsaturated FA content with oleic acid comprising more than 97% of them. Total phytosterol content of 100 g of cooked burgers represented 6% of the daily recommended dose, necessary to decrease cholesterol and heart disease risk and was not affected by any of the factors analyzed. This study showed that low-fat beef burgers formulated with pre-emulsified vegetable and fish oils (10%), tocopherols, and phytosterols could be considered potentially functional meat products.

The good sensory results obtained for flavor, texture and overall acceptability of oils formulated burgers showed that the presence of this unsaturated fatty acids rich oil did not adversely affect the product leading to an innovative and healthier product.

REFERENCES

American Oil Chemists' Society -AOCS (1998). Recommended Practice Ce 7-87. In: D. Firestone (Ed.), Official Methods and Recommended Practices of the American Oil Chemists' Society, fifth ed., American Oil Chemists' Society, Champaign.

Andrés, S.C., Zaritzky, N.E., & Califano, A.N. (2009). Innovations in the development of healthier chicken sausages formulated with different lipid sources. *Poultry Science*, *88*, 1755-1764.

Arici, M., Tasan, M., Gecgel, U., & Ozsoy, S. (2002). Determination of FA Composition and Total Trans FA of Turkish Margarines by Capillary GLC. *Journal of the American Oil Chemists' Society*, *79*, (5), 439- 441.

Ayo, J., Carballo, J., Serrano, J., Olmedilla-Alonso, B., Ruiz-Capillas, C., & Jiménez-Colmenero, F. (2007). Effect of total replacement of pork back fat with walnut on the nutritional profile of frankfurters. *Meat Science, 77,* 173-181.

Bobe, G., Zimmerman, S., Hammond, E.G., Freeman, G., Lindberg, G.L., & Beitz, D.C. (2004). Texture of butters made from milks differing in indices of atherogenicity. Iowa State University Animal Industry Report 2004. A.S. Leaflet R1902.

Bourne, M.C. (1978). Texture profile analysis. *Food Technology*, 32(7), 62-66, 72.

Brennan, J.G., Bourne, M.C. (1994). Effect of lubrication on the compression behaviour of cheese and frankfurters. *Journal of Texture Studies*, 25, 139-150.

Cáceres, E., García, M. L., & Selgas, M. D. (2006). Design of a new cooked meat sausage enriched with calcium. *Meat Science, 7,* 368-377.

Campo, M. M., Nute, G. R., Hughers, S. L., Enser, M., Wood, J. D., & Richardson, R. L. (2006). Flavor perception of oxidation in beef. *Meat Science*, *72*, 303-311.

Candogan, K., Kolsarici, N. (2003). Storage stability of low-fat beef frankfurters formulated with carrageenan or carrageenan with pectin. *Meat Science, 64*(2), 207-214.

Channon, H. A., & Trout, G. R. (2002) Effect of tocopherol concentration on rancidity development during frozen storage of a cured and an uncured processed pork product. *Meat Science, 62,* 9-17.

Chouliara, E., Karatapanis, A., Savvaidis, I. N., & Kontominas, M. G. (2007) Combined effect of oregano essential oil and modified atmosphere packaging on shelf-life extension of fresh chicken breast meat, stored at 4°C. *Food Microbiology, 24,* 607–617.

Carballo, J., Fernández, P., Barreto, G., Solas, M.T., & Jiménez-Colmenero, F. (1996). Morphology and texture of bologna sausage as related to content of fat, starch and egg white. Journal of Food Science, 61 (3), 652-655.

Lanari Vila, C. (1988). Modificaciones de la textura y coloración superficial de carnes bovinas refrigeradas y congeladas. Tesis doctoral. Facultad de Ciencias Exactas, UNLP.

López-López, I, Cofrades, S, Ruiz-Capillas, C, & Jiménez-Colmenero, F (2009). Design and nutritional properties of potential functional frankfurters based on lipid formulation, added seaweed and low salt content. *Meat Science*, 83, 255-262.

De Caterina, R., Liao, J. K., & Libby, P. (2000). Fatty acid modulation of endothelial activation. *American Journal of Clinical Nutrition*, 71, 213S-223S.

de Jong, N., Ros, M. M., Ocké, M. C., & Verhagen, H. (2008). A general postlaunch monitoring framework for functional foods tested with the phytosterol/-stanol case. *Trends in Food Science & Technology, 19,* 535-545.

Djordjevic, D., McClements, D. J., & Decker, E. A. (2004). Oxidative stability of whey protein-stabilized oil-in-water emulsions at pH 3: potential w-3 fatty acid delivery systems (Part B). *Journal of Food Science, 69* (5), C356-C362.

El-Magoli, S. B., Laroia, S. & Hansen P. M. T. (1996). Flavor and texture characteristics of low fat ground beef patties formulated with whey protein concentrate. Meat Science, 42 (2) 179-193.

Estévez, M., & Cava, R. (2006) Effectiveness of rosemary essential oil as an inhibitor of lipid and protein oxidation: contradictory effects in different types of frankfurters. *Meat Science, 72,* 348-355.

Estévez, M., Ramírez, R., Ventanas, S., & Cava, R. (2007). Sage and rosemary essential oils versus BHT for the inhibition of lipid oxidative reactions in liver paté. *Food Science and Technology - Lebensmittel-Wissenschaft und - Technologie (lwt), 40,* 58–65.

FDA-CFSAN (2003). U.S. Food and Drug Administration. Center for Food Safety and Applied Nutritions. Cook it safely. Available from: http//:www.foodsafety.gov /acrobat/f99broch.pdf. Accessed May 15, 2003.

Foegeding, E. A., Lanier, T. C., & Hultin, H. O. (1996) Characteristic of edible muscle tissues. In O. R Fennema, *Food Chemistry* (pp. 879-942), New York: Marcel Dekker, Inc.

Folch, J., Lees, M., & Sloane Stanley, G. H. (1957). A simple method for isolation and purification of total lipids from animal tissue. *Journal of Biological Chemistry, 226,* 497-509.

Georgantelis, D., Blekas, G., Katikou, P., Ambrosiadis, I., & Fletouris, D.J. (2007). Effect of rosemary extract, chitosan and α-tocopherol on lipid oxidation and colour stability during frozen storage of beef burgers. *Meat Science, 75,* 256-264.

Hernández-Hernández, E., Ponce-Alquicira, E., Jaramillo-Flores, M. E., & Guerrero Legarreta, I. (2009). Antioxidant effect rosemary (*Rosmarinus officinalis L.*) and oregano (*Origanum vulgare L.*) extracts on TBARS and colour of model raw pork batters. *Meat Science, 81,* 410–417.

Hu, M., McClements, D. J., & Decker, E. A. (2003). Impact of whey protein emulsifiers on the oxidative stability of salmon oil-in-water emulsions. *Journal of Agricultural and Food Chemistry, 51,* 1435-1439.

Jeremiah, L. E. (2001). Packaging alternatives of fresh meats using short- or long term distribution. *Food Research International, 34,* 749–772.

Jiménez-Colmenero, F. (2000). Relevant factors in strategies for fat reduction in meat products. *Trends in Food Science & Technology, 11,* 56-66.

Jiménez-Colmenero, F. (2007). Healthier lipid formulation approaches in meat-based functional foods. Technological options fro replacement of meat fats by non-meat fats. *Trends in Food Science & Technology, 18,* 567-578.

Kanner, J. (1994). Oxidative processes in meat and meat products: Quality implications. *Meat Science, 36,* 169-189.

Khalil, A. H (2000). Quality characteristics of low-fat beef patties formulated with modified corn starch and water. *Food Chemistry, 68,* 61-68.

Kim, J. M, & Lee, C. M. (1987). Effect of starch of textural properties of surimi gel. Journal of Food Science, 52 (3), 722-725.

Lee, S., Decker, E. A., Faustman, C., & Mancini, R. A. (2005). The effects of antioxidant combinations on color and lipid oxidation in *n-3* oil fortified ground beef patties. *Meat Science, 70,* 683-689.

Lee, S., Faustman, C., Djordjevic, D., Faraji, H., & Decker, E. A. (2006a). Effect of antioxidants on stabilization of meat products fortified with *n-3* fatty acids. *Meat Science, 72,* 18-24.

Lee, S., Hernández, P., Djordjevic, D., Faraji, H, Hollender, R., Faustman, C., & Decker, E.A. (2006b). Effect of antioxidants and cooking on stability of fatty acids in fortified meat products. *Journal of Food Science, 71(3)*, 233-238.

Li, J., & Yeh, A. (2002). Functions of starch in formation of starch/meat composite during heating. *Journal of Texture Studies*, 33, 341-366.

Li, J., & Yeh, A. (2003). Effects of starch properties on rheological characteristics of starch/meat complexes. *Journal of Food Engineering*, 57, 287–294.

Li, J., Yeh, A. & Fan, K. (2007). Gelation characteristics and morphology of corn starch/soy protein concentrate composites during heating. *Journal of Food Engineering, 78*, 1240–1247.

Linares, M. B., Berruga, M. I., Bórnez, R., & Vergara, H. (2007). Lipid oxidation of lamb meat: Effect of the weight, handling previous slaughter and modified atmospheres. *Meat Science, 76*, 715–720.

Linares, M. B., Bórnez, R., & Vergara, H. (2006). Effect of the type of stunning on lipid oxidation and colour of light lamb meat. In: Proceedings of the 52th International Congress on Meat Science and Technology (pp. 185–186). Dublin, Ireland.

Linares, M. B., Bórnez, R., & Vergara, H. (2008). Effect of stunning systems on meat quality of Manchego suckling lamb packed under modified atmospheres. *Meat Science, 78*, 279–287.

Lu, G. H., & Chen, T. C. (1999). Application of egg white and plasma powders as muscle food binding agents. *Journal of Food Engineering, 42*, 147-151.

Mensink, R. P. Zock, P. L., Kester, A. D., & Katan, M. B. (2003). Effects of dietary fatty acids and carbohydrates on the ratio of serum total to HDL cholesterol and on serum lipids and apolipoproteins: a meta-analysis of 60 controlled trials. *American Journal of Clinical Nutrition, 77*, 1146-1155.

Morrison, W. R., & Smith, L. M. (1964). Preparation of fatty acid methyl esters and dimetylacetals from lipids with boron fluoride methanol. *Journal of Lipids Research, 5*, 600-608.

Muchenje, V., Dzama, K., Chimonyo, M., Strydom, P. E., Hugo, A., & Raats, J. G. (2009). Some biochemical aspects pertaining to beef eating quality and consumer health: a review. *Food Chemistry*, 112, 279-289.

Ostlund, R. E. (2007). Phytosterols, cholesterol absorption and healthy diets. *Lipids, 42*, 41-45.

Paneras, E. D., & Bloukas, J. G. (1994). Vegetable oils replace pork back fat for low-fat frankfurters. *Journal of Food Science, 59 (4)*, 725-728, 733.

Pappa, I. C., Bloukas, J. G., & Arvanitoyannis, I. S. (2000). Optimization of salt, olive oil and pectin level for low-fat frankfurters produced by replacing pork backfat with olive oil. *Meat Science, 56*, 81-88.

Park, J., Rhee, K. S., & Ziprin, Y. A. (1990). Low-fat frankfurters with elevated levels of water and oleic acid. *Journal of Food Science, 55 (3)*, 871-872, 874.

Park, J., Rhee, K. S., Keeton, J. T., & Rhee, K. C. (1989). Properties of low-fat frankfurters containing monounsaturated and omega-3 polyunsaturated oils. *Journal of Food Science, 54 (3)*, 500-504.

Pelser, W. M, Linssen, J. P. F., Legger, A., & Houben, J. H. (2007). Lipid oxidation in *n-3* fatty acid enriched dutch style fermented sausages. *Meat Science, 75*, 1-11.

Pennisi Forrel, S.C., Ranalli, N., Zaritzky N.E., Andrés, S.C., & Califano, A.N. (2010). Effect of type of emulsifiers and antioxidants on oxidative stability, colour and fatty acid profile of low-fat beef burgers enriched with unsaturated fatty acids and phytosterols. *Meat Science*, 86, 364-370.

Piñero M.P., Parra, K., Huerta-Leidenz, N., Arenas de Moreno, L., Ferrer, M., Araujo S., & Barboza, Y. (2008). Effect of oat´s soluble fibre (β-glucan) as a fat replacer on physical, chemical, microbiological and sensory properties of low-fat beef patties. *Meat Science*, 80, 675-680.

Rojas, M. C., & Brewer, M. S. (2008). Effect of natural antioxidants on oxidative stability of frozen, vacuum-packaged beef and pork. *Journal of Food Quality, 31(2),*173-188.

Sante´-Lhoutellier V., Engel E., Gatellier P. H. (2008). Assessment of the influence of diet on lamb meat oxidation. *Food Chemistry, 109,* 573–579.

Sarıçoban, C., Yılmaz, M. T. & Karakaya, M. (2009). Response surface methodology study on the optimisation of effects of fat, wheat bran and salt on chemical, textural and sensory properties of patties. *Meat Science 83*, 610–619.

Scientific Committee on Food (2002). General view of the Scientific Committee on Food on the long-term effects of the intake of elevated levels of phytosterols from multiple dietary sources, with particular attention to the effects on β-carotene. European Commission Health & Consumer Protection Directorate-General (expressed on 26 September 2002, SCF/CS/NF/DOS/20 ADD 1 Final 3 October 2002). Available from http://ec.europa.eu/food/fs/sc/scf/ out143_en.pdf, accessed: Nov 20, 2009.

Serdaroğlu, M. (2006). Improving low fat meatball characteristics by adding whey powder. *Meat Science, 72,* 155-163.

Siegel, D. G., Church, K. E., & Schmidt, G. R. (1979). Gel structure of non-meat proteins as related to their ability to bind meat pieces. *Journal of Food Science, 44* (5) 1276-1279/1284.

Smit, L. A., Mozaffarian, D., & Willet, W. (2009). Review of fat and fatty acid requirements and criteria for developing dietary guidelines. *Annals of Nutrition & Metabolism, 55,* 44-55.

Soldatou, N., Nerantzaki, A., Kontominas, M. G., & Savvaidis, I. N. (2009). Physicochemical and microbiological changes of "Souvlaki" – A Greek delicacy lamb meat product: Evaluation of shelf-life using microbial, colour and lipid oxidation parameters. *Food Chemistry, 113 (1),* 36-42.

Stanley, D. W. (1983). Relation of structure to physical properties of food. In: *Physical properties of foods.* Peleg, M; Bagley, E.B. Eds. AVI Publishing Co. Inc. Westport Connecticut. 422-473.

Subhadra, B., Lochmann, R., Rawles, S., & Chen, R. (2006). Effect of dietary lipid source on the growth, tissue composition and hematological parameters of largemouth bass (Micropterus salmoides). *Aquaculture*, 255, 210–222.

Ulbricht, T. L. V., & Southgate, D. A. T. (1991). Coronary Heart-Disease – 7 Dietary Factors. *Lancet,* 338 (8773), 985-992.

USDA: USDA National Nutrient Database for Standard Reference (releases 20 and 21; release numbers change as new versions are released): nutrient data laboratory home page. 2009. http://www.ars.usda.gov/nutrientdata.

Velioğlu, H. M., Velioğlu, S. D., Boyacı, I. H., Yılmaz, İ., & Kurultay, Ş. (2010). Investigating the effects of ingredient levels on physical quality properties of cooked hamburger patties using response surface methodology and image processing technology. *Meat Science, 84*, 477-483.

WHO. (2004). Global strategies on diet, physical activity and health. WHA57. http://www.who.int/gb/ebwha/pdf_files/WHA57/A57_R17-en.pdf. Accessed: 4/7/2008.

Wolmarans, P. (2009). Background paper on global trends in food production, intake and composition. *Ann Nutr Metab*, *55,* 244–272.

Yilmaz, İ., Şimşek, O., & Işikh, M. (2002). Fatty acid composition and quality characteristics of low-fat cooked sausages made with beef and chicken meat, tomato juice and sunflower oil. *Meat Science, 62,* 253-258.

Zorrilla, S.E., Rovedo, C. O., & Singh, R. P. (2000). A new approach to correlate textural and cooking parameters with operating conditions during double-sided cooking of meat patties. *Journal of Texture Studies*, 31, 499-523.

In: Agricultural Research Updates. Volume 2
Editor: Barbara P. Hendriks

ISBN: 978-1-61470-191-0
© 2012 Nova Science Publishers, Inc.

Chapter 8

POTENTIALS AND LIMITATIONS OF HUSBANDRY PRACTICE IN SUSTAINABLE SYSTEMS TO SECURE ANIMALS' MINERAL NUTRITION

Marta López-Alonso[1], Marta Miranda[2] and Isabel Blanco-Penedo[3]

[1] Universidade de Santiago de Compostela, Departamento de Patoloxía Animal, Facultade de Veterinaria. Lugo, Spain
[2] Universidade de Santiago de Compostela, Departamento de Ciencias Clínicas Veterinarias, Facultade de Veterinaria. Lugo, Spain
[3] Swedish University of Agricultural Sciences, Department of Clinical Sciences, Faculty of Veterinary Medicine and Animal Science. Uppsala, Sweden

ABSTRACT

Principles of sustainable systems are based on a vision where animals should be part of an agricultural system that is environmentally sound, animal friendly and considering the whole system rather than only optimizing its parts. Sustainable systems such as integrated, low-input and organic farming use ecologically sound management strategies with the potential to benefit and respect the physiological and behavioral needs of livestock. This chapter focuses on the most critical obstacles to meet mineral requirements in ruminants in sustainable systems when a high degree of home-grown feed and constrained lower concentrate and mineral supplementation in the ration is promoted; and identifies the effects of different feeding regimes on mineral nutrition such as winter feeding and grazing intensity, fertilizer and pesticide-free pasture management, forage diversity and the evidences that mineral metabolism might be negatively affected by parasite infections that severely affect sustainable systems. The document addresses all these factors that are likely to exert a potentially adverse effect to meet nutritional requirements for animals and current research on strategies to improve animals' mineral nutrition and means of minimizing mineral disorders by specific husbandry practices.

Keywords: sustainable and organic systems, mineral nutrition, livestock, ruminants, husbandry practices.

INTRODUCTION

Minerals, both macro (Ca, P, K, Mg, N, Na and S) and trace elements (Co, Cu, Fe, I, Mn, Mo, Se, Zn among others), are essential to maintain animal health and productivity. Optimal nutrition, with adequate mineral levels, guarantees proper functions of the organism, amongst which the most important are structural, physiological, catalytic and regulatory functions (Suttle, 2010).

In sustainable systems such as integrated, low-input and organic farming the main source of minerals for livestock is the soil itself and the waste materials from the farm which are recycled onto the pasture. The available mineral concentrations in the different soil types are dependent on the concentrations in the parent rock and on the chemistry of the soil (Sundrum, 1997; Hayashida et al., 2004; Suttle, 2010), and as a consequence, forages in some circumstances may be deficient or imbalanced in some trace elements. In rare cases acute deficiency or toxicity may occur and remedial actions may be complex and difficult. On the contrary, subclinical deficiencies or toxicities, which are very difficult to diagnose but associated with low productivity, are very common worldwide (Coonan et al., 2002; Blanco-Penedo et al., 2009).

Traditionally, these mineral imbalances have been relatively easy to manage in conventional agriculture where animals routinely receive mineral supplements in the concentrate feed. In contrast to conventionally managed farms, in organic and other sustainable systems the main (or even the only) source of livestock feed is locally produced and the use of mineral supplements is restricted. This means organic farms must practice a highly efficient soil and forage management (Kuusela and Khalili, 2002) as well as to ensure the quality of the purchased organic feed if they are to achieve a mineral balanced dietary intake for livestock (Bard, 2006).

This chapter first analyses the concept of agricultural sustainability and the role of trace minerals on animal health. The chapter secondly addresses the main obstacles to meet mineral requirements in sustainable systems when a high degree of home-grown feed and constrained lower concentrate and mineral supplementation in the ration is promoted and discuss the effect of good management farming and husbandry practices to increase the supply of minerals for livestock.

CONCEPT OF AGRICULTURAL SUSTAINABILITY AND ORGANIC FARMING

The word "sustainable" is derived from the Latin, *sustinere*, meaning to keep in existence, implying permanence or long-term support. In the context of agricultural production, Ikerd (1993) defines a sustainable agriculture as "capable of maintaining its productivity and usefulness to society over the long run. ...it must be resource-conserving, economically viable and socially supportive and commercially competitive, and environmentally sound". Sustainability of the agricultural sector has everything to do with biodiversity. Biodiversity plays a key role in all farming processes and at all levels: soils, crops, livestock, and the farming environment. Improving on farm biodiversity leads to better nutrient use efficiency, increased disease resistance, more reliable production, resilient production systems, and varied landscapes. Not surprisingly biodiversity has become a focus of agricultural policy development.

During the last years, sustainable agriculture has progressed from a focus primarily on a low-input, organic farming approach with a major emphasis on small fruit or vegetable production farms—often described as "Low Input Sustainable Agriculture"—to the current situation where sustainability is an important part of mainstream animal and plant production units (Wagner, 1999). Many ``alternative'' approaches have been developed with respect to issues of sustainability, these include integrated pest management, integrated crop management, low-input agriculture, low-input sustainable agriculture, low external input sustainable agriculture, agroecology, permaculture, biodynamic farming and organic farming (for review see Rigby and Cáceres, 2001). The focus here is on organic farming and their relationship with the concept of sustainability. Organic farming is underpinned by a set of guiding principles -the principles of health, ecology, fairness and the principle of care-, drawn up by the International Federation of Organic Agricultural Movements (IFOAM, 1998). There is no real dispute that sustainable agriculture and organic farming are closely related terms. There is however disagreement on the exact nature of this relationship. Despite the variety of definitions of organic farming, the general agreements regarding what is necessary to produce organically are in stark contrast to the debates and arguments that rage regarding the nature of agricultural sustainability (Rigby and Cáceres, 2001). Organic farms utilise practices designed to meet the physiological and behavioural requirements of livestock and involve maintaining animals on good-quality, organically-grown feedstuffs (IFOAM, 2002; Lund, 2006).

Ruminants have served and will continue to serve a valuable role in sustainable agricultural systems. They are particularly useful in converting vast renewable resources from rangeland, pasture, and crop residues or other by-products into food edible for humans. With ruminants, land that is too poor or too erodible to cultivate becomes productive (Oltjen and Beckett, 1996). Also, nutrients in by-products are utilized and do not become a waste-disposal problem. There is large evidence about the efficiency with which ruminants convert humanly edible energy and protein into meat or milk in sustainable systems, and in addition, the protein resulting from ruminant livestock production is of higher quality with a greater biological value than protein in the substrate feeds.

TRACE MINERALS ON ANIMAL HEALTH

Despite the role of trace minerals in animal health being well established (for review see Suttle, 2010), they are a commonly forgotten source of nutrients in animal feedstuffs. Their physiological role is often underestimated and their presence in the feed in adequate quantities taken for granted. However, they are necessary to maintain body function, to optimise growth and reproduction and to stimulate immune response and therefore determine the health status (Suttle, 2010). Indeed, it is difficult to realize the impact of insufficient trace minerals, as symptoms of deficiency or mineral unbalances may not always be evident. However, a slight deficiency of trace elements can cause a considerable reduction in performance and production.

Many trace elements have very specific but often multiple roles. In example Se, it has been known for a considerable time to be necessary for growth and fertility in animals and for the prevention of a variety of diseases. More recently it has been established to form an integral part of a number of enzymes (selenoproteins) most of which function as antioxidants in the cellular cytoplasm in a range of situations (Tame, 2008). As well as individual trace elements having several functions, several trace minerals may be involved in a single function. For example, Se, Zn and Cu are all involved in immune function (Meglia, 2004). Rather than consider the role of each trace element it would be more helpful to consider the role of trace elements as a whole in some of the more common problems encountered in livestock.

Trace Element Requirements

Minerals must be provided in optimal concentrations and according to requirements that change during the rapid growth and development of the animal and the production cycle. It is rather difficult to justify the term 'requirements' for minerals in the same way as it is for energy, protein or amino acids. Requirements for minerals are hard to establish and most estimates are based on the minimum level required to overcome a deficiency symptom and not necessarily to promote productivity (Close, 2006).

Many authorities—in example the INRA en France, ACR in the United Kingdom or FEDNA in Spain—have recommended mineral requirements to ensure that the production of native livestock is not impaired by dietary mineral imbalances, however, agreement is rare. The recommendations for livestock in the USA are perhaps more up to date being revised in 2000 and 2001 (NRC, 2000, 2001). Weiss (2002) suggests that these should be regarded as minimum requirements – as they do not included safety margins. Mineral recommendations should include a safety margin to take account of the presence of antagonists (Underwood and Suttle, 1999; López-Alonso et al., 2004; Blanco-Penedo et al., 2006, 2009). For example, in ruminants Cu uptake is inhibited particularly by Mo but also by S and to a lesser extent by Fe (Underwood and Suttle, 1999) and high levels of Ca in the feed inhibit the uptake of Zn. It is also established that higher levels of Cu are required in the presence of high levels of Zn and that animals under stress require higher levels of Cu and Zn. Moreover, when determining mineral requirements and supplementation, consideration must be given to the quantity and type of raw ingredients and their inherent mineral content, the processing of the diet, the storage and environmental conditions, as well as the inclusion and content of other minerals (Close, 2006).

Sources of Trace Elements

In organic farming livestock normally obtain most of their minerals from the feeds and forages that they consume, and their mineral intakes are influenced by the factors that determine the mineral content of plants and their seeds. The concentrations of all minerals in plants depend largely on plant genotype, soil environment, climate and stage of maturity (Suttle, 2010). Micronutrient concentrations are generally higher in the surface soil and decrease with soil depth. In spite of the high concentration of most micronutrients in soils, only a small fraction is available to plants (Gupta et al., 2008) and soil pH is one of the most important factors affecting the availability of micronutrients to plants. Trace metal concentrations in pasture also varied seasonally (Socha et al., 2002; Griffiths et al., 2007; Suttle, 2010). Leguminous species are generally much richer in macro-elements than grasses growing in comparable conditions (Suttle, 2010). The trace elements, notably I, Cu, Zn, Co and Ni, are also generally higher in leguminous than in gramineous species grown in temperate climates, with Cu and Zn higher in mixed than in pure grass swards (Hopkins et al., 1994).

There is a considerable amount of data collected over a long period of time on the trace element levels in forages, particularly from ryegrass based swards. However, the over-riding feature of this data is the very wide range of values for each element. A tenfold or greater range between the lowest and highest values is not uncommon making the interpretation of the data extremely difficult (Tame, 2008). The problem is that no reference is made to the stage of growth, the ratio of leaf to stem to flower/seed head, at which the samples were harvested or to the weather conditions or light intensity under which the samples were harvested. It has been known for some time that values are different in leaf, stem and seed and even differ between young and old leaves (Terry et al, 2000). Following with the example of Se, a silage sample from a sward cut prior to heading will be expected to have a higher Se level than a sward cut when it is in flower. Many organic farmers take only one cut of silage and to ensure that they have a sufficient quantity leave the sward to "bulk up"; this usually means that the grasses have made the switch from vegetative to reproductive growth and this will have consequences for Se content. There is also evidence that the form in which Se is available has a strong influence on the level of Se in the plant (Terry et al, 2000). It is likely that variations in the content of other trace minerals will be dependent also on stage of growth, ratio of leaf to stem or the weather conditions.

Only a small number of studies to date in which grasses, legumes and "pasture weeds" (herbs) have been grown under the same conditions and analyzed for minerals and trace elements are available to allow a proper comparison (Weller and Bowling, 2002; Harrington et al., 2006; van Eckeren et al. 2006). Weller and Bowling (2002) have also shown that there were differences in the major minerals confirming that white clover and chicory had much higher levels of Ca, Mg and Na and ribgrass plantain had much higher levels of Ca and Mg. Surprisingly, timothy was lower than ryegrass in both Ca and Na. Clover was also lower in P and K. They also highlighted significant seasonal variations on the Na content, ryegrass (30%) and chicory (93%) showing higher and ribgrass (58%) lower Na concentrations in May-June compared to July-October. There also appeared to be a seasonal variation in magnesium content in ribgrass plantain.

The Harrington study was conducted in New Zealand and samples were taken mid-summer though there is no comment about growth stage. Levels of both macro and micro nutrients were compared between ryegrass, Yorkshire fog, white clover, chicory, narrow leaved plantain, broad-leaved dock, and dandelion. The chicory and narrow leaved plantain had significantly higher levels of P, S and Na, chicory also had higher levels of Mg and the narrow leaved plantain had higher levels of Ca. Dandelion had significantly higher levels of P Mg and Na. With regard to micro-nutrients, chicory had significantly higher levels of Cu, Zn and B, while narrow leaved plantain had higher levels of Cu and Co. Dandelion had higher levels of Cu, zn and B. Yorkshire fog had a significantly higher content of Mo at a level likely to interfere with cu absorption. In the same study the levels of Se were highest in white clover but also higher in the "weeds" than in the ryegrass though the difference between the "weeds" and ryegrass was not significant.

The van Eckeren et al. (2006) study compared ryegrass, white clover, chicory and plantain and was conducted on two different soil types, sandy and clay, though plantain was only grown on the sandy soils. This study showed that Ca levels were very much higher in white clover and chicory on both soils and in plantain on the sandy soils. Sodium was higher in both chicory and plantain on the sandy soils and in chicory on the clay soils. Of the trace minerals Cu was much higher in the chicory on both soils types than in the grass or clover, Zn was higher in the chicory and the plantain than in grass and clover on the sandy soils and higher in the chicory on the clay soils. On the clay soils the chicory had higher contents of both Co and Se than either grass or clover. There were no significant differences on the sandy soils.

As indicated, stage of maturity and time of year can have an effect on the concentrations of trace elements in pasture species. Numerous papers have been written on this subject but the results are difficult to summarize due to the different conditions under which the experiments were carried out, the different dates on which the yields were recorded and the different locations of the experiments. Much of the data suggests that the concentration of Fe, B, Cu, Zn and Mo are high in April and May when pasture is growing vigorously. Trace elements are most active in young growth in spring and concentrations are high in the young tissue during this period. Concentrations of Fe, Cu and Zn fall slightly in June and reach a peak again in September and October. Herbage Mo and Mn falling gradually as the season progresses, whereas Se increases (Tame, 2008).

Assessment of Trace Element Status in Ruminants

In the light of the data detailed above, it seems clear that great care needs to be taken when assessing the trace element status of grazing ruminant livestock. In addition to the great variability on the mineral content in forages, in grazing systems it is not possible to do a precise and straightforward estimation of the animal mineral intake (as a function of the mineral content in the feed and the amount of feed consumed) as well as the mineral digestibility. Besides, it appears that marginal trace element status often does not result in clearly defined symptoms, which could help to guide mineral unbalance diagnosis.

Perhaps the best starting point to assess mineral status for organic and other sustainable systems is the forage, both grazed and conserved, because represents the main source of feed. However, we should take note of the stage of growth, weather conditions, time of year, composition of the sward etc., as these are all likely to affect trace mineral levels. Such analyses should give a good indication as to whether the animals are likely to have adequate, marginal or deficient levels of a particular trace element.

Assessment of the trace element status in the animal is much more difficult. We need to take into account its point in the production cycle, the level of stress imposed on the animal, the choice of analysis, the level of trace element antagonists as well as other trace elements in the feed and indeed the nature of any supplementary feeds used. For example blood Cu concentrations are not good indicators of the Cu status as not all the Cu is available to the animal and are not correlated with Cu concentrations in the liver (Clark et al., 1993) the best indicator of Cu status: cattle with low plasma Cu levels can have adequate liver Cu levels

(Mulryan and Mason, 1992). Cu concentrations are also highly influenced by antagonist (such as S, Mo, Fe or Zn), infection, trauma and stage of production (Underwood and Suttle, 1999, Puls, 1994) and hepatic Cu concentrations largely vary with stage in the production cycle, declining during the pre-parturient period reaching its lowest at calving and increasing post calving (Boland, 2003).

OBSTACLES IN GETTING MINERAL REQUIREMENTS IN SUSTAINABLE SYSTEMS

Traditionally, getting mineral requirements has been relatively easy on intensively managed farms. This is because mineral supplements are routinely incorporated into concentrates and this generally ensures that animals receive the required intake of minerals (Chládek and Zapletal, 2007). Concentrate rations are often formulated with large "safety margins" so that nutrient intakes largely exceed requirements without any negative consequence for animal health and productivity.

In contrast to intensively managed farms, sustainable systems including organic farms use locally or on-farm produced roughage as the main source of feed, with a very limited use of concentrate feed and mineral supplements. This type of diet has been associated with mineral deficiencies and imbalances in areas where soils have an inadequate mineral content and/or low mineral bioavailability (Owens, 2001; Weller and Bowling, 2002; Coonan et al., 2002; Owens and Watson, 2002; Govasmark et al., 2005; Scott, 2007; Tame, 2008; Blanco-Penedo et al., 2009) demonstrating the need of mineral supplementation in organic farms. Coonan et al. (2002) reported that mineral and trace element supplements were an essential part of a dietary regime for organic dairy cattle which were identified as at potential risk from Cu, Zn, iodine and Se deficiency. Similarly, in a recent study in organic beef cattle in North-West Spain, it was observed that animals receiving a high forage diet and no mineral supplementation suffered deficiencies of certain elements; these deficiencies were not observed in coterminous traditional farms that received supplemented concentrate diets (Blanco-Penedo et al., 2009).

In addition, mineral deficiencies or unbalances can be exacerbated at times when the natural farm resources are unable to guarantee an adequate level of nutrition. Conventional farming has numerous tools to deal with these situations, for example offering the animals a complementary winter feeding, improving pasture production by using fertilizers and pesticides or having a pharmacological control of parasite infections, all measures that directly or indirectly help to maintain/improve the animal mineral status. On the contrary, on the organic and other sustainable systems most of these farming practices are not allowed or are economically unviable, which forces farmers to focus on a highly efficient grazing management to get an adequate mineral status.

GOOD PRACTICES TO IMPROVE LIVESTOCK MINERAL STATUS

Getting animal mineral requirements in low-input and other sustainable systems needs for guaranteeing an adequate and balanced flow of minerals from the soil, the main source of minerals of the farm system, to the animal. An integral management of the farm should consider aspects related to the soil, pasture and animal to ensure an adequate mineral nutrition.

Soil Management

Topsoil is the capital reserve of every farm. To sustain agriculture means to sustain soil resources, because that's the source of a farmer's livelihood (Sullivan, 2010). Appropriate mineral nutrition needs to be present for soil organisms and plants to prosper. The balance between the different nutrients is equally important: the concentration of an individual nutrient in the soil can affect the uptake of another by the pasture herbage and ultimately cause chronic or acute deficiency in livestock. Maintaining the correct soil balance is more difficult in organic farming because the regulations do not permit the use of manufactured fertilizers for correcting such an imbalance except in extreme situations. A high level of management skill is needed to put an effective system of actions in place that will ensure an adequate supply of trace elements for livestock. Most of the actions are interdependent and interrelated and the effect of each action on the other must be carefully assessed so as to ensure a balance and to avoid extremes.

Soil Organic Matter

Organic matter (humus) is an important constituent of every soil because it is the most chemically and biologically active of all the soil phases. It is of particular importance in organic farming because the farmer is dependent on inherent soil properties to support crop and livestock growth. The organic matter content of a soil is dependent on climate and soil type: the organic matter content tends to be high in wet temperate regions and in heavy clay soils and low in arid regions with light, sandy or gravelly soils. Soil organic matter content is greater under permanent grassland than in arable land (MacNaeidhe, 2001).

Soil organic matter is strongly associated with the transfer of trace elements from the soil to the plant (Stevenson and Ardakani, 1972). Although trace elements may be fixed by the fully decomposed organic materials such as the humic acids, the incompleted decomposition of organic acids, polyphenols, amino acids, peptides, proteins and polysaccarides carry trace elements in loosely bound chelated form which are easily available to plants. From 98 to 99% of Cu, 84 to 99% of Mn and 75% of Zn are carried on organic complexes within the soil (MacNaeidhe, 2001). It is clear from the above that practices that build up the organic matter in a soil to the optimum level will also increase the mineral and trace element supply to livestock. A low livestock density (1.0 to 1.5 LSU/ha) prevents over grazing and a reduction of organic matter input to the soil. Excessive liming causes a reduction in soil organic matter, the soil pH should be at 6.5 or slightly below. Adequate moisture supply should be available to pasture herbage but over-drainage in pastureland should be avoided.

Lime Application and Soil pH

The soil pH affects the availability of trace elements for uptake by pasture species. Over time leaching by rainfall causes the soil pH to fall and this affects the balance of trace elements in the soil. The soil pH can be adjusted upwards by the application of lime. A soil pH of 6.5 is considered to be the optimum for a soil containing trace elements in well-balanced amounts. At soil pH values below 6.5 the availability of Mo and Se is reduced and the availability of Fe, Mn, Co, Zn and B is increased; the opposite is true at soil pH values above 6.5 (Fleming, 1965).

Soils that are high in Mo and Se should be limed to a pH which does not exceed 6.0 to reduce the risk of induced Cu deficiency or Se toxicity in livestock. The pH of soils which are low in Mn, Co, Zn and B should not be allowed to exceed 6.0. Over-liming of soils low in Co can result in Co deficiency and ill thrift in sheep. A high soil pH can also causes B deficiency, resulting in poor establishment of clover (Kehoe, 1981) which is essential for the provision of N in organic systems. Severe Zn deficiency results in poor pasture growth, which is due in part to poor utilization of N within the plant when Zn is low (MacNaeidhe, 1991).

Drainage

Free draining sandy or gravelly soils are low in trace elements. The low status of these soils is associated with low clay, oxide and organic matter content rather than with the leaching of trace elements from the soil profile (Shuman, 1979), these soil constituents containing important sites onto which the trace elements are bound. The exceptions are Se, I and B, which are relatively easily leached by rainfall. Application of farmyard manure and inclusion of species containing high concentrations of trace elements in the pasture can increase the supply of trace elements to grazing livestock. Various measures can be taken to control moisture loss in dry weather; these include the use of dams, sluices and sub-irrigation through existing drainage canals (Lucas, 1982).

On the contrary, poor drainage can increase the concentrations of some trace elements in soil and in pasture species (Walsh and Fleming, 1978). Special attention should be given to high Mo and Fe concentrations, which can induce Cu deficiency in livestock. In addition, poor drainage can increase soil ingestion by livestock and this can lead to induced deficiency of Cu as will be discussed later.

Nutrient Budgeting and Farmyard Manure

Nutrient management is a major challenge on organic farms: lower crop yields, poor animal performance and stocking rates on farms can occur as a result of poor nutrient management. Anecdotal evidence suggests that many farms that have entered organic farming have not completed nutrient budgets for their farms and prior to moving into the organic system of farming were returning little or no P or K to their soils; in some of these cases, crop yields are poor because soil reserves have dropped over time due to insufficient nutrient inputs (MacNaeidhe, 2001).

"Feed the soil to feed the plant" is often referred to as one of the principles of organic farming (McDonnell, 2010). It is absolutely essential that all nutrients removed in farm produced products are replaced. Nutrient advice should be formulated on the basis of soil fertility levels, stocking rate and crop nutrient requirements. It is also essential to know current soil fertility status of our soils. Nutrient management in organic farming systems should be based on regular soil nutrient analysis and nutrient budgets which are used to plan applications of manures, composts and permitted fertilisers (Sullivan, 2010).

A key component of building and maintaining soil fertility in organic farming is a readily available supply of organic manures (McDonnell, 2010). Manure recycling will help replenish soil reserves and ensure that annual crop nutrient requirements are satisfied.

Organic manures should be applied on a rotational basis depending on soil test results and crop nutrient requirements. Organic manure nutrient content can vary widely depending on the source of nutrients and it is advisable to have the nutrient content of manures checked through laboratory analysis. Manure application rates should be adjusted in relation to nutrient content and crop requirements. Apply manures in the spring time to maximise the recovery of N especially for manures with high N availabilities (for example cattle slurry). Where liquid manures are applied to tillage soils ensure that they are well agitated, spread evenly and rapidly incorporated.

Pasture Management and Grazing

Pasture is at the heart of organic livestock management. According to the organic normative, all ruminants must have access to pasture, which must be managed organically using strategies that improve soil and water while increasing the mineral nutrient value of the pasture (IFOAM, 1998).

Considering Trace Element Concentrations in Pasture Species
The inclusion of grass and herb species with high concentrations of trace elements in pasture has been proposed as an efficient tool to increase the intake of trace elements in livestock (Tame, 2008). Scientific research into individual herbs species, their mineral concentrations and their availability is still limited. In particular rough stalked meadow grass and clover have high trace element concentrations and should be included in grass seed mixtures. Chicory, sheep's sorrel, plantain, dandelion and Lotus species (trefoils) have all been investigated to some extent as sources of minerals in livestock diets, these last species because of their high tannin content could be useful in the control of parasite infections as will be discussed later.

A number of seed houses are now offering seed mixes which include a range of grasses, clovers, trefoils and herbs. While these mixes are generally significantly more expensive and take longer to establish the indications are that they are much longer lasting. If they also result in better trace element status in the livestock it is likely that the extra cost will be more than offset by a combination of less frequent re-seeding and better animal health. However, before any reliable recommendations can be made we need to understand much more about how the level of each trace element varies in different parts of the plant as well as how it varies with stage of growth through the season. If quantitative recommendations are to be made we must take into account the proportions of the various herbage plants within the sward or great care will need to be taken when sampling to ensure that the sample is truly representative of the sward.

Rotations
A correctly designed and implemented crop rotation is at the heart of organic crop production. It is well known that organic farming without crop rotation is effectively impossible in the long run because of detrimental effects on soil fertility, weeds and plant health (Finckh, 2011). In addition, rotations and cover crops may significantly contribute to erosion control, another important agricultural problem field. Especially grass-clover

mixtures play a crucial role in crop rotations with respect to nutrient management, soil organic matter accumulation and microbial activity and problem weed management. In addition, the role of oats and certain brassica crops for the reduction and management of weeds and fungal and nematode diseases should not be underestimated. Future research should concentrate on the identification of crops especially useful for rotations and possibly intercropping to enhance such beneficial effects in organic farming as well as in conventional farming.

Grazing Management

Regardless of the species or class of grazing animal, graze management is a key point in the organic systems, and this should emphasise on maximizing dry matter intake (DMI) from pasture (Sullivan, 2010). Efficient grazing practices can make the farm more profitable in several ways: reduce purchased grain costs, reduce mechanical forage harvesting and associated fossil fuel costs, reduce mechanical manure spreading costs and lower fertilizer costs as manure nutrients are returned to the soil. Numerous factors including plant, animal, and human, have an influence over how much pasture forage an animal will consume. The higher an animal's requirements are, based on production level, the more important maximizing intake becomes. Thus, lactating dairy cows are the kind and class of livestock that are most sensitive to factors influencing intake. Maximizing pasture DMI is not as critical with beef cattle, sheep, or growing dairy heifers as it is with lactating dairy cattle.

The management of the pastures, as well as the physical attributes of the pasture (too tall, too short, no clover, too much stubble, etc.), are the key to animal production. The components of pasture that we need to be concerned about are plant density, number of tillers/plant, the height of the grass, and species composition. Research from around the world has clearly shown that dry matter intake from pasture is the result of a relatively simple equation: *DMI = intake/bite x rate of biting x time spent grazing*. A high plant density results in a higher intake level because the animal can stand in one spot in a pasture and graze from many plants—a high biting rate. A thinner stand will result in a lower biting rate because the animal needs to spend more time walking around and looking for plants to graze.

In relation to the mineral nutrition, an essential point that must be taken into account within the grazing management is the amount of soil ingested when grazing. As previously indicated soil is the main source of minerals, both toxic and essentials, in agricultural not polluted areas (Blanco Penedo et al., 2009). This is because toxic metal concentrations in soils are up to 2-3 fold those in plants and it is assumed that ruminants ingest up to 18 % of their dietary dry matter when grazing (Thornton and Abrahams, 1983). Soil ingestion is directly related with the stocking rate. Soil ingestion was greater at high stocking rate (3 LSU/ha) on loam soil than at low stocking rate (1.6 LSU/ha) on sandy loam (MacNaeide, 2001) and can represent a problem when soils have an unbalanced mineral content; for example a high rate of Fe ingestion constitutes a high risk of induced Cu deficiency (Jarvis and Austin, 1983). It is well demonstrated that soil ingestion can be minimized by using low stocking rates and removing animals from the pasture in wet weather especially in the case of soils that are low in Cu.

Issues Related to the Concentrate and Mineral Supplements

Ideally, sustainable systems should provide as most as feed as possible from the own farm and land-based, reducing at maximum the purchased feed, usually concentrate feed, at moments when local or own-farm feedstuffs are not available (i.e. winter feeding) or to achieve a high level of production (i.e. milk production). Organic legislation is restrictive on this respect: organic standards for sheep and cattle require that the feed ration is 100% organic and that at least 60% of dry matter is from fresh or conserved forage.

When used in organic systems, purchased concentrate feed represents an important source of minerals for livestock. It is well known that most trace elements are present at higher in cereals than in forages, their bioavailablility being also higher in the former (Underwood and Suttle, 1999). In a recent study carried out in NW Spain it has been demonstrated that the inclusion of a low proportion of concentrate feed in the ration prevents cattle on organic farms from suffering mineral deficiencies (Blanco Penedo et al., 2009).

Mineral supplementation is allowed by the standards only when the requirements of the animals cannot be met by the practices of organic husbandry, such as grazing of pastures with mineral availability. At present the main strategy for improving trace element status in livestock is to analyse the forages and possibly other feeds and to seek derogation from the appropriate Sector Body to feed a mineral supplement to overcome any deficits highlighted by the analyses. Supplementation of trace elements in organic livestock is most commonly made through the purchased concentrate feed, rumen boluses and free access mineral blocks. There is also some use of trace element injections, particularly local inhibitory factors (such as soil Mo affecting Cu uptake) present; however these kind of mineral formulations have the disadvantage that the rumen is by-passed and consequently rumen microbes may not receive an adequate supply of trace elements for their critically important functions.

Mineral supplementation should be regarded as a short-term measure, as it is a largely conventional approach and not in keeping with organic philosophy, trying to devise more natural and sustainable soil and pasture strategies to increase the mineral content in the forage. If supplements were needed to maintain animal health and productivity these should be tailored to the specific farm needs (in example: for some specific minerals) and should involve minerals with a high biodisponibility. In conventional farms inorganic salts, such as sulphates, carbonates, chlorides and oxides have been the compounds most commonly used as mineral supplements. These salts are broken down in the digestive tract to form free ions and are then absorbed. However, free ions are very reactive and can form complexes with other dietary molecules, which are difficult to absorb. Large quantities of undigested minerals are then excreted and cause environmental pollution. For this reason, there is growing interest in organic farming for proteinated or chelated trace minerals. In this form, the trace elements are chemically bound to a chelating agent or ligand, usually a mixture of amino acids or small peptides. This makes them more bio-available and bio-active and provides the animal with a metabolic advantage that often results in improved performance. They can therefore be included at much lower levels without compromising performance, thus minimizing nutrient excretion and environmental impact (Close, 2006).

Parasite Problems in Organic Systems and Options for Control

In contrast to conventional farms, parasite control on organic farms is affected by several of the prescribed changes in management e.g. access to the outdoors in the summer and in most countries, a ban on preventive medication, including use of anti-parasiticides. Organic animal production relies heavily on grazing, and pasture or soil related parasites are thus of major importance. In addition, pastured livestock rely on the available vegetation for feeding, wherefore they may face potential problems of insufficient nutrition combined with high parasite transmission. This last fact could be of high relevance when considering mineral nutrition: It is well demonstrated that parasites can largely enhance mineral deficiencies (Suttle, 2010) and on the other hand, some minerals, including Co (used to synthesize vitamin B12) and Fe can play a key role in affecting ruminant susceptibility (Singh, 2010).

In a short to medium term perspective, integrated parasite control in sustainable systems may combine grazing management with biological control using nematophagous micro-fungi, selected crops like tanniferous plants, and if allowed, limited use of antiparasiticides.

At present, the non-chemotherapeutic control of pasture related infections is based mainly on grazing management strategies (Niezen et al., 1996). Preventive strategies, where young previously unexposed stock, are turned out on parasite-free pastures. Evasive strategies aim at avoiding disease producing infections of a contaminated area by moving to a clean area and may also be relevant. Securing sufficient exposure to induce immunity in young stock through an integrated approach based primarily on grazing management, perhaps supported by herd monitoring, seems more likely to be successful. An effective control of nematodes can also be achieved by repeated moving of the herd or alternate grazing with other species. In general, high stocking rates are linked with increased parasite loads; lowering the animal density serves two purposes: it reduces the amount of manure in a given area and the residual grazing height of the forage is often much higher, which significantly reduces the probability of parasite infection (80% of parasites live in the first 5 cm of forage aboveground) (Singh, 2010).

Bioactive forages can be characterised as forages containing metabolites that may reduce the establishment of incoming nematodes or reduce existing worm burdens. Initial work has focused on plants with a high content of condensed tannins, like sulla (*Hedysarum coronarium*) and the trefoils (*Lotus spp.*) (Niezen et al., 1995). Chicory, a plant with a low tannin content, has been shown to offer animals some protection from internal parasitism (Niezen et al., 1998), but is high in soluble carbohydrate and minerals which may enable the host to tolerate parasites. The use of bioactive forages easily incorporated in the crop rotation for grazing animals or in the diet of housed animals is highly compatible with principles of organic farming.

Control of nematodes by larvae-trapping fungi, or perhaps in the future by egg-destroying fungi, looks promising for ruminants and certain monogastric animals but delivery systems and practical dosing regimens integrated with grazing management have to be developed (Thamsborg and Roepstorff, 2010).

Reduction of Heavy Metal Input – A Task also for Organic Animal Husbandry

One of the main goals of organic systems is to avoid environmental pollution. Organic farming offers important advantages in terms of protecting an agricultural system against the overload of heavy metals: a high degree of integration between animal husbandry and land use and the low external inputs.

It is well established that in relatively unpolluted agricultural regions the main source of toxic metals to conventionally reared livestock are inorganic fertilizers and the mineral supplements given to the animals. Of particular concern is the contamination of pastures with Cd derived from superphosphate fertilizers and its subsequent fate in grazing livestock (Stark et al., 1998; Loganathan et al., 2008), in fact Cd enriched phosphate fertilizers are the most important exogenous sources of Cd to agricultural land worldwide (Nicholson et al., 1994). The very restrictive use of phosphate and other inorganic fertilizers in organic legislation has significantly contributed to decrease soil cadmium contamination (Linden et al., 2001). Sewage sludge application to agricultural land, which use is not allowed in organic agriculture, is also responsible for an increase in Cd and other toxic elements as Pb in crops (EFSA 2004 a,b).

The main sources of heavy metal inputs in the nutrient cycle of a farm are the mineral supplements in animal nutrition; in a lesser extend food baths with Cu-based agents, medicines, mineral bedding materials and other chemicals in stable equipment (Zn) contribute to metal exposure (Schumacher, 2004). Mineral supplements and pre-mixes generally contain higher toxic metal concentrations than the other main components of the ration (McBride, 1998), however the significance of their contribution to the total animal intake is difficult to establish, due to the low but very variable rate of inclusion within the animal diet. An important consideration from the environmental pollution point of view is that essential metals present in the supplements that are not absorbed by the animal become toxic residues when excreted by the animal in the manures. Cu and Zn in pig and poultry supplements are possibly the micronutrients that best exemplify the conflict between short-term animal welfare (that may involve providing supplements in feed) and the long term sustainability of soil fertility (Gustafson et al., 2007). Excess dietary Cu and Zn that is not absorbed by the animal is excreted in the faeces, producing Cu-enriched manures (Jondreville et al., 2003). When these manures are used as fertilizers in the pastures for cattle grazing or hay/forage production, they can lead to increased soil Cu concentrations (Poulsen, 1998; Bengtsson et al., 2003) and toxic effects in plants and micro-organisms (Coppenet et al., 2003) and in animals (López-Alonso et al., 2000, 2006; Miranda et al., 2006). Important measures should be adopted in terms of preventing soil contamination in organic systems by reducing the Cu- and Zn-contents in feedstuff. This is mostly caused by unknown contents in the home-grown feedstuffs and, sometimes, by the necessity to compensate bad hygiene management with measures that lead to Cu and Zn surplus in some farm cycles (Schumacher, 2004). A better choice of Cu and Zn compounds as well as the use of enzymes to improve the availability of trace elements could also be essential points.

CONCLUSION

The data presented in this chapter indicates that in organic and other livestock sustainable systems with a high degree of home-grown feed their mineral nutrition is highly dependent on local conditions. This means that a high level of management skill is needed to put an effective system of actions in place that will ensure an adequate supply of trace elements for livestock. Most of the actions are interdependent and interrelated and the effect of each action on the other must be carefully assessed so as to ensure a balance and to avoid extremes.

REFERENCES

Bard, PW. Requerimientos de formación para los veterinarios que trabajan en ganadería ecológica. Ganadería ecológica en el sur de Europa. 1ª Conferencia Internacional. Zamora, 2006, pp. 145-148.

Bengtsson, H; Öborn, I; Jonsson, S; Nilsson, I; Andersson, A. Field balances of some mineral nutrients and trace elements in organic and conventional dairy farming — a case study at Öjebyn, Sweden. *European Journal of Agronomy*. 2003, 20: 101–116.

Blanco-Penedo, I; Cruz, JM; López-Alonso, M; Miranda, M; Castillo, C; Hernández, J; Benedito, JL. Influence of copper status on the accumulation of toxic and essential metals in cattle. *Environment International*, 2006, 32, 901–906.

Blanco-Penedo, I; Shore, RF; Miranda, M; Benedito, JL; Lopez-Alonso, M. Factors affecting trace element status in calves in NW Spain. *Livestock Science*, 2009, 123(2/3): 198–208.

Chládek, G; Zapletal, D. A free-choice intake of mineral blocks in beef cows during the grazing season and winter. *Livestock Science*. 2007, 106: 41-46.

Clark, TW; Xin, Z; Du, Z; Hempken, RW.. A field trial comparing copper sulfate, copper proteinate and copper oxide as copper sources for beef cattle. *Journal of Dairy Science*. 1993, 76(Suppl.):462.

Close, WH. Trace mineral nutrition of pigs. Meeting production and environmental objectives. *EAAP Annual Meeting*: Antalya, Turkey, 2006.

Coonan, C; Freestone-Smith, C; Allen, J; Wilde, D. Determination of the major mineral and trace element balance of dairy cows in organic production systems. In: Kyriazakis, Zervas (Eds.) *Proceeding of Organic Meat and Milk from Ruminants*, Athens, EAAP Publication, 2002, 106: 181–183.

Coppenet, M; Golven, J; Simon, JC; Le Roy, M. Evolution chemique des soils en exploitations d'élevage intensif: example du Finistère. *Agronomie*. 2003, 13: 77–83.

EFSA (European Food Safety Authority). Opinion on the Scientific Panel on Contaminants in the Food Chain on a request from the Commission related to cadmium as undesirable substance in animal feed. Adopted on 2 June 2004. *The EFSA Journal*, 2004a, 72, 1-24.

EFSA (European Food Safety Authority). Opinion on the Scientific Panel on Contaminants in the Food Chain on a request from the Commission related to lead as undesirable substance in animal feed. Adopted on 2 June 2004. *The EFSA Journal*, 2004b,71, 1-20.

Finckh, *MR*. Crop rotation, crop succession and aspects of plant health. Available at www.wiz.uni-kassel.de/phytomed/crop_rotation.pdf, 2011.

Fleming, GA. Trace Elements in Plants with Particular Reference to Pasture Species. *Outlook on Agriculture. 1965,* 4 (6): 270-285.

Govasmark, E; Steen, A; Strom, T; Hansen, S; Singh, B R; and Bernhoft, A. Status of selenium and vitamin E on Norwegian organic sheep and dairy cattle farms. Acta Agriculturae Scandinavica.Section A, Animal Science, 2005, 55, 40-46.

Griffiths, LM; Loeffler, SH; Socha, MT; Tomlinson, DJ; Johnson, AB. Effects of supplementing complexed zinc, manganese, copper and cobalt on lactation and reproductive performance of intensively grazed lactating dairy cattle on the South Island of New Zealand. *Animal Feed Science and Technology.* 2007, 137: 69–83.

Gupta, UC; Kening, WU; Siyuan, L. Micronutrients in Soils, Crops, and Livestock. *Earth Science Frontiers*, 2008, 15(5): 110–125.

Gustafson, GM;Salomon, E; Jonsson, S.. Barn balance calculations of Ca, Cu, K, Mg, Mn, N, P, S and Zn in a conventional and organic dairy farm in Sweden. *Agriculture, Ecosystems & Environment.* 2007,119: 160–170.

Harrington, KC;Thatcher, A; Kemp, PD. Mineral composition and nutritive value of some common pasture weeds. *New Zealand Plant Protectio. 2006,* 59: 261-265.

Hayashida, M; Orden, EA; Cruz, EM; Cruz, LC; Fujihara, T. Effects of concentrate supplementation on blood mineral concentration of growing upgraded Philippine goats. *Animal Science Journal.* 2004, 75: 139-145.

Hopkins, A; Adamson, AH; Bowling, PJ. Response of permanent and reseeded grassland to fertilizer nitrogen. 2. Effects on concentrations of Ca, Mg, Na, K, S, P, Mn, Zn, Cu, Co and Mo in herbage at a range of sites. *Grass and Forage Science.* 1994, 49, 9–20.

IFOAM. *Basic Standards for Organic Production and Processing.* IFOAM. Tholey-Theley, Germany. 1998.

IFOAM. *IFOAM Norms. II. IFOAM Basic Standards for organic production and processing.* International Federation of Organic Movements, Tholey-Theley, Germany. 2002.

Ikerd, J. Two related but distinctly different concepts: organic farming and sustainable agriculture. *Small Farm Today,* 1993, 10, 30–31.

Jarvis, SC; Austin, AR. Soil and plant factors limiting the availability of copper to a beef suckler herd. *The Journal of Agricultural Science.* 1983, 101: 39-46.

Jondreville, C; Revy, PS; Dourmad, JY. Dietary means to better control the environmental impact of copper and zinc by pigs from weaning to slaughter. *Livestock Science.* 2003, 84: 147–156.

Kehoe, C. Mise en valeur de tour bieres residuelles en Irlande: essays sur l'implantation du trefle blanc. *Thesis, M.Sc.* University College, Dublin. 1981.

Kuusela, E; Khalili, H. Effect of grazing method and herbage allowance on the grazing efficiency of milk production in organic farming. *Animal Feed Science and Technology.* 2002, 98: 87-101.

Lindén A; Andersson K; Oskarsson A. Cadmium in organic and conventional pig production. *Arch Environ Contam Toxicol.,* 2001, 40, 425-31.

Loganathan, P; Hedley, MJ; Grace, ND. Pasture soils contaminated with fertilizer-derived cadmium and fluorine: livestock effects. *Reviews in Environmental Contamination and Toxicology,* 2008, 192, 29–66.

López Alonso, M; Prieto Montaña, F; Miranda, M; Castillo, C; Hernández, J; Benedito, JL. Interactions between toxic (As, Cd, Hg and Pb) and nutritional essential (Ca, Co, Cr, Mn,

Mo, Ni, Se, Zn) elements in the tissues of cattle from NW Spain. *BioMetals* 2004, 17: 389–97.

López-Alonso, M; Benedito, JL; Miranda, M; Castillo, C; Hernandez, J; Shore, RF. The effect of pig farming on copper and zinc accumulation in cattle in Galicia (North-Western Spain). *Veterinary Journal*. 2000, 160: 259-266.

López-Alonso, M; Crespo, A; Miranda, M; Castillo, C; Hernandez, J; Benedito, JL. Assessment of some blood parameters as potential markers of hepatic Cu accumulation in catte. *Journal of Veterinary Diagnostic Investigation*. 2006, 18: 70-74.

Lucas, RL. Organic Soils (Histosols), Formation, Distribution, Physical and Chemical Properties and Management for Crop Production. Research Report 435. Farm Science. Michigan State University, USA. 1982, 77pp.

Lund, V. Natural living – a precondition for animal welfare in organic farming. *Livestock Science*. 2006, 100: 71–83.

MacNaeidhe, FS. Pasture management and composition as a means of minimizing 102 mineral disorders in organic livestock. In: Hovi, M; Vaarst, M (Eds). *Positive health: preventive measures and alternative strategies*. Proceedings of the Fifth NAHWOA Workshop . RØdding, Denmark – November 11-13, 2001

MacNaeidhe, FS. The effect of zinc and nitrogen on pasture growth in a zinc deficient soil. *Proceeding of Irish Grassland and Animal Production Association*. 17th Annual Meeting. University College Dublin. 1991, 1-2.

McBride, MB. Growing food crops on sludge-amended soils: problems with the U.S. Environmental Protection Agency method of estimating toxic metal transfer. *Environment Toxicology Chemistry*, 1998, 17, 2274-2281.

McDonnell, J. Nutrient budgeting and management on organic farms. Available at www.teagasc.ie/.../NutrientBudgeting_GettingValueFromNutrients.pdf, 2010.

Meglia, GE., Nutrition and Immune Response in Periparturient Dairy Cows. *Doctoral Thesis*. Swedish University of Agricultural Sciences, Upsala, Sweden. 2004.

Miranda, M; Cruz, JM; López-Alonso, M; Benedito, JL. Variations in liver and blood copper concentrations in young beef cattle raised in north-west Spain: associations with breed, sex, age and season. *Animal Science*, 2006, 82:253–258.

Mulryan, G; Mason, J. Assessment of liver copper status in cattle from plasma copper and plasma copper enzymes. *Annales de Recherches Vétérinaires*. *1992*; *23: 233*-238.

Nicholson, FA; Jones, KC; Johnston, AE. Effect of phosphate fertilizers and atmospheric deposition on long-term changes in the cadmium content of soils and crops. *Environment Science and Technology*,1994, *28*, 2170–2175.

Niezen, JH; Charleston, WA; Hodgson, J; Miller, CM; Waghorn, TS; Robertson, HA. Effect of plant species *International Journal of Parasitology*. 1998, 28(5): 791-803.

Niezen, JH; Charleston, WAG; Hodgson, J; Mackay, AD; Leathwick, DM. Controlling internal parasites in grazing ruminants without recourse to anthelmintics: approaches, experiences and prospects. *International Journal of Parasitology*. 1996, 26: 983–992.

Niezen, JH; Waghorn, TS; Charleston, WAG; Waghorn, GC. Growth and gastrointestinal nematode parasitism in lambs grazing either lucerne (*Medicago sativa*) or sulla (*Hedysarum coronarium*) which contains condensed tannins. *The Journal of Agricultural Science*. 1995, 125: 281–289.

NRC (National Research Council). *Nutrient Requirements of Beef Cattle: 7th Revised Edition: Update 15*. National Academy Press, Washington, DC. *2000*.

NRC (National Research Council). *Nutrient Requirements of Dairy Cattle*, 7th edition. National Academy Press, Washington, DC. 2001.

Oltjen, JW; Beckett, JL. Role of ruminant livestock in sustainable agricultural systems. Journal *Animal Science*, 1996, 74: 1406-1409.

Owens, K. Copper, Cobalt and Molybdenum budgets on organic farms in Aberdeenshire. MSc Thesis, SAC, 2001.

Owens, K; Watson, C. Nutrient budgeting for trace elements: Examples from Scottish organic farms. In: Powell et al. (eds), UK Organic Research 2002: Proceedings of the COR Conference, 26-28 March 2002, Aberystwyth, pp. 147-148. Also available at: http://www.organic.aber.ac.uk/library/Nutrient%20budgeting%20for%20trace%20eleme nts.pdf

Poulsen, HD. Zinc and copper as feed additives, growth factors or unwanted environmental factors. *Journal Animal Feed Science*. 1998, 7: 135–142

Puls, R. Mineral levels in Animal Health. *Sherpa International, Clearbrook. , British Columbia, Canada, 1994.*

Rigby, D; Cáceres, D. Organic farming and the sustainability of agricultural systems. *Agricultural Systems*. 2001, 68: 21-40.

Schumacher, U. Reduction of heavy metal input – a task also for organic animal husbandry. In: Hovi, M; Zastawny, J.; Padel, S (Eds). *Enhancing animal health security and food safety in organic livestock production.* Proceedings of the 3rd SAFO Workshop 16-18 September 2004, Falenty, Poland, 2004.

Scott, P. Trace element deficiency in growing lambs. NADIS Health Bulletin. 2007. Available at: http://www.nadis.org.uk/EBLEX%20Bulletins/07-04Trace%20element %20deficiency.pdf

Shuman, LM. Zinc, manganese and copper in soil fractions. *Soil Science*. 1979, 127(1): 10-17.

Singh, A. Managing internal parasites in organic Livestock, 2010. Available at: http://ebookbrowse.com/managing-internal-parasites-in-organic-livestock-pdf-d17044196

Socha, MT; Tomlinson, DT; Rapp, CJ; Johnson, AB. Effect of nutrition on claw health. *Proceedings of the Society of Dairy Cattle Veterinarians, New Zealand Veterinary Association Conference*, Foundation for Continuing Education of the NZ Veterinary Association. Massey University, Palmerston North, NZ, 2002, 73–91.

Stark, BA; Livesey, CT; Smith, SR; Suttle, NF; Wilkinson, JM; Cripps, PJ. Implications of Research on the Uptake of PTEs from Sewage Sludge by Grazing Animals. Report to the Department of the Environment, Transport and the Regions (DETR) and the Ministry of Agriculture, Fisheries and Food (MAFF). WRc, Marlow, UK, 1998.

Stevensen, FJ; Ardakani, MS. Organic Matter Reactions Involving Micronutrients in Soils. In: Mortvedt, Giordani, Lindsay (Eds). *Ch.5. Micronutrients in Agriculture*. Soil Science Society. Amer.Inc. Wisconsin, USA. 1972, 79-114.

Sullivan, KH. Maximizing dry matter intake from pastures. Available at http://www.extension.org/article/19660, 2010.

Sundrum, A. Assessing animal welfare standards of housing conditions-possibilities and limitations. In: Sörensen, J., *Livestock Farming Systems. More than Food Production.* EAAP, Publ. 1997, 89, 238-246.

Suttle, NF. Mineral Nutrition of Livestock. 4th Edition, Cabi Publishing, UK. 2010.

Tame, MJ. Management of trace elements and vitamins in organic ruminant livestock nutrition in the context of the whole farm system. Institute of Organic Training & Advice: *Research Review: Management of trace elements and vitamins*, 2008.

Terry, N; Zayed, AM; de Souza, MP; Tarun, AS. Selenium in higher plants. *Annual Review of Plant Physiology and Plant Biology*. 2000, 51: 401-432.

Thamsborg, SM; Roepstorff, A. Running head: parasites and organic livestock parasite problems in organic livestock production systems and options for control. Available at http://orgprints.org/4509, 2010.

Thornton, I; Abrahams, P. Soil ingestion—a major pathway of heavy metals into livestock grazing contaminated land. *The Science of the Total Environment*. 1983, 28: 287-294.

Underwood, EJ; Suttle, NF. *The mineral nutrition of livestock*. (3rd edition), Cabi Publishing, UK. *1999*.

Van Ekeren, N; Wagenaar, JP; Jansonius, PJ. Mineral content of chicory (*Cichorium intybus*) and narrow leaf plantain (*Plantago lanceolata*) in grass-white clover mixtures. In: Helgadóttir, Á; Pötsch, EM (Eds). *Proceedings of the Final Meeting Quality Legume-Based Forage Systems for Contrasting Environments*. Gumpenstein, Austria, 2006: 121-123.

Wagner, WC. Sustainable agriculture: how to sustain a production system in a changing environment. *International Journal for Parasitology*, 1999, 29(1), 1-5.

Walsh,T; Fleming, GA. Trace elements in agriculture. Proceeding Symposium Soil, Plant Animal Interface. University of Aberdeen. 1978, 1-24.

Weiss, WP. Relationship of mineral and vitamin supplementation with mastitis and milk quality. *Proceedings of National Mastitis Council Annual Meeting, 2002,* 37-44.

Weller, RF; Bowling, PJ. The yield and quality of plant species grown in mixed organic swards. In: Kyriazakis, Zervas (Eds.) *Proceeding of Organic Meat and Milk from Ruminants*, Athens, EAAP Publication, 2002, 106: 177–180.

In: Agricultural Research Updates. Volume 2   ISBN: 978-1-61470-191-0
Editor: Barbara P. Hendriks   © 2012 Nova Science Publishers, Inc.

Chapter 9

TOXIC ELEMENT CONTAMINATION IN ANIMAL FEED

Marta López-Alonso[1] and Marta Miranda[2]

[1] Universidade de Santiago de Compostela, Departamento de Patoloxía Animal,
Facultade de Veterinaria. Lugo, Spain
[2] Universidade de Santiago de Compostela, Departamento de Ciencias Clínicas
Veterinarias, Facultade de Veterinaria. Lugo, Spain

ABSTRACT

Industrial and agricultural activities are responsible for polluting the environment
with toxic elements. In terms of potential adverse effects on animal and consequently
human health, cadmium, lead, but also mercury and the metalloid arsenic are amongst the
elements that have caused most concern; this is because they are readily transferred
through food-chains and their exposure can result in adverse effects on a great variety of
physiological and biochemical processes. Although episodes of lethal toxicity associated
to accidental exposure to very high doses of toxic elements have been largely described
in literature, the main concern for livestock is dietary exposure, in agricultural regions
diet being the main source of toxic elements for animals. This chapter reviews current
data on toxic element concentrations in animal feedstuffs and husbandry practices related
to toxic metal exposure, and analyse the effect of toxic element exposure on animal
health and residues in animal products.

Keywords: toxic metals; animal feed; routes of exposure; animal health; residues in animal
products.

INTRODUCTION

Anthropogenic activities, mainly related to industrial and agricultural development, have
been largely responsible for pollution of the environment with toxic metals, although some
contamination is also derived from natural geological resources. This is because elements that
are potentially toxic to animals and humans also have important industrial uses. In terms of
potential adverse effects on animal and consequently human health, cadmium, lead, but also

mercury and the metalloid arsenic are amongst the elements that have caused most concern. They naturally occur at trace levels in the foods and tissues of animals, and although very recent studies suggest that they might be essential dietary constituents for the normal development and health of livestock at ultratrace concentrations (for review see Suttle, 2010) it is well demonstrated that they adversely affect the animal or human health when amounts in the diet or water become excessive. Numerous episodes of lethal toxicity associated to accidental exposure to very high doses of toxic metals have been largely described in the literature, however the main concern of toxic metals in livestock is dietary exposure in animals raised in polluted environments. Dietary sub-lethal exposure to these elements can result in adverse effects on a variety of physiological and biochemical processes.

The prevention of toxic metal contamination in animal feed is an essential part of animal nutrition and management. Although feed contamination by toxic elements cannot be entirely avoided given the prevalence of these pollutants in the environment, there is a clear need for such contamination to be minimized with the aim of reducing both direct effects on animal health and indirect effects on human health. In this context, it is essential to establish the levels at which each toxic element becomes toxic to safeguard the feed that animals consume to optimize the animal health and minimize residues in food for human consumption.

TOXIC METAL CONCENTRATIONS IN FEED MATERIALS

Official information about metal content in feed materials is scarce outside of the EU, and although there are some data published on the scientific literature on toxic metal residues in animal feedstuffs (for detail review Kabata Pendias and Pendias, 2001 or NRC, 2005) they are generally based on samples collected at a very local scale and/or difficult to classify in categories. Furthermore, it is difficult to do a comprehensive review of toxic metal content in animal feed materials worldwide; because limited information is available about the nature of the samples, methods of chemical and statistical analysis and the way data are presented. Figures 1-4 summarize available data that can be reliable categorized both for feed materials and for commercially manufactured compound or complementary feeds.

Arsenic

Arsenic concentrations in agricultural non-polluted soils range from 0.1 to 40 mg/kg DM (Kabata-Pendias and Pendias, 2001). Mining and smelting activities, dating back centuries, distributed arsenic in the environment. A study carried out in New Zealand found far less soil arsenic (5.4 mg/kg DM) and no evidence of enrichment associated with farming (Longhurst et al., 2004). Arsenic uptake by plants largely varies depending on factors as the amount of soluble arsenic species in the soil, soil properties, redox and pH conditions and microbiological activity, as well as the plant species. Very limited information is available regarding arsenic speciation and metabolism in terrestrial plants and concentrations are regularly measured as total arsenic. Information on arsenic concentration in single components, apart from in fish and marine products and mineral supplements, is sparse (Figure 1) although generally arsenic concentrations in forages and crops grown on agricultural non-polluted soils remain below 0.5 mg/kg DM (Underwood and Suttle, 1999); arsenic concentrations up to 73 mg/kg DM were reported in the proximities of industrial smelting plants (Kabata-Pendias and Pendias, 2001). Unlike the main toxic metals cadmium and lead, the oral bioavailability of arsenic from the soil is considerably lower than from water or food, so it can be assumed that soil contamination of forages little contributes to total animal exposure. Within the feed components in animal diets, cereal grains content little arsenic (<0.2 mg/kg DM) and the highest arsenic residues are found in products of marine origin, as fish meal and fish oil, even though arsenic is present as relatively non-toxic organic forms (e.g. arsenobetaine and arsenosugars). Arsenic concentrations in mineral supplements are higher than in the main dietary components, however the low inclusion rate does not make it a significant contributing factor. Finally, drinking water can be a major source of inorganic arsenic in some arseniferous areas (e.g. aquifers under strongly reducing or oxidizing conditions, pH >8 or mining and geothermal areas; Mandal and Suzuki, 2002) with arsenic concentrations as high as 5 mg/l (in unpolluted fresh waters arsenic range between 1-10 µgl/l). The maximum acceptable concentrations of arsenic in drinking water in the European Union is 50 µg/l, and in the USA is 10 µg/l. The European Union has established permissible levels in animal feeds in 2 mg/kg DM, making exceptions for some forage by-products (4 mg/kg DM) (EC, 2002).

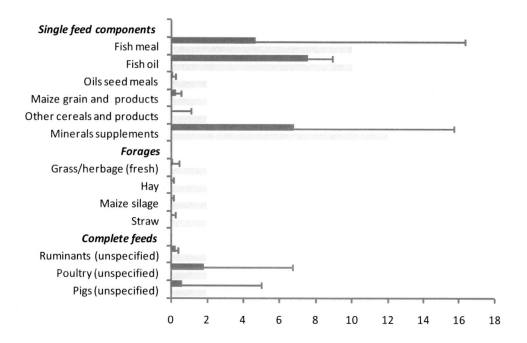

Figure 1. Total arsenic content in feed materials (mg/kg DM) expressed as mean values and maximum (in dark) and EU legislation on maximum concentrations in feed materials (in light).
Sources: Opinion on the Scientific Panel on Contaminants in the Food Chain on a request from the Commission related to arsenic as undesirable substance in animal feed (The EFSA Journal (2005) 180, 1-35). EU Directive 2002/32/EC of 7 May 2002 on undesirable substances in animal feed.

Cadmium

Cadmium is a highly reactive toxic element that is sparsely distributed in most agricultural ecosystems (Pinot et al., 2002). Cadmium concentrations in non contaminated soils range from 0.06 to 1.1 mg/kg DM (Kabata Pendias and Pendias, 2001) the main factor determining the cadmium content of soil being the chemical composition of the parent rock. The uptake of cadmium by plants is generally poor, and forages and crops on agricultural regions remains generally below 1 mg/kg DM (Figure 2), even though cadmium concentrations >10 mg/kg DM (1 fold higher) were observed in herbage growing close to industrial areas (Kabata Pendias and Pendias, 2001). Cadmium is strongly retained in the top-soil and cadmium uptake by plants depends on soil pH, plant species and the part of the plant sampled. In general, cadmium concentrations in forage crops are higher than in concentrate feed materials; this is associated at least in part, like other toxic metals, to soil physical contamination during harvesting or processing (EFSA, 2004a). Of particular concern is the contamination of pastures with cadmium derived from fertilizers and its subsequent fate in grazing livestock (Loganathan et al., 2008), so increased cadmium concentrations are regularly found in crops after superphosphate fertilizers (levels of cadmium vary from 5-134 mg/kg); in fact cadmium enriched phosphate fertilizers are the most important exogenous sources of cadmium to agricultural land worldwide (Nicholson et al., 1994). In New Zealand, soil cadmium is correlated with soil phosphorus, indicating the importance of phosphorus fertilizers as a major source of cadmium enrichment (Longhurst et al., 2004). In fact, it has been demonstrated that the regular use of superphosphates has led to increase in soil and pasture cadmium sufficient to raise kidney cadmium in grazing lambs above the maximum acceptable concentrations (> 1mg/kg FW) (Bramley, 1990; Morcombe et al., 1994). The ingestion of fertilizer and topsoil during grazing adds to the body burden of cadmium in the grazing animal (Stark et al., 1998). Sewage sludge application to agricultural land is also responsible for an increase in cadmium concentrations in crops. In Europe, guidelines aim to restrict the accumulation of cadmium in sludge-amended soils to <3 mg/kg DM (sampling depth, 20 cm) (EC, 2002). Mineral supplements and pre-mixes generally contain higher cadmium concentrations than the other main components of the ration (McBride, 1998), however the significance of their cadmium contribution to the total animal intake is difficult to establish, since the low but very variable rate of inclusion. The European Union has established permissible levels in animal feeds of vegetable origin in 1 mg/kg DM and of animal origin in 2 mg/kg DM (EC, 2002).

Lead

Lead poisoning is one of the most frequently reported causes of poisoning in farm livestock (Suttle, 2010). Soil lead varies widely (Kabata Pendias and Pendias, 2001), and although levels of to 20-40 mg/kg DM are normal, significantly higher lead concentrations (> 500 mg/kg DM) have been reported as a result of industrial and mining activities (Geeson et al., 1998) and application of sewage sludges (Hill et al., 1998; Gaskin et al., 2003; Tiffany et al., 2006). Lead is poorly taken by plants and lead residues in forages and agricultural crops are generally below 1.5 mg/kg DM (Figure 3) and rarely exceed 5 mg/kg DM (Underwood and Suttle, 1999); even though lead concentrations in excess of 600 mg/kg DM have been reported in herbage grown on mining soils or to which sewage sludge has been applied. The principal threat to livestock and the consumer comes from the soil, as livestock consume soil while grazing. In fact, lead concentrations in forages are generally higher than in concentrate feed materials; despite, the low lead uptake by plants indicates that in most cases forages are contaminated with soil or sewage sludges spread on the farmland (EFSA, 2004b). As for the other toxic metals, the processes of harvesting herbage for conservation as hay or silage often result in soil being picked up with the crops, and therefore elevated levels of lead may reflect spurious contamination. When lead-contaminated pasture is ensiled, downward migration of lead into the silo can lead to a fourfold increase in lead concentration in the lowermost layers (Coppock et al., 1988). Lead concentrations in mineral supplements and premixes can be higher compared to other feedstuffs, however their low proportion in the total diet makes their contribution to the total lead intake was low. The European Union has established permissible levels in animal feeds in 10 mg/kg DM, making exceptions with green fodder (40 mg/kg DM) (EC, 2002).

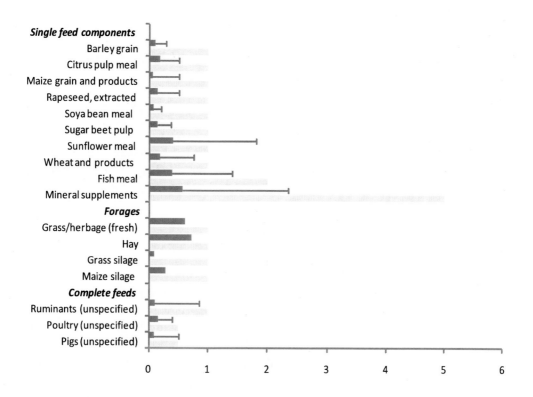

Figure 2. Cadmium content in feed materials (mg/kg DM) expressed as mean values and maximum (in dark) and EU legislation on maximum concentrations in feed materials (in light).
Sources: Opinion on the Scientific Panel on Contaminants in the Food Chain on a request from the Commission related to cadmium as undesirable substance in animal feed (The EFSA Journal (2004a) 72, 1-24). EU Directive 2002/32/EC of 7 May 2002 on undesirable substances in animal feed.

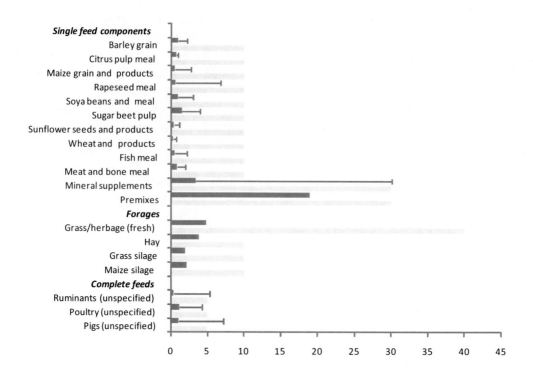

Figure 3. Lead content in feed materials (mg/kg DM) expressed as mean values and maximum (in dark) and EU legislation on maximum concentrations in feed materials (in light).
Sources: Opinion on the Scientific Panel on Contaminants in the Food Chain on a request from the Commission related to lead as undesirable substance in animal feed (The EFSA Journal (2004b) 71, 1-20). EU Directive 2002/32/EC of 7 May 2002 on undesirable substances in animal feed.

Mercury

The sources of mercury that pose threats to livestock or enter the food chain via livestock do not originate in the farm environment. Soils contain little mercury, the background, although not easy to determine due to the widespread mercury pollution, has been approximately estimated as 0.5 mg/kg DM; mercury concentrations exceeding this value should be considered as contamination from anthropogenic sources, mainly coal-burning power plants and other sources of fossil fuels combustion, waste incinerators and crematoria (Nriagu and Pacina, 1988). In general, mercury uptake by plants is very low and mercury residues are independent of soil mercury concentration, most of mercury found in the foliage have an atmospheric deposition origin (Kabata-Pendias and Pendias, 2001; Ericksen et al., 2003). Mercury concentrations in plants are very low, being very close or below the detection limits in most analyzed samples; this means that limited information exists in the literature on mercury residues in forages (Figure 4), being assumed that mercury exposure in farm animals feed almost entirely of forages in non mercury polluted areas is very low (ESFA, 2008), and anthropogenic mercury exposure accounting for most of mercury residues in grazing livestock (López-Alonso et al., 2003a). The only mercury-rich component in normal diets is likely to be fish meal and other fish feeds, mean total mercury concentrations being 1-2-fold higher than in other feed compounds. Relatively few data are available on the speciation of mercury in fish products; nevertheless the available data (ESFA, 2008) showed that is mainly present as methylmercury, the most toxic form of mercury. Mercury concentrations in mineral supplements and mineral feedstuffs are low and generally well below 0.1 mg/kg DM. The European Union has established permissible levels in animal feeds in 0.1 mg/kg DM, making exceptions with feedingstuffs produced by the processing of fish or other marine animals (0.5 mg/kg DM) (EC, 2002).

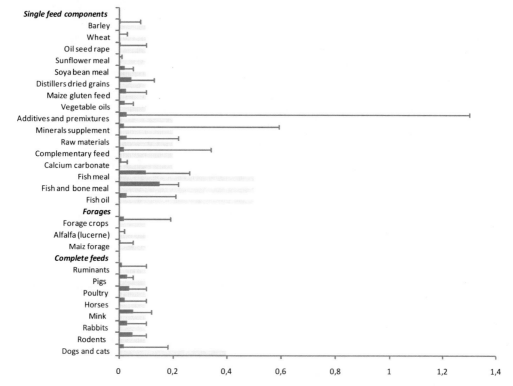

Sources: Mercury as undesirable substance in animal feed. Scientific opinion of the Panel on Contaminants in the Food Chain (The EFSA Journal (2008) 654, 1-74). EU Directive 2002/32/EC of 7 May 2002 on undesirable substances in animal feed.

Figure 4. Total mercury content in feed materials (mg/kg DM) expressed as mean values and maximum (bars) (in dark lines). EU legislation on maximum concentrations in feed materials (in light).

HUSBANDRY PRACTICES AND NUTRITIONAL MANAGEMENT RELATED TO ANIMAL TOXIC METAL EXPOSURE

An important point for consideration when evaluating livestock exposure to toxic metals is that dietary exposure will be highly conditioned by the nutritional management and other husbandry practices: minor dietary ingredients and/or supplements, soil ingestion or soil spurious contamination in foliages being responsible in many occasions of the main toxic metal exposure.

An essential issue that should be addressed is extensive versus intensive production systems. Although it is not possible to do general considerations because of the variety of husbandry practices worldwide in *extensive systems*—mainly represented by ruminants—animals receive a diet based on local products, usually from the own farm, supplemented with concentrate feeds to achieve the required level of production. Degree of feed supplementation is very variable worldwide, being low in areas with high agricultural land availability but also in sustainable systems. From a general point of view, in non-polluted agricultural areas, ingestion of soil when grazing or when consuming pastures or forages physically contaminated with soil, represents the main source of toxic metal exposure to extensively grown animals. Soil ingestion although highly variable (determined by seasonal, husbandry and weather-related factors) contributes up to 18 % of their dietary DM intake when grazing (Thornton and Abrahams, 1983) reaching maximum in out wintered animals. A recent study comparing toxic metal accumulation in cattle from different management practices has demonstrated that toxic metal residues are directly related to degree of extensification; animals that obtain most feed grazing have higher tissue toxic metal residues, despite toxic metal concentrations in feeds not being significantly higher (Blanco-Penedo et al., 2009). Finally, a special mention of the husbandry practices related to toxic metal exposure should be given to organic production systems: they arise to reduce environmental contamination, and some practices as the limit of use of inorganic fertilizers (e.g. phosphates with high cadmium content) or the limitation of inorganic mineral supplements in compound feeds, ensure that toxic metal concentrations in organic feeds are lower than in conventional ones (Blanco-Penedo et al., 2009).

On the contrary, in *intensive systems*—mainly represented by non-ruminant (pigs and poultry) but also highly intensive ruminant (feedlots or high intensive milk production cattle)—diets are almost entirely based on concentrate feed. The complete feed consist of a range of feed materials from the international market selected on the basis of price, availability, and the contribution that they make to the supply of nutrients required by the target animals at a specific level of production. Complete feeds are generally supplemented which additives and mineral supplements containing up to 2-3 fold toxic metal concentrations that the main ingredients; however, their contribution to the total metal intake is generally low because the low rate of inclusion, but can be significant for certain elements as arsenic and mercury, found at very low concentrations in forages and cereals, but very high in feeds of marine origin.

RISK OF TOXICITY FOR ANIMALS

Naturally occurring episodes of toxic metal poisoning have been largely described in previous literature, in most cases being associated to accidental exposure to high doses (single or repeated) of toxic metals that lead to acute or even lethal clinical manifestations (for review see for example NRC, 2005 or Suttle, 2010). However, the main concern of toxic metal toxicity in livestock is dietary exposure to relatively low levels of toxic metals caused by environmental contamination. These low levels of metal exposure are typically associated to subclinical toxicity, their effects on domestic livestock including mutagenicity, carcinogenicity, teratogenicity, inmunosuppression, poor body condition and impaired reproduction (Janicki et al., 1987; Bires et al., 1995) which cause clinical manifestations are not easy to identify, in the majority of cases being associated with a decrease in the animal productions. In this scenario it is a priority to establish the maximum tolerable level in the diet, defined as the dietary level that, when fed for a defined period of time, will not impair accepted indices of animal health or performance.

There is a vast literature on the toxicological effects of toxic metals on laboratory animals, but most of it has little or no nutritional significance because of the high dietary concentrations used during experiments. For livestock, toxicological database is very limited in terms of proper dose-response experiments and toxicological endpoints, which makes it impossible to properly establish what the concentrations of toxic metals are, that are not associated with risk in animals. In addition, animals that are very young, old, reproducing, sick, exposed to stressful environments or consuming nutritionally imbalanced diets can be especially sensitive to the toxic effects of metals (NRC, 2005).

Arsenic

Arsenic was once surpassed only by lead as a toxicological hazard to farm livestock, most incidents being associated to its use as rodenticide, parasitizide, insecticide and fungicides or even at low doses as growth promoters. However the incidence of arsenic toxicity has fallen markedly following the withdrawal of most arsenic-containing products in Europe, where clinical cases of arsenic toxicity associated to dietary exposure are rare. At high concentrations all livestock species are susceptible to the toxic effects of inorganic arsenic. Signs of chronic arsenic intoxication include depressed growth, feed intake, feed efficiency, and, for some species, convulsions, uncoordinated gait, and decreased hemoglobin (NRC, 2005).

The highest dietary arsenic tolerable concentrations, as well as the lowest level associated to toxicity, largely varies with animal species and arsenic compounds. NRC (2005) has suggested an arsenic maximum tolerable level for domestic animals of 30 mg/kg DM. The toxic dietary concentrations of arsenic are generally 2-3 fold greater than the concentrations normally found in animal feeds. In addition, marine organisms, which have been identified as the major sources of feed contamination with arsenic, accumulate the metalloid predominantly as non-toxic arsenobetaine and arsenocholine. Thus, except for localized areas where arsenic is extremely high in drinking water, major arsenic contamination by mining and smelting industries has occurred, the available data suggest that the non-toxic organic arsenic in feed materials do not pose a significant health risk to animals.

Cadmium

Clinical manifestations of cadmium toxicity largely varies depending on numerous factors as dose and time of exposure, species, gender, and environmental and nutritional aspects; in addition cadmium tissue accumulation is significantly influenced by dietary interactions with zinc and copper, but also with iron and calcium (López-Alonso et al., 2002b). Subsequently, large differences exist between the effects of a single exposure to a high concentration to cadmium or the most common low chronic dietary exposure. In general, clinical signs of cadmium toxicity in animals include kidney and liver damage, anaemia, retarded testicular development and infertility, enlarged joints, osteomalacia, scaly skin, hypertension, reduced growth and increased mortality (Underwood and Suttle, 1999; NRC, 2005).

Limits of tolerance, determined as minimum toxic doses or maximum tolerable levels, are hard to define because of the great variability on cadmium exposure and tissue deposition. Considering a quite constant dietary cadmium exposure during the productive life span, it is generally assumed that for most of the domestic species, cadmium concentrations up to 5 mg/Kg DM gross clinical symptoms are most likely to start. Pigs have been considered to be the most sensitive species and clinical signs of toxicity as growth retardation has been observed at concentrations >4.43 mg/kg DM (King et al., 1992). In this species the extensive use of copper and zinc as growth promoters largely increases the risk of adverse effects, as well as increasing cadmium accumulation in the liver and kidney (Ramberck et al., 1991; Rothe et al., 1992); these findings seem to be associated with the extraordinary capacity of the pig liver to respond to cadmium exposure with a significant increase in metallothionein synthesis (Henry et al., 1994). In ruminants, because to their great prevalence to copper deficiency, relatively low levels of cadmium in the diet (1 mg/kg DM) have shown to act as potent antagonists of copper metabolism, leading to secondary copper deficiency (Smith et al., 1999; Miranda et al., 2005). Cadmium intoxication in the other livestock species and in dogs and cats are rarely reported.

Lead

Lead poisoning is one of the most frequently reported causes of poisoning in farm livestock, with cattle the most commonly affected species because of their natural curiosity, licking and indiscriminate eating habits. Primary sources of lead exposure to animals include contaminated soils and lead-based products, especially batteries and older paints (Sharpe and Livesey, 2004; Miranda et al., 2006). Low levels of lead exposure cause subtle cardiovascular, hematological, and neurodevelopmental changes. Higher levels of exposure cause renal, gastrointestinal, hepatic, and immunological disturbances (NRC, 2005).

The effects of chronic lead dietary exposure on tissue lead in farm livestock are receiving increasing attention due to the subclinical effects of raised blood lead in young children (Suttle, 2010). A reliable estimate of dose-dependent effects of lead in livestock is not possible as the relevant information based on experimental studies is lacking. Like the other metals, the toxicity of lead depends on the chemical form, and it is generally recognized that following oral exposure, the toxicity of lead decreases in the following order: lead acetate>chloride>lactate> carbonate>sulfite> sulfate>phosphate. Young animals absorb lead more efficiently than older animals and have a lower tolerance (NRC, 2005). In pigs, 25 mg/kg DM from lead acetate results in decreased growth, but there is insufficient information to establish a maximum tolerable dose in this species. In ruminants, 250 mg/kg lead in the diet can be tolerated for several months without significant effects on performance; however, levels of lead in kidneys and bone become of concern if consumed by humans. Dogs tolerate 10 mg lead/kg DM without changes in functional indices in hematopoiesis or kidney function. In chickens slight but significant changes in growth and egg production occur with the addition of 1 mg Pb/kg DM as lead acetate, and 0.5 mg/kg of highly soluble lead source appears to be the maximum tolerable dose for chronic exposure in these species when dietary calcium levels are low; however, when dietary calcium levels are high, 100 mg Pb/kg DM is tolerated. It should be noted that the above maximum tolerable levels (NRC, 2005) are for highly available sources such as lead acetate, so animals are more likely to be able to tolerate higher levels of many other lead sources. Based on these toxicological data, the lead concentrations found in commercial feed are generally too low to induce clinical signs of toxicity. However, individual intoxications may result from ingestion of feed material originating from polluted areas, in fact incidental poisonings related to waste disposal are regularly reported. It is therefore recommended to monitor areas where sewage has been applied, but also farmland in close proximity to industrial activities likely to emit lead.

Mercury

While there is a large amount of data on mercury dose-response effects in laboratory animals, few and rather old data are available for livestock, mostly focused on clinical signs of toxicity observed in acute situations. Because of their differing bioavailabilities and tissue distributions, the toxicity profiles of organic mercury and inorganic mercury differ: accumulation of inorganic mercury in the kidneys causes changes in renal function, whereas the easy transport of methylmercury into the brain and across the placenta makes the nervous system and the fetus sensitive indicators for the organic form (NRC, 2005).

The livestock health risk to mercury dietary exposure is difficult to evaluate because most information on mercury residues in feedstuffs is given as total mercury concentrations and the methylmercury is the form of greatest toxicological concern. However, taking into account that the main source of mercury are marine products in which the predominant form is methylmercury, from a conservative position it could be assumed that all mercury in feedstuffs is methylmercury. The European Food Safety Authority (EFSA) has recently reviewed the risk of potential adverse effects of dietary mercury at concentrations allowed in the EU (EFSA, 2008). Within the domestic animal species cats are the most sensitive species to methylmercury toxicity and the current regulatory maximum limit for pets (0.4 mg/kg feed; see Figure 4) would not be low enough to protect the animal whose exclusively home-made diets are based on raw fish. Fur animals also seems to be quite sensitive to mercury, and although minks will be able to tolerate the maximum levels set for total mercury in complete animal feedstuffs in the EU, it cannot be excluded that the extensive use of offal from fish or other marine animals could result in neurotoxic effects in this species. For the other land animal species and poultry the mercury concentrations usually found in feedstuffs are well below the risk level for clinical toxicity.

RESIDUES IN ANIMAL PRODUCTS FOR HUMAN CONSUMPTION

Meat, milk and eggs are essential components of the human diet. Although toxic metal concentrations in muscle and other animal products as milk and eggs are generally low, offal such as liver and kidney often accumulate higher toxic metal concentrations than most other foods (López-Alonso et al., 2000) and can reach metal concentrations that might adversely affect human health even when animals are exposed to safe dietary levels (i.e. levels at or below their respective maximum tolerable levels). Consequently, in the case of animals raised to supply human food, acceptable concentrations of toxic elements in feeds must consider the health of those consuming food products derived from these animals as well as the health of the animal itself.

Transfer of dietary toxic metal to animal tissues is difficult to evaluate. This is because there is little information in the literature about experimental studies in livestock species given experimental doses of toxic metals at concentrations generally seen in animal diets. In general, the carry-over of an orally administered compound to animal tissues and products (milk, eggs) is dependent on the absorption, distribution, metabolism and excretion/deposition of the compound and its eventual metabolites; these biological processes largely vary depending on the chemical form, are dose and/or time dependent (especially for some bioacumulative elements like cadmium), and are influenced by other factors such as the interaction with other compounds (e.g. cadmium greatly interferes with essential elements as copper and zinc; López-Alonso et al., 2002)).

Arsenic

The carry-over of **arsenic** compounds from feeds to edible tissues of mammalian species and **poultry** is very low. Thus although from the human health point of view some caution is warranted, no major concerns on **arsenic** in animal feed seem necessary (NRC, 2005).

Very limited information is available on experimental studies on arsenic transfer to animal tissues. Arsenic residues in meat and edible offals of livestock (Woolson, 1975; Thatcher et al., 1985; Veen and Vreman, 1985; Vreman et al., 1986, 1988; Proudfoot et al., 1991) fed standard or control diets (< 2 mg/kg DM) are usually less than 0.01 mg/kg fresh weight. Similar arsenic residues are found in monitoring studies in animal products from various agricultural regions (Kramer et al., 1983; Vos et al., 1987; Jorhem et al., 1991; Salisbury et al., 1991; Kluge-Berge et al., 1992; López-Alonso et al., 2000, 2007). With increasing dietary arsenic exposure, arsenic residues significantly increase in all the analyzed tissues compared with control animals. However, the absolute arsenic residue levels vary significantly depending on animal species, arsenic compounds and duration of exposure (Vreman et al., 1986; 1988; Eisler, 1994).

Cadmium

After a long-term exposure to dietary cadmium target tissues are liver and kidney, cadmium deposition being dose and time dependent (Linden, 2002). In the muscle, on the contrary, cadmium deposition is very low and independent of the level of dietary cadmium exposure. Cadmium transfer to milk and eggs is very low or absent (<0.05%; Blüthgen, 2000). Cadmium residues in the liver and kidneys of livestock (Hansen and Hinesly, 1979; Sharma et al., 1979; Leach et al., 1979; Vreman et al., 1988; Smith et al., 1991; Lee et al., 1996; Hill et al.,1998; Linden et al.,1999; López-Alonso et al., 2000, 2007) fed standard diets (< 0.5 mg/kg DM) were in most cases below the EU maximum admissible levels for animal products (i.e. 0.05, 0.5 and 1 mg/kg fresh weigh for meat, liver and kidney respectively; EC,

2006). With increasing dietary cadmium exposure (1-5 mg/kg DM) cadmium residues in the liver and kidneys generally exceeded the cadmium permissible residues in all the farm animals and with diets containing > 5 mg/kg DM cadmium residues were one order of magnitude above these limits.

Lead

Lead does not accumulate in muscle during commonly encountered levels of lead exposure (Stark et al., 1998) and there is no risk to humans in consuming carcass meat from livestock exposed to lead. After a recent exposure, the highest lead residues are found in the liver and kidney, whereas after chronic exposure lead accumulates mainly in the bone. Lead concentrations in milk are usually much lower than blood levels. Results of experimental studies in livestock (Vreman et al., 1988; Hill et al., 1998; Phillips et al., 2003) fed diets containing levels of lead varying between 15-25 mg/kg DM indicate that although residues in the liver, and especially in the kidney, were generally higher than in the control animals, they remained below the maximum permissible levels for animal products (0.1 and 0.5 mg/kg fresh weight for meat and offal's, respectively; EC, 2006); in muscle lead residues were low and not significantly different from those in the control animals. At higher levels of lead dietary exposure (100 mg/kg DM) no significant changes in the tissue residues were found in the liver, kidney or muscle of sheep (Fick et al., 1976) and cattle (Dinius et al., 1973) although tissue levels significantly increased when dietary levels were 10 times higher. Blüthgen (2000) reported a carry-over percentage from feed to milk of 0.1–1%.

Mercury

After a chronic dietary exposure the highest mercury levels are present in the skin, nails, hair and feathers. Within the edible tissues, kidneys contain the highest residues, approximately 100-fold those in other tissues including liver or meat (Clarkson, 1992).

Data on mercury transfer into livestock animal tissues is very limited and no dose-response studies at the mercury concentrations usually found in feedstuffs are available. Data from biomonitoring studies in livestock from relatively unpolluted areas (Korsrud et al., 1985; Vos et al., 1986; Jorhem et al., 1991; Niemi et al., 1991; Salisbury et al., 1991; Kluge-Berge et al., 1992; Falandysz, 1993a,b; Raszyk et al., 1996; Ulrich et al., 2001; López-Alonso et al., 2003b, 2007) indicates that total mercury concentrations in meat and meat products are generally below 10-20 µg/kg fresh weight, being below the quantification limit (generally 1-5 µg/kg fresh weight) in many liver and muscle samples. Based on these data, mercury residues in meat products do not appear to pose a relevant risk for human health, in fact none of the EU or other countries have established a maximum mercury concentration for meat or meat products. However, the NRC (2005) points out that the levels of dietary and water mercury that are tolerated by livestock would result in tissue levels higher than 50 µg/kg, and consequently, standards for mercury levels in feed and water supplied to animals intended for human consumption in countries where maximum concentrations in animal feeds have not been established, should be based on tissue residue levels and not animal health concerns.

FUTURE TRENDS AND RESEARCH NEEDS

As indicated, information on toxic metal concentrations in animal feedstuffs out of the EU is sparse, quite old, and restricted to very isolated areas. Monitoring programs at a large scale, including standardized sampling and analytical procedures, to evaluate toxic metal concentrations in feedstuffs and animal products for human consumption are encouraged. These will allow to get more comprehensive information on toxic metal residues in feedstuffs worldwide, to identify contaminated feed commodities avoiding entering the food chain, to adopt measures to reduce toxic metal contamination in feedstuffs and to evaluate temporal trends of toxic metal contamination in agriculture. Because of the great prevalence of cadmium and lead in the terrestrial environment, these monitoring programs should include all categories of feedstuffs; in fact cadmium and lead residues above the regulatory limits are frequently found in feedstuffs from polluted, specially industrialized and mining areas, but also following the disposal of waste and sewage sludges. For arsenic and mercury, and within the exception of very located areas, most residues came from feeds of fish origin, which makes easier to control toxic residues entrance in the food chain.

Prevention of the adverse effects of toxic metals on the health of animals and consumers requires the setting of limits on toxic metal exposure to animals. Current information on maximum tolerable levels in livestock species as well as on the transfer to animal tissues is rather imprecise and based on experimental studies with animals exposed to toxic metal compounds and doses that do not resemble dietary exposure. New dose-response studies in real farm conditions are essential to establish regulatory maximum toxic metal concentrations in animal feedstuffs at an international level.

In addition, and because of the great variability in toxicity among arsenical and mercurial compounds, precise information on the concentration of the different compounds in feedstuffs is essential to properly evaluate the risk of animals to dietary exposure. Prior to getting this information, official methods should be developed and validated. For marine products the concentration of inorganic arsenic needs to be determined as a prerequisite for a comprehensive assessment of the potential animal health risks; more information on release of inorganic from organoarsenic compounds such as arsenosugars is also needed. Because methylmercury is the form of greatest toxicological concern, but also the predominant in marine organisms, the analysis of methylmercury in feeds should be encouraged.

REFERENCES

Bires, J; Dianovsky, J; Bartko, P; Juhasova, Z. Effects on enzymes and the genetic apparatus of sheep after administration of samples from industrial emissions. *Biometals*, 1995, 8, 53-58.

Blanco-Penedo, I; Shore, RF; Miranda, M; Benedito, JL; López-Alonso, M. Factors affecting trace element status in calves in NW Spain. *Livestock Science,* 2009, 123,198-208.

Blüthgen, AH. Contamination of milk from feed. *Bulletin of the International Dairy Federation*, 2000, 356, 43–47.

Bramley, RGV. Cadmium in New Zealand agriculture. *New Zealand Journal of Agricultural Research*, 1990, 33, 505–519.

Clarkson, TW. Mercury, major issues in environmental health. *Environment Health and Perspectives*, 1992, 100, 31-38.

Coppock, RW; Wagner, WC; Reynolds, RD. Migration of lead in a glass-lined, bottom-loading silo. *Veterinary and Human Toxicology,* 1988, 30, 458–459.

Dinius, DA; Brinsfield, TH; Willians, EE. Effect of subclinical lead intake on calves. *Journal of Animal Science*, 1973, 37, 169.

EC Commission Regulation No 1881/2006 of 19 December 2006 setting maximum levels for certain contaminants in foodstuffs. *Journal of the European Commission* OJ, 2006, L364/5, 20.12.2006, p.20.

EC Directive 2002/32/EC of the European parliament and of the Council of 7 May 2002 on undesirable substances in animal feed. *Journal of the European Commission* OJ, 2002, L140/10, 30.05.2002, p.12.

EFSA (European Food Safety Authority). Mercury as undesirable substance in animal feed. Scientific opinion of the Panel on Contaminants in the Food Chain. Adopted on 20 Feburary 2008. *The EFSA Journal*, 2008, 654, 1-74.

EFSA (European Food Safety Authority). Opinion on the Scientific Panel on Contaminants in the Food Chain on a request from the Commission related to cadmium as undesirable substance in animal feed. Adopted on 2 June 2004. *The EFSA Journal*, 2004a, 72, 1-24.

EFSA (European Food Safety Authority). Opinion on the Scientific Panel on Contaminants in the Food Chain on a request from the Commission related to lead as undesirable substance in animal feed. Adopted on 2 June 2004. *The EFSA Journal*, 2004b,71, 1-20.

EFSA (European Food Safety Authority). Opinion on the Scientific Panel on Contaminants in the Food Chain on a request from the Commission related to arsenic as undesirable substance in animal feed. Adopted on 31 January 2005. *The EFSA Journal*, 2005180, 1-35.

Eisler, R. A review of arsenic hazards to plants and animals with emphasis on fishery and wildlife resources. In: Nriagu JO, editor. *Arsenic in the Environment. Part II: Human Health and Ecosystem Effects*. John Wiley & Sons, Inc, 1994.

Ericksen, JA; Gustin, MS; Schorran, DE; Johnson, DW; Lindberg, SE; Coleman, JS. Accumulation of atmospheric mercury in forest foliage. *Atmospheric Environment*, 2003, 37, 1613-1622.

Falandysz, J. Some toxic and essential trace metals in cattle from the northern part of Poland. *The Science of the Total Environment*, 1993b, 36, 177-191.

Falandysz, J. Some toxic and essential trace metals in swine from Northern Poland. *The Science of the Total Environment*, 1993a, 136, 193-204.

Fick, KR; Ammerman, CB; Miller, SM; Simpson, CF; Loggins, PE. Effects of dietary lead on performance, tissue mineral composition and lead absorption in sheep. *Journal of Animal Science,* 1976, 42, 515-523.

Gaskin, JW; Brobst, RB; Miller, WB; Tollner, EW. Long-term biosolids application effects on metal concentrations in soil and Bermudagrass forage. *Journal of Environmental Quality*, 2003, 32, 146–152.

Geeson, NA; Abrahams, PW; Murphy, MP; Thornton, I. Fluorine and metal enrichment of soils and pasture herbage in the old mining areas of Derbyshire, UK. *Agriculture, Ecosystems and the Environment,* 1998, 217–231.

Hansen, LG; Hinesly, TD. Cadmium from soils amended with sewage sludge: effects and residues in swine. *Environment Health and Perspectives,* 1979, 28, 51-57.

Henry, RB; Liu, J; Choudhuri, S; Klaassen, CD. Species variation in hepatic metallothionein. *Toxicology Letters,* 1994, 74, 23-33.

Hill, J; Stark, BA; Wilkinson, JM; Curran, MK; Lean, IJ; Hall, JE; Livesey, CT. Accumulation of potentially toxic elements by sheep given diets containing soil and sewage sludge. 2. Effect of the ingestion of soils treated historically with sewage sludge. *Animal Science,* 1998, 67, 87-96.

Janicki, K; Dobrowolski, J; Krasnicki, K. Correlation between contamination of the rural environment with mercury and occurrence of leukaemia in men and cattle. *Chemosphere,* 1987, 16, 253-257.

Jorhem, L; Slorach, S; Sundstrom, B; Ohlin, B. Lead, cadmium, arsenic and mercury in meat, liver and kidney of Swedish pigs and cattle in 1984-88. *Food Additives and Contaminants,* 1991, 8, 201-212.

Kabata-Pendias, A; Pendias, H. *Trace Elements in Soils and Plants.* (3rd edition). CRC Press, Boca Raton, FL, US. 2001.

King, RH; Brown, WG; Amenta, VCM; Shelley, BC; Handson, PD; Greenhill, NB; Willcock, GP. The effect of dietary-cadmium intake on the growth-performance and retention of cadmium in growing pigs. *Animal Feed Science and Technology,* 1992, 37, 1-7.

Kluge-Berge, S; Skjerve, E; Sivertsen, T; Godal, A. Lead, cadmium, mercury and arsenic in Norwegian cattle and pigs. *Proceedings of the 3rd World Congress Foodborne Infections and Intoxications.* Berlin, 1992, pp. 745-748.

Korsrud, GO; Meldrum, JB; Salisbury, CD; Houlahan, BJ; Saschenbrecker, PW; Tittiger, F. Trace element levels in liver and kidney from cattle, swine and poultry slaughtered in Canada. *Canadian Journal of Comparative Medicine,* 1985, 49, 159-163.

Kramer, HL; Steiner, JW; Vallely, PJ. Trace element concentration in the liver, kidney and muscle of Queensland cattle. *Bulletin of Environmental Contamination and Toxicology,* 1983, 30, 588-594.

Leach, RM; Wang, KWL; Baker, DE. Cadmium and the food chain: The effects of dietary cadmium on tissue deposition in chicks and haying hens. *Journal of Nutrition,* 1979, 109, 437.

Lee, J; Rounce, JR; Mackay, AD; Grace, ND. Accumulation of cadmium with time in Rommey sheep grazing ryegrass-white clover pasture: Effect of cadmium from pasture and soil intake. *Australian Journal of Agricultural Research,* 1996, 47, 877-894.

Linden, A. Biomonitoring of cadmium in pig production. *Doctoral Thesis.* Swedish University of Agricultural Sciences. 2002.

Linden, A; Olsson, IM; Oskarsson, A. Cadmium levels in feed components and kidneys of growing/finishing pigs. *Journal of AOAC International,* 1999, 82, 1288-1297.

Loganathan, P; Hedley, MJ; Grace, ND. Pasture soils contaminated with fertilizer-derived cadmium and fluorine: livestock effects. *Reviews in Environmental Contamination and Toxicology,* 2008, 192, 29–66.

Longhurst, RD; Roberts, AHC; Waller, JE. Concentrations of arsenic, cadmium, copper, lead and zinc in New Zealand pastoral topsoils and herbage. *New Zealand Journal of Agricultural Research*, 2004, 47, 23–32.

López-Alonso ,M; Benedito, JL; Miranda, M; Castillo, C; Hernández, J; Shore, RF. Interaction between toxic and essential trace metals in cattle from a region with low levels of pollution. *Archives of Environmental Contamination and Toxicology*, 2002, 42, 165-172.

López-Alonso ,M; Benedito, JL; Miranda, M; Castillo, C; Hernández, J; Shore, RF. Arsenic cadmium, lead, copper and zinc in cattle from Galicia, NW Spain. *The Science of the Total Environment*, 2000, 246, 237-248.

López-Alonso ,M; Benedito, JL; Miranda, M; Castillo, C; Hernández, J; Shore, RF. Mercury concentrations in cattle from NW Spain. *The Science of the Total Environment*, 2003b, 302, 93-100.

López-Alonso ,M; Benedito, JL; Miranda, M; Fernández ,JA; Castillo, C; Hernández, J; Shore, RF. Large-scale spatial variation in mercury concentrations in cattle in NW Spain. *Environmental Pollution*, 2003a, 125,173-181.

López-Alonso, M; Miranda, M; Castillo, C; Hernández, J; García-Vaquero, M, Benedito, JL. Toxic and essential metals in liver, kidney and muscle of pigs at slaughter in Galicia north-west Spain. *Food Additives and Contaminants*, 2007, 24, 943-954.

Mandal, BK; Suzuki, KT. Arsenic around the world: a review. *Talanta*, 2002, 58,201-235.

McBride, MB. Growing food crops on sludge-amended soils: problems with the U.S. Environmental Protection Agency method of estimating toxic metal transfer. *Environment Toxicology Chemistry*, 1998, 17, 2274-2281.

Miranda, M; López-Alonso, M; Castillo, C; Hernández, J; Benedito, JL. Effects of moderate pollution on toxic and trace metal levels in calves from a polluted area of northern Spain. *Environment International*, 2005, 31,543-548.

Miranda, M; López-Alonso, M; Garcia, P; Velasco, J; Benedito, JL. Long-term follow-up of blood lead levels and haematological and biochemical parameters in heifers that survived an accidental lead poisoning episode. *Journal Veterinary Medicine, A*, 2006, 53, 305-310.

Morcombe, PW; Petterson, DS; Masters, HG; Ross, PJ; Edwards, JR. Cadmium concentrations in kidneys of sheep and cattle in Western Australia. 1. Regional distribution. *Australian Journal of Agricultural Research*, 1994, 45, 851–862.

Nicholson, FA; Jones, KC; Johnston, AE. Effect of phosphate fertilizers and atmospheric deposition on long-term changes in the cadmium content of soils and crops. *Environment Science and Technology*,1994, *28*, 2170–2175.

Niemi, A; Venäläinen, ER; Hirvi, T; Hirn, J; Karppanen, E. The lead, cadmium and mercury concentrations in muscle, liver and kidney from Finnish pigs and cattle during 1987-1988. *Zeitschrift für Lebensmittel-Untersuchung und -Forschung*, 1991,192, 427-429.

NRC (National Research Council). *Mineral tolerance of animals* (second revised edition) The National Academies Press, Washington, D.C. 2005.

Nriagu, JO; Pacyna, JM. Quantitative assessment of worldwide contamination of air, water and soils by trace metals. *Nature*, 1988, 333, 134-139.

Phillips, C; Gyori, Z; Kovacs, B. The effect of adding cadmium and lead alone or in combination to the diet of pigs on their growth, carcass composition and reproduction. *Journal of the Science of Food and Agriculture*, 2003, 83, 1357-1365.

Pinot, F; Kreps, SE; Bachelet, M; Hainault, P; Bakonyi, M; Polla, BS. Cadmium in the environment: sources, mechanisms of biotoxicity. *Reviews in Environmental Health*, 2002, 15, 299–323.

Proudfoot, FG; Jackson, ED; Hulan, HW; Salisbury, CDC. Arsanilic acid as a growth promoter for chicken broilers when administered via either the feed or drinking water. *Canadian Journal of Animal Science*, 1991, 71,221-226.

Rambeck, WA; Brehm, HW; Kollmer, WE. The effect of high dietary copper supplements on cadmium residues in pigs. *Zeitschrift fur Ernährungswissenschaft*, 1991, 30, 298-306.

Raszyk, J; Gajduskva, V; Ulrich, R; Nezveda, K; Jarosova, A; Sabatova, V; Docekalova, H; Salava, J; Palac, J; Scjöndorf, J. Evaluation of the presence of harmful pollutants in fattened pigs. *Veterinarni Medicina*, 1996, 9,261-266. [in Czech].

Rothe, S; Kollmer, WE; Rambeck, WA. Dietary factors influencing cadmium retention. *Revue de Médecine Vétérinaire*, 1992, 143, 255-260.

Salisbury, CDC; Chan, W; Saschenbrecker, PW. Multielement concentrations in liver and kidney tissues from five species of Canadian slaughter animals. Journal of AOAC International, 1991, 74,587-591.

Sharma, RP; Street, JC; Verma, MP; Shupe, JL. Cadmium uptake form feed and its distribution to food products of livestock. *Environment Health and Perspectives*, 1979, 28, 59.

Sharpe, RT; Livesey, CT. An overview of lead poisoning in cattle. *Cattle Practice*, 2004, 12, 199-203.

Smith, RM; Griel, LC; Muller, LD; Leach, RM; Baker, DE. Effects of dietary cadmium chloride on tissue, milk, and urine mineral concentrations of lactating dairy cows. *Journal of Animal Science*, 1991, 69, 4088-4096.

Stark, BA; Livesey, CT; Smith, SR; Suttle, NF; Wilkinson, JM; Cripps, PJ. Implications of Research on the Uptake of PTEs from Sewage Sludge by Grazing Animals. Report to the Department of the Environment, Transport and the Regions (DETR) and the Ministry of Agriculture, Fisheries and Food (MAFF). WRc, Marlow, UK, 1998.

Suttle, NF. Mineral Nutrition of Livestock. 2010, 4th Edition, Cabi Publishing, UK.

Thatcher, CD; Meldrum, JB; Wikse, SE; Whittier, WD. Arsenic toxicosis and suspected chromium toxicosis in a herd of cattle. *Journal of the American Veterinary Medical Association*, 1985, 187, 179-182.

Thornton, I; Abrahams, P. Soil ingestion—a major pathway of heavy metals into livestock grazing contaminated land. *The Science of the Total Environment*, 1983, 28, 287-294.

Tiffany, ME; McDowell, LR; O'Connor, GA; Martin, FG; Wilkinson, NS; Percival, SS; Rabiansky, PA. Effects of residual and reapplied biosolids on performance and mineral status of grazing steers. *Journal of Animal Science,* 2006, 80, 260–269.

Ulrich, R; Raszyk, J; Napravnik, A. Variations in contamination by mercury, cadmium and lead on swine farms in the district of Hodonin in 1994 to 1999. *Veterinarni Medicina*, 2001, 46,132-139.

Underwood, EJ; Suttle, NF. *The mineral nutrition of livestock* (3rd edition), Cabi Publishing, UK. 1999.

Veen, NG; Vreman, K. Transfer of cadmium, lead, mercury and arsenic from feed into various organs and tissues of fattening lamb. *Netherlands Journal of Agricultural Science*, 1985, 34,145-153.

Vos, G; Hovens, JPC; Delft ,WV. Arsenic, cadmium, lead and mercury in meat, livers and kidneys of cattle slaughtered in The Netherlands during 1980-1985. *Food Additives and Contaminants*, 1987, 4,73-88.

Vos, G; Teeuwen, JJMH; Van Delft, W. Arsenic, cadmium, lead and mercury in meat, livers and kidneys of swine slaughtered in the Netherlands during the period 1980- 1985. *Zeitschrift für Lebensmittel-Untersuchung und -Forschung*, 1986, 183, 397-401.

Vreman, K; van der Veen, NG; van der Molen, EJ; de Ruig, WG. Transfer of cadmium, lead, mercury and arsenic from feed into tissues of fattening bulls: chemical and pathological data. *Netherlands Journal of Agricultural Science*, 1988, 36, 327-338.

Vreman, K; Veen, NG; Molen, EJ; Ruig, WB.Transfer of cadmium, lead, mercury and arsenic from feed into milk and various tissues of dairy cows: Chemical and pathological data. *Netherlands Journal of Agricultural Science*, 1986, 34,129-144.

Woolson, EA. Arsenical pesticides. *ACS Symposium Series,* 1975, 7, 1-176.

In: Agricultural Research Updates. Volume 2
Editor: Barbara P. Hendriks

ISBN: 978-1-61470-191-0
© 2012 Nova Science Publishers, Inc.

Chapter 10

IMPORTANCE OF INTRINSIC AND EXTRINSIC ATTRIBUTES IN THE PURCHASE OF BEEF IN CENTRAL-SOUTHERN CHILE: A MARKET SEGMENTATION STUDY

B. Schnettler[1]*, F. Llancanpán[1], V. Vera[2], N. Sepúlveda[1], J. Sepúlveda[3] and M. Denegri[3]

[1] Departamento de Producción Agropecuaria,
Facultad de Ciencias Agropecuarias y Forestales
Universidad de La Frontera. Casilla, Temuco, Chile
[2] Magíster en Gestión y Manejo Agropecuario,
Facultad de Ciencias Agropecuarias y Forestales.
Universidad de La Frontera
[3] Departamento de Psicología, Facultad de Educación y Humanidades
Universidad de La Frontera

ABSTRACT

The objective of this study was to evaluate the importance of 25 intrinsic and extrinsic attributes in the purchase of beef, to detect relations between the attributes and to distinguish the existence of consumer segments in the south of Chile. The study was based on a direct survey of 1,200 people in the Maule, Biobío and Araucanía Regions, Chile. Five dimensions were obtained which characterise the relations between the attributes (62.7% variance). Consumers attributed great importance to the price and intrinsic attributes related to the organoleptic quality, harmlessness and health care. Three segments were distinguished, with differences of gender, region and zone of residence, age, family size, education, socio-economic level and ethnic origin. The majority group (42.0%) valued intrinsic and extrinsic attributes, the second group (29.8%) attributed little importance to the production system and the third group (28.5%) gave less

* Corresponding author: Tel.: 56-45-325655; fax: 56-45-325634
E-mail address: bschnett@ufro.cl

importance to the organoleptic quality. The results are discussed in relation to studies conducted in developed countries.

Additional Keywords: Intrinsic attributes, extrinsic attributes, beef, purchase decision, developing countries.

INTRODUCTION

Knowledge of consumer opinion is a key factor in the process of adding value to the meat and understanding purchasing decisions (Destefanis *et al.*, 2008). When making a purchase, consumers base their selection on different attributes before deciding and acquiring the product. However, products are conceived as a set of attributes, each one of which gives an indication that contributes to the formation of consumer preferences. Therefore, the quality of products is not a one-dimensional concept, but one which requires a multi-attribute approach (Chocarro *et al.*, 2009). The attributes can be divided into intrinsic, those related to the meat's physical aspects, and extrinsic, those related to the product but which are not physically part of it (Steenkamp, 1989). It has been reported that the intrinsic attributes have a greater influence than the extrinsic attributes on the decision to purchase beef (Krystallis and Arvanitoyannis, 2006).

The most highly valued intrinsic attributes in the selection of meat are tenderness (Goodson *et al.*, 2002; Krystallis and Arvanitoyannis, 2006; Severiano-Pérez *et al.*, 2006; Barrera and Sánchez, 2009), freshness (Henson and Northern, 2000; Verbeke *et al.*, 2000; Bernués *et al.*, 2003ab; Verbeke and Ward, 2006), color (Carpenter *et al.*, 2001; Kubberod *et al.*, 2002; Robbins *et al.*, 2003; Killinger *et al.*, 2004; Krystallis and Arvanitoyannis, 2006), fat and cholesterol content (Grunert, 1997; Robbins *et al.*, 2003; Roosen *et al.*, 2003; Killinger *et al.*, 2004, Jackman *et al.*, 2009), presence of additives and hormones (Verbeke *et al.*, 2000; Alfnes, 2004; Schnettler *et al.*, 2008a), nutritional value (Bernués *et al.*, 2003a; Bredahl, 2003; Roosen *et al.*, 2003), flavor (Goodson *et al.*, 2002; Krystallis y Arvanitoyannis, 2006; Severiano-Pérez *et al.*, 2006; Barrera and Sánchez, 2009), juiciness (Goodson *et al.*, 2002; Kristallis and Arvanitoyannis, 2006; Severiano-Pérez *et al.*, 2006), maturation time (Bernués *et al.*, 2003b; Revilla and Vivar-Quintana, 2006) and aroma (Severiano-Pérez *et al.*, 2006).

Among the extrinsic attributes of greatest importance in beef are country of origin (Bernués *et al.*, 2003ab; Roosen *et al.*, 2003; Alfnes, 2004; Loureiro and Umberger, 2007; Schnettler *et al.*, 2008b, 2009), animal welfare (Bernués *et al.*, 2003b; McEachern and Reaman, 2005; María, 2006; Schnettler *et al.*, 2008a), the feed the animal receives (Henson and Northern, 2000; Bernués *et al.*, 2003ab; Olaizola *et al.*, 2005; Schnettler *et al.*, 2008a), production system and traceability (Bernués *et al.*, 2003ab; Loureiro and Umberger, 2007), brand name (Bredahl, 2003; Banović *et al.*, 2009) and packaging (Carpenter *et al.*, 2001). With respect to the price, while some investigations conclude that this extrinsic attribute is in second place in the decision to purchase (Hui *et al.*, 1995; Schnettler *et al.*, 2008b, 2009), others claim it to be the predominant factor (Yen and Huang, 2002; Barrera and Sanchez, 2009). Nevertheless, the importance assigned to the attributes of meat differs according to the consumer's age (Bernués *et al.*, 2003ab; Alfnes, 2004; Krystallis and Arvanitoyannis, 2006; María, 2006; Verbeke and Ward, 2006; Schnettler *et al.*, 2008a), gender (Beardsworth *et al.*,

2002; Kubberod *et al.*, 2002; Alfnes, 2004; María, 2006), education (Bernués *et al.*, 2003b; Olaizola *et al.*, 2005; María, 2006), area of residence (Alfnes, 2004; Verbeke and Ward, 2006) and ethnic group (Yen and Huang, 2002).

Despite the promotion of beef exports in Chile, the domestic market predominates and is highly competitive. In 2008, beef production reached 240,257 tons, while 4,505 tons were exported and, 89,936 tons were imported, mainly from Paraguay and Argentina, but also from Australia, Brazil, Uruguay and the United States (ODEPA[1], 2009a). Chilean beef not only competes with the imported meat, but also faces stiff competition from poultry and pork, which is more easily prepared and has a wider variety of presentations than beef (ODEPA, 2008). The increase in the demand for poultry and the decrease in the consumption of beef has been associated with consumers' heightened concern for health issues (Bernués *et al.*, 2003ab; Bredahl, 2003; Roosen *et al.*, 2003; Loureiro y Umberger, 2007), changes in the relative prices (Yen and Huang, 2002) and modifications in consumer preferences (Resurreccion, 2003).

Based on this background, the aim of this study was to evaluate the relative importance of intrinsic and extrinsic attributes to the purchase of beef, to identify the connections between the attributes and to distinguish consumer segments in the Regions of Maule, Biobío and Araucanía, Chile.

MATERIALS AND METHODS

A personal survey was administered to a sample of 400 people in the Region of Maule, 400 in Biobío and 400 in the Araucanía, Chile, who are responsible for meat purchases in their home. This number of consumers was obtained using the formula of simple random sampling for non-finite populations according to the population of the 2002 Census (INE[2], 2003), considering 95% confidence, 5% estimation error and maximum dispersion (Fernández, 2002). The surveys were conducted in cities with more than 100,000 inhabitants in each region, and therefore all the surveys in the Region of the Araucanía were applied in Temuco (245,347 inhabitants; 38°45'S, 73°03'W). In the Region of Bío-Bío, surveys were distributed by proportional representation in Chillán (161.953 inhabitants, 119 surveys; 36°60'S, 72°12'W), Concepción (216,061 inhabitants, 159 surveys; 36°77'S, 73°07'W) and Los Angeles (166,556 inhabitants, 122 surveys; 37°45'S, 72°36'W). The same procedure was used for the Region of Maule, in Talca (201,797 inhabitants, 251 surveys, 35°25'S, 71°40'W) and Curicó (119,585 inhabitants, 149 surveys, 34°58'S, 71°14'W). The instrument of data collection was a questionnaire with questions on the frequency of beef consumption and monthly food and meat expenditure for the home. The questionnaire included questions to classify of survey participants: gender, age, area of residence, number of family members; ethnic origin, occupation and education of the head of the household and the possession of 10 domestic goods, with these last two variables meant to determine socioeconomic group (Adimarc, 2004). The survey was applied in two supermarkets in each city between August

---

[1] Office of Agrarian Studies and Policies
[2] National Statistics Institute, Chile.

and December 2008, once the questionnaire had been validated with a pilot test of 10% of the sample from Temuco.

In order to determine the importance of different intrinsic and extrinsic attributes in the purchase of beef, a scale with three levels was used (3: very important, 2: important and 1: not important) similar to the used one by Bernués *et al.* (2003b). The intrinsic attributes evaluated were: flavor, juiciness, aroma, color, tenderness, freshness, nutritional content, fat content, marbling, cholesterol content, absence of additives and hormones and maturation time. The extrinsic attributes were: price; country of origin; region of production in Chile; respect for the environment during production; type of feed; breed; grading[3]; place where it was born, bred, fattened and slaughtered; suitable treatment prior to slaughtering, brand, seal of quality assurance; uniform quality over time and packaging. The results were analyzed using descriptive statistics (calculation of averages) and then a Principal Component Analysis was used to determine those factors that explain the relations between the attributes. The Factor Analysis (FA) is a class of multivariate statistical methods that allows the structure of interrelations (correlations) to be analyzed among a great number of variables with the definition of a series of common underlying dimensions, known as factors (Hair *et al.*, 1999). The extraction of factors was done with eigenvalues greater than 1 and varimax rotation. In order to determine the relevance of the factor analysis, a KMO (Kaiser-Meyer-Olkin) test and Bartlett's test of sphericity were carried out. KMO values close to one indicate that the FA is a suitable procedure. Bartlett's test is used to verify whether the correlation matrix between variables is an identity matrix; the greater the Chi^2 and the lower the degree of significance, the more unlikely it is that it is an identity matrix (Vizauta, 1998).

In order to determine consumer segments according to the importance of intrinsic and extrinsic attributes in the purchase of meat, a hierarchical cluster analysis was used, with Ward's method (Hair *et al.*, 1999) as a type of connector and the Euclidean distance squared as a measure of similarity between objects (Hair *et al.*, 1999). The number of clusters was taken on the basis of the R^2 obtained and from a strong increase produced in the Cubic Criterion of Clustering and Pseudo-F values. In order to describe the segments, Pearson's chi-square test was applied to the discrete variables and an analysis of variance to the importance values of the attributes. Those variables where the analysis of variance resulted in significant differences ($P<0.001$) underwent Tukey's multiple comparisons test. The SPSS 16.0 program (SPSS, 2007) for Windows was used.

RESULTS

The total sample of consumers surveyed (n = 1,200) contained a greater proportion of women, between 35 and 64, belonging to families of three or four members, with children between 5 and 12 or 13 and 17, with complete secondary and complete technical college or incomplete university, from middle and high socioeconomic groups, resident in urban areas and non-Mapuche (Table 1). In addition, the sample had family buyers that ate beef three

---

[3] Chilean law 19.162, called *"The Meat Law"*, enacted on August 29, 1992, contains 11 articles that establish an obligatory system of cattle classification, the grading of its meats and the nomenclature of its cuts. Grading qualifies carcass beef considering the classification (gender and age) and some parameters that evaluate its quality (weight of hot carcass, degree of fat cover and as a complementary requirement fat and muscle color as well as hygiene standards) with the order being determined by uniform categories (V-C-U-N-O).

times per week, with an average expenditure on groceries for the home of US $257.20 and US $44.40 on meat (17.3% of food expenditure)

Table 1. Description in percentage of the total sample in Maule, Biobío and Araucanía Regions, Chile

Sample	Composition	Total sample
Gender	Female	24.0
	Male	76.0
Age	< 35 years	14.0
	35-49 years	37.4
	50-64 years	38.2
	65 years or more	10.4
Family size	1-2 family members	29.5
	3-4 family members	47.8
	5 or more	22.8
Presence and age of the children	Without children	18.0
	Children < 5 years	11.5
	Children 5-12 years	22.7
	Children 13-17 years	27.3
	Children $\geq$ 18 years	20.5
Education	Without studies	0.2
	Elementary incomplete	3.2
	Elementary complete	7.8
	Secondary incomplete	5.8
	Secondary complete	26.0
	Technical college incomplete	12.7
	Technical college complete or university incomplete	25.4
	University complete or higher	18.8
Socioeconomic group	ABC1	27.5
	C2	38.6
	C3	22.2
	D	9.4
	E	2.3
Residence	Urban	86.6
	Rural	13.4
Ethnic origin	Mapuche	11.7
	Non-Mapuche	88.3
Beef consumption frequency	Daily	2.0
	three times a week	57.6
	Once a week	30.2
	Occasionally	10.2
Average monthly expenditure (US$)[1]	Foods	257.2
	Meat	44.4

[1]The national currency values (Chilean pesos) were converted to dollars using the average 2008 value ($522.46/US$)
December 2008.

In accordance with the three-level evaluation scale used, the attributes can be classified (Table 2) as follows:

Table 2. Importance of intrinsic and extrinsic attributes in the purchase of beef and results of the factorial analysis, Maule, Biobío and Araucanía Regions, Chile

Attribute	Total sample		Factor				
	Average	Standard deviation	1	2	3	4	5
Region of production in Chile	1.72	0.694	0.832	0.045	0.067	0.121	-0.048
Type of feed	1.96	0.667	0.792	0.080	0.134	0.249	0.050
Country of origin	1.81	0.727	0.786	0.118	0.048	0.065	-0.142
Breed	1.90	0.648	0.752	0.034	0.080	0.304	0.021
Respect for the environment during production	2.00	0.705	0.735	0.166	0.089	0.151	0.101
Place where it was born, bred, fattened and slaughtered	1.97	0.681	0.672	0.063	0.099	0.320	0.059
Suitable treatment prior to slaughtering	2.14	0.668	0.550	0.129	0.137	0.410	0.162
Flavor	2.61	0.491	0.075	0.911	0.072	0.088	-0.039
Juiciness	2.56	0.514	0.102	0.886	0.060	0.071	0.060
Aroma	2.64	0.482	0.083	0.876	0.112	0.083	0.032
Color	2.54	0.545	0.126	0.830	0.157	0.070	0.110
Tenderness	2.51	0.549	0.032	0.619	0.330	0.186	-0.213
Absence of additives and hormones	2.55	0.587	0.168	0.162	0.757	0.005	0.043
Fat content	2.64	0.529	0.021	0.080	0.717	0.075	0.100
Freshness	2.62	0.518	0.026	0.388	0.654	0.150	-0.139
Cholesterol content	2.55	0.621	0.135	-0.022	0.640	0.023	0.220
Nutritional content	2.38	0.632	0.170	0.253	0.629	0.129	0.115
Marbling	2.44	0.603	0.022	-0.003	0.606	0.100	-0.113
Uniform quality over time	2.26	0.602	0.224	0.150	0.068	0.765	-0.048
Packaging	2.10	0.652	0.310	0.116	0.124	0.765	0.141
Brand	1.98	0.651	0.358	0.046	0.104	0.747	0.052
Seal of quality assurance	2.30	0.603	0.213	0.172	0.064	0.738	-0.035
Grading (V-C-U-N-O)	2.42	0.611	0.090	0.176	0.150	0.789	0.005
Maturation time	2.29	0.666	0.164	0.055	0.083	0.771	0.092
Price	2.79	0.478	-0.022	0.022	0.149	0.073	0.912
Variance by factor (%)			17.070	15.507	11.986	11.506	6.694
Accumulated variance (%)			17.070	32.576	44.563	56.069	62.763

Kaiser-Meyer-Olkin measure of sampling adequacy = 0.869. Bartlett's test of sphericity: Approximate Chi2 = 1,505.794; gl = 300; Sig. = 0.000. Method of extraction: Principal components analysis. Method of rotation: Varimax normalization with Kaiser. The rotation converged in six iterations. December 2008.

Very Important Attributes (Importance Averages between 2.5 and 3.0)

Price, aroma, fat content, freshness, flavor, juiciness, absence of additives and hormones, cholesterol content, color and tenderness.

Important Attributes (Averages between 2.0 and 2.49)

Marbling, grading, nutritional content, seal of quality assurance, maturation time, uniform quality over time, suitable treatment of the animals prior to slaughter, packaging and respect for the environment.

Important Attributes (Averages between 1.7 and 1.99)

Brand; place where the animal was born, bred, fattened and slaughter; feed, breed, country of origin and region of Chile where the meat was produced.

FACTORS THAT EXPLAIN THE CHOICE OF BEEF

By means of a principal component analysis, five dimensions were obtained that make it possible to characterize the relations between the attributes evaluated. The factors obtained represent 62.7% of the accumulated variance (Table 2). The value of the KMO test of sample adequacy is considered good and the Bartlett's sphericity test was significant ($P \leq 0.001$) (Vizauta, 1998). All the attributes presented correlated positively with their respective factors. The factors obtained are:

Factor 1. Production System

This factor explains 17.1% of the variance. The factor is comprised of the following attributes: region of Chile where the meat was produced, type of feed, country of origin, breed, respect for the environment; place where the animal was born, bred, fattened and slaughtered and adequate treatment of the animals prior to slaughter.

Factor 2. Organoleptic Attributes

Explains 15.5% of the variance. This is composed of the attributes: flavor, juiciness, aroma, color and tenderness.

Factor 3. Health and Safety

This factor explains 12.0% of the variance. This is comprised of the attributes: absence of additives and hormones, fat content, freshness, cholesterol content, nutritional content and marbling.

Factor 4. Signs of Quality Associated with Marketing

This explains 11.5% of the variance. This is composed of the attributes: uniform quality over time, packaging, brand, seal of quality assurance, grading and maturation time of the meat.

Factor 5. Price

This explains 6.7% of the variance and is composed only of the attribute price.

Despite Factor 5 having been composed of only one attribute, it was decided not to omit it because the price was the attribute of greatest importance in the group evaluated.

CONSUMER SEGMENTS ACCORDING TO IMPORTANCE ASSIGNED TO THE ATTRIBUTES

Using a hierarchical cluster analysis, three consumer segments were obtained with statistically significant differences ($P \leq 0.001$) in the importance of all the attributes evaluated and in the monthly expenditure on food ($P \leq 0.05$) and meat ($P \leq 0.001$) for the home (Table 3). Significant differences were observed between the groups identified according to the participant's region, gender, age, number of family members, education, socioeconomic group, ethnic origin ($P \leq 0.001$) and area of residence ($P \leq 0.05$) (Table 4), but not in the case of presence and age of the children and the frequency of beef consumption ($P > 0.1$). The segments obtained are described next:

Group 1. Meat Consumers Concerned about Intrinsic and Extrinsic Attributes

The largest group was made up of 42.0% of the consumers surveyed (n = 504), who assigned the greatest importance to all the attributes evaluated, significantly higher than the remaining groups ($P \leq 0.001$), except in the importance assigned to the cholesterol content and price, attributes that did not differ statistically from Group 3. This group presented amounts spent on foods and meat statistically similar to Group 2 and higher than Group 3 (Table 3).

Table 3. Importance of intrinsic and extrinsic attributes in the purchase of beef in the groups obtained through the analysis of hierarchical clusters. Maule, Biobío and Araucanía Regions, Chile

Attribute	Group 1 (n = 504)	Group 2 (n = 357)	Group 3 (n = 339)	Levene's statistic	F	P-value
Region of production in Chile	2.11 a	1.25 c	1.66 b	2.528	219.762 **	0.000
Type of feed	2.37 a	1.49 c	1.84 b	2.512	271.855 **	0.000
Country of origin	2.23 a	1.36 c	1.65 b	2.456	219.162 **	0.000
Breed	2.22 a	1.45 c	1.88 b	2.648	197.289 **	0.000
Respect for the environment during production	2.51 a	1.48 c	1.78 b	1.236	416.433 **	0.000
Place where it was born, bred, fattened and slaughtered	2.44 a	1.37 c	1.89 b	1.444	464.015 **	0.000
Suitable treatment prior to slaughtering	2.62 a	1.59 c	2.01 b	1.654	452.396 **	0.000
Flavor	2.90 a	2.78 b	2.01 c	1.024	913.109 **	0.000
Juiciness	2.85 a	2.69 b	2.01 c	1.333	549.761 **	0.000
Aroma	2.93 a	2.77 b	2.09 c	1.115	728.707 **	0.000
Color	2.80 a	2.68 b	2.01 c	1.059	371.573 **	0.000
Tenderness	2.78 a	2.52 b	2.10 c	2.621	207.971 **	0.000
Absence of additives and hormones	2.81 a	2.38 b	2.34 b	2.630	97.249 **	0.000
Fat content	2.82 a	2.48 b	2.52 b	2.639	60.318 **	0.000
Freshness	2.83 a	2.62 b	2.31 c	2.555	119.024 **	0.000
Cholesterol content	2.66 a	2.40 b	2.55 ab	2.648	18.165 **	0.000
Nutritional content	2.56 a	2.29 b	2.22 b	2.836	36.294 **	0.000
Marbling	2.82 a	2.48 b	2.52 b	2.845	26.507 **	0.000
Uniform quality over time	2.53 a	1.99 c	2.14 b	2.563	111.072 **	0.000
Packaging	2.41 a	1.70 c	2.06 b	2.452	158.97 **	0.000
Brand	2.26 a	1.56 c	2.01 b	2.450	153.364 *	0.000

Table 3. (Continued)

Attribute	Group 1 (n = 504)	Group 2 (n = 357)	Group 3 (n = 339)	Levene's statistic	F	P-value
Seal of quality assurance	2.61 a	2.01 b	2.15 b	2.554	145.371 **	0.000
Grading (V-C-U-N-O)	2.64 a	2.25 b	2.27 b	2.777	62.406 **	0.000
Maturation time	2.47 a	2.08 b	2.24 b	2.852	40.722 **	0.000
Price	2.83 a	2.71 b	2.82 a	2.883	7.764 **	0.000
Monthly expenditure (US$)						
Food	268.7 a	263.9 a	232.9 b	2.889	5.331*	0.005
Meet	47.2 a	49.3 a	34.9 b	2.789	12.861**	0.000

* Significance at 0.05%. ** Significance at 0.001%. Different letters indicate statistically significant differences according to the Tukey test of multiple comparisons (p ≤ 0.05).

Group 1. Meat consumers concerned about intrinsic and extrinsic attributes

Group 2. Meat consumers indifferent to the production system

Group 3. Beef consumers less concerned about organoleptic qualities

December 2008.

In relation to the total sample of survey consumers, Group 1 had the greatest proportion of people from the Region of the Araucanía (45.0%), men (30.6%), families with one or two members (36.9%), rural residents (16.9%), with complete university studies (22.0%), from socioeconomic groups ABC1 (31.3%) and of Mapuche origin (16.7%) (Table 4).

Table 4. Demographic characteristics with statistical differences between the groups obtained through the analysis of hierarchical conglomerates. Maule. Biobío and Araucanía Regions. Chile

	Group 1 (n = 504)	Group 2 (n = 357)	Group 3 (n = 339)	Pearson Chi-Sq.	P-value
Region					
Maule	34.3	25.2	40.3		
Biobío	20.6	33.1	52.5	172.500	0.000
Araucanía	45.0	41.7	7.1		
Gender					
Female	30.6	30.0	8.0	66.645	0.000
Male	69.4	70.0	92.0		
Age					
< 35 years	11.7	21.3	9.7		
35-49 years	37.5	33.6	41.3	25.566	0.000
50-64 years	40.3	36.4	36.9		
55 years or more	10.5	8.7	12.1		
Family size					
1-2 family members	36.9	35.0	12.7		
3-4 family members	43.8	43.1	58.4	65.300	0.000
5 or more	19.2	21.8	28.9		
Residence					
Urban	83.1	89.9	88.2	9.336	0.009
Rural	16.9	10.1	11.8		
Education					
Without studies	0.4	0	0		
Elementary incomplete	4.8	2.5	1.8		
Elementary complete	8.9	5.9	8.3		
Secondary incomplete	6.0	7.0	4.4		
Secondary complete	25.4	27.7	25.1	50.896	0.000
Technical college incomplete	8.3	11.5	20.4		
Technical college complete or university incomplete	24.2	24.6	28.0		
University complete or higher	22.0	20.7	12.1		
Socioeconomic Group					
ABC1	31.3	30.8	18.3		
C2	32.5	35.6	50.7		
C3	21.4	23.5	21.8	42.207	0.000
D	11.1	8.4	8.0		
E	3.6	1.7	1.2		
Ethnic origin					
Mapuche	16.7	9.2	6.8	22.100	0.000
Non-Mapuche	83.3	90.8	93.2		

P-value is the asymptotic significance (bilateral) obtained in Pearson Chi-Square Test.
Group 1. Meat consumers concerned about intrinsic and extrinsic attributes
Group 2. Meat consumers indifferent to the production system
Group 3. Beef consumers less concerned about organoleptic qualities
December 2008.

Group 2. Meat Consumers Indifferent to the Production System

This is made up of 29.8% of the consumers (n = 357). The people in this group gave low importance to the Factor 1 attributes, significantly lower than Groups 1 and 3. The importance assigned to the Factor 2 attributes was intermediate, differing statistically from Groups 1 and 3. This group gave significantly less importance than Group 1 to all the Factor 3 attributes, similar to Group 3 in the majority of the attributes except freshness. Group 2 attributed less importance to the Factor 4 attributes, significantly lower than Groups 1 and 3, except in the case of the seal of quality assurance, grading and maturation time, in which it did not differ from Group 3. This group afforded significantly less importance to price. The amounts spent on food and meat for the home were similar to Group 1 and significantly higher than Group 3 (Table 3). This group had a higher presence of people from the Araucanía (41.7%), men (30.0%), those under 35 (21.3%) and families with one or two members (35.0%) (Table 4).

Group 3. Beef Consumers Less Concerned about Organoleptic Qualities

This group was made up of 28.5% of the consumers (n = 339), who gave intermediate importance to the Factor 1 attributes, significantly more than Group 2. Although the average values are within the range of important attributes, this group was the one that gave significantly less importance to the Factor 2 attributes. Group 3 gave high importance to the price, similar to Group 1 and significantly higher than Group 2. The expenditure on food and meat were lower than the other groups by a broad margin (Table 3). Group 3 was made up of a greater proportion of people from the regions of Maule (40.3%) and Biobío (52.5%), women (92.0%), families with three or four members (58.4%) and five or more (28.4%), with complete technical education (20.4%) and from the C2 stratum (50.7%) (Table 4).

DISCUSSION

This study focuses on evaluating the relative importance of intrinsic and extrinsic attributes in the purchase of beef in central-southern Chile, a developing country in South America. Generally, the composition of the group of "very important attributes" makes it possible to confirm the greater influence that the intrinsic attributes (aroma, fat content, freshness, flavor, juiciness, color and tenderness) exert on the choice when purchasing beef. According to Krystallis and Arvanitoyannis (2006), the decision to purchase is related mainly to factors associated with enjoyment during consumption. Also, the presence of the attributes absence of chemical additives and hormones and cholesterol content in this group tallies with research that reports consumer concerns over health care in the decision to purchase beef, which is in line with studies in developed countries (Bernués et al., 2003ab; Bredahl, 2003; Roosen et al., 2003; Loureiro and Umberger, 2007).

The only extrinsic attribute within this group was the price, the one that obtained the greatest average importance value (2.79). This result agrees with studies conducted in Spain (Barrera and Sanchez, 2009) and the United States (Yen and Huang, 2002), in which price took first place in importance among the attributes considered in the purchase of beef.

However, this is contrary to what Hui *et al.* (1995) found in the United States and Schnettler *et al.* (2008b, 2009) found in Chile, where price was secondary. This contradiction of previous studies in Chile may be related to the period of time in which the survey was applied (August-December 2008), when the worldwide economic crisis which began in the United States was being widely broadcast. To this can be added the higher price of the different cuts of beef in 2008. By way of an example, according to figures cited by ODEPA (2009b), between 2004 and 2008 the price per kilogram of round increased in real terms by 25.6%, loin by 37.3% and stew beef by 23.4%. The second attribute in importance was the aroma, which confirms the results reported by Severiano-Perez *et al.* (2006) concerning the importance of this attribute in the quality of the meat relating to its freshness. The presence of fat content and color in the group of attributes of greatest importance accounts for the relevance of the meat's appearance when consumers are making their choice, as was the case with studies conducted in developed countries concerning color (Carpenter *et al.*, 2001; Kubberod *et al.*, 2002; Robbins *et al.*, 2003; Killinger *et al.*, 2004; Krystallis and Arvanitoyannis, 2006) and fat content (Robbins *et al.*, 2003; Roosen *et al.*, 2003; Killinger *et al.*, 2004). The importance of the fat content lies in the consumer preference for lean meats (Robbins *et al.*, 2003; Roosen *et al.*, 2003; Killinger *et al.*, 2004), while the color is used by consumers as a freshness indicator (Kubberod *et al.*, 2002; Robbins *et al.*, 2003; Killinger *et al.*, 2004). Additionally, the importance attributed to freshness agrees with the high valuation of this attribute in developed countries (Bernués *et al.*, 2003; Verbeke *et al.*, 2000; Verbeke and Ward, 2006). In these studies freshness was associated with the product's safety (Henson and Northern, 2000), constituting one of the main attributes of quality at the time of purchase in the supermarket, anticipating a pleasant experience during consumption (Bernués *et al.*, 2003b). The importance of flavor (Godson *et al.*, 2002; Krystallis and Arvanitoyannis, 2006; Severiano-Pérez *et al.*, 2006; Barrera and Sánchez, 2009) and juiciness (Godson *et al.*, 2002; Kristallis and Arvanitoyannis, 2006; Severiano-Pérez *et al.*, 2006) to consumer preferences are also aspects that will determine whether the consumption experience is enjoyable or not. The average importance obtained by the absence of additives and hormones in this investigation matches previous studies that report the rejection by consumers of their use in meat production (Verbeke *et al.*, 2000; Alfnes, 2004; Schnettler *et al.*, 2008a), associated with a growing concern for food safety. The presence of tenderness among the attributes of greatest importance corroborates results of studies in developed countries that indicate the attribute of palatability as most important and the primary determinant of quality, which positively affects consumer acceptance (Goodson *et al.*, 2002; Krystallis and Arvanitoyannis, 2006; Barrera and Sánchez, 2009).

The second group of attributes according to importance comprised intrinsic and extrinsic attributes. Among these, the importance assigned to marbling would be congruent with the relevance given to tenderness since this variable is used to predict palatability in beef (Jackman *et al.*, 2009). The importance of grading agrees with the high utility consumers assign to the presence of this information on the labeling, according to results from a previous study in Chile (Schnettler *et al.*, no date). The secondary importance of the nutritional content among the attributes evaluated agrees with the results of studies in developed countries (Bernués *et al.*, 2003a; Bredahl, 2003; Roosen *et al.*, 2003) that accredit this to consumers' in-depth knowledge of the subject. However, in the case of the area studied here, this aspect will have to be explored further with new investigations. Despite the maturation time improving the quality of the meat (tenderness and flavor), especially of beef (Bernués *et al.*, 2003b;

Revilla and Vivar-Quintana, 2006), this attribute is in the second group of importance. This is similar to what Bernués *et al.*, (2003b) found in Mediterranean countries where it is considered a cue for evaluating freshness in meat. The relevance assigned to the seal of quality assurance and the uniform quality over time agrees with consumers' growing need to have information about the quality of the product, both in developed countries (Verbeke *et al.*, 2000; Verbeke and Ward, 2006) and in Chile (Schnettler *et al.*, no date). Although since 1990 the literature has indicated the importance of animal welfare in the formation of consumer preferences (McEachern and Reaman, 2005, Maria, 2006), in this study the suitable treatment of the animal prior to slaughter was in the second group of attributes, in agreement with the results of investigations that put animal welfare as a secondary factor in the decision to purchase beef (Bernués *et al.*, 2003; Davidon *et al.*, 2003; Schnettler *et al.*, 2009). The presence of packaging in this group confirms to the secondary importance of this attribute in meat detected previously in Chile (Schnettler *et al.*, 2008b), which may be connected to this attribute affecting the intention to purchase, but not the satisfaction during consumption (Carpenter *et al.*, 2001). The importance assigned by survey participants to respect for the environment during the breeding and fattening of the animal is in line with consumers' increasing concern in developed countries for the impact of intensive production on the environment (Bernués *et al.*, 2003ab).

The group of "least important" attributes was formed entirely by extrinsic characteristics. Although it was to be expected that brand would be catalogued as an attribute of low importance, it is worthy of note that it takes first place in this group. Despite this result being contrary to the conclusions drawn in the studies by Bredahl (2003) in Denmark and Banović *et al.* (2009) in Portugal with respect to the high importance of brand name as a sign of quality in beef and its potential in differentiating the product (Banović *et al.*, 2009), it is plausible if one considers that to date the consumers of the regions in the study usually buy meat with no brand name due to scarcity and few data about brands of beef in Chile. The low importance assigned to the place where the animal was born, bred, fattened and slaughtered tallies with consumers' low interest in information referring to traceability (Verbeke *et al.*, 2000; Verbeke and Ward, 2006; Schnettler *et al.*, no date). However, this result can be attributed to consumers not requiring technical information, preferring information that is easily understood and processed quickly (Van Rijswijk *et al.*, 2008). The low importance attributed to the feeding of the animal runs contrary to results from previous studies that indicate this as an attribute relevant to the consumer (Henson and Northern, 2000; Bernués *et al.*, 2003ab; Olaizola *et al.*, 2005; Schnettler *et al.*, 2008a) associated with the safety of the product (Henson and Northern, 2000). If it is considered that in a previous study in Temuco it was determined that consumers value grassland feeding positively and reject feeding with broiler litter (Schnettler *et al.*, 2008a), it is possible to suggest that consumers are not concerned about feeding unless the type is indicated to them specifically. As for breed, in spite of the fact that cattle producers consider it one of the main factors in obtaining quality meat, those surveyed considered this attribute unimportant, similar to what has been reported in developed countries (Bernués *et al.*, 2003a). This can be accredited to consumers not knowing the differences between breeds or not having access to this information. This illustrates how the perception of quality differs between producers and consumers, thus demonstrating gaps in the information given to consumers and in the feedback to meat producers (Corcoran *et al.*, 2001). The low average importance of origin is in opposition to the results of studies conducted in both European countries (Bernués *et al.*, 2003ab; Roosen *et al.*, 2003; Alfnes, 2004; Loureiro and Umberger, 2007) and in Chile (Schnettler *et al.*, 2008b, 2009). Similar to what occurred with the importance of price, this may be a sign of consumers' changing priorities according to the economic period and the dynamic behavior of the markets. In addition to this, the region of Chile where the meat was produced registered the least average value among the attributes evaluated, which would indicate either a low preference or local identification with the origin of the meat in contrast to that determined by Bernués *et al.* (2003b) in Europe.

Through the use of a Principal Component Analysis, 25 attributes evaluated were associated in five dimensions. In general, the attributes of greatest importance were concentrated in the dimensions "price" (Factor 5), "organoleptic attributes" (Factor 2) and "health and safety" (Factor 3). The dimension "signs of quality associated with marketing" had the greatest presence of attributes considered important (Factor 4), while the dimension "production system" (Factor 1) gathered in the least important attributes. Factor 3 highlights the association of marbling with attributes related to health and safety. This seems to indicate that consumers perceive the presence of fat negatively and that they do not recognize the positive aspects associated with this attribute, such as flavor and tenderness, in agreement with Grunert (1997).

The use of the cluster analysis made it possible to distinguish three consumer segments regarding the importance assigned the attributes evaluated: "Meat consumers concerned about intrinsic and extrinsic attributes" (Group 1, 42.0%), "Meat consumers indifferent to the production system" (Group 2, 29.8%) and "Meat consumers less concerned about organoleptic quality" (Group 3, 28.5%). Although it was possible to confirm gender differences in the importance assigned to the attributes, the results are contradictory. The greater presence of men in the Group "Meat consumers indifferent to the production system", which assigned the least importance to the Factor 1 attributes, agrees with the lower concern detected in men regarding the effect of intensive production systems on the environment (Alfnes, 2004), animal welfare (Maria, 2006) and the origin of the meat (Alfnes, 2004). Nevertheless, the higher presence of men in the Group "Meat consumers concerned about intrinsic and extrinsic attributes" contradicts these authors' results. The majority proportion of women in the Group "Meat consumers less concerned about organoleptic quality" may be related to women's lower consumption of beef (Beardsworth *et al.*, 2002; Kubberod *et al.*, 2002; Schnettler *et al.*, 2008c), which is also congruent with the lower expenditure on meat in this group even though it is formed by larger families. With respect to age, the higher presence of people under 35 in the Group "Meat consumers indifferent to the production system" is opposed to the results of Maria (2006) in Spain and Schnettler *et al.* (2008a) in Chile regarding young people's heightened sensitivity towards animal welfare. However, this tallies with the lower valuation of the origin of beef in that age segment according to what was found by Alfnes (2004) in Norway. Again, the results of group composition according to educational level are contradictory. Thus, the greater presence of people with university studies in the Group "Meat consumers concerned about intrinsic and extrinsic attributes" agrees with the results of Bernués *et al.* (2003b) and Maria (2006) with respect to a higher value being placed on environmental aspects and animal welfare by consumers with higher education. Nevertheless, this is in opposition to the lower valuation of the origin of the meat in this type of consumer according to Alfnes (2004). As far as the socioeconomic level is concerned, the greater presence of people from the ABC1 stratum in the Group "Meat consumers concerned about intrinsic and extrinsic attributes" agrees with the results of

Olaizola *et al.* (2005) in Spain regarding which consumers with higher education levels and income value in large measure the attributes that differentiate meat on the basis of the production system. The greater presence of people residing in rural areas in the Group "Meat consumers concerned about intrinsic and extrinsic attributes" agrees with the results of Alfnes (2004) that the proximity of these consumers to livestock activity creates an appreciation for aspects connected to the production system. This same group contains the biggest proportion of people of Mapuche origin, confirming the importance of this demographic variable in consumer preferences (Yen and Huang, 2002). However, because this latter point may be related to the greater rurality of this group, this aspect will require further investigation.

The different regional composition among the groups agrees with the studies by Henson and Northern (2000), Bernués *et al.* (2003a) and Schnettler *et al.* (2009) that report on the importance of cultural differences in consumer preferences. It should, however, be noted that the Group "Meat consumers concerned about intrinsic and extrinsic attributes" and the Group "Meat consumers indifferent to the production system" had a larger presence of people from the Araucanía, which may be related to the greater presence of people who live in rural areas in Group 1.

The attributes considered most important in the selection of beef were those that determine their intrinsic quality and enjoyment during consumption, fundamental to a repeat purchase. This suggests the need for producers and slaughterhouses to optimize the organoleptic quality of the meat. Yet the existence of a consumer segment concerned about intrinsic and extrinsic attributes (Group 1; 42.0%) signifies the opportunity to advance towards the development of beef differentiated by the production system used, for which the efforts of all the actors in the meat production chain must come together.

With respect to the consumer sample, it should be pointed out that this is not representative of the country's population distribution. But the consumer distribution in this survey was similar to the sample obtained by Schnettler *et al.* (2006) and Schnettler *et al.* (2008) in supermarket consumer studies. Therefore, although the results and conclusions in this study may not be applicable to the total population, they might be for those consumers that purchase beef in supermarkets. The highest proportion of women surveyed clearly reflects the notion that in Chile women are more likely to do the shopping in the supermarket than are men, just as in developed countries (Harvey, Erdos and Chalinor, 2001; Verbeke and Vackier, 2004).

CONCLUSION

In the purchase of beef, consumers in the main cities of the Regions of Maule, Biobío and the Araucanía, Chile, afford high relevance to price and the intrinsic attributes related to organoleptic quality, health and safety. The principal component analysis made it possible to associate the 25 attributes evaluated in five dimensions corresponding to production system, organoleptic attributes, health and safety, sign of quality associated to marketing and price.

Three consumer segments were identified, with differing valuations of the attributes and different demographic profiles as far as gender, region and area of residence, age, family size, socioeconomic level, education and ethnic origin. The largest segment (42.0%) values the meat's intrinsic as well as extrinsic attributes. The second group (29.8%) gives low importance to the attributes related to the production system. The third group (28.5%) gives minor importance to the intrinsic attributes associated with the organoleptic quality of the meat.

ACKNOWLEDGMENTS

The results presented here were obtained as part of studies financed by Fondecyt Project 1080146 and Project FIA PIT-2007-009.

REFERENCES

Adimark, 2004. Mapa socioeconómico de Chile. Disponible en http://www.adimark.cl/medios/estudios/informe_mapa_socioeconomico_de_chile.pdf (Leído el 20 de octubre de 2005).

Alfnes, F., 2004. Stated preferences for imported and hormone-treated beef: application of a mixed logit model. *Eur Rev Agric Econ.* 31(1), 19-37.

Banović, M., Grunert, K.G., Barreira, M.M., Fontes, M.A., 2009. Beef quality perception at the point of purchase: A study from Portugal. Food Quality and Preference 20(4), 335-342.

Barrera, R., Sánchez, M., 2009. Consumption frequency and degree of abstraction: A study using the laddering technique on beef consumers. *Food Qual Prefer.* 20(2), 144-155.

Beardsworth, A., Bryman, A., Keil, T., Goode, J., Haslam, C., Lancashire, E., 2002. Women, men and food: the significance of gender for nutritional attitudes and choices. *Br Food J.* 104(7), 470-491.

Bernués, A., Olaizola, A., Corcoran, K., 2003a. Extrinsic attributes of red meat as indicators of quality in Europe: an application for market segmentation. *Food Qual Prefer.* 14(4), 265-276.

Bernués, A., Olaizola, A., Corcoran, K., 2003b. Labelling information demanded by European consumers and relationships with purchasing motives, quality and safety of meat. *Meat Sci.* 65(3), 1095-1106.

Bredahl, L., 2003. Cue utilization and quality perception with regard to branded beef. *Food Qual Prefer.* 15(1), 65-75.

Carpenter, C., Cornforth, D., Whittier, D., 2001. Consumer preferences for beef color and packaging did not affect eating satisfaction. *Meat Sci.* 57(4), 359-363.

Corcoran, K., Bernués, A., Manrique, E., Paccioli, M.T., Baines, R., Boutonnet, J.P., 2001. Current consumer attitudes toward lamb and beef in Europe. Options Méditerranéennes a46, 75-79.

Chocarro, R., Cortiñas, M., Elorz, M., 2009. The impact of product category knowledge on consumer use of extrinsic cues-A study involving agriffods products. *Food Qual Prefer.* 20(3), 176-186.

Davidson, A., Schröder, M.J.A., Bower, J.A., 2003. The importance of origin as a quality attribute for beef: results from a Scottish consumer survey. *Int J Cons Stud.* 27(2), 91-98.

Destefanis, G., Brugiapaglia, A., Barge, M.T., Dal Molin, E., 2008. Relationship between beef consumer tenderness perception and Warner-Bratzler shear force. *Meat Sci.* 78(3), 153-156.

Fernández, A., 2002. Investigación y técnicas de mercado. Primera edición. Editorial Esic, Madrid, España. 273 pp.

Goodson, K., Morgan, W., Reagan, J., Gwartney, B., Courington, S., Wise, J., Savell, J., 2002. Beef customer satisfaction: factors affecting consumer evaluations of clod steaks. *J Animal Sci.* 80(2), 401-408.

Grunert, K.G., 1997. What's in a steak? A cross-cultural study on the quality perception of beef. *Food Qual Prefer.* 8(3), 157-174.

Hair, J., Anderson, R., Tatham, R., Black, W., 1999. Análisis cluster. Análisis Multivariante. p. 407-454. Ed. Prentice Hall Internacional. Inc. España.

Harvey, J., Erdos, G., Chalinor, S., 2001. The relationship between attitudes, demographic factors and perceived consumption of meats and other proteins in relation to the BSE crisis: a regional study in the United Kingdom. *Health, Risk & Society*, 3, 181-197.

Henson, S., Northen, J., 2000. Consumer assessment of the safety of beef at the point of purchase: a pan-European study. *J Agric Econ.* 51(1), 90-105.

Hui, J., Mclean-Meyinsee, P.E., Jones, D., 1995. An empirical investigation of importance of ratings of meat attributes by Louisiana and Texas consumers. *J Agric Appl Econ.* 27, 636-643.

INE (Instituto Nacional de Estadísticas, Chile), 2003. Censo 2002. Resultados Volumen I: Población; País – Región. Instituto Nacional de Estadísticas, Gobierno de Chile, Santiago, Chile. 246 pp.

Jackman, P., Sun, D.W., Du, Ch., Allen, P., 2009. Prediction of beef eating qualities from colour, marbling and wavelet surface texture features using homogenous carcass treatment. Pattern Recognition 42(5), 751-763.

Killinger, K.M., Calkins, C.R., Umberger, W.J., Feuz, D.M., Eskridge, K.M., 2004. Consumer visual preference and value for beef steaks differing in marbling level and color. *J Anim Sci.* 82, 3288-3293.

Krystallis, A., Arvanitoyannis, I., 2006. Investigating the concept of meat quality from the consumers' perspective: The case of Greece. *Meat Sci.* 72(1), 164-176.

Kubberod, E., Ueland, O., Rodbotten, M., Westad, F., Risvik, E., 2002. Gender specific preferences and attitudes toward meat. *Food Qual Prefer.* 13(5), 285-294.

Loureiro, M.L., Umberger, W.L., 2007. A choice experiment model for beef: What US consumer responses tell us about relative preferences for food safety, country-of-origin labeling and traceability. *Food Policy*, 32(4), 496-514.

María, G.A., 2006. Public perception of farm animal welfare in Spain. *Livestock Science*, 103(3): 250-256.

Mceachern, M.G., Seaman, C., 2005. Consumer perception of meat production. *Br Food J.* 107(8), 572-593.

Odepa, 2008. Antecedentes de la carne bovina en Chile en el año 2007. Disponible en https://www.odepa.gob.cl/odepaweb/servlet/contenidos.ServletDetallesScr;jsessionid=0A 9312DBA97BD8FA2E3EF4ACF07EF3C4?idcla=2&idcat=&idn=2079 (Leído el 30 de julio de 2008).

Odepa, 2009a. Boletín Estadístico de Comercio Exterior N° 52. Disponible en http://www.odepa.gob.cl/odepaweb/servicios-informacion/ComexTrim/Bol-Trimestral-52.pdf (Leído el 20 de enero de 2009).

Odepa, 2009b. Estadísticas y precios / Series de precios / Avance mensual. Disponible en http://www.odepa.gob.cl/odepaweb/servlet/sistemas.precios.ServletPreciosScr;jsessionid =870A3EBD3D8748319A16F7A705751869 (Leído el 23 de febrero de 2009).

Olaizola, A., Whebi, Z., Manrique, E., 2005. Quality perception and consumer attitudes to "specific quality beef" in Aragón, Spain. *Span J Agric Res*. 3(4), 418-428.

Resurreccion, A.V.A., 2003. Sensory aspects of consumer choices for meat and meats products. *Meat Sci*. 66(1), 11-20.

Revilla, I., Vivar-Quintana, A.M., 2006. Effect of breed and ageing time on meat quality and sensory attributes of veal calves of the "ternera de Aliste" quality label. *Meat Sci*. 73(2), 189-195.

Robbins, K., Jensen, J., Ryan, K.J., Homco-Ryan, C., Mckeith, F.K., Brewer, M.S., 2003. Consumer attitudes towards beef and acceptability of enhanced beef. *Meat Sci*. 65(2), 721-729.

Roosen, J., Lusk, J.L., Fox, J.A., 2003. Consumer demand for and attitudes toward alternative beef labeling strategies in France, Germany and the UK. *Agribusiness*. 19(1), 77-90.

Severiano-Pérez, P., Vivar-Quintana, A.M., 2006. Determination and evaluation of the parameters affecting the choice of veal meta of the "Ternera de Aliste" quality appellation. Meat Sci 73(3), 491-497.

Schnettler, B., Vidal, R., Silva, R., Vallejos, L., Sepúlveda, N., 2008a. Consumer perception of animal welfare and livestock production in the Araucanía Region, Chile. *Chilean J Agric Res*. 68(1), 80-93.

Schnettler, B., Ruiz, D., Sepúlveda, O., Sepúlveda, N., 2008b. Importance of the country of origin in food consumption in a developing country. *Food Qual Prefer*. 19(4), 372-382.

Schnettler, B., Silva, R., Sepúlveda, N., 2008c. El consumo de carne en el sur de Chile y su relación con las características sociodemográficas de los consumidores. *Revista Chilena de Nutrición*. 35(1), 262-271.

Schnettler, B., Vidal, R., Silva, R., Vallejos, L., Sepúlveda. N., 2009. Consumer Willingness to Pay for Beef Meat in a Developing Country: The Effect of Information Regarding Country of Origin, Price and Animal Handling Prior to Slaughter. *Food Qual Prefer* .20(2), 156-165.

Schnettler, B., Silva, R., Sepúlveda, N., (Aceptada). Utilidad y Aceptación de Información en el Etiquetado de la Carne Bovina en Consumidores del sur de Chile. Chilean J Agric Res.

Steenkamp, J.B.E.M., 1989. Product quality: an investigation into the concept and how it is perceived by consumers. Royal Van Gorcum, Assen, The Netherlands. 288 pp.

Van Rijswijk, W., Frewer, K., Menozzi, D., Faioli, G., 2008. Consumer perceptions of traceability: A cross-national comparison of the associated benefits. *Food Qual Prefer*. 19(5), 452-464.

Verbeke, W., Ward, R., Viaene, J., 2000. Probit analysis of fresh meat consumption in Belgium: exploring BSE and television communication impact. *Agribusiness*. 16(2), 215-234.

Verbeke, W., Vackier, I., 2004. Profile and effects of consumer involvement in fresh meat. *Meat Sci*. 67, 159-168.

Verbeke, W., Ward, R., 2006. Consumer interest in information cues denoting quality, traceability and origin: An application of ordered probit models to beef labels. *Food Qual Prefer.* 17(6), 453-467.

Vizauta, B., 1998. Análisis estadístico con SPSS para Windows. Volumen II Estadística. 2ª edición. McGraw Hill/Interamericana de España, S.A.U., Madrid, España. 358 pp.

Yen, S., Huang, Ch., 2002. Cross-sectional estimation of U.S. demand for beef products: a censored system approach. *J Agric Res Econ.* 27, 320-334.

In: Agricultural Research Updates. Volume 2
Editor: Barbara P. Hendriks

ISBN: 978-1-61470-191-0
© 2012 Nova Science Publishers, Inc.

Chapter 11

TEN-YEAR ASSESSMENT OF AGRICULTURAL MANAGEMENT AND LAND-USE PRACTICES ON PESTICIDE LOADS AND RISK TO AQUATIC BIOTA OF AN OXBOW LAKE IN THE MISSISSIPPI DELTA, USA

Richard E. Lizotte, Jr., Scott S. Knight, Martin A. Locke and R. Wade Steinriede, Jr.

USDA-ARS National Sedimentation Laboratory,
Oxford, MS

ABSTRACT

The current chapter examined the combined influence of changing row crop production, implementation of agricultural Best Management Practices (BMPs), and enrollment of 112 ha into Conservation Reserve Program (CRP) on pesticide contamination and potential risk to lake aquatic biota in a 914-ha Beasley Lake watershed from 2000-2009. A suite of six current-use herbicides, five current-use insecticides, and two legacy insecticides were measured in lake surface water sampled approximately monthly from 2000-2009. Relative risk of these pesticides to lake aquatic biota was assessed using individual toxicity quotients (TQs), mixture pesticide toxicity index (PTI) scores based upon acute (48-96h) LC/EC50 values, and acute restricted-use pesticide levels of concern (LOCs) (LC/EC50 x 1, 0.5, and 0.1) for freshwater crustaceans (*Daphnia* sp.), insects (*Culex* sp.), fish (*Lepomis* sp.), and algae (*Psuedokirchneriella* sp.). During the ten-year study period, row-crop production shifted from primarily cotton in 2000-2001 to predominantly soybean in 2002-2004, 2006, 2008, and 2009 with milo and corn dominant in 2007. Reduced tillage BMPs were implemented in 2001 and CRP enrollment began in 2003. From 2000-2009, most individual pesticide concentrations were frequently <0.1 µg/L, with the exception of atrazine. Greatest herbicide concentrations occurred for triazine herbicides atrazine and cyanazine. Greatest insecticide concentrations occurred for methyl parathion and bifenthrin. Greatest legacy compound concentrations occurred for the organochlorine insecticide, *p,p'*-DDT. Temporally, peak lake water concentrations of current-use herbicides, current-use insecticides, and legacy compounds occurred during 2000, 2002, and 2002, respectively. Lowest lake water concentrations of current-use herbicides, current-use insecticides, and legacy compounds occurred during 2005. Results of the pesticide risk assessment showed greatest risk would be to crustaceans, primarily from the pyrethroid bifenthrin, during 2000-2002, with decreasing risk to fish and aquatic insects and minimal risk to algae. Although most individual pesticides were below LOCs, PTIs indicated increased risk from pesticide mixtures to aquatic fauna. Temporally, relative risk to lake aquatic biota decreased from greatest potential risk in 2000-2002 to minimal risk in 2005-2006 with infrequent risk to aquatic fauna in 2007-2009. Overall, lake water pesticide contamination decreased annually until 2005-2006 and increased again in 2007-2009 due, in part, to a shift in row crop from reduced tillage soybeans to conventional-till milo and corn in 2007. Concomitantly, changes in land-use with implementation of BMPs, CRP and crop type reduced the frequency and duration of risk of pesticides to lake aquatic biota.

INTRODUCTION

The Mississippi River watershed is the single largest drainage basin in North America encompassing approximately 2.9 million km^2 and includes some of the most intensive agricultural regions of the continent [1]. The lower Mississippi River alluvial plain (i.e. the Delta) is a predominantly agricultural region comprising the southern portion of the Mississippi River basin that extends over 1100 km from southeastern Missouri to Louisiana at the Gulf of Mexico and encompasses 18,130 km^2 [2]. The Delta is comprised of numerous aquatic systems that have been physically isolated from their respective main river channels such as bayous, sloughs and oxbow lakes [3]. Climatic conditions in the Delta provide for a long growing season with average annual rainfall amounts between 114 and 152 cm y^{-1} [4], enhancing the proliferation of insect and weed pests which require frequent pesticide use for control [5]. Associated with intensive pesticide use in the Delta is a high potential for transport into nearby aquatic systems such as lakes, rivers and streams. These aquatic systems often receive pesticide laden effluent from agricultural fields, primarily during storm events [6]. Historically, Delta oxbow lakes were once valued for their productivity and recreational use, especially fishing, but many oxbow lakes have become impaired from sediment, nutrient and pesticide contamination from nearby agricultural practices [6]. This has resulted in the degradation of environmental quality and ecological diversity in Delta oxbow lakes and their recreational popularity has decreased.

Beginning in 1994, a Mississippi Delta Management Systems Evaluation Area (MSEA) project was initiated with a primary objective of developing alternative farming systems providing combinations of region-specific Best Management Practices (BMPs) [2]. The project was essential in piloting the protection of surface and ground water resources and improving the environmental quality of aquatic systems in the Delta [7, 8]. The project included utilizing varying levels of physical BMPs (e.g. ponding of water using slotted board risers) and cultural (e.g. reduced tillage) and monitoring changes in environmental conditions (e.g. lake water quality) within Beasley Lake (Table 1), one of three Delta oxbow lake watersheds included in the study [7]. When the MSEA project concluded in 2002, one of three original watersheds, Beasley, was selected as a benchmark watershed for the

Conservation Effects Assessment Project (CEAP). Begun in 2003, CEAP is a US National Watershed Research Project conducted by the USDA Agricultural Research Service (ARS) and USDA Natural Resources Conservation Service (NRCS) to assess long-term changes in integrated farm management practices and BMPs on environmental quality at the watershed scale. In addition to BMPs initiated in Beasley Lake watershed during the MSEA project, additional practices were implemented in 2003 including changes from conventional to reduced tillage throughout the watershed, vegetated drainage ditches, and inclusion of the NRCS Conservation Reserve Program (CRP). Beasley Lake, an agriculturally impacted oxbow lake located in the Delta, was chosen as a CEAP watershed due to an extensive long-term data base (approximately 8 y) available on the contributions and understanding of BMP effectiveness at the watershed scale [7, 9]. Recently Locke et al. [10] reported some success in restoring productivity and recreational value in Beasley Lake. Lake improvements were measured as decreased sediment, nutrient and pesticide loads with concomitant increases in water clarity, primary productivity and fisheries productivity from 1995-2005 [7, 10].

Table 1. Agricultural best management practices in the
Beasley Lake watershed implemented from 1996-2009

Year	Location	Best Management Practice
1996	Edge-of-field	In-field impoundment using slotted board risers [7]
1996	Edge-of-field	In-field impoundment using slotted pipes [7]
1996	Edge-of-field	Vegetated filter strips using fescue (*Festuca arudinacea* Schreb.) [10]
1996	Edge-of-field	Vegetated filter strips using switchgrass (*Panicum virgatum*) [10]
2001	Watershed-wide	Reduced tillage [10]
2003	Edge-of-field	Constructed wetland [24]
2003-2004	Watershed-wide	Conservation Reserve Program cottonwood trees (*Populus deltoides* Bartr. ex. Marsh.) [10]
2006	Edge-of-field	Quail buffer

The purpose of this chapter is to examine the combined influence of changing row crop production, implementation of agricultural BMPs, and enrollment of 114 ha into CRP on pesticide contamination and potential risk to lake aquatic biota in Beasley Lake watershed from 2000-2009. Relative pesticide risk to lake aquatic biota was assessed using individual toxicity quotients (TQs), mixture pesticide toxicity index (PTI) scores based upon acute (48-96h) LC/EC50 values, and acute restricted use pesticide levels of concern (LOCs) (LC/EC50 x 0.1) for freshwater crustaceans (*Daphnia* sp.), insects (*Culex* sp.), fish (*Lepomis* sp.), and algae (*Psuedokirchneriella* sp.).

MATERIALS AND METHODS

Study Site Description

Beasley Lake watershed (Sunflower County, Mississippi, USA: latitude 33°24'15N, longitude 90°40'05W) was selected as a CEAP watershed beginning in 2003 (Figure 1). The watershed drainage area is approximately 915 ha and the lake has a surface area of about 25-30 ha. An unusual feature of the watershed is a 5.5-m change in elevation from the highest point in the watershed to the lake whereas the topography of most Delta oxbow lake watersheds include a <3-m change [11]. Approximately 150 ha of the watershed is non-arable wetland with hardwood forest and herbaceous riparian vegetation. The 722 ha of arable land in the watershed has been primarily farmed in cotton (*Gossypium hirsutum* L.), corn (*Zea mays* L.), soybeans (*Glycine max* [L.] Merr.) and occasionally milo (*Sorghum bicolor* L.). Cropping patterns within watershed have varied over the 10-y observation period (Table 2). From 2000 to 2002, an average of 401 ha cotton, 200 ha soybeans, and 11 ha corn were planted. From 2003 to 2006, an average of 99 ha cotton, 410 ha soybean, and 1 ha corn were planted. Beginning in 1995, only structural BMPs (e.g. grassed buffers, slotted pipes) were used in the watershed. These BMPs were followed with the implementation of reduced tillage in specific regions of the watershed in 2001 and reduced tillage throughout the entire watershed in 2003. Most recently, a 2003 enrollment into CRP consisted of 91 ha planted with cottonwood trees (*Populus deltoides* Bartr. ex. Marsh.) [10].

Table 2. Land-use and row crop production in hectares in the Beasley Lake watershed 2000-2009

Year	Cotton	Sorghum	Soybean	Wheat	Corn	Milo	CRP	Fallow	Forest
2000	514	0	155	0	34	0	0	18	150
2001	584	0	37	0	0	94	0	6	150
2002	106	0	407	98	0	101	0	9	150
2003	144	0	398	0	0	49	90	40	150
2004	119	0	432	0	0	0	112	58	150
2005	81	280	243	0	4	0	112	0	150
2006	52	0	557	0	0	0	112	0	150
2007	0	0	22	0	228	359	112	0	150
2008	0	0	609	0	0	0	112	0	150
2009	0	0	376	0	70	0	112	163	150

Figure 1. Aerial photograph of Beasley Lake agricultural watershed in the Mississippi Delta, USA.

SAMPLE COLLECTION AND PESTICIDE ANALYSIS

Surface aqueous samples of 4 L were collected approximately monthly in Beasley Lake from 2000-2009 at the midpoint of the lake (Figure 1). Aqueous samples were extracted and preserved on site with the addition of 4 g of KCl and 400 mL of pesticide grade ethyl acetate and manually mixed for one minute. Samples were placed on wet ice and transported to the USDA-ARS National Sedimentation Laboratory, Oxford, MS for pesticide analysis. Upon arrival, samples were stored at 4°C (usually <24 h) prior to pesticide analyses via gas chromatograph (GC) using a method developed by Smith et al. [12]. Sample preparation involved partitioning via separatory funnel and discarding the water phase. The pesticide-grade ethyl acetate phase was dried using anhydrous Na_2SO_4 and concentrated to near dryness via rotary evaporation. The extract was taken up in approximately 5-mL pesticide-grade hexane, subjected to silica gel column chromatography clean-up, and concentrated to 1 mL under ultra high purity (UHP) dry nitrogen for GC analysis. Two Agilent model HP 6890 GCs each equipped with dual Agilent HP 7683 ALS autoinjectors, dual split-splitless inlets, dual capillary columns, an Agilent HP Kayak XA Chemstation were used to conduct all pesticide analyses [12, 13]. One Agilent HP 6890 was equipped with two Agilent HP micro electron capture detectors (µECDs) and the other Agilent HP 6890 with one Agilent HP µECD, one Agilent HP nitrogen phosphorus detector (NPD), and an Agilent HP 5973 mass selective detector (MSD). Present and past pesticide usage in the Beasley Lake watershed area resulted in 13 pesticides being targeted for analysis (Table 3). The primary analytical column was an Agilent HP 5MS capillary column (30 m x 0.25 mm i.d. x 0.25 µm film thickness). Column oven temperatures were as follows: initial at 85°C for 1 min, ramp at $25°C$ min^{-1} to 190°C, hold at 190°C for 25 min, ramp at 25°C to 230°C, and hold for 30 min. Carrier gas was UHP helium at 28 cm sec^{-1} average velocity with the inlet pressure at 8.64 psi and inlet temperature at 250°C. Micro ECD temperature was 325°C with a constant gas flow of 40 mL min^{-1} UHP nitrogen. The autoinjector was set at 1.0-µL injection volume, fast mode. When necessary, pesticide residues were confirmed with an Agilent HP 1MS capillary column (30 m x .25 mm i.d. x 0.25-µm film thickness) and/or with the MSD (e.g. positive identification of a particular pesticide). Extraction efficiencies of all fortified samples analyzed for quality assurance/quality control were ≥ 90%. Level of detection for aqueous analyses ranged from 0.01-0.001 µg L^{-1} [12].

Calculation of Pesticide TQ, PTI, and LOC Values

A toxicity ranking system used in several recent studies [14, 15, 16] was utilized for the current chapter to assess relative risk of agricultural pesticides to a range of aquatic organisms representing various trophic levels. Toxicity data were obtained for each pesticide for four groups of organisms, crustaceans (*Daphnia* sp.), insects (*Culex* sp.), fish (*Lepomis* sp.), and algae (*Pseudokirchneriella* sp.) from USEPA Ecotoxicology Database (ECOTOX) [17]. Qualifying data were obtained for 10 pesticides for *Daphnia* sp., *Culex* sp., and *Lepomis* sp., and nine pesticides for *Pseudokirchneriella* sp. (Table 4). A toxicity quotient (TQ) value for qualifying pesticides in each sample for each group of organisms, as described by Munn et al. [14] and Belden et al. [16], was calculated using equation 1:

$$TQ_{ax} = PC_a \div EC_{ax} \text{ (eq. 1)}$$

Where TQ_{ax} is the toxicity quotient of pesticide a, for species x, PC_a is the concentration of pesticide a, and EC_{ax}, is the toxic effects concentration of pesticide a, for species x (e.g., LC50 or EC50) (Table 4). A total pesticide toxicity index (PTI) value for all pesticides in each sample, according to Munn et al. [14] and Belden et al. [16], was determined using equation 2:

$$PTI = \sum TQ_{a, b, c.., x} \text{ (eq. 2)}$$

Where $TQ_{a, b, c.., x}$ is the toxicity quotient for pesticide a, b, c,...etc., for species x, in a sample. The PTI is based upon the assumption of an additive toxicity model (concentration addition, CA) where co-occurring pesticides are expected to act cumulatively on organisms [18]. Although a CA model is relatively simplistic and may not strictly apply to all complex pesticides mixtures, especially when the mixture contains pesticides with differing modes of action [19], it is a useful tool in assessing relative risk of complex mixtures and has been shown to predict effects concentrations within a factor of two [18, 19]. Levels of concern (LOCs) were assessed according to Battaglin and Fairchild [14] where TQ and or PTI values ≥1.0 indicate probable toxicity, values ≥0.5 indicate potential toxicity, and values ≥0.1 indicate limited toxicity.

Table 3. Class, use and application of pesticides examined in Beasley Lake 2000-2009

Pesticide	Class	Use	Application Rate	Time of Application
Herbicide				
Trifluralin	Dinitroanaline	Annual grasses Broad leaf weeds	1.1-2.8 kg ha⁻¹ 0.6-1.1 kg ha⁻¹	October-December April-May
Pendimethalin	Dinitroanaline	Annual grasses Broad leaf weeds	0.8-1.7 kg ha⁻¹ 0.6-1.1 kg ha⁻¹	October-December May-June
Atrazine	Triazine	Grassy weeds Broad leaf weeds	1.1-3.4 kg ha⁻¹ 1.1-3.4 kg ha⁻¹	April-May April-May
Cyanazine	Triazine	Grassy weeds Broad leaf weeds	1.3-2.2 kg ha⁻¹ 0.6-1.8 kg ha⁻¹	April-May May-June
Alachlor	Acetanilide	Annual grasses Broad leaf weeds	2.2-3.4 kg ha⁻¹ 2.2-3.4 kg ha⁻¹	April April

Table 3. (Continued)

Pesticide	Class	Use	Application Rate	Time of Application
Metolachlor	Acetanilide	Annual grasses Broad leaf weeds	1.7-2.8 kg ha⁻¹ 1.7-2.8 kg ha⁻¹	April April
Insecticide				
Methyl Parathion	Organophosphate	Boll weevils	0.3-1.7 kg ha⁻¹	May-August
Chloropyrifos	Organophosphate	Broad spectrum insecticide	0.2-1.1 kg ha⁻¹	May-July
Bifenthrin	Pyrethroid	Broad spectrum insecticide	0.07-0.11 kg ha⁻¹	May-August
λ-cyhalothrin	Pyrethroid	Broad spectrum insecticide	0.028-0.045 kg ha⁻¹	May-August
Fipronil	Pyrazole	Broad spectrum insecticide	0.17 kg ha⁻¹	April-May
Legacy Insecticide				
Dieldrin	Organochlorine	Broad spectrum insecticide	Banned in USA	No longer applied
p,p'-DDT	Organochlorine	Broad spectrum insecticide	Banned in USA	No longer applied

Data from NASS [39].

Table 4. Concentration effects ($\mu g\ L^{-1}$) of targeted restricted use pesticides for crustaceans, insects, fish, and algae [17] in assessing relative risk to aquatic biota in Beasley Lake from 2000-2009 (--, no data available for that pesticide)

Pesticide	Crustacean *Daphnia* sp. 48h LC50	Insect *Culex* sp. 48h LC50	Fish *Lepomis* sp. 96h LC50	Algae *Pseudokirchneriella* sp. 96h EC50
Trifluralin	215.2	1643.2	114.7	447.5
Pendimethalin	--	--	--	16.8
Atrazine	27452.1	803.3	26606.4	75.1
Cyanazine	--	--	--	16.7
Alachlor	--	--	--	6.7
Metolachlor	13000	1000	5656.9	56.5
Methyl Parathion	8.3	2.4	3111.7	150
Chlorpyrifos	0.5	2.7	5.7	5900
Bifenthrin	0.3	0.5	0.4	--
λ-Cyhalothrin	0.6	1.6	0.8	--
Fipronil	15.6	1.8	45.6	108
Dieldrin	79.5	62.2	9.6	--
p,p'-DDT	4.2	87	4.6	--

DATA ANALYSIS

Descriptive statistical analysis was conducted on pesticide concentrations to present the annual distribution of pesticide levels in Beasley Lake water. Box-whisker plots were used to present median (line), 25th-75th percentiles (box), and 10th-90th (whiskers) percentiles for each pesticide (Figures 2-4). Relative pesticide risk was qualitatively assessed using categorical LOCs (previously described) at 1.0, 0.5, and 0.1 levels and percent of annual samples which exceeded these levels for pesticide TQs and or PTIs. Pearson Product Moment correlation analysis was conducted on mean and maximum annual concentrations of each pesticide versus land-use to assess the influence of land-use changes on observed changes in pesticide levels in Beasley Lake water. Pearson Product Moment correlation analysis was also conducted on qualifying TQs and PTIs (where LOCs ≥ 0.1) versus land-use to assess the influence of land-use changes on changes in relative risk of pesticides to aquatic biota in Beasley Lake water.

RESULTS AND DISCUSSION

During the 10-y study period, significant changes in land-use occurred. For agricultural BMPs, two watershed-wide and one edge-of-field BMPs were implemented (Table 1). Watershed-wide reduced tillage BMPs were implemented in 2001; edge-of-field vegetated constructed wetland (0.5 ha); and watershed-wide CRP enrollment (114 ha planted cottonwood trees) began in 2003 and was fully implemented by 2004. These additional BMPs were implemented in conjunction with previous BMPs incorporated in 1994-1998 (Table 1). Row-crop production shifted from primarily cotton in 2000-2001 to predominantly soybean in 2002-2004, 2006, 2008, and 2009 with milo and corn dominant in 2007 (Table 2). Cotton production steadily declined from its peak in 2001 until by 2007 none of the watershed was planted in cotton. Soybean production increased more than 10-fold from 2001 to 2002 and remained above 400 ha for all years except 2005 and 2007. Corn and milo production fluctuated almost annually with greatest production occurring in 2007 at 226 and 358 ha, respectively. In the last 10 years there have been numerous small-scale studies assessing the effectiveness of a variety of BMPs on pesticide mitigation in the Delta. Such studies have included examining vegetated filter strips [20], vegetated drainage ditches [21, 22, 23], and constructed wetlands [24, 25]. Fewer studies have attempted to assess pesticide mitigation with watershed-wide BMPs, and additionally, land-use [8, 9, 10]. The latter assessments [8, 9, 10] however, were more qualitative than quantitative assessments.

A total of 111 monthly lake water samples were collected from 2000-2009. Nearly every sample (109 of 111, 98%) detected at least one of the 13 pesticides assessed; >80% of all samples detected at least two pesticides; approximately 70% of all samples detected at least five pesticides; and approximately 10% of all samples detected at least 10 pesticides. Measured pesticide concentrations in Beasley Lake water varied widely within and among years (Figures 2-4). In this chapter, most individual pesticide concentrations were frequently <0.1 μg L^{-1}, with the exception of the current-use herbicide, atrazine. In general, pesticides were detected most frequently and in greater concentrations from 2000-2002 and least frequently and in lower concentrations from 2004-2006. Lowest current-use herbicide concentrations occurred for the dinitroanalines, trifluralin and pendimethalin, with all concentrations below 0.05 µg L^{-1} (Figure 2a-b). Both dinitroanalines have a relatively high affinity for sorption to soil [26] and trifluralin has been measured in sediment laden runoff at concentrations >5-fold from experimental cotton fields in the Delta [27]. This, in part, accounts for the relatively low concentrations of these two pesticides measured in the water column. Intermediate herbicide concentrations were observed for the acetanilides, alachlor and metolachlor, with concentrations frequently below 0.05 and 0.5 µg L^{-1}, respectively (Figure 2e-f). Greatest herbicide concentrations occurred for the triazines atrazine and cyanazine with median concentrations ranging from <0.001-0.310 and <0.001-0.216 µg L^{-1}, respectively and maximum concentrations ranging from 0.125-3.539 and <0.001-1.339 µg L^{-1}, respectively (Figure 2c-d). Concentrations of metolachlor, atrazine and cyanazine in Beasley Lake were similar to those of other Delta oxbow lakes in Mississippi [6, 9].

For current-use insecticides, greatest insecticide concentrations occurred for bifenthrin and methyl parathion with median concentrations ranging from <0.001-0.037 and <0.001-0.035 µg L^{-1}, respectively and maximum concentrations ranging from <0.001-1.408 and 0.015-0.163 µg L^{-1}, respectively (Figure 3a and c). Intermediate insecticide concentrations were observed for λ-cyhalothrin and chlorpyrifos, with median concentrations ranging from <0.001-0.016 and <0.001-0.004 µg L^{-1}, respectively and maximum concentrations ranging from 0.007-0.119 and 0.001-0.045 µg L^{-1}, respectively (Figure 3b and d). Lowest concentrations were observed for the pyrazole, fipronil, with nearly all concentrations ≤0.01 µg L^{-1} (Figure 3e). Levels of organophosphate insecticides measured in Beasley Lake were similar to those from surface waters in other agricultural watersheds within the Mississippi Delta [6, 28, 29, 30] and California [31]. Both methyl parathion and chlorpyrifos were infrequently found together in mixtures (33 of 111 samples) reflecting changes in organophosphate applications in the watershed. Lake water pyrethroid levels were comparatively similar to those found in other watersheds having row-crop agriculture [32, 33]. Pyrethroid bifenthrin levels were consistently below 0.05 µg L^{-1}(8 of 111 samples) with only one sample on 2/20/2007 with measured concentrations above 0.1 µg L^{-1} (1.408 µg L^{-1}). These results are similar to results from studies in other Delta oxbow lakes [6, 11, 13]. Pyrethroid λ-cyhalothrin concentrations were typically below aqueous worst-case expected environmental concentrations of 0.1 µg L^{-1} [34], with only 1 of 111 samples exceeding this level. Most concentrations of this pyrethroid were ≥10-fold below the aqueous worst-case expected environmental concentration. Levels of the pyrazole fipronil in Beasley Lake water was comparable with concentrations found in surface waters with agricultural land-use throughout the United States [35]. Analysis of legacy organochlorine insecticides dieldrin and *p,p'*-DDT indicated relatively low residual concentrations in Beasley Lake water during the study period. Lowest concentrations were observed for dieldrin ranging from <0.001-0.029 µg L^{-1} and greatest measured concentrations were *p,p'*-DDT, ranging from <0.001-0.374 µg L^{-1} (Figure 4a-b). These measured levels of residual legacy compounds are similar to concentrations measured in other Mississippi Delta watersheds [6, 11, 13, 36]. Results of our study continue to support the conclusion that, although banned from use in the United States for more than 25 years, organochlorine pesticides remain persistent [37] and that Mississippi Delta watershed soils and sediments will continue to release these compounds to the aqueous phase into the 21st century [36].

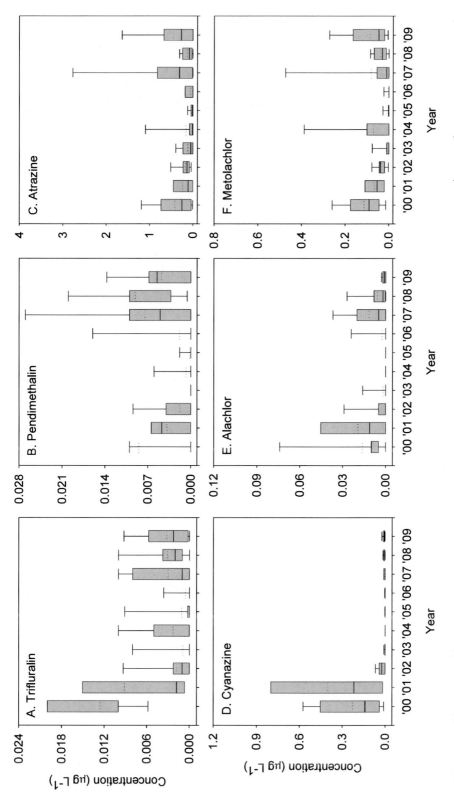

Figure 2. Box-Whisker plots of current-use herbicide concentrations (A-F) in Beasley Lake water 2000-2009. Whiskers, 10th and 90th percentiles; boxes 25th and 75th percentiles; solid lines, median; dotted lines, mean.

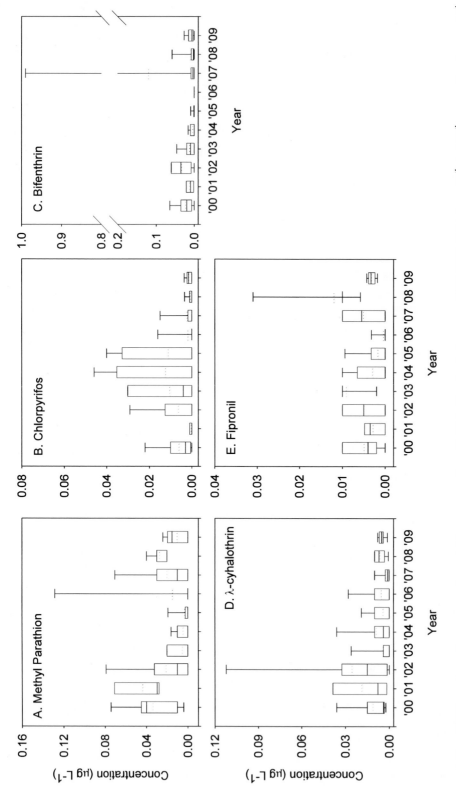

Figure 3. Box-Whisker plots of current-use insecticide concentrations (A-E) in Beasley Lake water 2000-2009. Whiskers, 10th and 90th percentiles; boxes, 25th and 75th percentiles; solid lines, median; dotted lines, mean.

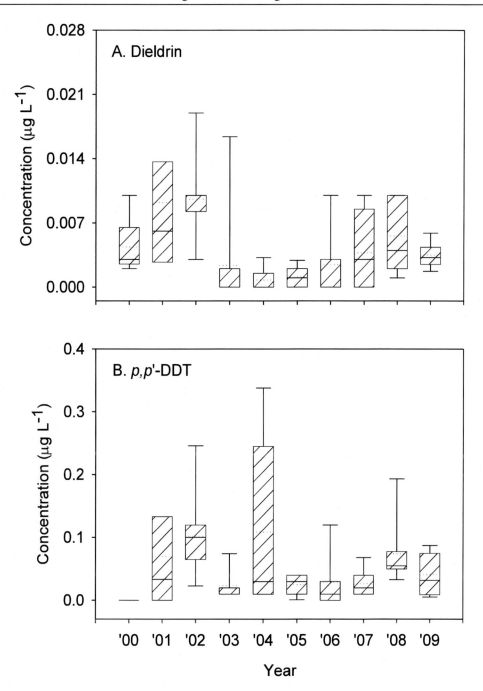

Figure 4. Box-Whisker plots of legacy insecticide concentrations (A-B) in Beasley Lake water 2000-2009. Whiskers, 10th and 90th percentiles; boxes, 25th and 75th percentiles; solid lines, median; dotted lines, mean.

Land-use changes in Beasley Lake watershed were significantly (P <0.05) associated with changes in pesticide levels of 10 of 13 pesticides examined in this chapter (Table 5). Land-use changes produced a significant positive correlation with 8 of 11 current-use pesticides and implementation of the BMP, CRP produced a significant negative association with six current-use pesticides and one legacy pesticide (Table 5). Decreased cotton production in the watershed from 2000-2009 (Table 2) had a significant positive correlation with decreases in lake water concentrations of herbicides trifluralin, cyanazine, and alachlor (Table 5, Figure 2a, d, e) as well as the insecticide methyl parathion (Table 5, Figure 3a). Trifluralin, cyanazine, and methyl parathion have been viewed as predominantly cotton pesticides in the Mississippi Delta [9, 30, 38, 39] although other pesticides such as alachlor may be used locally for watershed-specific weed control. While cyanazine concentrations were strongly correlated with cotton production ($r > 0.9$), additional non-crop factors may also affect such relationships. The measured decrease in cyanazine also coincides with the cancellation of herbicide reregistration by USEPA in December 1999 [40]. Permitted application of existing herbicide stocks was allowed until December 2002 after which cyanazine was not to be used. Changes in watershed corn production resulted in significant positive correlation with herbicides pendimethalin, atrazine, and metolachlor as well as with the pyrethroid insecticide befenthrin. The associations of corn and atrazine and metolachlor were expected since these herbicides are commonly used for this crop [39]. Associations with pendimethalin and bifenthrin were due to localized applications to treat corn for specific pests in the watershed. A similar pattern for milo was observed resulting in associations with both atrazine and bifenthrin. Implementation of CRP was significantly associated with decreases in a variety of current-use and legacy compounds. Significant negative correlations with CRP included herbicides trifluralin, alachlor and cyanazine (Table 5), although association with the latter herbicide should include mitigating factors described previously. CRP was also negatively associated with insecticides methyl parathion, bifenthrin and λ-cyhalothrin with both pyrethroids exhibiting higher correlation coefficients (r -0.747 to -0.879) than more water-soluble herbicides. Additionally, CRP was the only land-use variable significantly associated with a legacy compound, dieldrin ($r = -0.766$) indicating a decreased aqueous detection of this legacy insecticide with the implementation of this BMP. No associations were observed for land-use and current use insecticides chlorpyrifos (a broad spectrum insecticide) and fipronil. Fipronil is more commonly used on rice crops [41], a significant crop within the Delta [2, 38]. Although rice was not planted in Beasley Lake watershed during the study years (2000-2009), 41 ha rice was planted in the watershed in 1998 and concentrations measured in lake water are likely due to residual levels remaining in the soil.

Assessment of relative risk of agricultural pesticide mixtures to aquatic organisms representing various trophic levels as PTI in the current chapter showed greatest potential risk to crustaceans (as *Daphnia* sp.) and lowest potential risk to algae (as *Pseudokirchneriella* sp.) (Figure 5a-d). PTIs for *Daphnia* sp. ranged from <0.01 to 0.48 (Figure 5a) with approximately 23% (26 of 111) of samples having PTIs >0.1, most of which occurred from 2000-2004. As a result, LOCs were between 0.1 and 0.5, with no values greater than 0.5, indicating a potential for limited toxicity for this group of organisms [14]. For samples with LOCs >0.1, a significant fraction of potential limited toxicity to *Daphnia* sp. could come from a single pyrethroid, bifenthrin. For crustaceans, bifenthrin TQs closely matched PTIs for all years except 2004 (Figure 6a). PTI values for aquatic insects (as *Culex* sp.) ranged from <0.01 to 0.362 (Figure 5b) with approximately 10% (11 of 111) of samples having PTIs >0.1, most of which occurred from 2000-2002. Again LOC values ranged between 0.1-0.5, suggesting limited toxicity for aquatic insects. Also similar to *Daphnia* sp., samples with LOCs >0.1 for *Culex* sp. showed bifenthrin TQs closely matching PTIs for all years except 2001 (Figure 6b). Fish PTIs (as *Lepomis* sp.) ranged from <0.01 to 0.36 (Figure 5c) with approximately 17% (19 of 111) of samples having PTIs >0.1, with most occurring from 2000-2002. As with previous aquatic animal groups examined, *Lepomis* sp. LOCs were between 0.1-0.5 indicating limited potential toxicity for fish in Beasley Lake surface water. For fish, as with crustaceans, samples having LOCs >0.1, bifenthrin TQs closely matched PTIs for all years except 2004 (Figure 6c). Algae PTIs ranged from <0.01 to 0.09 with no samples having PTIs >0.1 and, as a result, no samples reaching any LOC category [14]. An increasing number of studies within the last ten years have attempted to assess risk of pesticide mixtures in water bodies receiving runoff from agricultural catchments using some form of pesticide toxicity index [14, 16, 18, 41]. Several of these studies have focused on watersheds within the intensively cultivated region of the Midwestern United States, where corn and/or soybeans are the predominant row crop. Within such watersheds, algae PTIs were greatest, sometimes exceeding 1.0 LOCs [14, 16], indicating potential for likely toxicity, with fish PTIs <0.01 indicating unlikely potential for toxicity. In contrast, Mize et al. [41] within watersheds of southeastern Louisiana observed greatest PTIs for aquatic invertebrates and Smith and Cooper [13] within oxbow lake watersheds of the Delta suggested fish might be more vulnerable than algae to pesticide runoff. The latter two studies are in agreement with the current chapter and suggest that insecticides rather than herbicides are the greater risk to aquatic biota in the Delta and that insecticides would have a greater potential impact on crustacean populations. This is corroborated by results of Moore et al. [11] and Lizotte et al. [42] utilizing the crustacean *Hyalella azteca* in laboratory exposures of whole lake water from Beasley Lake in 2001 and 2005 where significant growth inhibition was observed in crustaceans exposed to water from 2001 but not 2005.

Groups of aquatic organisms having LOCs >0.1 (indicating limited toxicity) were included in assessments to associate Beasley Lake watershed land-use changes with varying potential pesticide risk. Pearson Product Moment correlations showed significant (P <0.05) associations between aquatic animal PTIs or bifenthrin TQs and corn, milo and CRP (Table 6). Daphnia sp. PTI values were positively associated with milo production (r = 0.669) and negatively associated with CRP (r = -0.781). Bifenthrin TQs for crustaceans were positively correlated with both corn (r = 0.763) and milo (r = 0.798) production and negatively correlated with CRP (r = -0.752). Similar patterns of associations were observed for Culex sp. PTIs and bifenthrin TQs with land-use resulted in greatest correlation coefficients occurring between PTI and CRP (r = -0.845) and bifenthrin TQ and milo production (r = 0.791). Associations between Lepomis sp. PTIs and bifenthrin TQs with land-use also followed a similar pattern with greatest correlation coefficients occurring for bifenthrin TQ and corn production (r = 0.818) and PTI and CRP (r = -0.754). Previous studies often qualitatively or implicitly associate land-use or farm management practices with risk or changes to aquatic biota [41, 43, 44, 45]. Most of these types of studies focused on land-use and/or BMPs at the sub-watershed or field plot scale, not at a larger or broader watershed/catchment scale [43, 45]. Vondracek et al. [46] synthesized nine studies that assessed land-use changes, including agricultural landscapes, at the watershed level with changes in fish and macroinvertebrate assemblages, however the focus was on habitat variables and not on potential contaminants (i.e. pesticides). To our knowledge, no published studies have attempted to quantitatively correlate agricultural land-use changes and concomitant pesticide mixture levels in surface water with potential risk to aquatic biota. Smiley et al. [44] recommended hypothesis-driven ecological assessments of ecological responses to agricultural land-use changes (i.e. conservation practices), however such research is limited to categories of before and after and/or control versus treatment(s) and does not readily address land-use changes and ecological responses over extended periods of time at the watershed level.

Table 5. Pearson Product Moment correlation coefficients (*r*)
land-use (hectares) and pesticide loads (mean and maximum concentrations)
in Beasley Lake 2000-2009

Pesticide	Value	Cotton	Soybean	Corn	Milo	CRP	Fallow
Trifluralin	Mean	**0.888**	**-0.550**	-0.042	-0.064	**-0.733**	-0.051
	Maximum	**0.803**	**-0.569**	-0.100	0.120	**-0.650**	-0.135
Pendimethalin	Mean	0.046	-0.184	**0.515**	0.332	-0.065	-0.004
	Maximum	-0.369	-0.040	**0.704**	**0.579**	0.243	-0.111
Atrazine	Mean	0.085	**-0.601**	**0.847**	**0.627**	-0.109	0.321
	Maximum	-0.085	**-0.555**	**0.888**	**0.724**	0.133	0.236
Cyanazine	Mean	**0.934**	**-0.599**	-0.153	0.005	**-0.750**	-0.184
	Maximum	**0.913**	**-0.593**	-0.157	0.030	**-0.717**	-0.182
Alachlor	Mean	**0.788**	**-0.691**	0.233	0.374	**-0.704**	-0.360
	Maximum	**0.731**	-0.457	0.112	0.104	**-0.685**	-0.326
Metolachlor	Mean	0.400	**-0.502**	0.454	0.223	-0.323	0.380
	Maximum	-0.036	-0.391	**0.643**	0.469	0.182	0.317
Methyl Parathion	Mean	**0.672**	-0.384	0.031	0.197	**-0.686**	-0.345
	Maximum	0.220	0.019	0.055	0.143	-0.310	-0.418
Chlorpyrifos	Mean	-0.045	0.084	-0.300	-0.250	0.088	-0.009
	Maximum	-0.063	-0.066	-0.043	-0.027	0.072	-0.171
Bifenthrin	Mean	0.260	-0.225	0.109	0.303	**-0.747**	-0.088
	Maximum	-0.036	-0.452	**0.761**	**0.791**	-0.140	-0.233
λ-Cyhalothrin	Mean	**0.530**	-0.157	-0.350	-0.017	**-0.879**	-0.199
	Maximum	0.391	-0.045	-0.380	0.009	**-0.800**	-0.243
Fipronil	Mean	-0.057	0.254	-0.122	-0.048	-0.050	-0.065
	Maximum	-0.289	0.416	-0.010	-0.043	0.148	-0.325
Dieldrin	Mean	0.438	-0.210	-0.126	0.174	**-0.766**	-0.133
	Maximum	**0.524**	-0.334	-0.125	0.295	**-0.613**	-0.227
p,p'-DDT	Mean	-0.110	0.328	-0.332	-0.073	-0.124	0.041
	Maximum	0.012	0.287	-0.353	-0.084	-0.092	-0.031

Bold values are statistically significant correlation coefficients, $P < 0.05$.

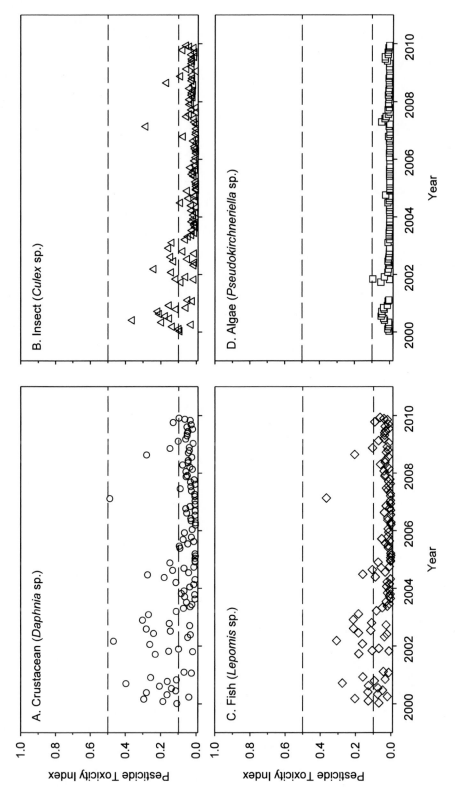

Figure 5. PTI values of aquatic organisms (A–D) representing relative risk of pesticides in Beasley Lake 2000-2009. Dashed lines at 0.1 and 0.5 represent levels of concern (LOCs) indicating probable toxicity and potential toxicity, respectively.

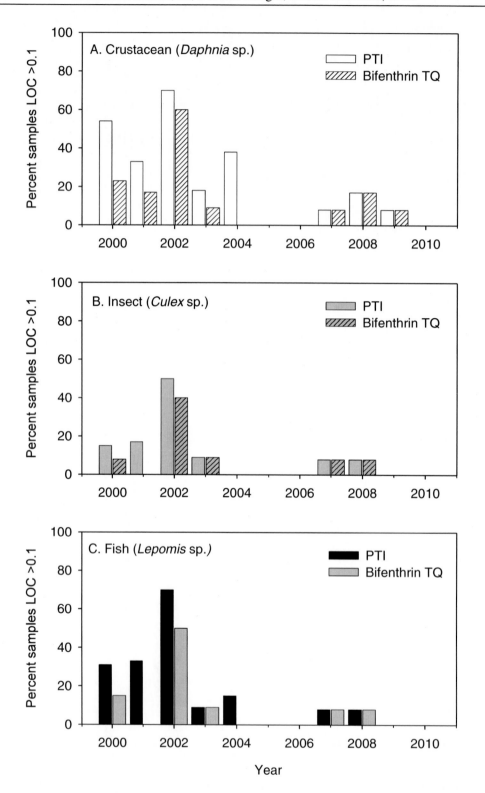

Figure 6. Percent samples exceeding LOC values of 0.1 for PTI and bifenthrin TQ in aquatic organisms (A-C) indicating risk of limited toxicity in Beasley Lake water 2000-2009.

Table 6. Pearson Product Moment correlation coefficients (*r*) land-use (hectares) and aquatic organism PTIs or bifenthrin TQs (mean and maximum values) in Beasley Lake 2000-2009

Organism	Index	Value	Cotton	Soybean	Corn	Milo	CRP	Fallow
Daphnia	PTI	Mean	0.317	-0.136	-0.136	0.158	-0.781	-0.132
		Maximum	0.083	-0.344	0.469	0.669	-0.414	-0.271
	Bifentrhin TQ	Mean	0.270	-0.226	0.106	0.304	-0.752	-0.104
		Maximum	-0.051	-0.444	0.763	0.798	-0.130	-0.234
Culex	PTI	Mean	0.420	-0.257	0.005	0.254	-0.845	-0.184
		Maximum	0.034	-0.311	0.567	0.717	-0.373	-0.327
	Bifentrhin TQ	Mean	0.280	-0.229	0.104	0.302	-0.759	-0.107
		Maximum	-0.035	-0.450	0.759	0.791	-0.142	-0.235
Lepomis	PTI	Mean	0.305	-0.123	-0.103	0.184	-0.783	-0.115
		Maximum	0.112	-0.408	0.540	0.731	-0.410	-0.289
	Bifentrhin TQ	Mean	0.271	-0.224	0.103	0.303	-0.754	-0.103
		Maximum	-0.118	-0.423	0.818	0.734	-0.068	0.054

Bold values are statistically significant correlation coefficients, $P < 0.05$.

CONCLUSION

Results of the current chapter provide a better understanding of the role of land-use and BMPs in determining risk to aquatic biota in agricultural watersheds. Land-use affects both pesticide types and combinations of pesticide mixtures occurring in agricultural watersheds [47], and as a result, the vulnerability of aquatic biota in water bodies receiving runoff from these watersheds. Overall, Beasley Lake surface water pesticide contamination decreased annually until 2005-2006 and increased again in 2007-2009 due in part, to a shift in row crop from reduced tillage soybeans to conventional-till milo and corn in 2007. Concomitantly, changes in land-use with implementation of BMPs, CRP and crop type reduced the frequency, magnitude, and duration of pesticide risk to lake aquatic biota. Such results have far-reaching implications for aiding researchers in modeling risk of pesticides to aquatic biota under dynamic agricultural landscapes [47, 48].

REFERENCES

[1]   Pereira, W. E. & Hostettler, F. D. (1993). Nonpoint source contamination of the Mississippi River and tributaries by herbicides. *Environmental Science and Technology*, *27*, 1542-1552.

[2]   Locke, M. A. (2004). Mississippi Delta management systems evaluation area: overview of water quality issues on a watershed scale. In M. Nett, M. Locke, D. Pennington, (eds.), *Water Quality Assessments in the Mississippi Delta, Regional Solutions, National Scope* (pp. 1-15). American Chemical Society, Oxford University Press, Chicago.

[3]   Knight, S. S. & Welch, T. D. (2004). Evaluation of watershed management practices on oxbow lake ecology and water quality. In M. Nett, M. Locke, D. Pennington, (eds.), *Water Quality Assessments in the Mississippi Delta, Regional Solutions, National Scope* (pp. 119-133). American Chemical Society, Oxford University Press, Chicago.

[4]   Snipes C. E., Evans, L. P., Poston, D. H. & Nichols, S. P. (2004). Agricultural practices of the Mississippi Delta. In M. Nett, M. Locke, D. Pennington, (eds.), *Water Quality Assessments in the Mississippi Delta, Regional Solutions, National Scope* (pp. 43-60). American Chemical Society, Oxford University Press, Chicago.

[5]   Moore, M. T., Lizotte, Jr., R. E., Cooper, C. M., Smith , Jr., S. & Knight, S. S. (2004). Survival and growth of Hyalella azteca exposed to three Mississippi oxbow lake sediments. *Bulletin of Environmental Contamination and Toxicology*, *72*, 777-783.

[6]   Cooper, C. M., Smith, Jr., S. & Moore, M. T. (2003). Surface water, ground water and sediment quality in three oxbow lake watersheds in the Mississippi Delta agricultural region: pesticides. *International Journal of Ecology and Environmental Science*, *29*, 171-184.

[7]   Cullum, R. F., Knight, S. S., Cooper, C. M. & Smith, Jr., S. (2006). Combined effects of best management practices on water quality in oxbow lakes from agricultural watersheds. *Soil Tillage Research*, *90*, 212-221.

[8]   Smith, Jr., S., Cooper, C. M., Lizotte, Jr., R. E., Locke, M. A. & Knight, S. S. (2007). Pesticides in lake water in the Beasley Lake watershed, 1998-2005. *International Journal of Ecology and Environmental Science, 33*, 61-71.

[9]   Zablotowicz, R. M., Locke, M. A., Krutz, L. J., Lerch, R. N., Lizotte, R. E., Knight, S. S., Gordon, R. E. & Steinriede, R. W. (2006). Influence of watershed system management on herbicide concentrations in Mississippi Delta oxbow lakes. *Science of the Total Environment, 370*, 552-560.

[10]   Locke, M. A., Knight, S. S., Smith, Jr., S., Cullum, R. F., Zablotowicz, R. M., Yuan, Y. & Bingner, R. L. (2008). Environmental quality research in the Beasley Lake watershed, 1995-2007: succession from conventional to conservation practices. *Journal of Soil and Water Conservation, 63*, 430-442.

[11]   Moore, M. T., Lizotte, Jr., R. E., Knight, S. S., Smith, Jr., S. & Cooper, C. M. (2007). Assessment of pesticide contamination in three Mississippi delta oxbow lakes using *Hyalella azteca. Chemosphere, 67*, 2184-2191.

[12]   Smith, Jr. S., Cooper, C. M., Lizotte, Jr., R. E., Locke, M. A. & Knight, S. S. (2007). Pesticides in lake water in the Beasley Lake watershed, 1998-2005. *International Journal of Ecology and International Sciences, 33*, 61-71

[13]   Smith, Jr., S. & Cooper, C. M. (2004). Pesticides in shallow groundwater and lake water in the Mississippi Delta MSEA. In M. Nett, M. Locke, D. Pennington, (eds.), *Water Quality Assessments in the Mississippi Delta, Regional Solutions, National Scope* (pp. 91-103). American Chemical Society, Oxford University Press, Chicago.

[14]   Battaglin, W. & Fairchild, J. (2002). Potential toxicity of pesticides measured in midwestern streams to aquatic organisms. *Water Science and Technology, 45*, 95-103.

[15]   Munn, M. D., Gilliom, R. J., Moran, P. W. & Nowell, L. H. (2006). *Pesticide Toxicity Index for Freshwater Aquatic Organisms* (2nd Edition). Scientific Investigations Report 2006-5148, Washington, DC: National Water Quality Assessment Program, US Department of the Interior, US Geological Survey.

[16]   Belden, J. B., Gilliom, R. J., Martin, J. D. & Lydy, M. J. (2007). Relative toxicity and occurrence patterns of pesticide mixtures in streams draining agricultural watersheds dominated by corn and soybean production. *Integrated Environmental Assessment and Management, 3*, 90-100.

[17]   US Environmental Protection Agency (USEPA). ECOTOX Database. 2010 [2010 09 23]. Available from: URL: http://cfpub.epa.gov/ecotox/quick_query.htm

[18]   Deneer, J. W. (2000). Toxicity of mixtures of pesticides in aquatic systems. *Pest Management Science, 56*, 516-520.

[19]   Belden, J. B., Gilliom, R. J. & Lydy, M. J. (2007). How well can we predict the toxicity of pesticide mixtures to aquatic life? *Integrated Environmental Assessment and Management, 3*, 364-372.

[20]   Blanche, S. B., Shaw, D. R., Massey, J. H., Boyette, M. & Smith, M. S. (2003). Fluometuron adsorption to vegetative filter strip components. *Weed Science, 51*, 125-129.

[21]   Moore, M. T., Bennett, E. R., Cooper, C. M., Smith, Jr., S., Shields, Jr., F. D., Milam, C. D. & Farris, J. L. (2001). Transport and fate of atrazine and lambda-cyhalothrin in an agricultural drainage ditch in the Mississippi delta, USA. *Agriculture, Ecosystems and Environment, 87*, 309-314.

[22] Cooper, C. M., Moore, M. T., Bennett, E. R., Smith, Jr., S. & Farris, J. L. (2002). Alternative environmental benefits of agricultural drainage ditches. *Verhandlungen des Internationalen Verein Limnologie*, *28*, 1678-1682.

[23] Bennett, E. R., Moore, M. T., Cooper, C. M., Smith, Jr., S., Shields, Jr., F. D., Drouillard, K. G. & Schulz, R. (2005). Vegetated agricultural drainage ditches for the mitigation of pyrethroid-associated runoff. *Environmental Toxicology and Chemistry*, *24*, 2121-2127.

[24] Moore, M. T., Cooper, C. M, Smith, Jr., S., Cullum, R. F., Knight, S. S., Locke, M. A. & Bennett, E. R. (2007). Diazinon mitigation in constructed wetlands: influence of vegetation. *Water Air and Soil Pollution*, *184*, 313-321.

[25] Moore, M. T., Cooper, C. M, Smith, Jr., S., Cullum, R. F., Knight, S. S., Locke, M. A. & Bennett, E. R. (2009). Mitigation of two pyrethroid insecticides in a Mississippi delta constructed wetland. *Environmental Pollution*, *157*, 250-256.

[26] Larson, S. J., Capel, P. D., Goolsby, D. A., Zaugg, S. D. & Sandstrom, M. W. (1995). Relations between pesticide use and riverine flux in the Mississippi River basin. *Chemosphere*, *31*, 3305-3321.

[27] Willis, G. H., McDowell, L. L., Murphree, C. E., Southwick, L. M. & Smith, Jr., S. (1983). Pesticide concentrations and yields in runoff from silty soils in the lower Mississippi Valley. *Journal of Agricultural and Food Chemistry*, *31*, 1171-1177.

[28] Senseman, S. A., Lavy, T. L., Mattice, J. D., Gbur, E. E. & Skulman, B. W. (1997). Trace level pesticide detections in Arkansas surface waters. *Environmental Science and Technology*, *31*, 395-401.

[29] Gruber, S. J. & Munn, M. D. (1998). Organophosphate and carbamate insecticides in agricultural waters and cholinesterase (ChE) inhibition in common carp (*Cyprinus carpio*). *Archives of Environmental Contamination and Toxicology*, *35*, 391-396.

[30] Thurman, E. M., Zimmerman, L. R., Scribner, E. A. & Coupe, Jr., R. H. (1998). Occurrence of cotton pesticides in surface water of the Mississippi embayment. USGS Fact Sheet FS-022-98.

[31] Pedersen, J. A., Yeager, M. A. & Suffet, I. H. (2006). Organophosphorus insecticides in agricultural and residential runoff: field observations and implications for total maximum daily load development. *Environmental Science and Technology*, *40*, 2120-2127.

[32] Starner, K., White, J., Spurlock F. & Kelley, K. (2008). Assessment of pyrethroid contamination of streams in high-use agricultural regions of California. pp. 72-83. In: Gan, J., Spurlock, F., Hendley, P. & Weston, D. (eds), Synthetic Pyrethroids: Occurrence and Behavior in Aquatic Environments. ACS Symposium Series 199, ACS, Washington, DC

[33] Weston, D. P. & Lydy, M. J. (2010). Urban and agricultural sources of pyrethroid insecticides to the Sacramento-San Joquin delta of California. *Environmental Science and Technology*, *44*, 1833-1840.

[34] Maund, S. J., Hamer, M. J., Warinton, J. S. & Kedwards, T. J. (1998). Aquatic ecotoxicology of the pyrethroid insecticide lambda-cyhalothrin: considerations for higher-tier aquatic risk assessment. *Pesticide Science*, *54*, 408-417.

[35] Gunasekara, A. S., Truong, T., Goh, K. S., Spurlock, F. & Tjeerdema, R. S. (2007). Environmental fate and toxicology of fipronil. *Journal of Pesticide Science*, *32*, 189-199.

[36] Cooper, C. M. (1991). Persistent organochlorine and current use insecticide concentrations in major watershed components of Moon Lake, Mississippi, USA. *Archiv für hydrobiologie, 121*, 103-113.

[37] Willis, G. H. & McDowell, L. L. (1982). Review: pesticides in agricultural runoff and their effects on downstream water quality. *Environmental Toxicology and Chemistry, 1*, 267-269.

[38] Coupe, R. H., Thurman, E. M. & Zimmerman, L. R. (1998). Relation of usage to the occurrence of cotton and rice herbicides in three streams of the Mississippi Delta. *Environmental Science and Technology, 32*, 3673-3680.

[39] National Agricultural Statistics Service (NASS). Agricultural chemical use database. 2011 [2011 02 09]. Available from: URL: http://www.pestmanagement.info/nass/

[40] US Environmental Protection Agency (USEPA). Pesticide reregistration status for triazines. 2010 [2011 02 07]. Available from: URL: http://www.epa.gov/pesticides/reregistration/status_triazines.htm

[41] Mize, S. V., Porter, S. D. & Demcheck, D. K. (2008). Influence of fipronil compounds and rice-cultivation land-use intensity on macroinvertebrate communities in streams of southwestern Louisiana, USA. *Environmental Pollution, 152*, 491-503.

[42] Lizotte, Jr., R. E., Knight, S. S. & Cooper, C. M. (2010). Toxicity evaluation of a conservation effects assessment program watershed, Beasley Lake in the Mississippi delta, USA. *Bulletin of Environmental Contamination and Toxicology, 84*, 422-426.

[43] Rice, P. J., Hapeman, C. J., McConnell, L. L., Sadeghi, A. M., Teasdale, J. R., Coffman, C. J., McCarty, G. W., Abdul-Baki, A. A. & Starr, J. L. (2007). Evaluation of vegetable production management practices to reduce the ecological risk of pesticides. *Environmental Toxicology and Chemistry, 26*, 2455-2464.

[44] Smiley, P. C., Shields, Jr., F. D. & Knight, S. S. (2009). Designing impact assessments for evaluating ecological effects of agricultural conservation practices on streams. *Journal of the American Water Resources Association, 45*, 867-878.

[45] Dunn, A. M., Julien, G., Ernst, W. R., Cook, A., Doe, K. G. & Jackman, P. M. (2011). Evaluation of buffer zone effectiveness in mitigating the risks associated with agricultural runoff in Prince Edward Island. *Science of the Total Environment, 409*, 868-882.

[46] Vondracek, B., Blann, K. L., Cox, C. B., Nerbonne, J. F., Mumford, K. G., Nerbonne, B. A., Sovell, L. A. & Zimmerman, J. K. H. (2005). Land-use, spatial scale, and stream systems: lessons from an agricultural region. *Environmental Management, 36*, 775-791.

[47] Lydy, M., Belden, J., Wheelock, C., Hammock, B. & Denton, D. (2004). Challenges in regulating pesticide mixtures. *Ecology and Society, 9*, 1-15.

[48] Reichenberger, S., Bach, M., Skitschak, A. & Frede H-G. (2007). Mitigation strategies to reduce pesticides inputs into ground- and surface water and their effectiveness; a review. *Science of the Total Environment, 384*, 1-35.

Reviewed by

Jennifer L. Bouldin, Assistant Research Professor, Director Ecotoxicology Research Facility, Arkansas State University, State University, Arkansas 72467, USA; and Robert Kröger, Assistant Professor, Department of Wildlife, Fisheries and Aquaculture, Mississippi State University, Mississippi State, Mississippi 39762, USA.

In: Agricultural Research Updates. Volume 2
Editor: Barbara P. Hendriks

ISBN: 978-1-61470-191-0
© 2012 Nova Science Publishers, Inc.

Chapter 12

CHARACTERIZATION OF THE SPATIAL CIRCUITS OF AGROFORESTRY SYSTEMS BELONGING TO INNOVATIVE FARMERS IN THE CAPIM RIVER POLE, PARÁ, BRAZIL

J. S. R. Oliveira, O. R. Kato, T. F. Oliveira and A. M. Silveira
Municipality of Irituia, Irituia, PA, Brazil

ABSTRACT

In the eastern Amazon Capim River pole, farming is a major economic activity, in which the slash-and-burn system prevails in creating small farms. In the last two decades, some smallholders have changed the landscape of their Family Production Units (FPU) by increasing their small farms and backyards to Agroforestry Systems (AFSs), in order to ensure food safety. Currently, the AFSs, in addition to the numerous environmental services they conduct in the region, already allow some smallholders to enter their products into the consumer market. This paper presents the structural characteristics related to the spatial arrangements of the circuit space and the main product marketing difficulties, which include the difficulty to obtain credit in order to produce, agricultural extension services – there is not enough human material to meet regional demand, the perishable nature of the products, the middleman, and the low prices and lack of specific marketing policies at this pole. From the results of the statistical analyses, as well as by the perceptual maps from expert knowledge this study presents a fuzzy system for the study of the marketing/development of the AFS in the studied area.

Keywords: Spatial Circuit; AFS, Innovative farmers, Regional system.

1. INTRODUCTION

The Northeastern area of Pará State, Brazil, is one of the oldest settlement areas of the eastern Brazilian Amazon. This area has been extensively modified regarding the original landscape, since currently only 15% of primary forests exist in the area (Pereira & Vieira, 2001).

For centuries the production system used in this area was the slash-and-burn system, which prepares the land for the creation of small farms. This system contributes to soil degradation, resulting in low soil fertility due to nutrient losses, harmful emissions into the atmosphere and fire hazards, besides becoming predatory when the demographic pressure on the land leads to land use in modules below their viability limit (Hébette 2004, Moran 1990, Hölscher et al. 1997a; Nepstad et al. 1999).

The slash-and-burn practice is due, in part, to the absence of policies aimed at the family farm sector, as the specific characteristics of the area are not considered, as well as economic development projects that were conceived and implemented from top to bottom (top-down development), based on policies designed and implemented by national and international technocratic agencies, without the participation of the local communities (Santos, 2005; Costa, 1998).

Given this situation, the public policy of the Socio-environmental Development Program of Family Production in the Amazon (Programa de Desenvolvimento Socioambiental da Produção Familiar) – PROAMBIENTE – was implemented by the Ministry of Environment (MMA). This program was created by the leading rural social movements in the Amazon, in partnership with nongovernmental organizations, and was then incorporated as a public policy of the Federal Government in the Multiyear Plan from 2004 to 2007.

According to the MMA, the PROAMBIENTE program aims to promote balance between the conservation of natural resources and the family rural production by means of environmental management of the land, by the integrated planning of the production units and by the provision of environmental services. The PROAMBIENTE program has as its priority family farmers and people from traditional communities. The Poles, all located in the Amazon, involve about 4,000 families.

The Capim River Pole is a component of the 14 Poles belonging to the PROAMBIENTE program, and is located in the middle region of Northeastern Pará, composed by the municipalities of Irituia, Concórdia do Pará, São Domingos do Capim and Mãe do Rio.

Some farmers at this Pole have made use of experiences accumulated by the interaction with the environment. As an example, they have expanded their family production units (FPU), which are the traditional sites that are technically known as Agroforestry Systems (AFS). These systems, initially designed to ensure the subsistence of the family – food and nutrition security, FNS – have been modified into other configurations due to their integration into the local and regional consumer market. The hallmark of an AFS is a regional species diversification, mainly of fruit (Oliveira et al., 2010).

The production of regional and exotic fruits is largely found in these AFS, in different stages of diversification, including the açaí (*Euterpe oleracea Mart*), cupuaçu (*Theobroma grandiflorum*), acerola (*Malpighia glabra* L), pineapple (*Ananas comosus* L. Merril), orange (*Citrus sinensis*), guava (*Psidium guajava* L), pupunha (*Bactris gasipaes*), caju (*Anacardium ocidentale*), cocoa (*Theobroma cocoa*) and passion fruit (*Passiflora edulis* Sims).

Regarding the spatial arrangements of these AFS, few or no crops use chemical industrial products, such as fertilizers, pesticides, herbicides, among others. Another highlight is the creation of environmental services characterized by the elimination of the use of fire, reforestation, watershed protection, soil protection against erosion and an increase in the local biodiversity.

However, these positive factors presented by the AFS are not synonymous with the assurance of product marketing, since several factors are necessary for marketing, such as low-interest credit, technical assistance and quality extension, access to new technologies and the creation of marketing channels with regard to the products generated by the social actors (Oliveira et al, 2010; Hespanhol, 2008).

It is noteworthy that in the municipalities of this study more than half of the population still resides in the fields, but with each passing year this reality changes more and more (IBGE, 2009), mainly due to the young population that moves to the cities in search of better opportunities.

In this sense, farmers have shown concerns regarding the future, since, without young people, the continuation of their actions will be compromised.

In this context, this study seeks to identify and classify the characteristics of the AFS in the Capim River Pole, located in the eastern Brazilian Amazon, based on the products grown/processed by these AFS and the main difficulties in their marketing, which are: the difficulty to obtain credit to produce the marketed products, the rural extension services (since the human laborers of the area are not sufficient to meet regional demands), the perishability of the marketed products, the middlemen, the low prices and the lack of specific marketing policies at this Pole.

2. MATERIAL AND METHODS

2.1. Pole Characterization

This study was conducted in the municipalities of Irituia, Concórdia do Pará, São Domingos do Capim and Mãe do Rio, that compose the Capim River Pole, in the northeastern area of the Pará State, Brazilian Eastern Amazon (Figure 1).

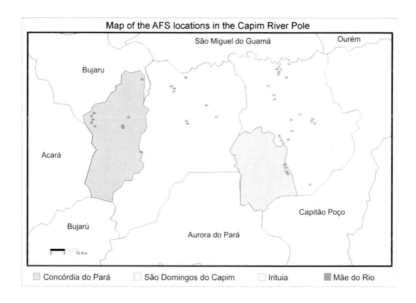

Figure 1. Location of the studied areas.

At the Capim River Pole, 417 families were enrolled in the PROAMBIENTE program, spatially distributed among the participant municipalities. In order to identify, locate and map the rural properties a GPS was used (Global Positioning System Garmin II Plus), facilitating georeferencing and the elaboration and validation of questionnaires created in this study.

The questionnaire formulation was undertaken with the farmers, taking into account their real situation, their experience locus. To characterize the spatial circuit of the AFS at this Pole, we used the theoretical Hart basis (1980) regarding agroecosystems and basic concepts. The characterization of the market stage aspect of these systems took place by analysing François (2000) and Ricci et al (2000), regarding long and short circuits, and the evolution of marketing systems was analysed using the theoretical basis of Kriesberg & Steele (1974).

2.2. Statistical Analysis and Perceptual Data Mapping

2.2.1. Statistical Analysis

The prices of the marketed products showed great variability and did not fit known theoretical distributions. Thus, we considered modal values.

The C test was used to measure the association between the variables by means of frequency, as was the correspondence analysis, that, along with the cluster analysis (K-means method), was used to determine the domains of the variables involved in the problem, generating the knowledge necessary to build the rule base and fuzzy sets of variables of the used fuzzy inference system (FIS).

The Correspondence Analysis is a multivariate technique that allows a graphical analysis in a multidimensional space of the association between variables by means of a contingency table for the calculation of inertia, which is the pondered sum of all the distances from the centroid divided by the sum of all the cells in the table (Hair et al, 1998).

To investigate the adequacy and best interpretation of the correspondence analysis, we used the contingency coefficient C and the chi-square residual analysis. The contingency coefficient C is indicated for the determination of the magnitude of the association of the measured variables, arranged in a contingency table. On the other hand, the residue analysis informs the importance of each cell of the contingency table by comparing the results with the standard probability of the normal curve (Ayres et al, 2007).

The K-means statistical technique, also known as a grouping or cluster analysis, was used for the analysis of the evolution of marketing systems of the studied municipalities. The goal of this technique is to associate variable arguments in groups, or clusters, so that the degree of similarity is high among members of a same group and small between different groups (Fávero, et al, 2009).

3. RESULTS AND DISCUSSION

3.1. Characterization of the Interviewed Families

Among the families enrolled in the PROAMBIENTE program at the Capim Pole, 360 families were correctly georeferenced and 53 were interviewed, with the criterion of choice being the AFS that already produce with the goal of direct insertion into the produce market. Table 1 shows the frequency distribution of the interviewed farmers, by municipality.

Table 1. Interviewed farmers and their respective municipalities

Municipality	Frequency	%
C. do Pará	12	22.60
Irituia	29	54.70
Mãe do Rio	4	7.50
S.D. do Capim	8	15.10
Total	**53**	**100.00**

It was observed that the interviewed families by municipality depend on three different situations: their historical siutation, the adoption of the AFS and the produce market. The municipalities of São Domingos, Irituia and Mãe do Rio are included in the first situation. At São Domingos and Irituia, most farmers reported that their grandfathers had already implemented the AFS (naming them farms). The same was true of farmers at the municipality of Mãe do Rio, corroborating the fact that this municipality was separated from Irituia since the interviewed families were located near Irituia´s oldest territorial area.

The municipality of Concórdia do Pará, located near the city of Tomé-Açu, is inserted in the second and third situations, where Japanese-Brazilian farmers practice AFS on a commercial basis. This context has motivated many families to make adjustments to their agroforestry systems according to their financial conditions and in line with the consumer market, since several types of products, such as fruit, have provided higher incomes to families practicing this system in neighboring municipalities.

3.2. Characteristics of the Spatial Circuit of the Pole

The spatial circuit of the Capim River Pole posesses specific physical, biotic and socio-economic components, as well as structural characteristics related to the spatial arrangements of its components and their functional characteristics.

3.2.1. Characterization of the Regional Structure of the Pole

The municipalities of the area are interconnected by the Belém - Brasília Federal Highway (BR-010), the PA 253 State Highway (commonly known as the Orange Highway, that crosses the territory of Irituia up to São Domingos do Capim) (Figure 3), the PA 252 (that connects Mãe do Rio to Concórdia do Pará), the PA 127 (that connects the BR-010 to the BR-316, cutting through the entire territory of São Domingos do Capim) and the PA 140 (that connects Concordia do Pará to the BR-316), as well as municipal branches of neighboring ramifications .

The hydrographic network formed by the basins of the Capim, Guama, Bujaru, Mãe do Rio and Irituia rivers and numerous stream micro-basins, once the only means of transportation and communication in the region, are still used quite often. The interviewed farmers, however, make greater use of the highways, where they primarily use buses, followed by boats, personal vehicles, middleman cars (trucks and vans), and other types of transportation (motorcycles, bicycles and animals).

3.2.1.1. Physical Components

The physical components of a region interact and form processes with energy, water and soil flows (Hart ,1980).

According to the Brazilian Institute of Geography and Statistics (IBGE), the prevailing climate in the Capim River Pole is the peculiar hot and humid climate of the Amazon region, with total rainfall exceeding 2,500 mm/year, distributed in two distinct periods: a rainy season (winter) from January to July, when there is higher rainfall incidence, of approximately 80.00% of the total rain, and a dry season from August to December (summer), when rainfall is less frequent, of about 20.00%. The thermal variation is very small, with minimum temperatures ranging from 22° C to 23° C and maximum temperatures ranging from 30° C to 34° C.

In the dry season, the rivers Irituia, Mãe do Rio and Bujaru decrease their water levels, which in the recent past was considered natural. However, this is now currently presented as a problem caused by the deforestation of headwaters and riparian areas, contributing to siltation processes.

Regarding topography, the region presents flat or slightly wavy areas. The exception is the Taperuçu community, in the São Domingos do Capim municipality, that presents altitudes of up to 80 m.

The predominant soil type is medium-textured oxisol, also featuring strips of laterite deposits. The terrain is high and tertiary, and the barrier formation is composed of sandstone, siltstone and clay. The predominant pedogenetic units are characterized, in general, as presenting low natural fertility.

3.2.1.2. Biotic Components

According to Hart (1980), different spatial and chronological arrangements of the biotic components that form the local fauna and flora are present in a certain region. The biotic processes differ according to the degree of change that humans have created.

In this sense, with the modification of the natural ecosystems of the Pole, input and output increments are considerable.

The Capim River Pole used to be dominated by firm land, floodplains and the characteristic wetlands that form the Amazon rainforest, with a rich biodiversity. The abundance of forest fauna and rivers rich in fish were important sources of protein in the diet of the local population. The opening of roads and exploitation of natural resources has contributed to the considerable loss of these resources.

3.2.1.3. Socioeconomic Components

The socioeconomic components of a region fall into three distinct sectors, primary, secondary and tertiary (Hart, 1980).

A) Primary Sector

In relation to agricultural products in general, those that provide income and part of family maintenance of the Capim River Pole are the cultivation of cassava (*Manihot esculenta*), rice (*Oryza sativa*), corn (*Zea mays*) and cowpea-beans [*Vigna unguiculata* (L) Walp].

Semi-permanent and permanent crops in monocrops and AFS also exist, such as: black pepper (*Piper nigrum* L), coconut (*Cocos nucifera*), orange (*Citrus sinensis*), banana (*Musa X paradisíaca* L), açaí (*Euterpe oleracea Mart*), cupuaçu (*Theobroma grandiflorum*), soursop (*Anona muricata* L), taperebá (*Spondias mombin L.*), guava (*Psidium guajava L*), pineapple (*Ananas sativa, Lindl*), cashew (*Anacardium Occidental*), peach palm (*Bactris gasipaes*); paricá (*Schizolobium amazonicum*), mahogany (*Swietenia macrophylla King*), Brazil nut (*Bertholletia excelsa*) and others.

The livestock system includes the raising of small, medium and large animals. However, the prominent farming activity of family farmers at the Pole is raising bovine livestock, in the form of extensive systems with few cattle heads and of mixed use (milk and meat), which in the last 15 years has become very evident due to credit access, especially from the Constitutional Fund of the North (FNO) – and the Support Program for Family Farming (PRONAF).

B) Secondary Sector

Relative to the secondary sector, we have the processing units of the primary products, their processing/marketing and, consequently, their value aggregation.

Specifically, in this region the main processing units are flourmills that transform the cassava fruit into several types of products, such as flour water, dried tapioca, tucupi and starch.

Also noteworthy is the transformation of homemade fruit into pulps, in all municipalities, with emphasis on açaí pulpers, as well the manual production of soursop and cupuaçu pulp.

At the municipality of Irituia, an agroindustry is established, that produces large scale fruit pulps, such as cupuaçu, taperebá, muruci (*Byrsonima crassifolia*), acerola (*Malpighia glabra L*) and tucumã (*Astrocaryum aculeatum*), plus raw material utilized in crafts, such as fibers, seeds and vines, among others.

C) Tertiary Sector

The tertiary sector of the Capim River Pole takes into account credit services, teaching/research/training, rural extension, transportation, organization, education, health and marketing of the products produced in the area.

At the Pole, the financial organizations are the Amazon bank (Banco da Amazônia - BASA) and the Brazilian Bank (Banco do Brasil - BB). Concerning the teaching/research aspect, we have the presence of the Brazilian Agricultural Research Corporation (EMBRAPA), the Federal University of Pará (UFPA), the Federal Rural University of the Amazon (UFRA) and the National Service of Rural Learning (SENAR). Extension services are carried out officially by the Technical Assistance and Rural Extension Company (EMATER) and by subcontractors.

Regarding transportation in the area, there is an extensive hydrographic network formed by the basins of rivers Capim, Guama, Bujaru, Mãe do Rio and Irituia, besides the road network linked by federal, state and municipal highways.

The organizational structure in the region is composed of unions, syndicates and work groups (Oliveira, 2006). Most of them are legalized, but others are still in the process of legalization.

These organizations have different purposes such as representing the category to government organs, facilitating credit access, defending the working class, defending the environment and providing different types of services. They seek to carry out their activities independently, but share the same difficulties and struggle with the lack of participation, financial resources and illiteracy of their leaders.

Among these organizations are some older syndicates, leftovers of the late 1960s, such as the Workers Union of Rural Irituia and São Domingos do Capim (Sindicatos dos Trabalhadores e Trabalhadoras Rurais de Irituia e São Domingos do Capim - STTR's).

Relationships also exist between the Unions, The Federation of Agricultural Workers of the State of Pará (Federação dos Trabalhadores na Agricultura do Estado do Pará - FETAGRI, subcontractors from Ater (Empresas Terceirizadas de Ater – EMATER), churches (mainly Catholic) and the local city halls.

These organizations have provided assistance to the community, including social, logistical and political support. Today, these organizations have pursued numerous achievements through fights, marches and walks.

Health services are restricted to emergency/urgency services. In more complex cases, patients are taken to the larger municipalities or to Belém, the state capital. Specifically regarding education, there are schools that serve urban and rural students in all districts.

Finally, the marketing system is based on the "mediation" exercised by traders (trading posts), middlemen and also directly at local markets, neighboring cities and outside the state.

3.3. Marketing Characteristics of the Pole

In order to characterize the marketing system of the Pole, the following variables were considered: *Municipality, transportation, education, institutionalism, circuits, difficulty in marketing* and *marketing strategies*, described below:

1) *Municipality*; refers to municipalities of the Pole: Concórdia do Pará; Irituia; Sâo Domingos do Capim and Mãe do Rio.
2) *Transportation*; refers to the way farmers circulate their production: 1 – Company Transportation (bus and boat), 2 - Private transportation (car, motorbike / bicycle / animal) and 3 – The middleman´s transportation.
3) *Education*; mentions the stratification of the study years of the farmers: 1 - illiterate, 2 - basic education and 3 - high school and higher education.
4) *Institutionalism*; Refers to the level of relationships that farmers have with the institutions at the Pole: 1 - Low 2 - Medium and 3 - High.
5) *Circuits* consider the number of intermediaries between production and consumption, and, particularly, where the sale occurs: 1 - Community, 2 – inside the municipality and 3 - outside the municipality (nearby city/other states).

6) *Difficulties in marketing* refer to: 1 - middleman, 2 – product packaging for transport and 3 – others (road condition/shipping value).
7) *Marketing strategies*: 1 - Unprocessed and 2 - Processed.

3.3.1. Main Products of the Pole

The identification of major products, even though a mix of cultures makes up the AFS, was established as the criteria for selecting those farmers who excel in marketing. According to the interviewed farmers, the main products were: Açaí, Cupuaçu, Pupunha, Cocoa and Cashew nuts.

The following modal values were considered for the calculation of the profit obtained from the sale of these products: R$ 70.00/bag of Açaí, R$ 3.00/ kg Cupuaçu; R$ 5.00/ bunch of Pupunha; R$ 5.00/kg of cocoa, and R$ 1.00/pound of Cashew Nuts. Table 2 lists the main products, the number of farmers, the amount produced, the percentage of consumption, the percentage of sale and the profit in Reais (R$) per municipality.

By observing Table 2 we can see that the municipality of Irituia has the largest number of farmers practicing agroforestry systems, but the city of Concórdia do Pará is the only one in which all farmers answered that they cultivate the above identified crops.

Table 2. Main Products, number of farmers, amount produced, consumed and sold, by municipality

Municipality	Farmers	Amount	% Cons	% Sold	Profit(R$)
AÇAÍ (R$ 70.0/bag)					
C. do Pará	12	765	18.60	81.40	52,650.00
Irituia	29	928	35.90	64.10	65,160.00
Mãe do Rio	4	18	85.00	15.00	1,260.00
S. D. do Capim	8	636	25.00	75.00	44,520.00
CUPUAÇU(R$ 3.00/kg)					
C. do Pará	12	3020	16.20	83.80	10,260.00
Irituia	29	11111	35.70	64.30	34, 353.00
Mãe do Rio	4	600	52.50	47.50	2,040.00
S. D. do Capim	8	790	58.60	41.40	2,390.00
PUPUNHA(R$ 5.00/bunch)					
C. do Pará	12	9185	26.00	74.00	45,925.00
Irituia	29	3855	40.00	60.00	19,375.00
Mãe do Rio	4	100	20.00	80.00	500.00
S. D. do Capim	8	110	83.30	16.70	3, 700.00
COCOA (R$ 5.00/kg)					
C. do Pará	12	4680	0.90	99.10	23,400.00
Irituia	29	0	0.00	0.00	0.00
Mãe do Rio	4	0	0.00	0.00	0.00
S. D. do Capim	8	930	10.00	90.00	4,650.00

Table 2. (Continued)

Municipality	Farmers	Amount	% Cons	% Sold	Profit(R$)
CASHEW NUT(R$ 1.00/kg)					
C. do Pará	12	1260	0.00	100.00	1,260.00
Irituia	29	1500	6.90	93.10	1,500.00
Mãe do Rio	4	2140	0.00	100.00	2,140.00
S. D. do Capim	8	0	0.00	0.00	0.00

3.3.2. Transportation

The roads present in the Pole are not always in good trafficability conditions, which significantly hampers the marketing of some products. In the case of the buses, usually company buses, product trampling by the passengers is not rare.

Regarding waterways, private boats with capacity for over 10 tons represent 70%, while 10% of these are canoes and 20% are company boats or freighters. In the more distant areas some perishable products often suffer problems.

Table 3 shows the frequency distribution of the types of transportation used for marketing products at the Pole. In most cases the transportation used for this purpose is company transportation, in 58.49% of cases.

Table 3. Percentage frequency (%) of the type of transportation used for product marketing in the municipalities of the Pole

Transportation	C. do Pará	Irituia	Mãe do Rio	S. D. do Capim	Total
Company	50.00	58.62	50.00	75.00	58.49
Private	16.67	34.48	50.00	12.50	28.30
Middleman	33.33	6.90	0.00	12.50	13.21

The C association test between *Transportation* and *Municipality* was not significant, with $p = 0.000$, indicating that there was no association between these two variables.

3.3.2. Education

Concórdia do Pará showed the highest relative percentage of farmers with high school and higher education, 58.33%. Irituia showed a prevalence of basic education, with a relative percentage of 86.21%. To test the association between *education* and *municipality*, a C test was used, resulting in $p = 0.000$, statistically significant. Thus, we conducted a correspondence analysis, with the residual values of the χ^2 test shown in Table 4. It is possible to evaluate which associations are significant by using the residue analysis, considering $\alpha = 5\%$. Thus, residue values greater than 1.96 represent significant associations.

Table 4. χ2 residues of the correspondence analysis between education and municipality

Municipality	Illiterate	Basic education	High school and higher education
C. do Pará	0.94	-2.92	2.63
Irituia	-0.14	3.13	-3.19
Mãe do Rio	-0.42	-1.91	0.16
S. D. do Capim	-0.61	0.47	0.23

* Values in bold represent significant associations considering $\alpha = 5\%$.

Figure 2 is a symmetry graph, showing the symmetry of the association between *education* and *municipality*, with component 1, the abscissa axis, representing 93.83% of the total variance and component 2 representing 6.17%. Both yielded 100% of the explanation of the analysis.

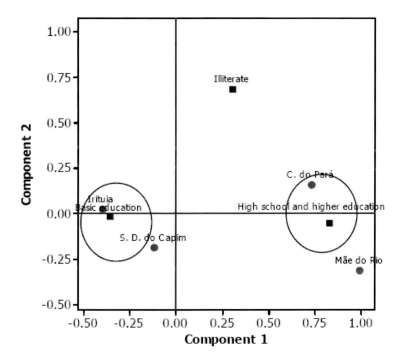

Figure 2. Representation of the symmetry of the association between education and municipality.

3.3.3. Institutionalism

Institutionalism refers to the level of relationship that the farmers have with the institutions at the Pole: 1 - Low 2 - Medium and 3 - High. The relationship with institutions at Concórdia do Pará presented the highest partial frequency, of 67.70%. In the case of São Domingos do Capim, the relationship with institutions was of 87.50%, the highest among the farmers of the surveyed municipalities.

The association between *institutionalism* and *municipality* was significant, with p = 0.001. According to the Correspondence Analysis, two components explain 100% of the data variability, with the first axis accounting for 69.35%.

Table 5. χ2 residues of the Correspondence Analysis between institutionalism and municipalities

Municipalities	Low	Medium	High
C. do Pará	-1.60	3.13	-1.48
Irituia	3.25	-1.66	-2.00
Mãe do Rio	0.29	-0.24	0.03
S.D. do Capim	-2.79	-1.18	4.49

* Values in bold represent significant associations considering α = 5%.

These significant associations can be observed in the symmetry graph shown in Figure 3.

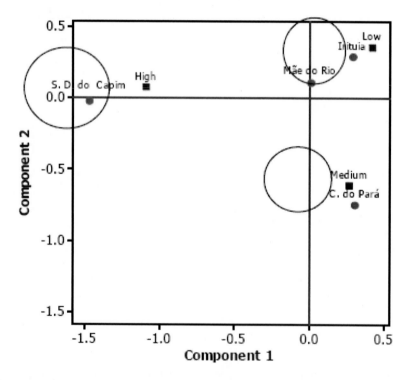

Figure 3. Representation of the symmetry of the association between institutionalism and municipality.

There are two possible genuine explanations for these situations: The largest number of courses offered by the PROAMBIENTE program in São Domingos Capim and in Concórdia do Pará (Nascimento, 2009), and the fact that the geographic location of Concórdia is near to the municipality of Tomé-Açu, that has been working with AFS in a tecnified manner since the 70's.

The PROAMBIENTE program trainings conducted by the Soci-environmental Foundation of Northeastern Pará (Fundação Sócio Ambiental do Nordeste Paraense -

FANEP) are most evident. However, the relationship with institutions varies from municipality to municipality, and, depending on each family decision regarding the AFS, external information and knowledge are configured differently. However, there are cases where farmers have no relationship with institutions and complain about the lack of the same, especially governmental institutions. However, those who complain are usually those who do not participate in any organization.

3.3.3. Circuits

According to the interviewed farmers, the marketing channels of the area are created in the community, in the municipality and outside the municipality. Table 6 shows the frequency (%) of the type of circuit used for product marketing at the Pole.

Table 6. Frequency Percentage (%) of the type of circuit in the municipalities of the Pole

Circuit	C. do Pará	Irituia	Mãe do Rio	S. D. do Capim
Community	8.33	10.34	0.00	12.50
In the Municipality	33.33	37.93	50.00	50.00
Outside the municipality	58.33	51.72	50.00	37.50

The farmers of Concórdia do Pará market products mostly outside the municipality, especially in the municipality of Tomé-Açu, followed by inside the municipalitty and, to a lesser extent, in the communities where they are located. There are no reports of out of state sales. In Irituia, the situation is similar to, and the affected marketing municipalities are: Castanhal, Braga and Belém. There are reports of outside state sales.

In the case of the Mãe do Rio, the highlight is that 50% of the entire production is sold in the municipality itself, due to the fact that, within the cities of the Pole, it is the only one that has a specific place for farmers to market their production directly. The proximity to Paragominas makes the latter its main marketing municipality. The highlight of outside state sales is related to the cashew nut, which is sold directly to Fortaleza.

São Domingos do Capim also concentrates sales in in the municipality itself, even without a proper place for this activity. Its proximity to Castanhal makes the latter its main marketing municipality. There are no reports of outside state sales, despite the common knowledge that part of the açaí production in the Castanhal industries is intended for out of state marketing.

3.3.4. Diffficulties

The difficulties in marketing refer to: 1 – the middleman, 2 – product packaging for transport and 3 - others (road conditions/shipping values). Table 7 shows the frequency (%) of the type of difficulty in marketing products in the municipalities of the Pole.

Table 7. Frequency Percentage (%) of the difficulty in marketing products in the municipalities of the Pole

Difficulty	C. do Pará	Irituia	Mãe do Rio	S.D. do Capim
Middleman	16.67	48.28	50.00	62.50
Product packaging for transport	25.00	20.69	0.00	25.00
Others	58.33	31.03	50.00	12.50

Table 7 shows that the greatest marketing difficulty for farmers from Concórdia do Pará corresponds to "others" (road conditions/shipping values), with 58.33%, since most farmers sell in other municipalities. 50.00% use company transports. The second difficulty is the product packaging for transport and marketing. In the case of Irituia, the main difficulty is the middleman, followed by "others", since 58.33% use company transports and the road conditions are poor. In the Mãe do Rio, the main difficulties are the middleman and "others", and in São Domingos do Capim the greatest difficulty is the middleman.

The association between *difficulty* and *circuit* was significant, considering $\alpha = 10\%$. According to correspondence analysis, two components explain 100%. Table 8 and the graph shown in figure 4 shows the significant associations.

Tabela 8. $\chi2$ Residues of the Correspondence Analysis between Difficulty and Circuit

Municipality	Community	In the municipality	Outside the municipality
Middleman	**1.73**	-0.04	-1.76
Product packaging for transport	1.64	-0.94	-0.90
Others	-2.62	0.94	**2.00**

* Values in bold represent significant associations considering $\alpha = 10\%$.

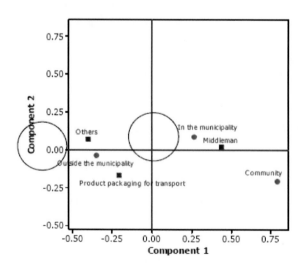

Figure 4. Representation of the symmetry of the association between difficulty and municipality.

3.3.5. Marketing Strategies

Research has shown that the products of agroforestry systems are classified regarding marketing in processed and unprocessed forms.

Among the municipalities of the Pole, Irituia showed the highest percentage regarding unprocessed items, which is contradictory, since the information obtained in interviews claims that, with the introduction of electricity in the district, the acquisition of industrial pulpers enabled açaí processing, addingaggregating value to this product, thus improving family income.

Those who turn cupuacu into pulp guarantee that it is economically viable if this product is marketed mainly in the offseason. At Concórdia, farmers now sell to the National Supply Company (Companhia Nacional de Abastecimento - CONAB) and for school meals. At Mãe do Rio this fruit is also marketed in the form of pulp, especially at Irituia and São Domingos that, besides the pulp, also markets the fruit in the form of sweets and chocolates.

Only the farmers of Concórdia do Pará and São Domingos answered that they produce cocoa and market it unprocessed. This production is sold to export companies and buyers acquire this product in two forms, dry and in some cases "wet" beans. The case of buying "wet" beans occurs only in São Domingos, where farmers only remove the kernels from the husk and sell them without the drying process.

Almost all the pupunha production is marketed fresh, although many farmers already recognize that this fruit can be transformed into flour and similar products. Only at Concórdia do Pará and Irituia do some farmers sell this fruit cooked by the roadside. The cashew nut production at the Pole is marketed in its fresh form.

To evaluate the marketing strategies utilized, the variable total percentual sale of processed products (TPS) was calculated, by multiplying the sale percentages by the modal values (SPMV) (fifth column of Table 2), pondered by the marketing strategy: 1 - unprocessed and 2 – processed, and added to the main products of the municipality. The following sets were constructed (R$):

Small: TPS <5650; Medium: TPS $\leq$ 5650 <7650, and Large: TPS $\geq$ 7650

The association between *TPS* and *municipality* was significant, with p = 0.017. According to the correspondence analysis, two components explain 100% of the variability.

Table 9 presents the χ^2 residues of the correspondence analysis between the *TPS* and *municipality*, with the graphical representation shown in figure 5.

Table 9. χ2 residues of the Correspondence Analysis between TPS and municipality

Municipality	Small	Medium	Large
C. do Pará	-0.54	-1.31	1.98
Irituia	0.12	2.04	-2.24
Mãe do Rio	2.55	-1.43	-1.30
S.D. do Capim	-1.50	-0.21	1.85

*Values in bold represent significant associations considering α = 10%.

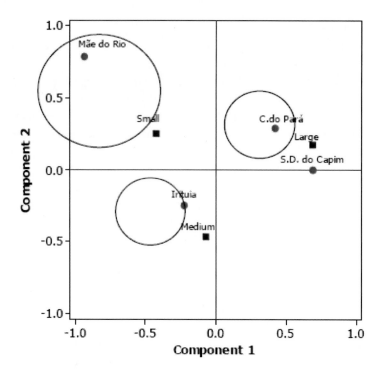

Figure 5. Symmetry graph representing the association between TPS and municipality.

3.4. Evolution of the Marketing System

The Kriesberg & Steele diagram (1974) formed the basis for the study of the evolution of the marketing situation of the farmers at the Rio Capim Pole. According to these authors, three types of economics are analysed in order to understand the evolution of the marketing system: traditional, transition and market economies, and in order to understand them the evolution depends on politicalm economic and socio-cultural forces.

The authors do not define or reflect on who is better or worse, but stress that, in all three cases, there are typical market problems. In the case of farmers at the Rio Capim Pole, these are based on the following variables: transportation, circuit, difficulty, education and institutionalism that are known to permeate political, economic, social and cultural rights.

Thus, to understand the evolution of the marketing system, in addition to the statistical techniques used previously, a K-means clustering analysis was also used, considering four clusters.

Cluster 1 consisted of farmers who have less formal education and some relation to institutions, whose major difficulty is product packaging, with the circuit basically restricted to the local communities.

The notable features of the farmers in cluster 2 are their higher schooling and higher relationships with the institutions. They have little relation to the middlemen regarding difficulties in transporting their products and the marketing circuits are not restricted merely to the community.

Farmers in cluster 3 have a good relationship with the institutions and the education is well distributed. The circuits are held in the communities, in the municipalities and outside the municipalities. The most widely used form of transportation is private transportation, followed by the middleman transport.

In cluster 4, education and relationships with institutions are well distributed. The major marketing difficulties are the road conditions and product packaging. Product transportation is conducted by company and private transportation. The circuits are limited to the local communities and the local municipality.

Figures 6(a) - 6(c) show the distributions of Total Sales (R$) Total Profit (R$), SPMV (R$) and TPS (R$) per cluster. The farmers in clusters 2 and 3 show higher total profit, however, the distribution of the total profit in cluster 2 is more homogeneous. This and the mapping from the perceptual knowledge of the researchers provided a basis for creating the rules of the fuzzy system for the study of marketing evolution: stage 1 - more evolved, stage 2-evolved; stage 3 – transition and stage 4 - less evolved.

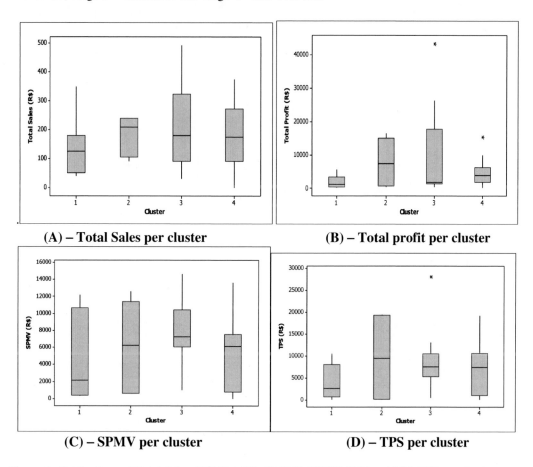

(A) – Total Sales per cluster   **(B) – Total profit per cluster**

(C) – SPMV per cluster   **(D) – TPS per cluster**

Figure 6. distributions of Total Sales (R$) Total Profit (R$), SPMV (R$) and TPS (R$) per cluster.

To develop the fuzzy solution we used the Fuzzy Logic Toolbox belonging to the computer software Matlab 7.0. The input variables of the system are the same as those used in the statistical analysis of the AFS, i.e.: transportation, circuit, difficulty, education, institutionalism, and the output variable was the marketing stage of the AFS.

The fuzzy sets of the system inputs and outputs were determined from the results of the conducted statistical analysis, as well as by the perceptual maps from expert knowledge. Thus, we determined the following fuzzy sets and linguistic terms for the variables involved in the problem solving: regarding *transportation* only difficult and easy classifications were considered, regarding *circuits,* low (community), medium (inside the municipality) and high (outside the municipality) classifications were considered; regarding *difficulty,* low (conditioning), medium (others) and high (middlemen) classifications were considered; regarding *education* low (illiterate), medium (basic education) and high (high school and upper education) classifications were considered; regarding *institutionalism,* low, medium and high classifications were considered; and regarding *TPS* small, medium and large classifications were considered.

For the output variable *marketing stage,* four classifications were defined: fourth, third, second and first, in which singular relevance values (singletons) of 0.5, 1.5, 2.5 and 3.5 were assigned.

The set of rules that process implications of the "if" (If), "then "(Then) type between the input and output variables, set in the exploratory data analysis and expert knowledge were:

- If (*Transport* = difficult and (*Circuit* = Low) and (*Difficulty* = high) and (education = low) and (*Institutionalism* = medium) and (*TPS* = small), then farmers are in the 4th marketing stage;
- If (*Transport* = difficult) and (*Circuit* = Low) and (*Difficulty* = medium) and (*education* = low) and (*Institutionalism* = medium) and (*TPS* = medium) then farmers are in the 3rd marketing stage.
- If (*Transport* = easy) and (*Circuit* = High) and (*Difficulty* = low) and (*education* = low) and (*Institutionalism* = medium) and (*TPS* = medium) then farmers are in the 2nd marketing stage.
- If (*Transport* = difficult) and (*Circuit* = medium) and (*Difficulty* = low) and (*education* = high) and (*Institutionalism* = high) and (*TPS* = large), then farmers are in the 1st marketing stage.

The final output of the system is a real number, resulting from the deffuzyfication of the fuzzy output generated by the inference machine from the implementation of the rule base for the provided input values, as shown in Figure 7.

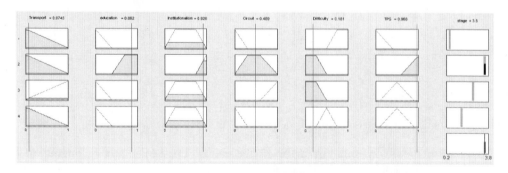

Figure 7. Operation of the fuzzy system in the marketing stages of the Capim River Pole.

CONCLUSION

This research identified the AFS characteristics at the Capim River Pole, located in the eastern Brazilian Amazon, based on products grown/processed by these AFS and the main difficulties in their marketing: the difficulty to obtain credit, rural extension services (since there is not enough human labor to meet regional demands), the perishability of the products, the presence of the middleman, the low price and lack of specific marketing policies at the Pole.

This set of factors that is reflected in the absence of marketing policies is not enough to undermine the expansion of agroforestry systems, but has been a limiting factor in the addition of more areas to family production units, or agroecosystems.

The circuits at a local scale show the insertion into the consumer market in the local communities, at a regional scale in different municipalities in the northeastern Pará area and even at a national scale.

The appeal of the current modern society for environmentally friendly products has not been translated into differentiated prices, which shows greater dissemination of the AFS products. However, in a timid manner some advances in this direction have been made, as in the purchase of products that are part of school meals in Concórdia do Pará, marketed at the organic market in Belém, the state capital, by farmers from Irituia.

Profits from product marketing do not have just an economic connotation, since the generation of environmental services, the social relations between farmers and the enhancement of local products, which can not always be measured, are significant gains that involve the entire chain of families and locations where the AFS are implanted.

To analyze the evolution of the marketing systems the defined variable was the total profit obtained, and we verified that farmers who have higher education levels and better relationships with institutions, and, therefore, easier access to information, are the ones in the best marketing stage, according to the classifications established in this study.

In the intermediary stages are those farmers that have some schooling and some relationship with institutions, and, finally, farmers in the lowest stage are those that are paid less, have less education and have more relationships with the middlemen, that show a marketing circuit restricted only to their local communities.

After the statistical analysis of the variables involved in the mapping of the AFS, the proposed solution using fuzzy logic to identify the marketing/development stage of the AFS of the studied area was proven adequate, and can be adapted and exported to other realities of a similar nature.

REFERENCES

[1]   Ayres, M.; Ayres, M, Jr; Ayres, D.L.; Santos, A. S. Bioestat 5.0. Aplicação estatisticas nas áreas de ciências biologicas e médicas. Instituto de Desenvolvimento Sustentável Marirauá-IDSM/MCT/CNPq. Belém, Para, 364p 2007,

[2]   Fávero, L. P; Belfione, P.; Silva, F. L ; Chan, B.L. *Análise de dados: Modelagem multivariada para tomada de decisão. Rio de Janeiro: Elsevier,* 640p, 2009.

[3]   Costa, F. de A. A diversidade estrutural e desenvolvimento sustentável: novas de política de planejamento agrícola para Amazônia. In: Perspectivas do desenvolvimento sustentável (uma contribuição para Amazônia 21) Tereza Ximenes (org.) Belém: UFPA/NAEA; Associação de Universidades Amazônicas, 1998

[4]   Matworks. Fuzzy Toolbox User's Guide: for use with MATLAB. Natick, MA: The MathWorks, Inc., 2001.

[5]   Turban, E.; Aronson, J. E. Decision Support Systems and Intelligent Systems. New Jersey: 6th edition. Prentice -Hall, 2001.

[6]   Wang, Li-Xin. A Course in Fuzzy Systems and Control. New Jersey: Prentice-Hall International, Inc., 1997.

[7]   Hair, J. F.; Anderson, R. E.; Tatham, R. L.; Black, W. C. Multivariate data analysis with readings. 5. ed. Upper Saddle River: Prentice Hall. 730p. 1998.

[8]   Hart, D. Robert. Agroecossistemas: conceitos básicos. Centro Agronômico Tropical de Investigacion y Enseñanza. Turrialba, Costa Rica. 1980. 201p.

[9]   Hébette, Jean. Cruzando a fronteira: 30 anos de estudo do campesinato na Amazônia. Vol. 1 – EDUFPA / UFPA. Belém, 2004.

[10]   Hespanhol, Antonio Nivaldo. Desafios da geração de renda em pequenas propriedades a questão do desenvolvimento rural sustentável no Brasil. *In: Desenvolvimento Territorial E Agroecologia*. Alves, Adilson Francelinoi, CORRIJO, Beatriz Rodrigues e CANDIOTTO, Luciano Zanetti (orgs) 1ª edição. São Paulo: Expressão Popular. 2008. 256p.

[11]   Hölscher, D.; Sá, T. D. De A .; Bastos, T. X.; Denich, M.; Fölster, H. Evaporation from young secondary vegetation in eastern Amazonia. Journal of Hydrology, 193: 293-305. 1997b.

[12]   Kriesberg, M & Steele,M. Mejoramento de los sistemas de comercializacion en los paises en dessarrollo. Um enfoque para La identificacion de problemas y El fortalecimento de la asistencia técnica. IICA. San Jose, Costa Rica, 1974.p6.

[13]   Moran, E. F. A ecologia humana das populações da Amazônia. Petrópolis: Vozes, 1990. 368p.

[14]   Nepstad, D. C.; Moreira, A . G.; Alencar, A. A. Flames in the rain forest: origins, impacts and alternatives to Amazonian fire. The Pilot Program to Conserve the Brazilian Rain Forest, Brasília, 1999.

[15]   Oliveira, J. S. R., Kato, O. R., Oliveira, T. F E Queiroz, J. C. B. Evaluation of sustainability in Eastern Amzon under proambiente program. Agroforest System. Pará, 2010.

[16]   Pereira, C. A. & Vieira, I. C. G.. A importância das florestas secundárias e os impactos de sua substituição por plantios mecanizados de grãos na Amazônia. Revista Interciência aug. 2001, vol.26 no8.

[17]   Revista Observatório Europeu Leader. Inovação no Meio Rural. FRANÇOIS, Martine.Comercializar os produtos locais através dos circuitos curtos. Caderno n°7. Julho de 2000. P 11-52.

[18]   Revista Observatório Europeu Leader. Inovação no Meio Rural. RICCE, Carlo. Comercializar os produtos locais através dos circuitos curtos e longos. Caderno n°7. Julho de 2000. P 53-80.

[19]   Santos, B. S (org.). Produzir para viver: os caminhos da produção não capitalista. 2ª edição. Rio de Janeiro: Civilização Brasileira, 2005. 514p.

[20] Santos, M. Metamorfoses do Espaço Habitado. Fundamentos teóricos e metodológicos da geografia. Ed. Hucitec. São Paulo, 1988.

[21] Sommer,R. Water and nutrient balance in deep soils under shifting cultivation with and without burning in the Eastern Amazon. Göttingen, Cuvillier, Tese Doutorado, 2000, p. 240.

In: Agricultural Research Updates. Volume 2
Editor: Barbara P. Hendriks

ISBN: 978-1-61470-191-0
© 2012 Nova Science Publishers, Inc.

Chapter 13

ADVANCED SYSTEMS FOR NITROGEN REMOVAL FROM EFFLUENTS PRODUCED IN THE FISH CANNING INDUSTRY

J. L. Campos, M. Figueroa, N. Morales, C. Fajardo, J. R. Vázquez-Padín, A. Mosquera-Corral and R. Méndez

Department of Chemical Engineering. School of Engineering.
Rua Lope Gómez de Marzoa s/n. University of
Santiago de Compostela,
Santiago de Compostela, Spain

ABSTRACT

Discharge of effluents produced in the fish-canning industry contributes significantly to the contamination of the environment in the littoral zones where they are discharged. These effluents have salinity similar to sea water, high organic matter content, and high protein concentration. Firstly, solids and oil are separated by physicochemical methods and, then, anaerobic digestion is generally applied to remove organic matter from these wastewaters. However the generated effluent contains high levels of ammonium concentration due to protein degradation, producing effluents characterized by low C/N ratios.

The post-treatment of these effluents by conventional nitrification-denitrification processes is not economically feasible since the addition of an external carbon source is needed. Therefore, the application of processes such as anammox or autotrophic denitrification, where ammonia and reduced sulphur compounds are used as electron donor, respectively, instead of heterotrophic denitrification can be a feasible alternative to remove nitrogen from these effluents.

Keywords: Anaerobic digestion, Anammox, autotrophic denitrification, CANON, COD/N ratio, fish canneries, nitrogen, SHARON.

1. INTRODUCTION

The presence of high Chemical Oxygen Demand (COD), organic nitrogen and solids concentrations characterizes the wastewater produced in fish canning factories (Veiga et al., 1994). The volume and concentration of wastewater from these factories depends mainly on the raw fish composition, additives used, processing water source and the unit process (Table 1).

These effluents are generally treated by biological processes although a suitable pre-treatment (screening, settling and/or dissolved air flotation (DAF)) is necessary in order to remove suspended solids and fats, oils and greases and to avoid their possible negative effects on the biological system operation (Chowdhury et al., 2010). The biological treatment can be carried out under either aerobic or anaerobic conditions. To carry out an aerobic process, air must be continuously supplied to convert 50% of the organic matter, measured as COD, into CO_2 and 50% into sludge which is a waste product and, therefore, it costs lots of money to get rid of it. Contrary, during the anaerobic process 90% of COD is converted into CH_4, which can be used as fuel, and only very little COD is converted into sludge but this process must operate at temperatures around 35 °C which implies an important requirement of energy.

Taking into account the operational costs, the anaerobic processes are recommended for the treatment of the fish cannery effluents since their high level of COD generates an amount of methane which can compensate the energetic cost to maintain the temperature of the reactor. Nevertheless, the drawback of this technology is that organic nitrogenous compounds present, such as proteins, aminoacids or urea, are mainly reduced to ammonia which is not further degraded in anaerobic conditions. On the other hand, fish cannery wastewater also contains high sulphate concentration, due to the use of seawater during different processes, which under anaerobic conditions is reduced into sulphide by sulphate reducing bacteria (Aspé et al., 1997). Then, the biogas produced in the anaerobic bioreactor treating fish canneries effluents usually contains small amount of hydrogen sulphide (SH_2) and ammonia (NH_3) and they must be removed from the biogas for its utilization for energy generation.

The presence of both SH_2 and NH_3 in the liquid can also have negative effects on anaerobic digestion since these compounds are inhibitory. For methanogenic bacteria, the inhibitory levels of SH_2 found by different authors (Rinzema and Lettinga, 1988; McCartney and Oleszkiewicz, 1993; Kroiss and Plahl-Wabnegg, 1983; Visser, 1995) are very different and depend on the kind of biomass, being granular sludge less sensitive to SH_2 than flocculent sludge. To avoid the possible negative effects of SH_2 during the treatment of fish canneries effluents (0.6-2.7 g $SO_4^{-2} \cdot L^{-1}$) Omil et al. (1995) proposed to control pH to maintain the concentration of this compound under inhibitory levels.

Free ammonia causes inhibitory effects on anaerobic digestion at values as low as 0.08-0.10 g $NH_3 \cdot L^{-1}$ (Braun et al., 1981) but, if the biomass is adapted, the process can operate under stable conditions even in presence of 1.2 g $NH_3 \cdot L^{-1}$ (Guerrero et al., 1997). The presence of high sodium and/or chloride concentrations is also inhibitory for anaerobic wastewater treatment (Rinzema et al., 1988). However, Omil et al. (1995) showed it was possible to operate under high saline conditions if the methanogenic biomass was previously adapted. In fact, Guerrero et al. (1997) and Aspé et al. (1997) were able to treat effluents from fish canneries with chloride concentrations of 7.5 and 16.2 g $Cl^- \cdot L^{-1}$, respectively.

In spite of the presence of the quoted compounds, anaerobic digesters have been successfully used for years in many fish cannery factories (Nyns, 1994). In order to maintain a stable performance in anaerobic reactors high concentrations of biomass must be retained inside. This can be achieved by promoting the formation of dense bacterial granules as in UASB (Upflow Anaerobic Sludge Blanket) or EGSB (Expanded Granular Sludge Bed) reactors or by promoting the attachment of bacteria to carriers as in AF (Anaerobic filter). When fish cannery effluents were applied to these systems organic loading rates (OLR) ranged from 1 to 8 g $COD \cdot L^{-1} \cdot d^{-1}$ were removed and COD removal efficiencies between 80-95% were achieved (Guerrero et al., 1997; Balslev-Olesen et al., 1990; Méndez et al., 1992; Palenzuela-Rollón et al., 2002; Mosquera et al., 2001; Mosquera et al., 2003). During anaerobic digestion mostly of proteins are also degraded into ammonia which causes that the effluent generated has a low COD/N ratio (Mosquera-Corral et al., 2005; Vázquez-Padín et al., 2009a). Therefore, the conventional technology for nitrogen removal based on nitrification and denitrification processes can not be applied since organic matter may be insufficient to achieve complete denitrification. Thus, other sustainable solutions, such as partial nitrification, Anammox and autotrophic denitrification, should be applied in order to fulfil the nitrogen disposal requirements (Kleerebezem and Méndez, 2002; Fajardo et al., 2011; Dapena-Mora et al., 2006; Vázquez-Padín et al., 2009a). In the next sections both processes and technologies related to nitrogen removal from this kind of effluents are described and their possible applications are evaluated.

2. NITROGEN REMOVAL ALTERNATIVES

Ammonium present in wastewaters can be removed by physicochemical or biological processes. The selection of the best alternative is generally based on cost-effectiveness issues which, in practice, are determined by the nitrogen concentration of the wastewater. Three concentration ranges can be distinguished (Mulder, 2003):

a) Diluted wastewater with ammonium concentration up to 100 mg $N \cdot L^{-1}$ (e.g. municipal wastewater). In this range biological processes are the most economical option.

b) Concentrated wastewater with ammonium concentrations in the range of 100-5000 mg $N \cdot L^{-1}$ (e.g. fish canneries effluents). In this case, both physicochemical and biological methods can be applied. However, an economical evaluation showed the treatment by biological processes is cheaper (2.3–4.5 euros per kg N removed) than the physicochemical methods (4.5–11.3 euros per kg N removed) (Van Dongen et al., 2001).

c) Concentrated wastewater with ammonium concentrations higher than 5000 mg $N \cdot L^{-1}$ (e.g. urine). In this range physicochemical methods are technically and economically feasible. The main physicochemical processes applied are air stripping, precipitation and selective ion exchange which are focused on N-recovery instead of nutrient removal (Maurer et al., 2006).

Table 1. Characteristics of the effluents generated by fish canneries

Parameter (g·L⁻¹)	Mosquera et al. (2001)	Omil et al. (1995)	Vidal et al. (1997)	Figueroa et al. (2008)	Guerrero et al. (1997)	Mosquera et al. (2003)	Artiga et al. (2008)	Aspé et al. (1997)	Veiga et al. (1994)
COD_t	2.9-4.0	12.36-55.40		4.8-18.5	120.85	50-56	8-26	6.0	11.7-25.0
COD_s	1.1-1.3	9.71-16.50	5.0-6.3	4.1-15.8	96.65		7-25		11.0-24.0
Grease + oil							0.5-1.7		
TKN	0.60-0.65		0.48-0.80			5-6	1.2-4.0	0.54	
NH_4^+-N	0.38-0.41			0.6-2.0	6.39	2-3	0.2-0.7	0.09	0.1-0.2
Protein	0.70-0.84	3.34-19.20	3.0-5.0		33.82	15-18			7.0-12.6
pH	7.57-7.86		6-7	6.6-7.5			6.5	4.5	6.0-6.5
TSS			11.0-22.0	1.3-2.0	17.93		1.0-2.1	6.5	0.4-1.1
Cl^-		7.8-19.5	7.1-9.1		10.7		9.3-41.5	10.8-18.5	1.6-7.6
SO_4^{-2}		0.59-2.70	0.98-3.30					2.40	0.2-0.6

On the other hand when biological processes are considered, the COD/N ratio of wastewater will determine the most suitable process to carry out nitrogen removal:

1) COD/N >20: in this case, the assimilation of nitrogen by heterotrophic bacteria is sufficient to remove nitrogen.
2) 20<COD/N<5: removal of nitrogen by assimilation and nitrification-denitrification pathway.
3) COD/N<5: in this case, the nitrification-denitrification processes are not suitable since an additional carbon source is needed. Nitrogen removal by "nitrite-route" processes such as partial nitrification-denitrification, partial nitrification-Anammox or nitrification-autotrophic denitrification should be implemented to optimize nitrogen removal.

2.1. Nitrification-Denitrification

The combination of nitrification-denitrification processes is generally applied to remove nitrogen from municipal wastewater and most of industrial wastewaters. Nitrification implies oxidation of ammonia into nitrate under strict aerobic conditions and is conducted in two sequential stages (Equations 1 and 2): ammonia into nitrite (ammonia oxidation) and nitrite into nitrate (nitrite oxidation). Each stage is performed by different bacterial genera which use ammonia or nitrite as energy source, molecular oxygen as electron acceptor and carbon dioxide as carbon source (Ahn, 2006).

$$NH_4^+ + 1.5\ O_2 \rightarrow NO_2^- + H_2O + 2\ H^+ \tag{1}$$

$$NO_2^- + 0.5\ O_2 \rightarrow NO_3^- \tag{2}$$

Denitrification is performed by heterotrophic bacteria under anoxic conditions (absence of oxygen but presence of oxidized compounds). Nitrate formed during nitrification is reduced into gaseous dinitrogen by these bacteria using organic matter as carbon and energy source (Equation 3):

$$8\ NO_3^- + 5\ CH_3COOH \rightarrow 8\ HCO_3^- + 6\ H_2O + 2\ CO_2 + 4\ N_2 \tag{3}$$

Since nitrification and denitrification are carried out under different conditions, these processes have to be separated in space (for continuous systems) or time (for batch systems). In continuous systems, two tanks are used (Figure 1). Generally, wastewater is fed into the denitrifying reactor (first tank) and afterwards nitrification is carried out in the second tank. A stream from the aerobic tank containing nitrate and/or nitrite is recirculated to the first unit to carry out denitrification. Therefore, nitrogen removal efficiency depends on the recycling ratio (ratio between recycle and inlet flows) (Equation 4):

$$\eta = \frac{R}{R+1} \cdot 100 \tag{4}$$

where η is the nitrogen removal percentage (%) and R is the recycling ratio.

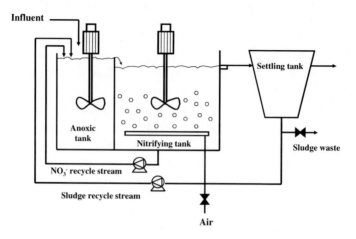

Figure 1. Tanks configuration to carry out nitrification-denitrification processes in continuous systems.

In batch systems, the reactor is operated in cycles (Figure 2). During the filling stage, influent containing organic matter and ammonia is fed to the system. Then the reactor is mixed in order to denitrify nitrate remaining from the previous cycle. During aeration phase, organic matter is consumed and ammonia oxidized into nitrate. In a fourth stage, both mixing and aeration are switched off to promote sedimentation of sludge. Finally, part of the liquid contained in the reactor is withdrawn.

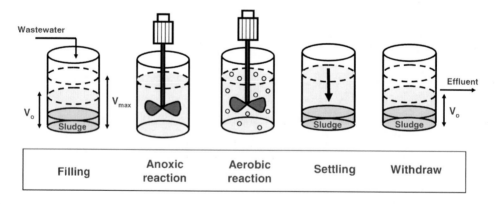

Figure 2. Cycle distribution to carry out nitrification-denitrification processes in batch systems.

In this case, nitrogen removal efficiency depends on the volumetric exchange ratio (ratio between the influent volume treated during one cycle and the useful volume of the reactor) (Equation 5):

$$\eta = 100 \cdot (1 - VER) \tag{5}$$

where VER is the volumetric exchange ratio.

2.2. Partial Nitrification

Nitrification and denitrification processes are suitable to remove ammonia from wastewater when its COD/N ratio is high, but when the amount of COD available in the effluent is not high enough to complete the denitrification process the addition of COD is necessary, which supposes an increase of costs. In those cases, partial oxidation of ammonia into nitrite (Figure 3) would suppose a decrease of both oxygen and organic matter requirements of 25% and 40%, respectively, and only 60% of sludge is generated compared to the full oxidation into nitrate (Van Kempen et al., 2001).

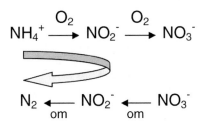

Figure 3. Scheme of the partial nitrification process (om: organic matter).

In order to remove nitrogen by means of the partial nitrification process, aerobic and anoxic tanks can be provided in a post-denitrification configuration where wastewater is fed into the partial nitrifying unit and its effluent enters into the denitrifying reactor. This configuration is very simple, easy to control and no recycling is needed. However addition of an external carbon source (methanol, acetic acid…) to the anoxic tank is necessary in order to complete denitrification (Figure 4).

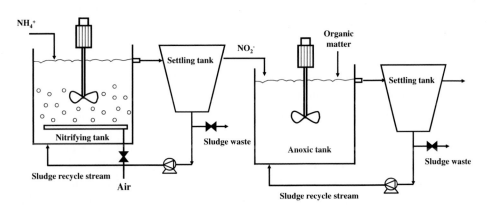

Figure 4. Tanks configuration to carry out partial nitrification-denitrification processes.

To obtain partial nitrification, the ammonium has to be converted into nitrite by ammonia oxidizing bacteria (AOB) while the oxidation of nitrite to nitrate carried out by nitrite oxidizing bacteria (NOB) has to be avoided. For this purpose, firstly, the system should be operated under conditions where AOB grow faster than NOB. And, then, biomass should be purged at least at a rate similar to the growth rate of NOB in order to promote their wash-out from the system but lower than the growth rate of AOB to allow their retention inside the system. This last condition can be expressed in terms of the solids retention time (SRT) by equation 6:

$$\frac{1}{\mu_{NOB}} > SRT > \frac{1}{\mu_{AOB}} \tag{6}$$

where $1/\mu_{AOB}$ and $1/\mu_{NOB}$ are the inverse of the AOB and NOB growth rates and gives minimum solids retention time (d) needed to avoid the wash-out of AOB and NOB, respectively; SRT (d) is calculated by equation [7].

$$SRT = \frac{X_r \cdot V_r}{Sludge\ purge\ rate} \tag{7}$$

where X_r is the biomass concentration inside the reactor (g $VSS \cdot L^{-1}$); V_r the volume of the reactor (L) and; Sludge purge rate the amount of biomass purged per unit of time (g $VSS \cdot d^{-1}$).

To achieve this aim, factors affecting to both reactions must be changed in such a way that the growth rate of ammonia-oxidizers is more enhanced than that of nitrite-oxidizers. Both growth rates depend on temperature, dissolved oxygen and free ammonia concentrations (Park and Bae, 2009) (Equation 8):

$$\mu = \mu_{max}(T) \cdot \frac{C_{O_2}}{K_{O2} + C_{O_2}} \cdot \frac{C_S}{K_S + C_S \cdot (1 + \frac{C_{NH_3}}{K_{INH_3}})} - b \tag{8}$$

where μ_{max} is the maximum growth rate (d^{-1}), C_{O2} the dissolved oxygen concentration (mg $O_2 \cdot L^{-1}$), K_{O2} the oxygen affinity constant (mg $O_2 \cdot L^{-1}$), C_S the substrate concentration (NH_4^+ for AOB and NO_2^- for NOB) (mg $N \cdot L^{-1}$), K_S the substrate affinity constant (mg $N \cdot L^{-1}$), C_{NH3} the free ammonia concentration which inhibits both ammonia- and nitrite oxidizers (mg $N \cdot L^{-1}$); K_{INH3} the free ammonia inhibition constant (mg $N \cdot L^{-1}$) and b the decay coefficient (d^{-1}).

Since these factors affect in different way to both AOB and NOB, the conditions where AOB grow faster than NOB can be selected ($\mu_{AOB} > \mu_{NOB}$). According to this, several parameters could be controlled in the nitrifying tank to reach partial nitrification:

1) Temperature: When the nitrifying systems are operated between 10 to 20 °C no nitrite build-up is registered due to the higher growth rate of NOB in this range of temperatures. However this tendency changes at temperatures around 25 °C due to the fact that ammonia oxidation has an activation energy (68 $kJ \cdot mol^{-1}$) higher than that of nitrite oxidation (43 $kJ \cdot mol^{-1}$) (Wiesmann, 1994). From an economic point of view, this strategy would be only feasible when effluents have already such temperature, for example effluents of anaerobic digesters.

Selection of AOB by controlling the operation temperature is the strategy chosen by the SHARON technology (Single reactor High Ammonia Removal Over Nitrite) (Hellinga et al., 1998; van Dongen et al., 2001; Mosquera-Corral et al., 2005). This technology consists essentially in a chemostat without biomass retention (solids retention = hydraulic retention time), operating at temperatures between 25-40°C and at a HRT of 1 day to promote the wash-out of nitrite-oxidizers while ammonia-oxidizers are retained.

2) Dissolved oxygen (DO): Since the oxygen affinity constant of nitrite-oxidizers is (1.1 mg $O_2 \cdot L^{-1}$) higher than that of ammonia-oxidizers (0.3 mg $O_2 \cdot L^{-1}$), a decrease of the dissolved oxygen in the reactor would exert a higher effect on the former one (Wiesmann, 1994). Therefore, nitrite generation would be favoured at low dissolved oxygen concentrations (Garrido et al., 1997). However, this operational strategy also implies the decrease of the ammonia-oxidizers activity and a control system for oxygen is required when the influent characteristics are not constant.

3) Free ammonia: The values of the free ammonia inhibition constant for both ammonia- and nitrite-oxidizers are 116 and 0.52 mg NH_3-$N \cdot L^{-1}$, respectively (Wiesmann, 1994). This supposes that AOB are more resistant to NH_3 than NOB and, therefore, partial nitrification could be achieved by maintaining free ammonia levels in the reactor which only cause the inhibition of nitrite-oxidizers. Albeit, the maintaining of a certain concentration of free ammonia in the system means that the effluent does not fulfil the disposal requirements.

2.3. Anammox Process

Broda (1977) predicted, using thermodynamic calculations, the existence of chemolitoautotrophic bacteria capable to oxidize ammonium using nitrite as electron acceptor. That prediction would be experimentally confirmed two decades later by Mulder et al. (1995) in a denitrifying pilot plant, treating wastewaters from a yeast plant. Anammox bacteria convert ammonium together with nitrite (electron acceptor) directly to dinitrogen gas in the absence of any organic carbon source, following the reaction described in equation 9 (Strous et al., 1998). In this process a small amount of nitrate is also produced in the anabolism of anammox bacteria.

$$NH_4^+ + 1.32\ NO_2^- + 0.066\ HCO_3^- + 0.13\ H^+ \rightarrow$$
$$1.02\ N_2 + 0.26\ NO_3^- + 0.066\ CH_2O_{0.5}N_{0.15} + 2\ H_2O \qquad (9)$$

The combination of partial nitrification and anammox processes allows minimizing oxygen and organic matter requirements during the treatment of wastewaters with low COD/N ratios. In a first unit 50% of inlet ammonia can be oxidized into nitrite under aerobic conditions (Figure 5). Later, anammox process can be carried out in UASB (Upflow Anaerobic Sludge Blanket) or IC (Internal Circulation) reactors similar to those used during anaerobic treatment at high loading rates.

Anammox bacteria belong to the phylum *Planctomycetes* and their yield coefficient is low (Y= 0.038 g VSS$\cdot$(g NH_4^+-N)$^{-1}$). This low sludge production reduces the management costs but makes the start-up of anammox reactors quite long. For this reason, systems with high biomass retention capacity must be used (granular systems (Strous et al., 1998); biofilm reactors (Fernández et al., 2008) or membrane reactors (Trigo et al., 2006)). Another characteristic of these microorganisms is the decrease of their activity in the presence of nitrite and, therefore, its concentration inside the reactor should be maintained at low levels to avoid the destabilization of the system (Strous et al., 1999; Dapena-Mora et al., 2004; 2007). In this sense the knowledge of kinetic parameters such as the maximum specific consumption rate and the inhibition constant for nitrite could be useful in order to avoid an overload of the system.

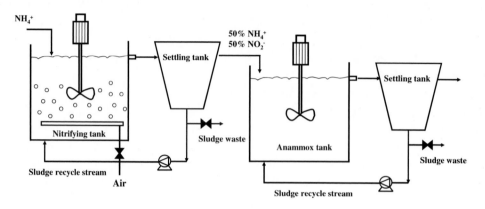

Figure 5. Tanks configuration to carry out partial nitrification and anammox processes.

2.4. CANON Process

Both partial nitrification and anammox processes can be simultaneously carried in a single-stage system under limiting oxygen conditions (Figure 6). In this system, AOB consume oxygen and generate both nitrite and an anoxic environment for anammox microorganisms. Then, ammonia is directly converted into nitrogen gas with nitrite as intermediate product (small amounts of nitrate are also produced) according to equation 11 (Strous, 2000). Different acronyms were used to define this process: OLAND (Oxygen-Limited Aerobic Nitrification and Denitrification) (Windey et al., 2005), aerobic deammonification (Wett, 2006) and CANON (Completely Autotrophic Nitrogen removal Over Nitrite) (Sliekers et al., 2002; 2003). The two former names are based on the idea that the own ammonia-oxidizers carried out the denitrification process. However, nowadays, it is known that anammox bacteria are responsible for the denitrification process, the last acronym being the most suitable to define the process.

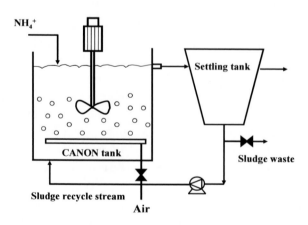

Figure 6. CANON process.

$$2.5\ NH_4^+ + 2.1\ O_2 \rightarrow 1.15\ N_2 + 0.2\ NO_3^- + 3.6\ H_2O + 2.8\ H^+ \tag{11}$$

Biofilm reactors are a suitable technology to develop the CANON process. AOB can grow in the outer part of the biofilm and produce nitrite and consume oxygen to provide anoxic conditions in the inner part of the biofilm. In this anoxic zone, ammonium (left by AOB) and nitrite (produced during partial nitrification) have to be present in order to allow the growth of anammox bacteria (Vázquez-Padín et al., 2009b).

2.5. Autotrophic Denitrification

Denitrification with sulphur compounds as electron donor is an alternative to heterotrophic denitrification for wastewater with high nitrogen concentration and low organic matter content (Campos et al., 2008). This process is carried out by autotrophic denitrifiers, such as *Thiobacillus denitrificans* and *Thiomicrospira denitrificans*. The energy required by these microorganisms is derived from oxidation-reduction reactions with elements such as hydrogen or various reduced-sulphur compounds (H_2S, S, $S_2O_3^{2-}$, $S_4O_6^{2-}$, SO_3^{2-}) acting as the electron donors. Autotrophic denitrifiers utilize inorganic carbon compounds (e.g., CO_2, HCO_3^-) as their carbon source for growth. Therefore, compared with heterotrophic conventional denitrification, autotrophic denitrification has two clear advantages: (1) no need for an external organic carbon source, e.g., methanol or ethanol, which decreases the cost of the process; and (2) less biomass production, which minimizes the handling of sludge (Claus and Kutzner, 1985; Zhang and Lampe, 1999). The common values of maximum growth rates for the autotrophic denitrifying microorganisms, 0.11 to 0.20 h^{-1} (Claus and Kutzner, 1985; Oh *et al.*, 2000), are similar to the heterotrophic ones (0.062-0.108 h^{-1} (Wiesmann, 1994)) while the biomass yields are lower for the autotrophic denitrifying microorganisms (0.40-0.57 g VSS (g NO_3^--N)$^{-1}$ (Claus and Kutzner, 1985; Oh *et al.*, 2000)) compared to 0.8-1.2 g VSS (g NO_x^--N)$^{-1}$ (Wiesmann, 1994) for the denitrifying heterotrophic ones.

The autotrophic denitrification rate depends on the oxidation state of the sulphur compound present in the wastewater (Table 2). The oxidation rates of thiosulphate are around 212 mg NO_3^--N·(g VSS·d)$^{-1}$ which are ten times higher than that of 22 mg NO_3^--N·(g VSS·d)$^{-1}$ reported elsewhere using elemental sulphur (Beristain *et al.*, 2006).

Table 2. Oxidation-reduction reactions with different sulphur compounds

Reaction	$\Delta G^{o\prime}$ (kJ·mol^{-1})	Reference
$1.25\ S^{2-} + 2\ NO_3^- + 2\ H^+ \rightarrow$ $1.25\ SO_4^{2-} + N_2 + H_2O$	-972.8	Reyes-Avila *et al.* (2004)
$5\ S^{2-} + 2\ NO_3^- + 12\ H^+ \rightarrow$ $5\ S^o + N_2 + 6\ H_2O$ $5\ S^o + 6\ NO_3^- + 8\ H_2O \rightarrow$ $5\ H_2SO_4 + 3\ N_2 + 6\ OH^-$	-1151.38 -1833.96	Wang *et al.* (2005)
$S^o + 1.2\ NO_3^- + 0.4\ H_2O \rightarrow$ $SO_4^{2-} + 0.6\ N_2 + 0.8\ H^+$ $S_2O_3^{2-} + 1.6\ NO_3^- + 0.2\ H_2O \rightarrow$ $2\ SO_4^{2-} + 0.8\ N_2 + 0.4\ H^+$	-547.6 -765.7	Beristain *et al.* (2006)

Autotrophic denitrifying bacteria have an optimum pH range of 7-8 (Oh *et al.*, 2000; Claus and Kutzner, 1985). In this range of pH values the end products of denitrification are N_2 and sulphate while at pH values below 7 the denitrification is incomplete and intermediate products such as nitrite and/or elemental sulphur are detected. At pH values under 6 or over 9 the complete inhibition of denitrification is observed (Moon *et al.*, 2004; Oh *et al.*, 2000). Sulphur oxidizing bacteria can grow under mesophilic environments (25-35 °C), their optimum value of temperature being around 35 °C. When temperature is higher than 40 °C (Oh *et al.*, 2000) or lower than 15 °C (Yamamoto-Ikemoto *et al.*, 2000), the autotrophic denitrification rate is negligible.

It has been observed that the S/N ratio of the feeding plays an important role on the autotrophic denitrification. Oh *et al.* (2000) tested different S/N ratios with thiosulphate as electron donor and nitrate as electron acceptor and they found that at ratios below of 6.51 g $S \cdot (g\ N)^{-1}$ (sulphur limitation) the denitrification was only carry out to nitrite. In the case of S/N ratios higher than the stoichiometric one, the sulphur compound is only oxidized to elemental sulphur (Gadekar *et al.*, 2006). The control of the inlet S/N ratio is required to drive the process to the products of interest.

3. APPLICATION OF NITROGEN REMOVAL TECHNOLOGIES

Mosquera et al. (2003) proposed the use of the combination of an anoxic filter and a nitrifying activated sludge system to treat an effluent of a fish cannery anaerobic digester. These authors observed that the efficiency of the system was mainly limited by the low organic content of the wastewater and only when the COD/N ratio was higher than 4 a complete denitrification was achieved. Such system was able to remove nitrogen with an efficiency of 60%, treating a nitrogen loading rate (NLR) up to 0.22 NH_4^+-$N \cdot L^{-1} \cdot d^{-1}$. Similar results were obtained by Figueroa *et al.* (2008). These authors used an aerobic granular system which was able to treat OLRs up to 1.7 g $COD \cdot L^{-1} \cdot d^{-1}$ and NLRs up to 0.18 g $N \cdot L^{-1} \cdot d^{-1}$ with average removal efficiencies of 95% and 40%, respectively. In spite of ammonia was only oxidized into nitrite, the nitrogen removal efficiency of this system was also limited by availability of organic matter in the feeding.

The SHARON-Anammox technology was applied to treat the effluent from fish canneries anaerobic digesters (Vázquez-Padín et al., 2009a; Mosquera-Corral et al., 2005; Dapena-Mora et al., 2006). Vázquez-Padín et al. (2009a) operated the SHARON unit at a HRT of 1 d and at pH value of 7.5 but the system did not achieved an stable operation in terms of ammonia oxidation efficiency and the value of NO_2^--N/NH_4^+-N molar ratio of the effluent ranged between 0.15 and 3.4. These obtained variable concentrations were caused by the changes of the organic matter content. The anammox reactor was fed with the effluent of the SHARON reactor at a NLR ranging between 0.34 and 0.67 g $N \cdot L^{-1} \cdot d^{-1}$. When the NO_2^--N/NH_4^+-N molar ratio of the SHARON effluent was lower than 1.3 (nitrite limiting conditions) an efficiency of 100% regarding to NO_2^- removal was achieved but the overall percentage of nitrogen removal was between 40-80%. However when the reactor was fed with an influent with a NO_2^--N/NH_4^+-N molar ratio higher than 1.3, the system turned unstable due to nitrite accumulation (Dapena-Mora *et al.*, 2004). As the overall efficiency of the system was limited by unstable performance of the SHARON unit, the authors proposed the control of operational parameters such as dissolved oxygen or pH in the SHARON process to obtain an optimal nitrite to ammonium ratio or the application of other technologies to obtain a suitable effluent to fed the anammox reactor (Vázquez-Padín et al., 2010; Yamamoto *et al.,* 2008).

On the other hand, fish cannery wastewater also contains high sulphate concentration. Sulphate is reduced into hydrogen sulphide during anaerobic digestion and it must be removed from the biogas for its utilization for energy generation. Kleerebezem and Méndez (2002) proposed to separate H_2S from methane by absorption and then to use it as electron donor for elimination of nitrate in a post-denitrification step. Therefore, combination of nitrification and autotrophic denitrification seems to be a suitable option to remove nitrogen from the effluent of an anaerobic digester treating a fish canning (Fajardo *et al.,* 2011). These authors observed that the nitrifying reactor was able to treat an ammonia loading rate (ALR) around 0.3 g NH_4^+-$N \cdot L^{-1} \cdot d^{-1}$ with an efficiency close to 100% when this system was fed with this effluent. Nitrate generated was removed in an autotrophic denitrifying SBR which treated a maximum sulphide loading rate (SLR) and NLR of 200 mg $S^{2-} \cdot L^{-1} \cdot d^{-1}$ and 0.1 g NO_3^--$N \cdot L^{-1} \cdot d^{-1}$, respectively. This reactor maintained a removal efficiency of 100% for sulphide while only around 30% of the inlet nitrate was removed. This low removal efficiency respect to nitrate was attributed to the limitation of the sulphur source. The S/N ratio of the fish canneries effluents ranged from 0.8 to 1.6 g $S \cdot (g\ N)^{-1}$ (Table 1). Since 1.4 kg S^{-2}-$S \cdot (kg\ NO_3^-$-$N)^{-1}$ are needed to achieve a complete denitrification (Park and Yoo, 2009), sulphide present in the biogas could be not enough to remove all the nitrogen and S° should be added to the denitrifying system.

On the other hand, literature shows NLRs removed by autotrophic denitrifying systems using sulphide as sulphur source ranged between 0.1 and 0.6 g NO_3^--$N \cdot L^{-1} \cdot d^{-1}$ (Manconi *et al.,* 2006; Wang *et al.,* 2005; Vaiopoulou *et al.,* 2005; Gadekar *et al.,* 2006) while NLRs up to 2 g NO_3^--$N \cdot L^{-1} \cdot d^{-1}$ can be removed when S° is used as sulphur source. Moreover when sulphide is employed, the low S/N applied causes the formation of nitrite as an intermediate product, which decreases both nitrogen removal efficiency and quality of the effluent. Nevertheless, when autotrophic denitrification is carried out with S° no control of the S/N is needed to avoid the formation of nitrite. Therefore the use of elemental sulphur instead of sulphide could be an interesting option to improve the performance of the autotrophic denitrifying reactor. H_2S coming from biogas could be previously oxidized under microaerobic conditions into S° and then, add it (together with an additional amount of S°) to the denitrifying unit to remove completely nitrate.

The applied NLR and the achieved nitrogen removal efficiencies during the post-treatment of fish canning industries ranged from 0.02 to 0.45 g $N \cdot L^{-1} \cdot d^{-1}$ and from 20 to 80%, respectively (Table 3). When the nitrification/denitrification or partial nitrification/denitrification processes were used, the nitrogen removal efficiency was limited by the low amount of organic matter available to carry out the denitrification process. In order to improve the efficiency of this process, the raw effluent (without a previous anaerobic treatment) should be used (Artiga et al., 2008). These authors treated the raw effluent by means of a hybrid biofilm-suspended biomass membrane reactor which was able to treat OLRs up to 4 g $COD \cdot L^{-1} \cdot d^{-1}$ and NLRs up to 0.7 g $N \cdot L^{-1} \cdot d^{-1}$ with removal efficiencies of 94% and 75%, respectively. This strategy is very interesting in terms of nitrogen removal efficiency but its operational costs are very high and no energy is recovery as biogas during anaerobic digestion.

During the application of anammox and autotrophic denitrification processes, full nitrogen removal efficiency was not achieved because both systems were under nitrite and sulphide limiting conditions, respectively, in order to avoid possible inhibitory effects.

Table 3. Processes applied for the treatment of effluents from anaerobic digesters treating fish canning industry wastewater

Process	NLR $(g \, N \cdot L^{-1} \cdot d^{-1})$	Nitrogen removal (%)	Reference
Nitrification/Denitrification	0.06-0.16*	20-60	Mosquera et al. (2003)
Partial nitrification/ Denitrification	0.16 - 0.28	20 - 45	Figueroa et al. (2008)
Partial nitrification/Anammox	0.20 -0.45*	40 - 80	Vázquez-Padín et al. (2009a)
Nitrification/ Autotrophic denitrification	0.02 - 0.08*	13 - 33	Fajardo et al. (2011)

*Taking into account the total volume of the system needed.

Table 4. Estimation of operational costs to remove 1 kg of nitrogen

Process	Energy (€)	Substrate (€)	Sludge (€)	Total cost (€)	Reference
Conventional N/D + MeOH	0.39	0.75	0.27	1.41	Fernández (2010)
SHARON/anammox	0.18	-----	0.04	0.22	Fernández (2010)
Nitrification/Autotrophic denitrification with S°*	0.39	0.25**	0.16***	0.80	Park and Yoo (2009)
Partial nitrification/ Autotrophic denitrification with S°*	0.31	0.15****	0.13	0.59	Park and Yoo (2009)

* Considering the total external addition of S°

**0.1 €·kg S^{-1} and 2.5 kg S consumed· (kg $NO_3^- $-N)$^{-1}$

*** An average yield coefficient of 0.5 g VSS·(g NO_3^--N)$^{-1}$ was used for autotrophic denitrifying bacteria

****1.5 kg S consumed· (kg NO_2^--N)$^{-1}$

Adapted from Fernández, 2010.

The combination of SHARON (partial nitrification) and anammox processes is the cheapest option to remove nitrogen from the effluents of anaerobic digesters of fish canneries even whether it is compared to the combination of partial nitrification and autotrophic denitrification with S° (Table 4). The main advantages of the application of the SHARON/anammox processes are: a) Only around 60% of ammonia is oxidized into nitrite which mainly decreases the aeration costs; b) Ammonia is directly used as electron donor to reduce nitrite while if autotrophic or heterotrophic denitrification are applied the addition of sulphur or organic matter is needed; c) Sludge production is minimized. Nevertheless, the post-treatment of fish canneries effluents by SHARON-anammox processes can present some troubles due to the stability performance of the SHARON system and the slow start-up of the Anammox reactor (Vázquez-Padín *et al.*, 2009a; Fernández *et al.*, 2008). In this sense, the application of a nitrifying system followed by an autotrophic denitrifying system using S^o as electron donor would be more stable in terms of nitrogen removal efficiency.

CONCLUSION

After a suitable pre-treatment to remove both solids and oil/greases, fish cannery effluents are treated by anaerobic digestion which allows the removal of around 85% of the organic matter. The anaerobic digesters can operate under stable conditions spite of the high levels of NH_3, SH_2 and salts present this kind of effluents.

The anaerobic digesters generate effluents with low COD/N ratios which makes unfeasible to apply the nitrification-denitrification processes to remove nitrogen. Therefore, no heterotrophic denitrifying processes such anammox or autotrophic denitrification can be economical options to fulfil the disposal requirements.

ACKNOWLEDGMENTS

This work was also funded by the Spanish CICYT (TOGRANSYS project CTQ2008-06792-C02-01/PPQ and NOVEDAR_Consolider project CSD2007-00055) and Xunta de Galicia (project 10MDS265003PR).

REFERENCES

Ahn Y.H. (2006). Sustainable nitrogen elimination biotechnologies: A review, *Process Biochemistry*, 41 1709-1721.

Artiga P., García-Toriello G., Méndez R., Garrido J.M. (2008) Use of a hybrid membrane bioreactor for the treatment of saline wastewater from a fish canning factory, *Desalination*, 221 518–525

Aspé E., Martí M.C., Roeckel M. (1997). Anaerobic treatment of fishery wastewater using a marine sediment inoculum, *Water Research*, 31(9) 2147-2160.

Balslev-Olesen, P., Lynggaard-Jensen, A., Nickelsen, C., (1990). Pilot-scale experiments on anaerobic treatment of wastewater from a fish processing plant, *Water Science and Technology*, 22 (1–2) 463–474.

Beristain C.R., Sierra-Alvarez R., Rowlette P., Razo-Flores E., Gómez J. Field J. (2006) Sulfide Oxidation Under Chemolitoautotrophic Denitrifying Conditions, *Biotechnology and Bioengineering*, 95 (6) 1148-1156.

Braun, R., Huber, P., Meyrath, J., (1981). Ammonia toxicity in liquid piggery manure digestion. *Biotechnology, Letters*, 3 159–164.

Broda E. (1977) Two kinds of lithotrophs missing in nature. Zeitschrift für Allgemeine Mikrobiologie, 17 491-493.

Campos J.L., Carvalho S., Portela R., Mosquera-Corral A., Méndez R. (2008). Kinetics of denitrification using sulfur compounds: Effects of S/N ratio, endogenous and exogenous compounds, *Bioresource Technology*, 99 1293-1299.

Chowdhury P., Viraraghavan T. and Srinivasan A. (2010). Biological treatment processes for fish processing wastewater – A review, Bioresource Technology, 101 439–449

Claus G., Kutzner J. (1985). Physiology and kinetics of autotrophic denitrification by Thiobacillus denitrificans, *Applied and Microbiology Biotechnology*, 22 283-288.

Dapena-Mora A., Campos J.L., Mosquera-Corral A., Jetten M.S.M., Méndez R. (2004). Stability of the Anammox process in a gas-lift reactor and a SBR, *Journal of Biotechnology*, 110 159-170.

Dapena-Mora A., Campos J.L., Mosquera-Corral A., Méndez R. (2006) Anammox process for nitrogen removal from anaerobically digested fish canning effluents, *Water Science and Technology*, 53(12) 265-274.

Dapena-Mora A., Fernández I., Campos J.L., Mosquera-Corral A., Méndez R., Jetten M.S.M. (2007). Evaluation of activity and inhibition effects on Anammox process by batch tests based on the nitrogen gas production, *Enzyme and Microbial Technology*, 40(4) 859-865.

Fajardo C. (2011). Autotrophic denitrification of wastewater with high concentration of sulphur and nitrogen compounds. PhD Thesis, University of Santiago de Compostela. Spain.

Fernández I., Vázquez-Padín J.R., Mosquera-Corral A., Campos J.L., Méndez R. (2008). Biofilm and granular systems to improve Anammox biomass retention, *Biochemical Engineering Journal*, 42 308-313.

Fernández I. (2010). Towards the improvement of start-up and operation of Anammox reactors. PhD Thesis, University of Santiago de Compostela. Spain.

Figueroa M., Mosquera-Corral A., Campos J.L., Méndez R. (2008). Treatment of saline wastewater in SBR aerobic granular reactors, Water Science and Technology, 58(2) 479-485.

Gadekar S., Nemati M., Hill G.A. (2006). Batch and continuous biooxidation of sulphide by Thiomicrospira sp. CVO: Reaction kinetics and stoichiometry, *Water Research*, 40 2436-2446.

Garrido J.M., Van Benthum W.A.J., Van Loosdrecht M.C.M., Heijnen J.J. (1997). Influence of dissolved oxygen concentration on nitrite accumulation in a biofilm airlift suspension reactor, *Biotechnology and Bioengineering*, 53 168-178.

Guerrero L., Omil F., Méndez R., Lema J.M. (1997) Treatment of saline wastewaters from fish meal factories in an anaerobic filter under extreme ammonia concentrations, *Bioresource Technology*, 61(1) 69-78.

Hellinga C., Schellen A.A.J.C., Mulder J.W., Van Loosdrecht M.C.M., Heijnen J.J. (1998). The Sharon process: An innovative method for nitrogen removal from ammonium-rich waste water, *Water Science and Technology*, 37(9) 135-142.

Kleerebezem R., Méndez R. (2002). Autotrophic denitrification for combined hydrogen sulphide removal from and porst-denitrification, *Water Science and Technology*, 45(10) 349-356.

Kroiss H., Plahl-Wabnegg (1983) Sulfide toxicity with anaerobic wastewater treatment. In Proceedings of the European Symposium on Anaerobic Wastewater Treatment (AWWT), TNO Corporate Communications Department, The Hague, The Netherlands, 72-80.

McCartney D.M., Oleszkiewicz J.A. (1993) Competition between methanogens and sulfate reducers: effect of COD:sulfate ratio and acclimatization, *Water Environmental Research*, 65(5) 655-664.

Manconi I., Carucci A., Lens P., Rossetti S. (2006). Simultaneous biological removal of sulphide and nitrate by autotrophic denitrification in an activated sludge system, *Water Science and Technology*, 53(12) 91-99.

Maurer M., Pronk W., Larsen T.A. (2006). Treatment processes for source-separated urine, *Water Research*, 40 3151-3166.

Mendez, R., Omil, F., Soto, M., Lema, J.M., (1992). Pilot plant studies on the anaerobic treatment of different wastewater from a fish-canning factory, *Water Science and Technology*, 25 (1) 37–44.

Moon H.S., Ahn K.-H., Lee S., Nam K., Kim J.Y. (2004). Use of autotrophic sulphur-oxidizers to remove nitrate from bank filtrate in a permeable reactive barrier system, *Environmental Pollution*, 129 499-507.

Mosquera-Corral A., Sánchez M., Campos J.L., Méndez R., Lema J.M. (2001) Simultaneous methanogenesis and denitrification of pretreated effluents from a fish canning industry, *Water Research*, 35(2) 411-418.

Mosquera-Corral A., Campos J.L., Sánchez M., Méndez R., Lema J.M. (2003). Combined system for biological removal of nitrogen and carbon from a fish cannery wastewater, *Journal of Environmental Engineering*, 129 (9) 82-833.

Mosquera-Corral A., González F., Campos J.L., Méndez R. (2005). Partial nitrification in a SHARON reactor in the presence of salts and organic carbon compounds, *Process Biochemistry*, 40(9) 3109-3118.

Mulder A., Van de Graaf A.A., Robertson L.A., Kuenen J.G. (1995). Anaerobic ammonium oxidation discovered in a denitrifying fluidized-bed reactor, *FEMS Microbiology Ecology*, 16 177-183.

Mulder A. (2003) The quest for sustainable nitrogen removal technologies, *Water Science and Technology*, 48 67-75.

Nyns E.-J. (1994). The anaerobic treatment of wastewater from fishery operations. Case studies Nº 12 and 13. In a THERMIE programme action: A guide to successful industrial implementation of biomethanation technologies in the European Union. 64-68.

Oh S.-E., Kim K.-S., Choi H.-C., Cho J., Kim I.S. (2000). Kinetics and physiological characteristics of autotrophic denitrifying sulphur bacteria, *Water Research and Technology*, 42(3-4) 959-68.

Omil F., Méndez R., Lema J.M. (1995) Anaerobic treatment of saline wastewaters under high sulphide and ammonia content, *Bioresource Technology*, 54(3) 269-278.

Palenzuela-Rollon, A., Zeeman, G., Lubberding, H.J., Lettinga, G., Alaerts, G.J., (2002). Treatment of fish processing wastewater in a one- or two-step upflow anaerobic sludge blanket (UASB) reactor, Water Science and Technology, 45 (10) 207–212.

Park S., Bae W. (2009). Modeling kinetics of ammonium oxidation and nitrite oxidation under simultaneous inhibition by free ammonia and free nitrous acid, *Process Biochemistry*, 44 631-640.

Park J.Y., Yoo Y.J. (2009). Biological nitrate removal in industrial wastewater treatment: which electron donor we can choose, *Applied Microbiology and Biotechnology*, 82 415-429.

Reyes-Avila J., Flores-Razo E., Gómez J. (2004). Simultaneous biological removal of nitrogen, carbon and sulfur by denitrification, *Water Research*, 38 3313-3321.

Rinzema, A., Van Lier, J., Lettinga, G., (1988). Sodium inhibition of acetoclastic methanogens in granular sludge from a UASB reactor, *Enzyme and Microbial Technology*, 10 (1) 101–109.

Rinzema A, Lettinga G (1988) Anaerobic treatment of sulfate containing wastewater. In Biotreatment systems (ed DL Wise), vol II, 65-109, CRC Press Inc., Boca raton, USA.

Sliekers O., Derwort N., Campos-Gomez J.L., Strous M., Kuenen J.G. Jetten M.S.M. (2002). Completely autotrophic nitrogen removal over nitrite in a single reactor, *Water Research*, 36 2475-2482.

Sliekers A.O., Tirad K.A., Abma W., Kuenen J.G., Jetten M.S.M. (2003). CANON and Anammox in a gas-lift reactor, *FEMS Microbiology Letters*, 218 339-344.

Strous M., Heijnen J.J., Kuenen J.G., Jetten, M.S.M. (1998). The sequencing batch reactor as a powerful tool for the study of slowly growing anaerobic ammonium-oxidizing microorganisms, *Applied Microbiology and Biotechnology*, 50 589-596.

Strous M., Kuenen J.G., Jetten M.S.M. (1999). Key physiology of anaerobic ammonium oxidation, *Applied and Environmental Microbiology*, 65 3248-3250.

Strous M. (2000). Microbiology of anaerobic ammonium oxidation. PhD Thesis, Technical University of Delft, The Netherlands.

Trigo C., Campos J.L., Garrido J.M., Mendez R. (2006). Start-up of the Anammox process in a membrane bioreactor, *Journal of Biotechnology*, 126 475-487.

Vaiopoulou E., Melidis P., Aivasidis A. (2005). Sulfide removal in wastewater from petrochemical industries by autotrophic denitrification, *Water Research*, 39 4101-4109.

Van Dongen U., Jetten M.S.M., Loosdrecht M.C.M. (2001). The SHARON®-Anammox® process for treatment of ammonium rich wastewater, *Water Science and Technology*, 44 (1) 153-160.

Van Kempen R., Mulder J.W., Uijterlinde C.A., van Loosdrecht M.C.M. (2001). Overview: full scale experience of the SHARON process for treatment of rejection water of digested sludge dewatering, *Water Science and Technology*, 44 (1) 145–52.

Vázquez-Padín J.R., Figueroa M., Ferández I., Mosquera-Corral A., Campos J.L., Méndez R. (2009a). Post-treatment of effluent from anaerobic digesters by the Anammox process, *Water Science and Technology*, 65(5) 1135-1143.

Vázquez-Padín J.R., Fernández I., Figueroa M., Mosquera-Corral A., Campos J.L., Méndez R. (2009b). Applications of Anammox based processes to treat anaerobic digester supernatant at room temperature, *Bioresource Technology*, 100 2988-2994.

Vázquez-Padín J.R., Figueroa M., Campos J.L., Mosquera-Corral A., Méndez R. (2010) Nitrifying granular systems: A suitable technology to obtain stable partial nitrification at room temperatura, *Separation and Purification Technology*, 74 178-186.

Wang A., Du D., Ren N., van Groenestijn J.W. (2005). An innovative process of simultaneous desulfurization and denitrication by Thiobacillus denitrificans, *Journal of Environmental Science and Health*, 40 1939-1949.

Veiga M.C., Méndez R., Lema J.M. (1994). Anarobic filter and DSFF reactors in anaerobic treatment of a tuna processing wastewater, *Water Science and Technology*, 30(12) 425-432.

Vidal G., Aspé E., Martí M.C., Roeckel M. (1997). Treatment of recycled wastewaters from fishmeal factory by an anerobic filter, *Biotechnology Letters*, 19(2) 117-121.

Visser A (1995) The anaerobic treatment of sulfate-containing wastewater. PhD Thesis, Wageningen Agricultural University, The Netherlands.

Wett B. (2006). Solved upscaling problems for implementing deammonification of rejection water, *Water Science and Technology*, 53(12) 121-128.

Wiesmann (1994). Biological Nitrogen Removal from Wastewater. In: Fletcher A. (ed.), Advances in Biochemical Engineering Biotechnology, vol. 51. Spinger-Verlag, Berlín. 113-154.

Windey K., de Bo I., Verstraete W. (2005). Oxygen-limited autotrophic nitrification-denitrification (OLAND) in a rotating biological contactor treating high-salinity wastewater, *Water Research*, 39 4512-4520.

Yamamoto-Ikemoto R., Komori T., Nomura M., Ide Y., Matsukami T. (2000). Nitrogen removal from hydroponic culture wastewater by autotrophic denitrification using thiosulfate, *Water Science and Technology*, 42(3-4) 369-376.

Yamamoto T., Takaki K., Koyama T., Furukawa K. (2008). Long-term stability of partial nitritation of swine wastewater digester liquor and its subsequent treatment by Anammox, *Bioresource Technology*, 99 6419–6425.

Zang T.C., Lampe D.G. (1999). Sulfur:Limostone autotrophic denitrification processes for treatment of nitrate-contaminated water: batch experiments, *Water Research*, 33(3) 599-608.

In: Agricultural Research Updates. Volume 2   ISBN: 978-1-61470-191-0
Editor: Barbara P. Hendriks   © 2012 Nova Science Publishers, Inc.

Chapter 14

STORAGE TEMPERATURE EFFECTS ON SURVIVAL OF *ESCHERICHIA COLI* AND *ESCHERICHIA COLI* O157: H7 IN FRESH ORANGE JUICE [*CITRUS SINENSIS* (L.) OSBECK]

Andreana Marino[1]*, Antonia Nostro[1], Antonio Tomaino[1], Francesco Cimino[1] and Paola Dugo[2]

[1] Pharmaco-Biological Department, University of Messina,
Viale Annunziata, Messina, Italy
[2] Pharmaco-Chemical Department, University of Messina,
Viale Annunziata, Messina, Italy

ABSTRACT

Orange juice is the predominant juice manufactured by the beverage processing industry worldwide. However, this product is not free from microbiological spoilage problems, especially unpasteurized single-strength juice. This challenge study was undertaken to assess the effects of storage temperature on survival of *Escherichia coli* ATCC 25922 and *Escherichia coli* O157:H7 ATCC 35150 in fresh Sanguinello, Tarocco and Moro orange juice [*Citrus sinensis* (L.) Osbeck] varieties. Standard (ascorbic acid) and sensory (anthocyanins)-influencing quality parameters of these orange juice cultivars were monitored in order to detect the limiting quality factor. Microbial and nutrient analyses were conducted every week. The initial concentration (ca. 5×10^8 cfu ml^{-1}) of *E. coli* gradually decreased by about 3 $\log_{10}$ in 4 weeks in all varieties compared to the control samples. *Escherichia coli* O157:H7 cells were reduced in Sanguinello juice with the same trend as *E. coli* whereas the former decreased by about 5 $\log_{10}$ and 6 $\log_{10}$ in Moro and Tarocco juices respectively. Both strains seem to use ascorbic acid for the their survival, they significantly (P > 0.001) reduced the ascorbic acid concentrations in all

---
* Correspondence: Andreana Marino,
Pharmaco-Biological Department, University of Messina
98168 Messina, Italy
Tel.: 390906766441; FAX: 390906766438; e-mail: marino@pharma.unime.it

orange juice varieties with respect to variety controls. The strains significantly also decreased the anthocyanins content in Sanguinello ($P > 0.05$) and Moro ($P > 0.001$) juices but not in the Tarocco juice. The reduction of the bacterial number of each strain was highly correlated ($r > 0.99$) to ascorbic acid and anthocyanin degradation levels. Despite its physical and chemical properties and critical conditions it is subjected to during storage, the fresh orange juice can be considered a food suitable for the survival of *E. coli* and *E. coli* O157:H7. Moreover, the ascorbic acid could be held as a good control marker to monitor microbiological quality and safety of orange juice.

Keywords: *Escherichia coli*, *Escherichia coli* O157:H7, orange juice, ascorbic acid, anthocyanins.

INTRODUCTION

Orange juice is the predominant juice manufactured by the beverage processing industry worldwide (UN Food and Agriculture Organization, 1991; Bull *et al.* 2004).

Blood oranges [*Citrus sinensis* (L.) Osbeck] are commonly grown in the Mediterranean Basin, but thanks to the area's special pedoclimatic characteristics, the Sicilian ones express their particular qualities to the best effect. The most common in descending order are Moro, Sanguinello and Tarocco cultivars. These cultivars show a high aroma profile and an antioxidant activity due to the presence of ascorbic acid, flavonoids, hydrocinnamic acids and anthocyanins at higher levels than in blond orange varieties (Rapisarda *et al.* 2001; Galvano *et al.* 2004). The most striking feature of these cultivars is the red colour, of varying intensity and prevalence, caused by anthocyanins, found in the flavedo or juice vesicles, mainly consisting of cyanidin-3-glucoside and cyanidin-3-(6''-malonyl)-β-glucoside (Hillebrand *et al.* 2004; Maccarone *et al.* 1998).

Moreover, the ascorbic acid, the hydrocinnamic acids and the anthocyanins constitute useful markers to allow recognition and evaluation of nutritional quality in fresh and processed products (Lo Scalzo *et al.* 2004; Kelebek *et al.* 2008).

The majority of citrus fruits arrives to the market in the form of processed products, such as single-strength orange juice and frozen juice concentrates (Arias *et al.* 2002). Before pasteurization, fruit juices contain a microbial load representative of the organisms normally found on fruits during harvest and post-harvest. The source of microbial contamination of fresh fruit juices has not been fully established, but the surface of citrus fruits are known to harbour microbial populations (1.3 to 5.3 log cfu ml^{-1}), which can potentially contaminate the juices (Fellers, 1988; Bonaventura & Russo, 1993; Eleftheriadou *et al.* 1998; Pao & Brown, 1998).

Although alternate processes have been developed, almost all commercially produced orange juice is thermally processed because this is still the most cost-effective means to reduce microbial populations and enzyme activity (Perez-Cacho, 2008). Therefore, these products, especially unpasteurized single-strength juice, are not free of microbiological spoilage problems (Rapisarda *et al.* 2001; Arias *et al.* 2002). In addition, blood orange varieties present problems during storage because of their high susceptibility to chilling injury when stored below 8 °C, and to altered fragrance caused by the onset of unpleasant off-flavors (Rapisarda *et al.* 2001).

However, consumers demand juice products with characteristics similar to those of freshly squeezed juices (Anonymous, 2000; Tillotson, 2000; Bull *et al.* 2004; Perez-Cacho, 2008).

In fact, the marketing of organic citrus fruit has markedly expanded both because of increased consumer demand for healthy food products, which are free of synthetic chemical residues, and the resulting improvements in the production and distribution systems. However, since organic fruit are not treated with chemical fungicides, they suffer from relatively high rates of decay, which develops during storage and shelf-life (Porat *et al.* 2000). Wounds on the fruit surface create optimal conditions for the breakdown of juice components and a rapid increase in microorganisms (Del Caro *et al.* 2004). Molds, yeasts, and bacteria can cause spoilage and sensory defects in fruit juices (Jay & Anderson, 2001; Perez- Cacho, 2008).

A limited range of yeasts, moulds and aciduric bacteria are capable of growth at the low pH of orange juice, typically pH 3.3–4.0 (Bracket, 1997). Historically, acid foods such as fruit juices have been considered safe, however, recent foodborne disease outbreaks attributed to unpasteurised apple juices contaminated with pathogens such as *Salmonella typhimurium* (Centers for Disease Control and Prevention, 1966) and *Escherichia coli* O157:H7 have demonstrated that unpasteurized juice can be a vehicle for food-borne illness (Australian Department of Health and Ageing, 1999; Cook et al. 1998; McClure & Hall, 2000). *E. coli* O157:H7 is a major food-borne pathogen which can cause bloody diarrhea, hemolytic uremic syndrome (HUS), or thrombotic thrombocytopenic purpura (TTP). Various foods such as milk, ground beef, drinking water, apple cider, lettuce, melon, and spinach have been associated with this pathogen (Feng, 1995; USFDA, 2006).

AIM

The objective of this work was to evaluate the effects of storage temperature on survival of *Escherichia coli* ATCC 25922 and *Escherichia coli* O157:H7 ATCC 35150 in fresh Sanguinello, Tarocco and Moro orange juice [*Citrus sinensis* (L.) Osbeck] as also to evaluate the consequent variations in standard (ascorbic acid) and sensory(anthocyanins)-influencing quality parameters.

MATERIALS AND METHODS

Bacterial Strains and Growth Conditions

Escherichia coli ATCC 25922 and *Escherichia coli* O157:H7 ATCC 35150 were used as the test organisms. After two successive transfers of the test organisms in Tryptic Soy Broth (TSB; Oxoid, Basingstoke, UK) at 37 °C for 24 h, the activated cultures were inoculated into TSB and incubated at the above temperature for 18 h. The suspensions were adjusted to 1 x 10^9 cfu ml^{-1} using a spectrophotometer (Perkin Elmer, Lambda 15).

Preparation of Orange Juice Samples

Sanguinello, Tarocco and Moro oranges [*Citrus sinensis* (L.) Osbeck] were obtained from a local orange grove and were used after a short storage period at 4 °C (no longer than 1 week). The oranges were of typical commercial maturity. The fruit was disinfected by dipping in a NaOCl 0.5% solution for 1 min and in a 70% ethanol solution for 2 min, rinsed with sterile distilled water, air-dried and randomized. The disinfected oranges were peeled and squeezed by a domestic squeezer. The juices were centrifuged at 6000 rpm for 15 min at 15 °C. The supernatant was filtered twice through a Whatman # 2 filter paper chart and subsequently again filtered through a 0.45 µm sterile filter (Millipore, Bedford, USA) into sterile bottles under aseptic conditions.

Survival Studies in Fresh Orange Juice Cultivars

Each orange juice cultivar sample was split into two 150 ml portions. Each portion was inoculated with the prepared suspensions of *Escherichia coli* or *Escherichia coli* O157:H7 to achieve an initial population of ca. 8.0 log cfu ml^{-1}. The temperature tolerance of both strains was determined in the manner described above by inoculating the suspension of each strain into TSB (control). The samples were then stored at 4 °C. During the storage period, aliquots of 1 ml, diluted with a solution: Bacto Peptone 1 g l^{-1} (Difco) and NaCl 8.5 g l^{-1} (Carlo Erba), were included in tubes containing molten Trypticase Soy Agar (TSA, Oxoid) and poured into plates which were then incubated at 37 °C for 24-48 h. The viable counts were determined at different intervals of time (0, 1, 2, 3, 4 weeks) and counts were expressed as colony forming units (cfu) per ml. Each experiment was performed in triplicate.

Determination of Ascorbic Acid

Analysis of ascorbic acid was performed by an HPLC method injecting 10 µl aliquots of sample or standard onto a Zorbax ODS column (4,6 mm x 25 mm) coupled with a Zorbax C18 guard column. Ascorbic acid was eluted isocratically with 2% KH_2PO_4 (adjusted to pH 2,4 with phosphoric acid) at a flow rate of 0,5 ml/min. The eluate was monitored by UV detection at 245 nm. Each assay was performed in triplicate.

RP-HPLC

The RP-HPLC analyses were performed on a Shimadzu HPLC system equipped with two LC 10 AD *Vp* pumps, an SPD-M10 A*vp* UV-Vis detector, a SCL-10-A*vp* controller and a degasser DGU-14A. UV data were acquired and processed using Shimadzu Class vp5 software. The column used was a Restek Ultra C18, 250 x 2.1 mm, 5 µm particle size. Data were acquired using a photodiode array detector in the range 450-700 nm and the chromatograms were extracted at 518 nm. Elution was effected with a binary high pressure gradient at a flow rate of 0.2 ml min^{-1}. Injection volume was 2 µl. Depending on the anthocyanin content, different dilutions of the extracts were injected: Moro juices were diluted 1:10 in H_2O; Tarocco juices 1:5 in H_2O and the Sanguinello juices were diluted 1:1 in H_2O. Solvent A was Water/Formic acid, 9:1; solvent B was Water/Formic acid/Acetonitrile, 4:1:5. The percentage of solvent B was increased linearly, after an initial hold of 1 min, from 12 to 30 % in 25 min; then to 100% in a further 9 min. Each assay was performed in triplicate.

Statistical Treatment of Results

All statistical analyses were performed using Microsoft Excel 2000 and Mathcad 2001i Professional (MathSofth Engineering & Education, Inc.). The results were submitted to analyses of regression, variance (ANOVA) and correlation.

RESULTS

The results of this challenge study proved the survival of *E. coli* ATCC 25922 and *E. coli* O157:H7 ATCC 35150 in fresh Sanguinello, Tarocco, and Moro orange juice [*Citrus sinensis* (L.) Osbeck] varieties at a refrigerated temperature.

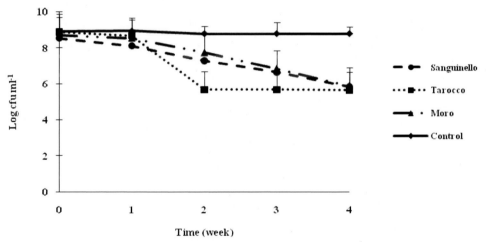

Control: inoculated medium. The results are expressed as mean log cfu ml^{-1} ± standard error (ES).

Figure 1. Survival of *E.coli* ATCC 25922 in Sanguinello, Tarocco and Moro orange juice cultivars during storage at 4 °C.

Escherichia coli showed good survival in all varieties of juice. The initial concentration (about 5 x 10^8 cfu ml^{-1}) decreased by about 3 log_{10} in 4 weeks in all varieties compared to the control samples (inoculated medium). *Escherichia coli* O157:H7 cells were reduced in Sanguinello juice in the same way as *E. coli* whereas the former decreased by about 5 log_{10} and 6 log_{10} in Moro and Tarocco juices respectively on the fourth week of observation compared to the control samples.

Escherichia coli and *E. coli* O157:H7 maintained the initial concentrations in control medium during the storage time (Figure 1-2).

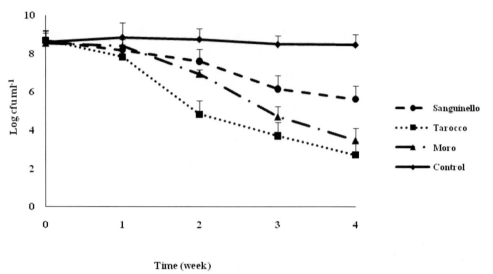

Control: inoculated medium. The results are expressed as mean log cfu ml^{-1} ± standard error (ES).

Figure 2. Survival of E.coli O157:H7 ATCC 35150 in Sanguinello, Tarocco and Moro orange juice cultivars during storage at 4 °C.

Ascorbic acid content of Sanguinello, Tarocco and Moro orange juices are given in table 1.

The Tarocco juice control (uncontaminated juice) showed a higher concentration of ascorbic acid (616.80 ± 28.57 µg ml^{-1}) compared to those of Sanguinello (496.4 ± 34.19 µg ml^{-1}) and Moro (428.30 ± 47.36 µg ml^{-1}) juice controls at the start time. A slight decrease in ascorbic acid content was detected in Sanguinello and Tarocco juice controls (-16% and -22% respectively), while levels in Moro juice control diminished by about 31% during the storage time.

The ascorbic acid concentration was reduced gradually in Sanguinello juices (ca. - 80%) and Tarocco juices (ca. -90%) contaminated with *E. coli* or *E. coli* O157:H7 whereas it dropped by about 70% and 92% in Moro juices contaminated with *E. coli* or *E. coli* O157:H7 respectively, as time progresses. Results of ANOVA analysis indicated that the degradation of ascorbic acid is significantly high (P >0.001) in Sanguinello, Tarocco and Moro contaminated juices compared to those in juice variety controls.

The anthocyanin contents of all the juice varieties are shown in tables 2-4.

The Moro juice control showed a higher concentration of anthocyanins (180.23 ± 1.30 µg ml^{-1}) compared to those of Tarocco (65.02 ± 1.58 µg ml^{-1}) and of Sanguinello (36.04 ± 1.06 µg ml^{-1}) juice controls at the start time. During storage of Sanguinello juice control, total anthocyanin concentrations decreased by about 43%, values for Tarocco were reduced by 54%, whereas the concentrations for Moro rose by 6% as time progresses.

Table 1. Degradation of ascorbic acid in contaminated and uncontaminated (control) orange juice cultivars stored for 4 weeks at 4 °C. The results are expressed as mean µg ml^{-1} ± standard deviations (SD)

Time	Sanguinello			Tarocco			Moro		
week	E. coli	E. coli O157:H7	Control	E. coli	E. coli O157:H7	Control	E. coli	E. coli O157:H7	Control
0	482.31±45.37	454.93±32.46	496.42±41.37	611.40±39.65	614.60±31.28	616.80±26.36	487.90±46.38	404.10±36.78	428.30±39.64
1	402.61±32.64	399.02±25.31	486.92±40.51	397.30±24.16	329.90±19.63	581.30±16.06	441.70±39.18	288.60±25.79	401.40±11.33
2	234.65±19.37	275.89±22.19	461.95±39.18	199.02±11.28	270.10±11.52	520.80±17.02	241.10±20.19	195.70±16.34	385.60±6.38
3	140.95±11.69	118.29±9.61	441.92±42.16	140.95±12.72	118.29±24.64	441.92±23.65	235.50±15.27	93.60±8.14	323.20±21.75
4	94.15±8.71	79.32±8.51	419.37±36.94	55.60±8.74	58.00±13.64	442.20±4.67	145.60±11.68	31.70±2.94	294.20±19.21

Control: uncontaminated juice.

Table 2. Degradation of total anthocyanins in contaminated and uncontaminated (control) orange juice cultivars stored for 4 weeks at 4 °C. The results are expressed as mean ppm ± standard deviations (SD)

Time	Sanguinello			Tarocco			Moro		
week	E. coli	E. coli O157:H7	Control	E. coli	E. coli O157:H7	Control	E. coli	E. coli O157:H7	Control
0	32.80±2.32	35.45±1.43	36.04±1.06	67.38±1.51	70.28±2.72	65.02±1.58	184.23±2.92	186.43±2.25	180.23±1.23
1	26.56±2.80	35.45±1.02	34.83±2.37	47.98±0.47	45.47±1.34	59.05±3.30	171.23±2.56	174.43±1.28	172.34±1.71
2	26.44±2.11	31.75±1.70	30.76±0.38	45.61±1.70	42.49±1.63	46.99±0.97	165.64±2.37	162.78±1.91	181.57±1.10
3	26.29±2.25	28.15±3.16	24.20±1.11	35.78±1.21	40.77±0.80	29.65±1.69	163.43±1.34	161.11±1.08	185.61±0.98
4	18.32±4.26	26.84±0.39	20.68±1.66	29.94±3.62	26.87±0.57	30.07±1.73	146.54±1.79	160.99±1.08	191.14±1.20

Control: uncontaminated juice.

Table 3. Degradation of cyanidin-3-(6''-malonyl)-β-glucoside in contaminated and uncontaminated (control) orange juice cultivars stored for 4 weeks at 4 °C. The results are expressed as mean ppm ± standard deviations (SD)

Time	Sanguinello			Tarocco			Moro		
week	E. coli	E. coli O157:H7	Control	E. coli	E. coli O157:H7	Control	E. coli	E. coli O157:H7	Control
0	7.97±1.85	8.20±1.28	8.16±2.04	19.95±2.55	20.11±2.39	20.45±1.45	66.45±3.12	64.56±3.21	65.06±2.75
1	6.43±1.67	7.81±2.05	7.71±1.95	15.10±2.77	13.95±2.92	19.26±1.39	58.81±1.57	59.34±2.84	61.00±1.78
2	6.24±1.91	7.15±1.82	7.03±1.70	14.35±1.08	13.51±1.92	15.79±0.36	49.26±2.53	49.23±1.97	52.41±2.48
3	5.72±1.82	6.82±0.72	5.92±1.62	10.78±2.24	12.15±2.61	8.69±1.15	49.99±2.01	49.77±3.83	54.06±2.08
4	4.16±0.58	6.47±0.89	5.44±0.86	10.11±2.46	9.10±1.36	9.43±1.78	43.36±2.06	49.04±1.62	66.27±1.95

Control: uncontaminated juice.

Table 4. Degradation of cyanidin-3-Oglucoside in contaminated and uncontaminated (control) orange juice cultivars stored for 4 weeks at 4 °C. The results are expressed as mean ppm ± standard deviations (SD)

Time	Sanguinello			Tarocco			Moro		
week	E. coli	E. coli O157:H7	Control	E. coli	E. coli O157:H7	Control	E. coli	E. coli O157:H7	Control
0	14.66±1.46	15.14±0.98	15.37±0.75	22.00±0.50	21.71±0.79	22.16±1.66	70.26±2.24	65.73±3.23	67.93±2.64
1	12.23±2.47	15.06±1.30	14.78±1.02	16.18±1.69	15.15±2.72	20.80±3.93	59.57±2.70	63.02±3.15	62.37±3.90
2	12.15±2.18	13.54±0.79	13.08±1.25	14.18±1.25	13.76±1.67	15.95±2.52	50.69±1.74	53.12±2.86	54.30±1.91
3	10.95±2.41	12.66±2.12	10.65±2.11	10.37±1.83	11.85±3.31	8.36±0.82	57.96±1.65	51.99±3.28	54.58±2.03
4	7.64±1.06	12.23±1.35	9.52±1.94	9.89±2.24	6.08±1.57	8.77±1.12	45.97±2.36	50.19±2.96	66.45±2.02

Control: uncontaminated juice.

In Sanguinello juice contaminated with *E. coli* the total anthocyanin concentration was reduced by 44%. In particular, cyanidin-3-(6''-malonyl)-β-glucoside and cyanidin-3-Oglucoside values decreased by about 48%. In the samples contaminated with *E. coli* O157:H7 cells the total content decreased by 24% and both the cyanidin values by about 20%.

In contaminated Tarocco juices with *E. coli* and *E. coli* O157:H7 the total content of anthocyanins decreased by about 55% and 62% respectively. In sample juice contaminated with *E. coli*, the cyanidin-3-(6''-malonyl)-β-glucoside and cyanidin 3-Oglucoside values reduced by about 49% and 55% respectively. In the sample contaminated with *E. coli* O157:H7 the cyanidin-3-(6''-malonyl)-β-glucoside and cyanidin 3-Oglucoside values decreased by about 55% and 72% respectively.

In contaminated Moro juices with *E. coli* ATCC 25922 and *E. coli* O157:H7 the total content of anthocyanins dropped by about 20% and 14% respectively. In the contaminated sample with *E. coli* ATCC 25922 the cyanidin-3-(6''-malonyl)-β-glucoside and cyanidin 3-Oglucoside values decreased by about 34%. In the contaminated sample with *E. coli* O157:H7 the cyanidin-3-(6''-malonyl)-β-glucoside and cyanidin 3-Oglucoside values diminished by about 24%.

ANOVA analysis showed that the degradation of total anthocyanin content is significantly higher in Moro (P > 0.001) than in Sanguinello (P> 0.05) and not significant in Tarocco contaminated juices compared to those in the control juices.

The reduction of bacterial number of each strain was highly correlated (r > 0.99) to ascorbic acid and anthocyanin degradation levels.

CONCLUSION

The dynamics of growth, survival and biochemical activity of microorganisms in food are the result of stress reactions in response to the changing of the physical and chemical conditions in the food microenvironment (Giraffa, 2004). Orange juice is a highly acidic food due to its high citrus acid content therefore not suitable for growth of most bacteria. Pathogenic and non-pathogenic *E. coli* can not grow in acidic foods but unfortunately they are able to survive in acidic products for varying periods of time, and this is dependent on a number of factors including, limiting nutrients, temperature and pH (McClure & Hall, 2000). *Escherichia coli* adapts to changes in the amounts of nutrients available in the growth medium. For example, decreases in the concentration of glucose, ammonium and inorganic phosphate (Pi) induce the Crp, GlnG/Nac, and PhoB regulons, respectively. The responses to nutrient limitation allow cells to scavenge for traces of the preferred nutrient and for alternative nutrients in order to maintain growth (Moreau, 2007). The high degradation of ascorbic acid detected in all orange juice varieties suggests that both *E. coli* can use it as a nutrient for survival. *Escherichia coli* can ferment L-ascorbate by gene products of the *ula* regulon. Three enzymes in the pathway that produce D-xylulose 5-phosphate have been functionally characterized: 3-keto-L-gulonate 6-phosphate decarboxylase (UlaD), L-xylulose 5-phosphate 3-epimerase (UlaE), and L-ribulose 5-phosphate 4-epimerase (UlaF) (Campos *et al.* 2007; Yew & Gerlt, 2002). However, at high levels of ascorbic acid, the oxygen scavenging effect and maintenance of low redox potential could improve viability of bacteria by protecting starved cells from protein oxidative damage (Dave & Shah, 1997; Moreau, 2007). The temperature had a significant effect on the growth kinetics (lag phase, growth rate) of *E. coli* O157:H7 with a lower growth temperature (15°C) signicantly lengthening the lag phase and slowing the growth rate from that observed at 37°C (Duffy *et al.* 1999). Several acid resistance systems potentially contribute to the survival of pathogenic *E. coli* in the different acid stress environments. Three known inducible systems have evolved for stationary phase acid resistance in *E. coli*. These systems, AR1, AR2 and AR3, provide a different level of protection with different requirements and induction conditions. These three systems protect against acid stress involved in food processing and facilitate the low infectious dose characteristic of *E. coli*, significantly contributing to the pathogenesis of this organism (Richard & Foster, 2003). The acid tolerance of E. coli O157:H7 at pH 3.5 is higher at 4 °C than at 21°C storage temperature (Uljas & Ingham, 1998). This might be related to changes in cell membrane fluidity when cells were exposed to a chilling temperature (Nair *et al.* 2005).

The survival of *E. coli* and *E. coli* O157:H7 in all varieties of orange juices tested was probably due to their high acid level and low temperature tolerance as also to their ability to ferment L-ascorbate. *Escherichia coli* survived well in all refrigerated juice varieties for 4 weeks. *Escherichia coli* O157:H7 survived a little better in Sanguinello than in Tarocco and Moro juice varieties. Both strains seem to use ascorbic acid for their survival, they strongly reduced the ascorbic acid concentrations in all orange juice varieties with respect to variety controls. The enhanced decay of ascorbic acid in Moro juice control could be ascribed to an interaction with anthocyanins present in high concentrations in this variety. In fact, direct condensation with anthocyanins may account for this decrease (Poei-Langston & Wrolstad 1981; Maccarone & Passerini 1990). The anthocyanins content decreased less, especially in Tarocco juice, probably because the strains can not synthesize enzymes like the β-D-Glucosidase involved in anthocyanins degradation (Barbagallo *et al.* 2007). The increased concentration of the anthocyanins in Moro juice control could be attributable to the activation of the enzymes involved in phenylpropanoid metabolism induced by the low temperature that stimulates anthocyanin biosynthesis (Kalt & McDonald, 1966).

Despite its physical and chemical properties and critical conditions it is subjected to during storage, the fresh orange juice can be considered a food suitable for the survival of *E. coli* and *E. coli* O157:H7. Moreover, the ascorbic acid could be held as a good control marker to improve microbiological quality and safety of orange juice.

REFERENCES

Anonymous. (2000). Quality drives growth. *Beverage Industry, 91* (9), 16-18.

Arias, C. R., Burns, J. K., Friedrich, L. M., Goodrich, R. M. & Parish, M. E. (2002). Yeast species associated with orange juice: evaluation of different identification methods. *Applied and Environmental Microbiology, 68*, 1955-1961.

Australian Department of Health and Ageing. (1999). Salmonellosis outbreack, South Australia. *Communicable Diseases Intelligence, 23* (3), 73.

Barbagallo, R. N., Palmeri R., Fabiano S., Rapisarda P. & Spagna, G. (2007). Characteristic of β-glucosidase from sicilian blood oranges in relation to anthocyanin degradation. *Enzyme and Microbial Technology, 41,* 570-575.

Bonaventura, S. & Russo, C. (1993). Refrigeration of blood oranges destined for transformation. *Fruit Processing, 10,* 284–289.

Bracket, R. E. Fruits, vegetables and grains. In: Doyle, M. P., Beuchat, L. R., Montville, T. J. editor. *Food microbiology fundamentals and frontiers.* Washington, DC: ASM Press; 1997; p. 121.

Bull, M. K., Zerdin, K., Howe, E., Goicoechea, D., Paramanandhan, P., Stockman, R., Sellahewa, J., Szabo, E. A., Johnson, L. & Stewart, C. M. (2004). The effect of high pressure processing on the microbial, physical and chemical properties of Valencia and Navel orange juice. *Innovative Food Science and Emerging Technologies, 5,* 135–149.

Campos, E., Montella, C., Garces, F., Baldoma, L., Aguilar, J. & Badia, J. (2007). Aerobic L-ascorbate metabolism and associated oxidative stress in *Escherichia coli. Microbiology, 153,* 3399-3408.

Centers for Disease Control and Prevention (CDC). (1996). Outbreak of *Escherichia coli* O157:H7 infections associated with drinking unpasteurized commercial apple juice-British Columbia, California, Colorado and Washington. *Morbidity and Mortality Weekly Report, 45,* 975.

Cook, K. A., Dobbs, T. E., Hlady, W. G., Wells, M. S., Barrett, T. J., Puhr, N. D., Lancette, G. A., Bodager, D. W., Toth, B. L., Genese, C. A., Highsmith, A. K., Pilot, K. E., Finelli, L. & Swerdlow, D. L. (1998). Outbreak of Salmonella serotype Hartford infections associated with unpasteurized orange juice. *Journal of American Medical Association, 280,* 1504–1509.

Dave, R. I. & Shah N. P. (1997). Viability of yoghurt and probiotic bacteria in yoghurts made from commercial starter cultures. *International Dairy Journal, 7,* 31–41.

Del Caro, A., Piga, A., Vacca, V. & Agabbio, M. (2004). Changes of flavonoids, vitamin C and antioxidant capacity in minimally processed citrus segments and juices during storage. *Food Chemistry, 84,* 99-105.

Duffy, G., Whiting, R. C. & Sheridan, J. J. (1999). The effect of a competitive microflora, pH and temperature on the growth kinetics of *Escherichia coli* O157:H7. *Food Microbiology, 16,* 299-307.

Eleftheriadou, M., Quantick, P., Nolan, M. & Akkeidou, D. (1998). Factors affecting quality and safety of freshly squeezed orange juice. *Dairy Food and Environmental Sanitation, 18,* 14-23.

Fellers, P. J. (1988). Shelf life and quality of freshly squeezed unpasteurized polyethylene-bottled citrus juices. *Journal of Food Science, 53,* 1699–1702.

Feng, P. (1995). *Escherichia coli* serotype, 0157:H7: novel vehicles of infection and emergence of phenotj-pic variants. *Emerging Infections, 1,* 16-21.

Galvano, F., La Fauci, L., Lazzarino, G., Fogliano, V., Ritieni, A., Cappellano, S., Battistini, N. C., Tavazzi, B. & Galvano, G. (2004). Cyanidins: metabolism and biological properties. *The Journal of Nutritional Biochemistry, 15,* 2-11.

Giraffa G. (2004). Studyng the dynamics of microbial populations during food fermentation. *FEMS Microbiology Reviews, 28,* 251-260.

Hillebrand, S., Schwarz, M. & Winterhalter, P. (2004). Characterization of anthocyanins and pyranoanthocyanins from blood orange [*Citrus sinensis* (L.) Osbeck] juice. *Journal of Agricultural and Food Chemistry, 52,*7331-7338.

Jay, S. & Anderson, J. Fruit juice and related products. In: Moir, C. J. M., Andrew-Kabilafkas, C., Arnold, G., Cox, B. M., Hocking, A. D., Jenson, I., editor. *Spoilage of Processed Foods: Causes and Diagnosis.* Marrickville: Southwood Press; 2001; pp. 187–197.

Kalt, W. & McDonald, J. E. (1996). Chemical composition of lowbush blueberry cultivars. *Journal of the American Society for Horticultural Science, 121,* 142-146.

Kelebek, H., Canbas, C. & Selli, S. (2008). Determination of phenolic composition and antioxidant capacity of blood orange juices obtained from cvs. Moro and Sanguinello (*Citrus sinensis* (L.) Osbeck) grown in Turkey. *Food Chemistry, 107,* 1710-1716.

Lo Scalzo, R., Iannoccari, T., Summa, C., Morelli, R. & Rapisarda, P. (2004). Effect of thermal treatments on antioxidant and antiradical activity of blood orange juice. *Food Chemistry, 85,* 41-47.

Maccarone, E., Rapisarda, P., Fanella, F., Arena, E. & Mondello, L. (1998). Cyanidin-3-(6''-malonyl)- β-glucoside). One of the major anthocyanins in blood orange juice. *Italian Journal of Food Science, 10,* 367-372.

McClure, P. J. & Hall, S. (2000). Survival of *Escherichia coli* in foods. *Journal of Applied Microbiology Symposium* Supplement, *88,* 61S-70S.

Moreau, P. L. (2007). The Lysine Decarboxylase CadA Protects *Escherichia coli* Starved of Phosphate against Fermentation Acids. *The Journal of Bacteriology, 189,* 2249–2261.

Nair, M., Manoj, K., Abouelezz, H., Hoagland, T. & Venkitanarayanan, K. (2005). Antibacterial Effect of Monocaprylin on *Escherichia coli* O157:H7 in Apple Juice. *Journal of Food Protection, 68,* 1895-1899.

Pao, S. & Brown, G. E. (1998). Reduction of microorganisms on citrus fruit surfaces during packinghouse processing. *Journal of Food Protection, 61,* 903-906.

Perez- Cacho, P. R. & Rouseff, R. (2008). Processing and Storage Effects on Orange Juice Aroma: A Review. *Journal of Agricultural and Food Chemistry, 56,* 9785–9796.

Porat, R., Avinoam, D., Weiss, B., Choen, L., Fallik, E. & Droby, S. (2000). Reduction of postharvest decay in organic citrus fruit by a short hot water brushing treatment. *Postharvest Biology and Technology, 18,* 151-157.

Rapisarda, P., Bellomo, S. E. & Intelisano, S. (2001). Storage Temperature Effects on Blood Orange Fruit Quality. *Journal of Agricultural and Food Chemistry, 49,* 3230-3235.

Richard, H. T. & Foster, J. W. (2003). Acid resistance in *Escherichia coli. Advances in Applied Microbiology, 52,* 167-86.

Tillotson, J. (2000). Are your juices and your company ready for the next century? Changing world, changing consumers, changing tastes. *Fruit Processing, 7,* 283–289.

U.N. Food and Agriculture Organization. (1991). *The world Market for Fruit Juice: Citrus and Tropical,* CCP 91/4 Rome.

Uljas, H.E. & Ingham, S. C. (1998). Survival of *Escherichia coli* O157:H7 in synthetic gastric fluid after cold and acid habituation in apple juice or trypticase soy broth acidified with hydrochloric acid or organic acids. *Journal of Food Protection, 61,* 939–947.

USFDA (2006). Everything added to food in the United States: a food additive database.

Yew, W. S. & Gerlt, J. A. (2002). Utilization of L-ascorbate by *Escherichia coli* K-12: assignments of functions to products of the *yjf-sga* and *yia-sgb* operons. *The Journal of Bacteriology*, *184*, 302–306.

In: Agricultural Research Updates. Volume 2   ISBN: 978-1-61470-191-0
Editor: Barbara P. Hendriks   © 2012 Nova Science Publishers, Inc.

Chapter 15

THE UNFOLDED PROTEIN RESPONSE (UPR) IN PLANT SEEDS

Johann Schernthaner[1], Jas Singh[1], Natalie Labbé[1], Dengqun Liao[2] and Frédéric Marsolais[2]

[1] Agriculture and Agri-Food Canada, Eastern Cereal and Oilseed Research Centre, Central Experimental Farm, Ottawa, Canada
.[2] Agriculture and Agri-Food Canada, Southern Crop Protection and Food Research Centre, London, Canada

ABSTRACT

While there are many studies reporting on the UPR in plant vegetative tissues, the number of publications investigating the effect of UPR on plant seeds is limited. However, as the few publications already show, the effect of UPR in seeds differs dramatically in the number as well as in the composition of genes that are affected regardless of whether UPR has been induced chemically or by the overexpression of recombinant proteins. Results shown so far indicate that, unlike in vegetative tissue, the UPR in embryogenic tissue affects components of the abscisic acid (ABA)-dependent transcriptome. Persistent UPR in developing seeds could thus have implications on seed maturation and seed stability. Here, we attempt to analyze the possible causes of this difference with respect to the specific physiological conditions that are present in the developing seed. Maize starchy endosperm mutants impaired in the accumulation of zeins, members of the prolamin family of seed storage proteins, exhibit several phenotypic features associated with the UPR, particularly those encoding defective zeins. Analysis of a common bean mutant lacking the 7S globulin phaseolin and major lectins revealed increased levels of cell division cycle protein 48 and ubiquitin, suggestive of enhanced endoplasmic retriuculum (ER) associated degradation. However, levels of the luminal binding protein (BiP) were actually reduced, along with those of rab1 GTPase, consistent with a decreased activity of the secretory pathway. These results indicated that BiP levels in seed may vary according to the rate of secretory traffic, and not necessarily as a component of UPR. A perspective integrating information from the different model systems of UPR in seed is presented.

INTRODUCTION

In eukaryotic cells, the ER has many functions such as protein folding, assembly, modification by glycosylation, transport, calcium homeostasis, lipid synthesis and protein storage. Under a variety of conditions, the ER can undergo a stress situation whereby a nascent protein synthesized on the rough ER will be misfolded and consequently not processed properly. Misfolding of nascent protein can lead to unbalanced ratios of subunits of hetero-oligomeric proteins, disturbances in calcium homeostasis, or overwhelm the capacity of the ER to keep up with the processing of newly synthesized proteins. If ER stress persists despite increased expression of genes that promote protein folding and removal of unfolded proteins from the ER, the apoptotic pathway is activated. The activation of ER stress-related pathways is referred to as the unfolded protein response (UPR) and is generally characterized by the up-regulation of a specific set of genes pertaining to protein folding, glycosylation, translocation, degradation, transport, sorting and cell wall and lipid metabolism. These responses safeguard the quality control processes in the ER [1], and the cellular mechanisms by which all eukaryotic cells preserve the homeostasis of the ER are well conserved [reviewed in 2-6].

While the molecular mechanisms underlying quality control of proteins and UPR have been thoroughly investigated in yeast and mammals, UPR response in plants still leaves many questions open. Although analogous pathways for quality control of proteins and UPR in plants have been identified, there are also distinctive differences. Global transcriptional analyses in *Arabidopsis* have identified genes that are up- or down-regulated upon treatment of plants with tunicamycin, an inhibitor of glycosylation, and dithiothreitol (DTT), an inhibitor of disulfide bond formation. These studies not only revealed the up-regulation of many ER stress response genes but also pointed to noted differences of the plant UPR compared to that of yeast and mammals [7-10].

Perhaps what differentiates the plant UPR most from that of mammals and yeast is that the plant signaling pathways play important roles not only in the ER stress response but also in other biological processes such as the response to biotic and abiotic stress. ER response to environmental stresses has received much attention and has brought about new insight in understanding the mechanisms of stress tolerance in plants [reviewed in 11]. For example, ER- and osmotic-stress responses in soybean caused by polyethylene glycol (PEG) treatment are to some degree integrated [12]. Similarly, Costa et al. reported that the PEG treatment of soybean suspension cells activated plant-specific, N-rich proteins that promoted cell death. When BiP was overexpressed in soybean, the leaves in the transgenic lines did not wilt and exhibited only a small decrease in water potential. During exposure to drought the stomata of the transgenic lines did not close as much as in the wild type, and the rates of photosynthesis and transpiration became less inhibited than in the wild type [13]. Interestingly, drought resistance in the BiP overexpressing lines was not associated with a higher level of the osmolytes proline, sucrose, and glucose. Instead, they had a lower level of the osmolytes and root weight as well as a lower mRNA abundance of several typical drought-induced genes such as *NAC2*, a seed maturation protein (*SMP*), a glutathione-S-transferase (*GST*), antiquitin, and protein disulphide isomerase 3 (*PDI-3*) than the wild type. The authors also studied the effect of drought on leaf senescence in soybean and tobacco. They found that BiP overexpressing tobacco and soybean showed delayed leaf senescence during drought. BiP antisense tobacco plants, conversely, showed advanced leaf senescence. The mechanism behind this is yet unknown [14]. Efforts to untangle the complex interactions are handicapped further by the fact that not all *BiP* genes are activated equally under abiotic stress and that there are also different phosphorylation states of BiP isoforms in response to different stimuli [15].

A signaling pathway that mediated salt stress response was described in *Arabidopsis* plants. This response was mechanistically related to ER stress responses described in mammalian systems. Specifically, the response involved processing and relocation of the ER membrane-associated transcription factor AtbZIP17 to the nucleus to activate stress response genes. Under salt stress conditions, the N-terminal fragment of AtbZIP17 was translocated to the nucleus. The N-terminal fragment bearing the bZIP DNA binding domain was also found to possess transcriptional activity that functions in yeast. AtbZIP17 activation directly or indirectly up-regulated the expression of several salt stress response genes, including the homeodomain transcription factor ATHB-7 [16]. The *Arabidopsis STT3a* gene encodes an essential subunit of the oligosaccharyltransferase complex that is involved in protein N-glycosylation. NaCl induces UPR in the ER and cell cycle arrest in root tip cells of *stt3a* mutant seedlings indicating that plant salt stress adaptation involves ER stress signal regulation of cell cycle progression. It appears that specific protein glycosylation is necessary for recovery from the UPR and for cell cycle progression during salt/osmotic stress. Recovery is associated uniquely with the function of the STT3a isoform [17]. The putative zinc transporter, ZTP29, was identified in *Arabidopsis* to be induced in root tissue by salt stress suggesting that ZTP29 is involved in the response to salt stress, perhaps through regulation of zinc levels required to induce the UPR pathway [18].

In mammalian cells, BiP is induced by glucose starvation. In *Arabidopsis* however, BiP transcripts decreased with sugar depletion and increased with sugar addition. Transcriptional analysis for β-glucuronidase (GUS) driven by either the BiP promoter or a core 35S promoter with UPR elements (UPRE) revealed that BiP is induced by sugar independent of the *cis*-element responsible for the UPR. The reason for this plant-specific regulatory pattern may be that higher sugar content leads to higher protein synthesis requiring more folding chaperones [19].

Programmed cell death (PCD) is a widespread response of plants against abiotic stress, such as heavy metal toxicity. Adamakis et al. investigated the effect of tungsten (W) toxicity in *Pisum sativum* roots. Transmission electron microscopy (TEM) and fluorescence microscopy revealed mitotic cycle arrest, protoplast shrinkage, disruption of the cytoskeleton, chromatin condensation and peripheral distribution in the nucleus of W-affected cells. These effects were suppressed by inhibitors of the 26S proteasome, caspases and ER stress. In addition, silencing of *DAD-1* and induction of *HSR203J*, *BiP-D*, *bZIP28* and *bZIP60* genes were also observed in W-treated roots implying that ER stress-unfolded protein response may be involved in W-induced PCD [20].

Heat-stressed *Arabidopsis* plants also triggered a stress response similar to UPR. Here, *Arabidopsis* seedlings mitigated stress damage by activating ER-associated transcription factors and an RNA splicing factor, IRE1b [21]. IRE1b splices the mRNA-encoding bZIP60, a basic leucine-zipper domain containing transcription factor associated with UPR in plants. Activation of bZIP60 leads directly to the induction of BiP.

Another variant is the anticipating ER stress response before pathogen-related gene induction. When a plant is attacked by a pathogen, it produces a range of defense-related proteins many of which are synthesized by the rough ER to be secreted from the cell or deposited in vacuoles. Under these conditions, ER-resident chaperones such as BiP are induced systemically throughout the plant. Furthermore, this induction occurs rapidly and precedes expression of genes encoding pathogenesis-related (PR) proteins. The underlying signal transduction pathway was shown to be independent of the signaling molecule salicylic acid and the UPR pathway. In addition, BiP induction was independent of PR gene induction [22]. It was proposed that the induction of BiP expression during plant-pathogen interactions is required as an early response to support PR protein synthesis on the rough ER and that a novel signal transduction pathway exists to trigger this rapid response.

The plant ER quality control system (ER-QC) and UPR have recently attracted considerable attention because many important nutritional plant proteins, such as the seed storage proteins in food crops are synthesized in the ER and the plant seed as a production platform for recombinant proteins is becoming increasingly attractive. Investigating the plant seed ER-QC has become even more urgent since differences in the UPR pathways between somatic cells and seed tissue have emerged. The maturation process of many plant seeds undergoes a phase of desiccation resembling environmental drought stress. In addition, during the seed-filling phase, storage proteins are synthesized and deposited in large amounts. This offers a unique opportunity to study the effect of stress on ER-QC *in vivo*. Knowledge of ER-QC also has practical implications as seeds have been considered an ideal platform for the production of heterologous proteins to add value to the crop and numerous attempts have been made to utilize plant seeds as a host for the manufacturing of valuable proteins such as therapeutic and industrial enzymes [23-25]. It has been shown repeatedly that targeting a recombinant protein to the secretory pathway results in increased yield [26]. However, the capacity of the seed to overexpress large quantities of ER-processed heterologous proteins has been shown to be diminished by the inability of some proteins to fold properly and maintain their native conformation, thus triggering UPR [27]. Moreover, the seed-specific ER stress caused by aggregating heterologous proteins seems to interfere with seed development in a manner similar to what has been observed in some maize zein mutations [28]. Although the folding chaperone BiP is involved in storage protein assembly, deposition and degradation [29,30], interestingly, overexpression as well as suppression of BiP alone in rice endosperm had an inhibitory effect on the accumulation of seed storage proteins and starch [31,32]. Overexpression of BiP in transgenic tobacco seedlings though alleviated tunicamycin-induced UPR and conferred tolerance to water stress while suppression of BiP had the opposite effect [33]. In antisense plants, the water stress stimulation of the antioxidative defenses was higher than in control plants, whereas in drought-stressed BiP-overexpressing lines, an induction of superoxide dismutase activity was not observed. It was suggested that overexpression of BiP in plants may prevent endogenous oxidative stress [34].

The obvious differences in ER stress response between seed and vegetative tissue demand closer studies of the ER-QC and the UPR pathways in seeds. Treatment of developing seeds with UPR inducing agents like tunicamycin is difficult but zygotic tissue can be generated *ex planta* in the form of somatic- or microspore-derived embryos (MDEs) which can be cultured in Petri dishes thus facilitating chemical or environmental treatments.

1. *BRASSICA* MDEs AS A MODEL SYSTEM FOR STUDYING THE UPR IN SEED TISSUE

Global transcriptional analysis of UPR in plants has so far been done mostly on plantlets, as they can be easily subjected to chemical inducers of UPR such as tunicamycin, L-azetidine-2-carboxylic acid (AZC) or DTT [7,10,35]. MDEs from *Brassica* have been shown to exhibit expression patterns similar to developing seed embryos [36-38]. While the chronic UPR observed in the maize mutations *floury 2* and *Mucronate* (*vide infra*) provides valuable insight pertaining to UPR pathways, chemical induction offers the possibility of determining common factors underlying the ER stress response. In seeds, it is important to differentiate between 'classical' UPR, caused by the accumulation of mutated proteins, and the 'anticipatory' UPR, which leads to a UPR-like accumulation of ER stress-related proteins following a pathogen attack or during normal seed development whereby large amounts of proteins are synthesized, transported, sorted and deposited [5]. *Brassica* MDEs produce many of the proteins unique to seed embryos like storage proteins, LEA proteins, oleosins, etc. [37-39]. Furthermore, the transcription of many embryo-specific genes in MDEs can be further enhanced by the addition of ABA [36,40]. This crucial plant hormone is indispensable for proper seed maturation which includes not only embryo formation and the filling of the seed with storage proteins, starch and lipids, but also the seed desiccation process, mediated by the expression of late embryogenesis abundant (LEA) proteins. LEA proteins are a family of hydrophilic proteins that form an integral part of desiccation tolerance of seeds [41,42]. LEA proteins have been also been postulated to play a protective role under different abiotic stresses [43-46].

In summary, embryonic tissue is physiologically fundamentally different from vegetative tissue. In order to study the key differences in the ER stress responses between vegetative and seed tissue, we attempted to analyze the transcriptome of MDEs during tunicamycin-induced ER stress.

2. RESULTS OF COMPARATIVE MICROARRAY ANALYSIS ON TUNICAMYCIN-TREATED *BRASSICA* MDEs

Microspore culture. Donor plants, *B. napus* cv. Topas were grown for a 16 hour photoperiod with full illumination and day/night temperatures of 10°C and 5°C respectively. Buds of 3-4 mm in length were selected, surface sterilized in bleach, and rinsed in sterile water. Sterile buds were crushed with a glass rod in 5 mL half-strength B5 medium. The suspension was filtered over 44 μm nylon filter by gravity. The eluate was spun down at 1,500 rpm for 3 min to gather microspores. The resulting pellet was washed two more times with half-strength B5 medium, and then suspended at 105 microspores per mL in NLN media. The 105 suspension was dispensed in 10 mL volume per 100 × 15 mm Petri dish. Plates were wrapped and placed in darkness at 32°C for three days and then at 25°C with gentle rotation (70 rpm). After one week, microspores were subcultured into larger 150 × 15 mm Petri dishes with a fourfold dilution into fresh NLN.

Tunicamycin treatment of embryos. At or just before torpedo stage (approximately day 14), microspores were treated with 5 μg/mL tunicamycin in dimethyl sulfoxide (DMSO) or an equivalent amount of DMSO as negative control for 5 hours. Embryos were harvested over a 250 μm nylon mesh and flash frozen in liquid nitrogen.

Table 1. Genes involved in ER quality control with a 2-fold or higher change in expression during tunicamycin-induced ER stress in *Brassica* microspore-derived embryos (MDEs). Hits corresponding to the A. th. plantlet studies by Martínez and Chrispeels [7] (M) , and Kamauchi et al. [10] (K), are indicated. Fold variances (FV) are rounded to whole numbers. Down-regulation is indicated by a (-)

ACI gene	Description	FV in MDEs	A.th. plantlets
Protein folding			
AT5G42020	BIP2	39	
AT5G28540	BIP1	21	M, K
AT1G56340	Calreticulin	15	K
AT1G09210	Calreticulin	4	M, K
AT5G61790	Calnexin	3	M, K
AT1G77510	ATPDIL1-2	10	K
AT2G32920	|ATPDIL2-3	6	M, K
AT2G47470	ATPDIL2-1	4	M, K
AT1G21750	ATPDIL1-1	4	
AT1G04980	ATPDIL2-2	4	K
AT4G24190	AtHSP90-7 (SHEPHERD)	14	M, K
AT5G58710	AtCYP20-1 (cyclophilin ROC7)	2	K
AT1G72280	AERO1	8	
AT1G71220	UGGT	2	
Chaperones			
AT4G16660	HSP70, putative	15	
AT2G25140	HSP98.7, CLPB-M, CLPB4	4	
AT1G79930	HSP91	3	
AT1G74310	HSP101, HOT1, ATHSP101	3	
AT5G50920	ATHSP93-V, HSP93-V, CLPC, DCA1, CLPC1	2	
AT5G03160	DNAJ heat shock N-terminal domain-containing protein	10	
AT3G62600	DNAJ heat shock family protein	7	
AT3G08970	DNAJ heat shock N-terminal domain-containing protein	7	
AT3G62600	DNAJ heat shock family protein	5	
AT4G12770	heat shock protein binding	3	
AT2G26890	KAM2 / heat shock protein binding	3	
AT1G50500	HIT1/ transporter	2	
AT5G37710	lipase class 3 family protein / calmodulin-binding heat-shock protein	2	
AT1G09180	ATSARA1A (secretion-associated Ras super family 1); GTP binding	6	

ACI gene	Description	FV in MDEs	A.th. plantlets
Co-chaperones			
AT5G22060	AtJ2	2	
AT3G08970	TMS1	7	
Glycosylation			
AT2G41490	GPT; UDP-N-acetylglucosamine-dolichyl-phosphate N-acetylglucosaminephosphotransferase	12	M, K
AT4G21150	RPN2 (HAP6)	3	
AT1G32210	AtDAD1	-2	
UPR activation			
AT3G10800	AtbZIP28	2	
AT1G42990	AtbZIP60	2	
AT5G24360	IRE1b	2	
AT3G14020	CCAAT-binding transcription factor (CBF-B/NF-YA)	4	
Translocation			
AT1G29310	SEC61 alpha subunit	13	K
AT2G34250	Sec61 alpha subunit	4	K
AT4G24920	SEC61 gamma subunit	4	M, K
AT1G27330	Similar to SERP1/RAMP4	4	M, K
AT3G51980	Similar to ER chaperone SIL1	10	K
AT5G03160	P58^{IPK}	10	K
Protein degradation			
AT4G21810	Derlin-2.1	19	M, K
AT4G29330	Derlin-1	6	
AT1G18260	SEL-1	5	K
AT4G28470	AtRPN1b/RPN1B (26S proteasome regulatory subunit S2 1B)	3	
AT2G46500	Phosphatidylinositol 3- and 4-kinase family protein / ubiquitin	3	M
AT3G02260	DOC1 binding / ubiquitin-protein ligase/ zinc ion binding	3	
AT4G05320	Polyubiquitin 10	3	
AT2G30110	ATUBA1 / ubiquitin activating enzyme	3	
AT4G24690	Ubiquitin-associated / TS-N domain-containing protein	3	
AT5G46210	CUL4 / ubiquitin-protein ligase	3	
AT5G17760	AAA-type ATPase family protein	5	
AT3G09840	CDC48 ATPase	3	
AT1G05910	CDC48-related	2	
AT3G62980	TIR1 / ubiquitin-protein ligase	2	
AT5G14950	ATGMII/GMII / alpha-mannosidase	2	
AT2G32730	26S proteasome regulatory subunit, putative	2	

Table 1. (Continued)

ACI gene	Description	FV in MDEs	A.th. plantlets
AT5G35080	Similar to OS-9	-2	M
Vacuolar			
AT1G78920	Vacuolar H+-Pyrophosphatase 2	4	M
Vesicle trafficking			
AT1G62020	coatomer protein complex, subunit alpha	27	K
AT1G09180	ATSARA1A / RAS Super Family 1; GTP binding	6	M
AT1G11890	ATSEC22 / (secretion 22); transporter	3	M, K
AT1G14010	Similar to emp24/gp25L/p24 family protein	2	K
Apoptosis			
AT5G47120	ATBI-1 / BAX Inhibitor 1	4	K
AT1G79340	ATMC4 / (Metacaspase 4); caspase/cysteine-type peptidase	3	
AT4G01090	Extra-large G-protein-related	2	
AT1G17470	ATDRG1 G-Protein 1; GTP binding	2	
Kinase			
AT3G59480	pfkB-type carbohydrate kinase family protein	20	
AT1G80460	NHO1 / carbohydrate kinase	9	
AT5G20050	protein kinase family protein	7	
AT1G26270	phosphatidylinositol 3- and 4-kinase family protein	6	
AT3G51850	CPK13 / calmodulin-dependent protein kinase/ kinase	5	
AT5G53450	ORG1 (OBP3-RESPONSIVE GENE 1); kinase	4	
AT3G58640	protein kinase family protein	4	
AT1G64460	phosphatidylinositol 3- and 4-kinase family protein	3	
AT4G23650	CDPK6 calmodulin-dependent protein kinase/ kinase	3	
AT4G35780	protein kinase family protein	3	
AT5G26570	PWD / Phosphoglucan water dikinase	3	
AT5G23450	ATLCBK1 / diacylglycerol kinase	3	
AT3G01510	5'-AMP-activated protein kinase beta-1 subunit-related	3	
AT3G57530	CPK32 / calmodulin-dependent protein kinase/ kinase	3	
AT2G03890	phosphatidylinositol 3- and 4-kinase family protein	3	
AT4G23320	protein kinase family protein	3	
AT3G13690	kinase family protein	3	
AT4G29130	ATHXK1 / ATP binding / hexokinase	3	
AT5G25930	leucine-rich repeat family protein / protein kinase family protein	3	
AT1G31650	ATROPGEF14 / Rho guanyl-nucleotide exchange factor	2	
AT3G26940	CDG1 / kinase	2	
AT2G26980	CIPK3 (CBL-INTERACTING PROTEIN KINASE 3)	2	

ACI gene	Description	FV in MDEs	A.th. plantlets
Transcription factor			
AT2G35930	U-box domain-containing protein	9	
AT3G24050	GATA transcription factor 1	3	M
AT5G12840	HAP2A / (embryo defective 2220); transcription factor	2	
AT2G38470	WRKY33 / (DNA-binding protein 33); transcription factor	3	M
AT4G31550	WRKY11 / transcription factor	4	
AT5G08790	ATAF2 / (Arabidopsis NAC domain containing protein 81)	2	M
AT1G22510	(C3HC4-type RING finger) family protein	37	
AT5G43530	helicase domain-containing protein / RING finger domain-containing protein	4	
AT5G65630	GTE7 (Global transcription factor group E 7)	6	
AT1G20980	SPL14 / transcription factor	4	
AT3G14020	CCAAT-binding transcription factor (CBF-B/NF-YA) family	4	
AT4G04890	PDF2 DNA binding / transcription factor	4	
AT3G49530	ANAC062 / transcription factor	4	
AT4G14365	(C3HC4-type RING finger) / ankyrin repeat family protein	-2	
AT5G59820	RHL41 / transcription factor/ zinc ion binding	-2	M
AT5G67300	ATMYB44/ATMYBR1	-3	
AT5G59820	RHL41 / transcription factor/ zinc ion binding	-2	
AT5G57390	AIL5 / transcription factor	-2	
AT1G21910	AP2 domain-containing transcription factor family protein	-2	
AT1G69180	CRC (CRABS CLAW)	-2	
AT2G42380	bZIP transcription factor family protein	-2	
AT5G59340	WOX2 (WUSCHEL-related homeobox 2)	-2	
AT5G17810	WOX12 (WUSCHEL-related homeobox 12)	-2	
AT1G28370	ATERF11/ERF11 (ERF domain protein 11)	-2	
AT4G37260	AtMYB73 / (myb domain protein 73)	-2	
AT3G50060	MYB77 / transcription factor	-2	
Regulatory protein			
AT3G07810	hnRNP, putative	3	
AT3G50670	U1SNRNP / (Spliceosomal protein U1A); RNA binding	2	
AT1G28060	snRNP family protein	2	
AT3G11500	snRNP-G, putative / Sm protein G, putative	-2	
AT4G25500	ATRSP35 / arginine/serine-rich splicing factor 35	3	
AT1G60200	splicing factor PWI domain-containing protein / RNA recognition motif (RRM)-containing protein	2	
AT5G01290	mRNA guanylyltransferase	3	
AT1G54080	UBP1A; mRNA 3'-UTR binding	2	
Sugar metabolism			
AT5G15650	RGP2 ; alpha-1,4-glucan-protein synthase (UDP-forming)	3	

Table 1. (Continued)

ACI gene	Description	FV in MDEs	A.th. plantlets
AT3G04240	SEC; transferase, transferring glycosyl groups	3	
AT1G23870	Arabidopsis thaliana trehalose-phosphatase/synthase 9;	3	
AT5G40390	SIP1; galactinol-sucrose galactosyltransferase	3	
AT5G15870	glycosyl hydrolase family 81 protein	2	
AT2G15490	UGT73B4; UDP-glycosyltransferase	2	
AT2G31750	UGT74D1; UDP-glycosyltransferase/ abscisic acid	2	
AT1G22400	ATUGT85A1; UDP-glycosyltransferase/ glucuronosyltransferase	2	
AT3G11540	SPY; transferase, transferring glycosyl groups	2	
AT4G34131	UGT73B3; UDP-glycosyltransferase/ abscisic acid	2	
AT1G18650	glycosyl hydrolase family protein 17	-3	
AT5G57560	TCH4; hydrolase, acting on glycosyl bonds / xyloglucan:xyloglucosyl transferase	-3	
AT3G23730	xyloglucan:xyloglucosyl transferase, putative	-3	
AT2G25630	glycosyl hydrolase family 1 protein	-2	
AT1G65310	ATXTH17; hydrolase, acting on glycosyl bonds	-2	
AT5G42100	ATBG_PPAP; glucan endo-1,3-beta-D-glucosidase	-2	
AT5G40390	SIP1; galactinol-sucrose galactosyltransferase	3	
AT1G22710	SUC2; carbohydrate transmembrane transporter / sucrose:hydrogen symporter/ sugar:hydrogen ion symporter	-2	
AT1G14360	ATUTR3 / UDP-Galactose transporter 3	8	
AT5G54860	integral membrane transporter family protein	7	
AT1G71960	ABC transporter family protein	6	
AT2G43240	nucleotide-sugar transmembrane transporter	4	
AT1G11890	SEC22 (secretion 22); transporter	3	
Cell wall			
AT3G22120	CWLP (cell wall-plasma membrane linker protein)	-3	
AT2G46150	similar to plant cell wall protein SlTFR88	-2	
Protease inhibitor			
AT2G43550	trypsin inhibitor, putative	-2	
AT1G73260	trypsin inhibitor family protein / Kunitz family protein	-2	
AT2G43510	ATTI1 / Trypsin inhibitor protein 1	-2	
AT3G12145	FLR1 (FLOR1); enzyme inhibitor	-3	
Plant defence			
AT1G66100	thionin, putative	-3	
AT2G43590	chitinase, putative	-2	
CYP450			

ACI gene	Description	FV in MDEs	A.th. plantlets
AT5G57220	CYP81F2 (cytochrome P450, family 81, subfamily F	4	
AT5G04660	CYP77A4 (cytochrome P450, family 77, subfamily A	3	
AT4G15396	CYP702A6 (cytochrome P450, family 702, subfamily A	-3	
Abscisic acid			
AT2G13540	CBP80, ABH1 | ABH1 (ABA hypersensitive 1)	2	
AT5G42100	ATBG_PAP / glucan endo-1,3-beta-D-glucosidase	-2	
AT4G34131	UDP-glycosyltransferase/ abscisic acid glucosyltransferase	2	
AT1G72770	HAB1 (Homology to ABI1); protein serine/threonine phosphatase	2	
AT4G23650	CDPK6 (Calcium-dependent proteinkinase 6)	3	
Hormone			
AT3G61830	ARF18 (auxin response factor 18); transcription factor	2	
AT4G25420.1	GA5 (GA REQUIRING 5); gibberellin 20-oxidase	-3	
Transaminase			
AT1G70580	AOAT2 (glutamate:glyoxylate aminotransferase 2)	16	
AT3G24090	Glutamine-fructose-6-phosphate transaminase 2	7	

Isolation of total RNA. Total RNA was extracted from approximately 80 mg embryos using Ambion RNA-4PCR kit with Ambion Plant Aid solution in extraction buffer. The DNAseI treatment performed was included with the kit and performed as per the manufacturer's guidelines. RNA was evaluated using a Bioanalyzer.

RT-PCR and microarray analysis. One μg total RNA was used in RT-PCR using Applied Biosystems High Capacity cDNA Reverse Transcription kit for first strand synthesis and gene specific primers in second strand synthesis with in-house Taq polymerase. For microarray analysis, RNA samples were prepared and labeled with Cy3 and hybridized on Agilent *Brassica* 4 × 44K arrays in single dye experiments using three different biological samples. Data were imported into GeneSpring for data analysis and Lowess normalization. The final sample set was determined by a *p* value < 0.05 and a 2- fold expression variance cut-off.

Tunicamycin treatment of *Brassica* MDEs induces the same UPR genes observed in similar experiments on *Arabidopsis* plantlets (7, 10). Comparisons between the *Arabidopsis* and the *Brassica* MDE microarrays are however somewhat made more difficult due to the much smaller probe set used in the *Arabidopsis* arrays as well as the differences in the transcriptomes of plantlets versus embryos.

The MDE arrays surfaced a lot more genes associated with ER stress (Table 1). Of the core set of the ER stress response genes, BiP1 and BiP2 were found but not BiP3 which was found in the *Arabidopsis* plantlet studies. Five PDIs were found, of which PDIL1-1 was not listed in the *Arabidopsis* studies. Calreticulin and calnexin were up-regulated as expected. In addition, a considerable number of heat shock proteins, co-chaperones were found as well as numerous components of the ER-associated protein degradation pathway (ERAD), genes involved in UPR regulation, kinases and transcription factors. Indeed, one of the most up-regulated genes was a C3HC4-type RING finger protein. Of the ER stress response pathway, bZIP28 and its interacting proteins, bZIP60 and the CCAAT-binding transcription factor

(CBF-B/NF-YA), as well as Ire1, the RNA splicing protein, were slightly up-regulated. These transcription factors activate BiP promoters through binding to the *cis*-elements of the plant-specific UPRE (P-UPRE) and ER stress-response element (ERSE) [9]. Highly up-regulated were also genes involved in protein and vesicle trafficking such as Sec6, SIL1 and the coatomer alpha subunit. The up-regulation of the BAX Inhibitor 1 gene points to a suppression of ER stress-induced cell death. In general, induction of apoptosis in plants is not fully understood yet and its initiation in plants seems to depend on more factors than in mammalian cells [11,47].

Somewhat surprising was the fact that typical embryo-specific genes such as storage protein genes, oleosins and late embryogenesis abundant (*lea*) genes were not particularly affected by tunicamycin treatment. However, this could be due to relatively low expression levels of these genes in MDEs. Our own experiments (unpublished data) and others [36,40] have shown that expression levels of embryo-specific genes can be increased in MDEs by the addition of exogenous ABA. Further experimentation using a combination of ABA and tunicamycin will generate more insight regarding this topic.

ABA signaling involves a highly complex web of interactions [see 48-50 for reviews]. The MDE array results do not clearly indicate that there is an immediately obvious disturbance in the ABA signaling cascade. The same is true for components involved in Ca^{2+} signaling, which is part of the ABA signaling network. Altogether, the tunicamycin-induced ER stress in *Brassica* MDEs revealed a great deal of UPR associated components many of which will have to be analyzed closer in order to fully understand their connection within the ER stress response pathways.

3. COMPARISON OF MDE RESULTS WITH THOSE FROM PUBLICATIONS REPORTING UPR DUE TO THE OVEREXPRESSION OF RECOMBINANT PROTEINS OR GENES INVOLVED IN UPR

While *Brassica* MDEs are an excellent model system to study embryonic tissue under various conditions and treatments, they do not represent a complete equivalent of a seed owing to a lack of an endosperm layer which is an integral part of a seed, and in monocots, comprises the major part of a seed.

Fortunately, several works regarding ER stress response in rice seeds have been published which allow comparison of ER stress response in these two systems. One study analyses the UPR elicited by the overexpression of the beta-amyloid (Ab) peptide in transgenic rice endosperm [27] while two more works describe the UPR caused by the recombinant overexpression/suppression of BiP, also in rice endosperm [31,51].

The study by Oono et al. revealed that the synthesis of Ab peptide in the ER lumen severely inhibited the synthesis and deposition of seed storage proteins. Starch synthesis was also negatively affected. Instead of normal protein bodies many small and abnormally appearing protein bodies were observed together with the accumulation of aggregated Ab peptide in the ER lumen. This resulted in a distinct seed phenotype of opaque and shrunken appearance due to incomplete grain filling. Interestingly, tetrazolium-based aleurone staining and germination tests proved the transgenic seeds to be more viable than dry wild type seeds.

Obviously, this can be explained by a higher water content in the transgenic seeds due to a lack of desiccation and therefore decreased dormancy.

Microarray analysis confirmed that UPR marker genes such as BiPs, PDIs and OsbZIP60 were up-regulated. The effects on grain phenotype were correlated with the expressed peptide causing ER stress rather than expression levels. The results indicated that even prolonged, severe UPR in the developing rice seeds did not lead to PCD, as has been observed in mammalian cells, but that the UPR-specific seed phenotype is the consequence of drastic changes in the transcriptional profile due to side effects of ER stress. In addition to storage proteins and starch, *lea* genes were also markedly down-regulated which would explain the decreased desiccation as LEA proteins are instrumental for the normal late-stage seed desiccation process.

While the reduction in storage proteins could be explained by a disturbance in the protein folding/trafficking/packaging processes, the reduction of starch and LEA proteins is somewhat more difficult to explain as they are not subjected to these processes and most LEA proteins do not pass the ER at all. Possible explanations are imbalances in the ABA and/or Ca^{2+} signaling. Most seed storage proteins and the major seed LEA proteins have ABA-responsive *cis* elements (ABRE) in their promoters and are induced by ABA. Loss of ABA due to ER stress could explain the down-regulation of the seed filling proteins. Because of the complicated web of ABA signaling, it is difficult to extract a coherent sequence of signaling events just from microarray experiments. Perhaps, a more detailed analysis of components in the ABA-dependent signaling pathway will give some answers. An analysis of the ABA-regulated transcriptomes in *Arabidopsis* was published recently [52]. A possible candidate that deserves a closer look is the Ca^{2+} signaling complex as it is at the basis of many ABA signaling pathways [see 53-55 for reviews]. In fact, the endoplasmic reticulum (ER) is a key organelle regulating intracellular calcium homeostasis. In addition, a large number of ABRE *cis* element-containing genes seem to be directly induced by calcium [56]. It is known that ER stress disturbs calcium homeostasis in mammalian as well as in plant cells. In plant cells, this could then in turn affect genes regulated by calcium and ABA. This might also explain why the overexpression/suppression of BiP alone in rice endosperm had a similar effect on seed phenotype as the UPR caused by the Ab peptide. In this experiment it was shown that BiP overexpression alone upset the folding chaperone balance thus triggering an ER stress response similar to that caused by the Ab peptide [31].

4. COMPARISON BETWEEN PROTEOMIC PROFILES OF STORAGE PROTEIN DEFICIENCY IN COMMON BEAN AND TRANSCRIPT PROFILES OF MAIZE STARCHY ENDOSPERM MUTANTS

Grains of cereal crops lacking or expressing defective major storage proteins constitute an additional model of UPR in seed. The maize *opaque 2* and other starchy and floury endosperm mutants are characterized by a defective synthesis of zeins, a group of prolamin storage proteins [57-59]. This deficiency has no impact on overall grain nitrogen content. The lack of lysine-poor zeins is compensated by increased levels of non-zein proteins. This almost doubles the content of lysine, an essential amino acid present at suboptimal levels for nutrition in cereal grain. Unfortunately, this desirable phenotype is accompanied by inferior agronomic characteristics associated with the starchy or floury endosperm phenotype, which disrupts the normal spatial distribution of soft and vitreous endosperm conferring grain hardness. Integration of modifier loci can restore the endosperm texture while keeping the high lysine trait. This strategy was used to develop nutritionally balanced 'Quality Protein Maize', grown in several developing countries [60]. Related high-lysine mutants have been isolated in barley [61].

The zeins are trafficked in the ER and then deposited into protein bodies [62]. Protein bodies are transferred directly to the vacuole by ER-vacuole trafficking, by-passing the Golgi apparatus, whereas globulins are trafficked through the secretory pathway. Interestingly, the lack of a 7S globulin β-conglycinin subunit in transgenic soybean results in a population of 11S globulin glycinin precursor being trafficked to protein bodies as in cereal prolamins [63]. The cereal prolamins are insoluble in aqueous solvent. Their accretion in the ER initiates a pathway whereby transport-incompetent insoluble protein can be trafficked directly to the vacuole. The ER-resident chaperone, BiP and PDI assist the process of accretion of zeins in protein bodies. The zein protein bodies are unique among prolamin protein bodies in remaining attached to tubular ER. The α- and γ-zeins are present at the core of protein bodies while the β- and δ-zeins are found at the periphery. The lack of zein subunits results in alterations to the structure of protein bodies. The soft endosperm texture may arise from a modified packing of protein bodies and starch granules. This hypothesis is supported by two different results. First, examination of starch granules in modified *opaque 2* has revealed an altered composition of amylopectin, increased extractability of granule-bound starch synthase I, and modified shape and adhesion between individual granules [64]. Secondly, one of the *opaque 2* modifiers results in increased levels of γ-zeins, which compensates for reduced levels of α-zeins [65]. This modifier locus is linked genetically to the 27 kDa γ-zein gene.

To circumvent the complex genetics associated with modifier loci, removal of specific zein subunits by RNA interference has been tried in transgenic maize as an approach to improve lysine content. Although this approach was successful at increasing lysine levels it also provoked a starchy endosperm phenotype [59]. By contrast, in barley, removal of the prolamin C-hordein by antisense RNA technology led to a modest increase in lysine content, without a deleterious grain phenotype [66,67]. Current approaches in maize focus on metabolic engineering of lysine biosynthesis and catabolism, avoiding the targeting of zeins for reduction due to the starchy endosperm phenotype [68,69].

Three of the high-lysine mutants of maize exhibit an UPR. They interfere with the processing of zeins. The *floury 2* and *Defective endosperm B30* loci encode mutated signal peptides [70,71]. *Mucronate* encodes a deletion in γ-zein resulting in a frameshift and mutated C-terminal segment [28]. These mutants have vastly increased levels of BiP in endosperm as compared with wild-type [72]. The defective zeins induce an UPR and ER quality control involving retrograde transport to the cytosol and degradation by the proteasome. The molecular responses of several starchy endosperm mutants, including *floury 2* and *Mucronate*, were later compared in a uniform genetic background by transcript profiling with an Affymetrix array [73]. The mutants displayed a pleiotropic phenotype, which complicates the identification of specific gene expression features associated with the UPR or other phenotypes. Common features between mutants included the activation of UPR-associated genes encoding ER chaperones or related to protein turnover, ubiquitin and polyubiquitin, as well as a general stress response involving an increase in the transcript levels of anti-microbial proteins and peroxidases. When comparing the different mutants, the increased transcript levels of ER chaperones were more pronounced in *floury 2* and *Mucronate*. However, the fact that UPR-associated transcripts were up-regulated even in starchy endosperm mutants not undergoing an UPR was intriguing. This finding led Vitale and Boston [5] to hypothesize that wild-type seeds experience an anticipatory UPR due to the high levels of secretory protein trafficking associated with the accumulation of storage protein. Transcript profiling by cDNA sequencing, complemented by proteomic profiling of *opaque 2*, later revealed a general increase in globulins, and proteins involved in secretory trafficking such as rab2 small GTPase, in the starchy endosperm mutants [57].

Genetic stocks of common bean (dry bean, *Phaseolus vulgaris*) are available which integrate recessive mutations resulting in very low levels of major seed proteins [74]. SARC1 integrates the lectin arcelin-1 from a wild accession of common bean. SMARC1N-PN1 lacks the 7S globulin phaseolin and the lectins, erythroagglutinating phytohemagglutinin and arcelin. They share a similar level of the parental background from the commercial cultivar Sanilac, of approximately 88%. The absence of the major seed proteins leads to a large increase in the total content of cysteine, by 70% and to a lesser extent methionine, by 10% [75]. This increase in sulfur amino acid content largely happens at the expense of an abundant non-protein sulfur amino acid, *S*-methyl-cysteine [75], which cannot substitute for methionine or cysteine in the diet [76]. While SARC1 and SMARC1N-PN1 share similar protein content, the lack of phaseolin and major lectins is associated with a two-fold increase in soluble extractible protein content. Protein composition of mature seed was compared between the two genotypes by a proteomic approach [77]. In one experiment, total protein was extracted in quadruplicate, separated by two-dimensional gel electrophoresis and a selected number of spots were excised, digested with trypsin and the resulting peptides submitted to liquid chromatography and tandem mass spectrometry. The relative volume of protein spots was quantified by image analysis to compare their abundance between genotypes. In a second experiment, total protein was extracted in triplicate, extracts were separated by SDS-PAGE, and lanes were excised, digested with trypsin and the resulting peptides submitted to liquid chromatography and tandem mass spectrometry. Proteins were quantified by spectral counting and the results analyzed using an in-house CellMapBase relational database or the SCAFFOLD software (Proteome Software, Portland, OR). These experiments were complemented by analyses of selective extracts of propanol-soluble, methanol-soluble and starch granule associated proteins. Proteins were identified by searching the UniProt database, Viridiplantae taxonomy, or a translated database of *P. vulgaris, Phaseolus coccineus* and *Glycine max* expressed sequence tag (EST) consensus available at the Legume Information System and a separate set of *P. vulgaris* and *Phaseolus angustissimus* ESTs [78] with Mascot (Matrix Science, Boston, MA). Due to the restricted sequence coverage of the target species, these searches were complemented for selected protein bands or spots by *de novo* sequencing with PEAKS Studio and search with the SPIDER algorithm (Bioinformatics Solutions, Waterloo, ON).

These analyses successfully identified several sulfur-rich proteins that are elevated in the absence of phaseolin and major lectins, including the 11S globulin legumin, albumin-2, albumin-1, defensin and Bowman-Birk type proteinase inhibitor 2. More recently, development of a seed specific EST resource from *P. vulgaris* enabled the isolation of cDNAs coding for legumin, albumin-2, albumin-1A and –B and defensin D1 whose protein levels are elevated [79]. In addition, the results obtained by proteomics highlighted changes in protein levels that are relevant to our understanding of UPR in seed. Like in the starchy endosperm mutants of maize, there was evidence for a pleiotropic phenotype in SMARC1N-PN1, including an apparent stress response, as evidenced by increased levels of sulfur-rich antimicrobial peptides, and induction of cytosolic peroxiredoxin and several members of the short chain dehydrogenase reductase family such as alcohol and formate dehydrogenase.

Surprisingly, the levels of the ER-resident chaperone BiP (ER HSC70-cognate binding protein) were actually reduced by 2-fold in the mutant line. Chaperones that were induced in the mutant were actually cytosolic, and include a peptidyl-prolyl *cis-trans* isomerase, HSP 90-1, and two HSP20 family members. The change in BiP levels was correlated with a decrease in a GTP binding protein homologous to *Arabidopsis* Rab1b implicated in ER to Golgi transport [80]. These results suggested that the decrease in BiP is related to a reduced secretory flux. This interpretation is supported by the already noted increased soluble protein content in SMARC1N-PN1. BiP has also been recently implicated as a positive regulator of the secretory pathway through its interaction with the KDEL receptor [81].

Increased levels of a cell division cycle protein 48 homolog (AAA+ ATPase or valosin-containing protein) and of a polyubiquitin present in methanol soluble extracts was interpreted as a possible indication of enhanced ER-associated degradation. However, upon further analysis, the cell division cycle protein 48 homolog is most similar to *Arabidopsis* cell division cycle protein 48A, implicated in the cell cycle [82], rather than ER-associated degradation [83].

5. CORRELATION AND EXTENSION OF PROTEOMIC DATA BY MICROARRAY DATA

These results were recently extended by performing a transcript profiling experiment. As a starting point, an assembly was generated from the developing seed ESTs [79] and other *Phaseolus vulgaris* ESTs available in GenBank (accessed on August 1, 2009; total number of 111,255 ESTs). Sequences were filtered using seqclean [84] and assembled using tgicl [85], yielding a total of 18,742 contigs or singletons. This assembly was used to design probes on a CustomArray 90K microarray (Mukilteo, WA). A total of 18,415 Unigenes were represented on the array. Most of these (88%) were represented by five unique 35- to 40-mer probes, with the remainder represented by one to four unique probes. The assembly was annotated by Blastx to the UniProt database, Viridiplantae taxonomy, by identifying the best hit with an informative annotation having an e-value smaller than or equal to 1^{-5}.

The array was used to profile transcripts from developing seeds at four developmental stages in SARC1 and SMARC1N-PN1. Developmental stages are designated after Walbot et al. [86]: stage IV – cotyledon, 25 mg seed weight; stage V – cotyledon, 50 mg seed weight; stage VI – maturation, 150 mg seed weight, corresponding to the most active phase of reserve accumulation; and stage VIII – maturation, 380 mg seed weight, corresponding to the onset of desiccation. Plants were grown in the field in London, Ontario, in 2009. Four biological replications were performed for each developmental stage. The results were analyzed with GeneSpring GX v. 11.5 (Agilent Technologies, Mississauga, ON). In Table 2, the fold change with respect to wild-type SARC1 after quantile normalization and baseline transformation is reported along with the ANOVA *p* value. A threshold of 1.54-fold was selected for transcripts present at different levels between genotypes. Genotypes were compared by one-way Analysis of Variance (ANOVA) at each developmental stage.

Globally, the data revealed a pleiotropic phenotype with regards to gene expression particularly at later developmental stages. At stages IV and V, 1.8 and 1.7% of transcripts were differentially expressed between genotypes, respectively. However, at stages VI and VIII, 8.4% and 10% of transcripts were differentially expressed, respectively. At the last two stages, there were about twice as many transcripts with reduced levels in SMARC1N-PN1 than those which were elevated. Among possible reasons that might explain the large differences in gene expression between genotypes is the necessity to maintain a balanced carbon to nitrogen ratio, despite a partial shift of nitrogen to the free amino acid pool in SMARC1N-PN1 [75]. It is possible that the lack of phaseolin, phytohemagglutinin and arcelin leads to a form of carbon excess. There was evidence in the transcript profiling data for alterations in several carbohydrate metabolic pathways (data not shown), beyond the previously identified starch and raffinose oligosaccharide biosynthesis [77]. There was also evidence in the transcript profiling data for changes in phytohormone metabolism and perception at the last two developmental stages (Table 2). This includes a stimulation of the bioactivation of jasmonic acid by amino acid conjugation and an increase in the jasmonic acid receptor (JAR1-like protein), bioactivation of auxin via amino acid conjugate hydrolysis which is stimulated by jasmonic acid [87], and a possible repression of the abscisic acid receptor (*Streptomyces* cyclase/dehydrase family protein). If true, these changes in phytohormone signaling pathways may have widespread consequences on gene expression.

Table 2. Selected transcripts having different levels between SARC1 and SMARC1N-PN1 lines of common bean having different seed protein composition. UniProt accession of top informative Blastx hit is indicated with its annotation

EST assembly accession	Function	UniProt accession	Stage	Fold change	ANOVA p value
FE897876.1	Superoxide dismutase	Q9M7R2	V	+2.8	1E-14
			VI	+3.3	2E-14
			VIII	+12.8	4E-14
FG233395.1	Pathogenesis-related protein 1	P25985	VIII	+10.8	4E-08
CL11179Contig1	17.9 kDa class II heat shock protein	P05477	IV	+1.7	8E-04
			V	+1.8	8E-04
			VI	+2.5	8E-04
			VIII	+6.4	9E-04
CL3399Contig1	18.5 kDa class I heat shock protein	P05478	V	+2.0	8E-04
			VI	+3.1	9E-04
			VIII	+6.3	9E-04
CL5587Contig1	HSP90-1	P27323	VI	+1.8	8E-07
			VIII	+2.7	8E-07
CL4975Contig2	Protein disulfide oxidoreductase	B9SIV3	IV	+1.6	2E-09
			V	+1.6	1E-09
			VI	+1.7	2E-09
			VIII	+2.0	2E-09

Table 2. (Continued)

EST assembly accession	Function	UniProt accession	Stage	Fold change	ANOVA p value
CL1852Contig1	γ-Interferon-inducible lysosomal thiol reductase	B6U3P7	VI	+1.9	2E-16
			VIII	+1.7	4E-16
AB020037.1	Calnexin	Q5NT70	VI	+1.8	3E-06
FE688047.1	Calreticulin	B9RM48	VI	+2.1	3E-08
CL11487Contig1	RAB1Y	Q40207	VIII	-1.7	2E-11
CL4012Contig1	Cysteine protease	Q84Y03	VI	+4.5	1E-15
			VIII	+14.1	3E-15
FE898615.1	Gly m Bd 30K allergen	C6TKC0	V	+1.6	1E-15
			VI	+5.1	2E-15
			VIII	+16.9	4E-15
CL1112Contig1	Vacuolar-processing enzyme	O24325	IV	-1.6	2E-14
			V	-1.7	1E-14
			VIII	-1.9	5E-14
CL2444Contig1	Jasmonic acid-amino acid-conjugating enzyme	A1BNG5	VI	+1.7	2E-04
CL8084Contig1	JAR1-like protein	B0VXR3	VI	+2.0	7E-04
PVUSE1NG_RP_1005_N03_23SEPT2008_004	Auxin conjugate hydrolase	Q0GXX4	VI	+1.9	4E-12
			VIII	+2.1	6E-12
CL7378Contig1	*Streptomyces* cyclase/dehydrase family protein	Q2A958	VIII	-2.1	7E-08

The transcript profiling data confirmed the changes in seed protein composition inferred from the proteomic data. Transcript levels of legumin, albumin-2, albumin-1A and −B, and defensin D1 were elevated in SMARC1N-PN1 (data not shown). Superoxide dismutase and pathogenesis-related protein 1 transcripts were elevated up to approximately 10-fold and may be representative of a physiological stress response (Table 2). Two phenotypes related to protein trafficking were confirmed by the transcript profiling data. The first one is the induction of cytosolic heat shock proteins, including 17.9 kDa class II and 18.5 kDa class I heat shock proteins, and HSP90-1. The second one is a repression of rab1 implicated as positive regulator of secretory trafficking. However, the transcript profiling data also revealed that transcripts of chaperones involved in the oxidation or reduction of disulfide bridges were increased in SMARC1N-PN1: protein disulfide oxidoreductase and γ-interferon-inducible lysosomal thiol reductase. This result is consistent with the increase in cysteine content and induction of cysteine-rich proteins in SMARC1N-PN1. Transcripts of lectin chaperones calnexin and calreticulin, involved in ER quality control of glycosylated proteins, were also elevated in SMARC1N-PN1. Transcripts of vacuolar, asparagine-specific cysteine proteases with similarity to soybean Gly m Bd 30K [88] were highly elevated in SMARC1N-PN1, up to 17-fold. This enzyme is likely to be involved in legumin and 2S albumin proteolytic maturation [89]. By contrast, transcripts of a related vacuolar-processing enzyme were reduced in SMARC1N-PN1. Interestingly, this enzyme has been implicated in phaseolin proteolytic digestion during seedling germination [90].

CONCLUSION

ER stress in seeds severely affects seed protein and carbohydrate quantity and with it grain quality. This needs to be taken into consideration when transgenic crops are used as bioreactors for the production of recombinant proteins. More detailed studies are needed in order to disentangle the causalities that comprise this side effect of ER stress response in seeds. In particular, focus should be on the effects of UPR on the ABA and calcium signaling pathways. Mutant grains of seed crops impaired in the accumulation of specific seed proteins constitute alternative models of the UPR in seed. The maize starchy endosperm mutants *floury 2*, *Mucronate* and *Defective endosperm B30* undergo a true UPR. Proteomic investigation of the common bean genotype SMARC1N-PN1, lacking phaseolin, phytohemagglutinin and arcelin, revealed that the levels of BiP may be reduced in seed in pace with secretory flux. Further investigation by transcript profiling highlighted a number of changes in chaperones, involved in disulfide bond formation and quality control of glycoproteins, and proteolytic enzymes, which may be specific to the type of storage proteins accumulated.

ACKNOWLEDGEMENTS

The authors are grateful to Steven Karcz at the Saskatoon Research Centre of Agriculture and Agri-Food Canada, for EST sequence assembly, and Patrick Chapman at the Southern Crop Protection and Food Research Centre of Agriculture and Agri-Food Canada, for assembly sequence annotation. We are indebted to staff at the Plant Biotechnology Institute of the National Research Council of Canada, Don Schwab, for array design and manufacturing, and Raju Datla and Cao Yongguo for assistance with array hybridization. Research performed in Frédéric Marsolais' laboratory was supported by the Ontario Bean Producers' Marketing Board, the Agricultural Adaptation Council funded by the Ontario Ministry of Agriculture, Food and Rural Affairs and Agriculture and Agri-Food Canada, the Pulse Research Network funded by the Agricultural Bioproducts Innovation Program of Agriculture and Agri-Food Canada, the Crop Genomics Initiative of Agriculture and Agri-Food Canada, and the Ontario Research Fund for Research Excellence of the Ontario Ministry of Research and Innovation. Research at Johann Schernthaner's lab was funded by Agriculture and Agri-Food Canada.

REFERENCES

[1]   Hubbard SC, Ivatt RJ. Synthesis and processing of asparagine-linked oligosaccharides. *Annu Rev Biochem* 1981;50:555-83.
[2]   Mori K. Tripartite management of unfolded proteins in the endoplasmic reticulum. *Cell* 2000, May 26;101:451-4.
[3]   Schröder M. The unfolded protein response. *Mol Biotechnol* 2006;34(2):279-90.
[4]   Ron D, Walter P. Signal integration in the endoplasmic reticulum unfolded protein response. *Nat Rev Mol Cell Biol* 2007;8:519-29.

[5]   Vitale A, Boston RS. Endoplasmic reticulum quality control and the unfolded protein response: Insights from plants. *Traffic* 2008, Sep;9:1581-8.

[6]   Urade R. The endoplasmic reticulum stress signaling pathways in plants. *Biofactors* 2009;35:326-31.

[7]   Martínez IM, Chrispeels MJ. Genomic analysis of the unfolded protein response in Arabidopsis shows its connection to important cellular processes. *Plant Cell* 2003, Feb;15:561-76.

[8]   Noh SJ, Kwon CS, Oh DH, Moon JS, Chung WI. Expression of an evolutionarily distinct novel BiP gene during the unfolded protein response in *Arabidopsis thaliana*. *Gene* 2003, Jun 5;311:81-91.

[9]   Iwata Y, Koizumi N. An *Arabidopsis* transcription factor, AtbZIP60, regulates the endoplasmic reticulum stress response in a manner unique to plants. *Proc Natl Acad Sci U S A* 2005, Apr 5;102:5280-5.

[10]   Kamauchi S, Nakatani H, Nakano C, Urade R. Gene expression in response to endoplasmic reticulum stress in *Arabidopsis thaliana*. FEBS J 2005, Jul;272:3461-76.

[11]   Liu JX, Howell SH. Endoplasmic reticulum protein quality control and its relationship to environmental stress responses in plants. *Plant Cell* 2010, Sep;22:2930-42.

[12]   Irsigler AS, Costa MD, Zhang P, Reis PA, Dewey RE, Boston RS, Fontes EP. Expression profiling on soybean leaves reveals integration of ER- and osmotic-stress pathways. *BMC Genomics* 2007;8:431.

[13]   Costa MD, Reis PA, Valente MA, Irsigler AS, Carvalho CM, Loureiro ME, et al. A new branch of endoplasmic reticulum stress signaling and the osmotic signal converge on plant-specific asparagine-rich proteins to promote cell death. *J Biol Chem* 2008, Jul 18;283:20209-19.

[14]   Valente MA, Faria JA, Soares-Ramos JR, Reis PA, Pinheiro GL, Piovesan ND, et al. The ER luminal binding protein (BiP) mediates an increase in drought tolerance in soybean and delays drought-induced leaf senescence in soybean and tobacco. *J Exp Bot* 2009;60:533-46.

[15]   Cascardo JC, Almeida RS, Buzeli RA, Carolino SM, Otoni WC, Fontes EP. The phosphorylation state and expression of soybean BiP isoforms are differentially regulated following abiotic stresses. *J Biol Chem* 2000, May 12;275:14494-500.

[16]   Liu JX, Srivastava R, Che P, Howell SH. Salt stress responses in Arabidopsis utilize a signal transduction pathway related to endoplasmic reticulum stress signaling. *Plant J* 2007, Sep;51:897-909.

[17]   Koiwa H, Li F, McCully MG, Mendoza I, Koizumi N, Manabe Y, et al. The STT3a subunit isoform of the Arabidopsis oligosaccharyltransferase controls adaptive responses to salt/osmotic stress. *Plant Cell* 2003, Oct;15:2273-84.

[18]   Wang M, Xu Q, Yu J, Yuan M. The putative *Arabidopsis* zinc transporter ZTP29 is involved in the response to salt stress. *Plant Mol Biol* 2010, Apr 1;73:467-79.

[19]   Tajima H, Koizumi N. Induction of BiP by sugar independent of a cis-element for the unfolded protein response in *Arabidopsis thaliana*. *Biochem Biophys Res Commun* 2006, Aug 4;346:926-30.

[20]   Adamakis ID, Panteris E, Eleftheriou EP. The fatal effect of tungsten on *Pisum sativum* L. root cells: Indications for endoplasmic reticulum stress-induced programmed cell death. *Planta* 2011, Feb 23.

[21] Deng Y, Humbert S, Liu JX, Srivastava R, Rothstein SJ, Howell SH. Heat induces the splicing by IRE1 of a mRNA encoding a transcription factor involved in the unfolded protein response in Arabidopsis. *Proc Natl Acad Sci U S A* 2011, Apr 11;108:7247-7252.

[22] Jelitto-Van Dooren EP, Vidal S, Denecke J. Anticipating endoplasmic reticulum stress. A novel early response before pathogenesis-related gene induction. *Plant Cell* 1999, Oct;11:1935-44.

[23] Stoger E, Ma JK, Fischer R, Christou P. Sowing the seeds of success: Pharmaceutical proteins from plants. *Curr Opin Biotechnol* 2005, Apr;16:167-73.

[24] Benchabane M, Goulet C, Rivard D, Faye L, Gomord V, Michaud D. Preventing unintended proteolysis in plant protein biofactories. *Plant Biotechnol J* 2008, Sep;6:633-48.

[25] Boothe J, Nykiforuk C, Shen Y, Zaplachinski S, Szarka S, Kuhlman P, et al. Seed-based expression systems for plant molecular farming. *Plant Biotechnol J* 2010, Jun;8:588-606.

[26] Takagi H, Saito S, Yang L, Nagasaka S, Nishizawa N, Takaiwa F. Oral immunotherapy against a pollen allergy using a seed-based peptide vaccine. *Plant Biotechnol J* 2005, Sep;3:521-33.

[27] Oono Y, Wakasa Y, Hirose S, Yang L, Sakuta C, Takaiwa F. Analysis of ER stress in developing rice endosperm accumulating beta-amyloid peptide. *Plant Biotechnol J* 2010, Aug;8:691-718.

[28] Kim CS, Gibbon BC, Gillikin JW, Larkins BA, Boston RS, Jung R. The maize *Mucronate* mutation is a deletion in the 16-kDa γ-zein gene that induces the unfolded protein response. *Plant J* 2006, Nov;48:440-51.

[29] Hatano K, Shimada T, Hiraiwa N, Nishimura M, Hara-Nishimura I. A rapid increase in the level of binding protein (BiP) is accompanied by synthesis and degradation of storage proteins in pumpkin cotyledons. *Plant Cell Physiol* 1997, Mar;38:344-51.

[30] Muench DG, Wu Y, Zhang Y, Li X, Boston RS, Okita TW. Molecular cloning, expression and subcellular localization of a BiP homolog from rice endosperm tissue. *Plant Cell Physiol* 1997, Apr;38:404-12.

[31] Yasuda H, Hirose S, Kawakatsu T, Wakasa Y, Takaiwa F. Overexpression of BiP has inhibitory effects on the accumulation of seed storage proteins in endosperm cells of rice. *Plant Cell Physiol* 2009, Aug;50:1532-43.

[32] Wakasa Y, Zhao H, Hirose S, Yamauchi D, Yamada Y, Yang L, et al. Antihypertensive activity of transgenic rice seed containing an 18-repeat novokinin peptide localized in the nucleolus of endosperm cells. *Plant Biotechnol J* 2010, Nov 16.

[33] Leborgne-Castel N, Jelitto-Van Dooren EP, Crofts AJ, Denecke J. Overexpression of BiP in tobacco alleviates endoplasmic reticulum stress. *Plant Cell* 1999, Mar;11:459-70.

[34] Alvim FC, Carolino SM, Cascardo JC, Nunes CC, Martinez CA, Otoni WC, Fontes EP. Enhanced accumulation of BiP in transgenic plants confers tolerance to water stress. *Plant Physiol* 2001, Jul;126:1042-54.

[35] Iwata Y, Koizumi N. Unfolded protein response followed by induction of cell death in cultured tobacco cells treated with tunicamycin. *Planta* 2005, Mar;220:804-7.

[36] Hays DB, Wilen RW, Sheng C, Moloney MM, Pharis RP. Embryo-specific gene expression in microspore-derived embryos of *Brassica napus*. An interaction between abscisic acid and jasmonic acid. *Plant Physiol* 1999, Mar;119:1065-72.

[37] Wakui K, Takahata Y. Isolation and expression of *Lea* gene in desiccation-tolerant microspore-derived embryos in *Brassica* spp. *Physiol Plant* 2002, Oct;116:223-30.

[38] Tsuwamoto R, Fukuoka H, Takahata Y. Identification and characterization of genes expressed in early embryogenesis from microspores of *Brassica napus*. *Planta* 2007, Feb;225:641-52.

[39] Boutilier KA, Ginés MJ, DeMoor JM, Huang B, Baszczynski CL, Iyer VN, Miki BL. Expression of the BnmNAP subfamily of napin genes coincides with the induction of *Brassica* microspore embryogenesis. *Plant Mol Biol* 1994, Dec;26:1711-23.

[40] Wilen RW, Mandel RM, Pharis RP, Holbrook LA, Moloney MM. Effects of abscisic acid and high osmoticum on storage protein gene expression in microspore embryos of *Brassica napus*. *Plant Physiol* 1990, Nov;94:875-81.

[41] Finkelstein D, Ewing R, Gollub J, Sterky F, Cherry JM, Somerville S. Microarray data quality analysis: Lessons from the AFGC project. *Plant Mol Biol* 2002, Jan;48:119-32.

[42] Hoth S, Morgante M, Sanchez JP, Hanafey MK, Tingey SV, Chua NH. Genome-wide gene expression profiling in *Arabidopsis thaliana* reveals new targets of abscisic acid and largely impaired gene regulation in the *abi1-1* mutant. *J Cell Sci* 2002, Dec 15;115:4891-900.

[43] Kiyosue T, Yamaguchi-Shinozaki K, Shinozaki K. Characterization of two cDNAs (ERD10 and ERD14) corresponding to genes that respond rapidly to dehydration stress in *Arabidopsis thaliana*. *Plant Cell Physiol* 1994, Mar;35:225-31.

[44] Brocard IM, Lynch TJ, Finkelstein RR. Regulation and role of the Arabidopsis *abscisic acid-insensitive 5* gene in abscisic acid, sugar, and stress response. *Plant Physiol* 2002, Aug;129:1533-43.

[45] Tunnacliffe A, Wise MJ. The continuing conundrum of the LEA proteins. *Naturwissenschaften* 2007, Oct;94:791-812.

[46] Dalal M, Tayal D, Chinnusamy V, Bansal KC. Abiotic stress and aba-inducible group 4 *LEA* from *Brassica napus* plays a key role in salt and drought tolerance. *J Biotechnol* 2009, Jan 15;139:137-45.

[47] Wang S, Narendra S, Fedoroff N. Heterotrimeric G protein signaling in the *Arabidopsis* unfolded protein response. *Proc Natl Acad Sci U S A* 2007, Mar 6;104:3817-22.

[48] Hirayama T, Shinozaki K. Perception and transduction of abscisic acid signals: Keys to the function of the versatile plant hormone ABA. *Trends Plant Sci* 2007, Aug;12:343-51.

[49] Wasilewska A, Vlad F, Sirichandra C, Redko Y, Jammes F, Valon C, et al. An update on abscisic acid signaling in plants and more. *Mol Plant* 2008, Mar;1:198-217.

[50] Cutler SR, Rodriguez PL, Finkelstein RR, Abrams SR. Abscisic acid: Emergence of a core signaling network. *Annu Rev Plant Biol* 2010, Jun 2;61:651-79.

[51] Wakasa Y, Yasuda H, Oono Y, Kawakatsu T, Hirose S, Takahashi H, et al. Expression of ER quality control-related genes in response to changes in BiP1 levels in developing rice endosperm. *Plant J* 2011, Mar;65:675-89.

[52] Choudhury A, Lahiri A. Comparative analysis of abscisic acid-regulated transcriptomes in *Arabidopsis*. *Plant Biol* 2011, Jan;13:28-35.

[53] White PJ, Broadley MR. Calcium in plants. *Ann Bot* 2003;92:487.

[54] Song WY, Zhang ZB, Shao HB, Guo XL, Cao HX, Zhao HB, et al. Relationship between calcium decoding elements and plant abiotic-stress resistance. *Int J Biol Sci* 2008;4:116-25.

[55] Roelfsema MR, Hedrich R. Making sense out of ca^{2+} signals: their role in regulating stomatal movements. *Plant Cell Environ* 2010, Mar 1;33:305-21.

[56] Kaplan B, Davydov O, Knight H, Galon Y, Knight MR, Fluhr R, Fromm H. Rapid transcriptome changes induced by cytosolic Ca^{2+} transients reveal ABRE-related sequences as Ca^{2+}-responsive *cis* elements in *Arabidopsis*. *Plant Cell* 2006, Oct;18:2733-48.

[57] Gibbon BC, Larkins BA. Molecular genetic approaches to developing quality protein maize. *Trends Genet* 2005, Apr;21:227-33.

[58] Azevedo RA, Arruda P. High-lysine maize: The key discoveries that have made it possible. *Amino Acids* 2010, Oct;39:979-89.

[59] Huang S, Frizzi A, Malvar TM. Engineering high lysine corn In Krishnan H, editor. Modification of seed composition to promote health and nutrition. Madison: *American Society of Agronomy, Crop Science Society of America, Soil Science Society of America*; 2009; 233-248.

[60] Vasal SK. Quality protein maize: overcoming the hurdles. In Kataki PK, Babu SC, editors. *Food systems for improved human nutrition: Linking agriculture, nutrition, and productivity.* Binghamton: Haworth Press; 2003; 193-228.

[61] Munck L. The case of high-lysine breeding. In Shewry PR, editor. Barley*: Genetics, biochemistry, molecular biology and biotechnology.* Wallingford: CABI International; 1992; 573-601.

[62] Herman EM. Endoplasmic reticulum bodies: Solving the insoluble. *Curr Opin Plant Biol* 2008, Dec;11:672-9.

[63] Kinney AJ, Jung R, Herman EM. Cosuppression of the α subunits of β-conglycinin in transgenic soybean seeds induces the formation of endoplasmic reticulum-derived protein bodies. *Plant Cell* 2001, May;13:1165-78.

[64] Gibbon BC, Wang X, Larkins BA. Altered starch structure is associated with endosperm modification in quality protein maize. *Proc Natl Acad Sci U S A* 2003, Dec 23;100:15329-34.

[65] Wu Y, Holding DR, Messing J. γ-Zeins are essential for endosperm modification in quality protein maize. *Proc Natl Acad Sci U S A* 2010, Jul 20;107:12810-5.

[66] Hansen M, Lange M, Friis C, Dionisio G, Holm PB, Vincze E. Antisense-mediated suppression of c-hordein biosynthesis in the barley grain results in correlated changes in the transcriptome, protein profile, and amino acid composition. *J Exp Bot* 2007;58:3987-95.

[67] Lange M, Vincze E, Wieser H, Schjoerring JK, Holm PB. Suppression of c-hordein synthesis in barley by antisense constructs results in a more balanced amino acid composition. *J Agric Food Chem* 2007, Jul 25;55:6074-81.

[68] Reyes AR, Bonin CP, Houmard NM, Huang S, Malvar TM. Genetic manipulation of lysine catabolism in maize kernels. *Plant Mol Biol* 2009, Jan;69:81-9.

[69] Frizzi A, Huang S, Gilbertson LA, Armstrong TA, Luethy MH, Malvar TM. Modifying lysine biosynthesis and catabolism in corn with a single bifunctional expression/silencing transgene cassette. *Plant Biotechnol J* 2008, Jan;6:13-21.

[70] Kim CS, Hunter BG, Kraft J, Boston RS, Yans S, Jung R, Larkins BA. A defective signal peptide in a 19-kD α-zein protein causes the unfolded protein response and an opaque endosperm phenotype in the maize *De*-B30* mutant. *Plant Physiol* 2004, Jan;134:380-7.

[71] Coleman CE, Lopes MA, Gillikin JW, Boston RS, Larkins BA. A defective signal peptide in the maize high-lysine mutant floury 2. *Proc Natl Acad Sci U S A* 1995, Jul 18;92:6828-31.

[72] Boston RS, Fontes EB, Shank BB, Wrobel RL. Increased expression of the maize immunoglobulin binding protein homolog b-70 in three zein regulatory mutants. *Plant Cell* 1991, May;3:497-505.

[73] Hunter BG, Beatty MK, Singletary GW, Hamaker BR, Dilkes BP, Larkins BA, Jung R. Maize opaque endosperm mutations create extensive changes in patterns of gene expression. Plant Cell 2002, Oct;14:2591-612.

[74] Osborn, Hartweck, Harmsen, Vogelzang, Kmiecik, Bliss. Registration of Phaseolus vulgaris genetic stocks with altered seed protein compositions. Crop Sci 2003;43:1570-1.

[75] Taylor M, Chapman R, Beyaert R, Hernández-Sebastià C, Marsolais F. Seed storage protein deficiency improves sulfur amino acid content in common bean (*Phaseolus vulgaris* L.): Redirection of sulfur from gamma-glutamyl-*s*-methyl-cysteine. *J Agric Food Chem* 2008, Jul 23;56:5647-54.

[76] Padovese R, Kina SM, Barros R, Borelli P, Lanfer Marquez UM. Biological importance of γ-glutamyl-*s*-methylcysteine of kidney bean (*Phaseolus vulgaris* L.). *Food Chemistry* 2001;73:291-7.

[77] Marsolais F, Pajak A, Yin F, Taylor M, Gabriel M, Merino DM, et al. Proteomic analysis of common bean seed with storage protein deficiency reveals up-regulation of sulfur-rich proteins and starch and raffinose metabolic enzymes, and down-regulation of the secretory pathway. *J Proteomics* 2010, Mar 27;73:1587-600.

[78] Vijayan P, Parkin IAP, Karcz SR, McGowan K, Vijayan K, Vandenberg A, Bett, KE. Capturing cold stress related diversity from a wild relative of common bean: Phaseolus angustissimus. *Genome* 2011; in press.

[79] Yin F, Pajak A, Chapman R, Sharpe A, Huang S, Marsolais F. Analysis of common bean expressed sequence tags identifies sulfur metabolic pathways active in seed and sulfur-rich proteins highly expressed in the absence of phaseolin and major lectins. *BMC Genomics* 2011; revision submitted.

[80] Batoko H, Zheng HQ, Hawes C, Moore I. A rab1 GTPase is required for transport between the endoplasmic reticulum and Golgi apparatus and for normal Golgi movement in plants. *Plant Cell* 2000, Nov;12:2201-18.

[81] Pulvirenti T, Giannotta M, Capestrano M, Capitani M, Pisanu A, Polishchuk RS, et al. A traffic-activated Golgi-based signalling circuit coordinates the secretory pathway. *Nat Cell Biol* 2008, Aug;10:912-22.

[82] Park S, Rancour DM, Bednarek SY. In planta analysis of the cell cycle-dependent localization of AtCDC48A and its critical roles in cell division, expansion, and differentiation. *Plant Physiol* 2008, Sep;148:246-58.

[83] Marshall RS, Jolliffe NA, Ceriotti A, Snowden CJ, Lord JM, Frigerio L, Roberts LM. The role of CDC48 in the retro-translocation of non-ubiquitinated toxin substrates in plant cells. *J Biol Chem* 2008, Jun 6;283:15869-77.

[84] Haas BJ, Delcher AL, Mount SM, Wortman JR, Smith RK, Hannick LI, et al. Improving the *Arabidopsis* genome annotation using maximal transcript alignment assemblies. *Nucleic Acids Res* 2003, Oct 1;31:5654-66.

[85] Pertea G, Huang X, Liang F, Antonescu V, Sultana R, Karamycheva S, et al. TIGR gene indices clustering tools (TGICL): A software system for fast clustering of large EST datasets. *Bioinformatics* 2003, Mar 22;19:651-2.

[86] Walbot V, Clutter M, Sussex IM. Reproductive development and embryogeny in *Phaseolus*. *Phytomorphology* 1972;22:59-68.

[87] Titarenko E, Rojo E, León J, Sánchez-Serrano JJ. Jasmonic acid-dependent and -independent signaling pathways control wound-induced gene activation in *Arabidopsis thaliana*. *Plant Physiol* 1997, Oct;115:817-26.

[88] Ogawa T, Tsuji H, Bando N, Kitamura K, Zhu YL, Hirano H, Nishikawa K. Identification of the soybean allergenic protein, Gly m Bd 30K, with the soybean seed 34-kDa oil-body-associated protein. *Biosci Biotechnol Biochem* 1993, Jun;57:1030-3.

[89] Gruis DF, Selinger DA, Curran JM, Jung R. Redundant proteolytic mechanisms process seed storage proteins in the absence of seed-type members of the vacuolar processing enzyme family of cysteine proteases. *Plant Cell* 2002, Nov;14:2863-82.

[90] Senyuk V, Rotari V, Becker C, Zakharov A, Horstmann C, Müntz K, Vaintraub I. Does an asparaginyl-specific cysteine endopeptidase trigger phaseolin degradation in cotyledons of kidney bean seedlings? *Eur J Biochem* 1998, Dec 1;258:546-58.

INDEX

B

F

G

H

I

J

K

L

Q

U

V

Z